Lecture Notes in Physics

For information about Vols. 1–151, please contact your bookseller or Springer-Verlag.

Vol. 152: Physics of Narrow Gap Semiconductors. Proceedings, 1981. Edited by E. Gornik, H. Heinrich and L. Palmetshofer. XIII, 485 pages. 1982.

Vol. 153: Mathematical Problems in Theoretical Physics. Proceedings, 1981. Edited by R. Schrader, R. Seiler, and D.A. Uhlenbrock. XII, 429 pages. 1982.

Vol. 154: Macroscopic Properties of Disordered Media. Proceedings, 1981. Edited by R. Burridge, S. Childress, and G. Papanicolaou. VII, 307 pages. 1982.

Vol. 155: Quantum Optics. Proceedings, 1981. Edited by C.A. Engelbrecht. VIII, 329 pages. 1982.

Vol. 156: Resonances in Heavy Ion Reactions. Proceedings, 1981. Edited by K.A. Eberhard. XII, 448 pages. 1982.

Vol. 157: P. Niyogi, Integral Equation Method in Transonic Flow. XI, 189 pages. 1982.

Vol. 158: Dynamics of Nuclear Fission and Related Collective Phenomena. Proceedings, 1981. Edited by P. David, T. Mayer-Kuckuk, and A. van der Woude. X, 462 pages. 1982.

Vol. 159: E. Seiler, Gauge Theories as a Problem of Constructive Quantum Field Theory and Statistical Mechanics. V, 192 pages. 1982.

Vol. 160: Unified Theories of Elementary Particles. Critical Assessment and Prospects. Proceedings, 1981. Edited by P. Breitenlohner and H.P. Dürr. VI, 217 pages. 1982.

Vol. 161: Interacting Bosons in Nuclei. Proceedings, 1981. Edited by J.S. Dehesa, J.M.G. Gomez, and J. Ros. V, 209 pages. 1982.

Vol. 162: Relativistic Action at a Distance: Classical and Quantum Aspects. Proceedings, 1981. Edited by J. Llosa. X, 263 pages. 1982.

Vol. 163: J.S. Darrozes, C. Francois, Mécanique des Fluides Incompressibles. XIX, 459 pages. 1982.

Vol. 164: Stability of Thermodynamic Systems. Proceedings, 1981. Edited by J. Casas-Vázquez and G. Lebon. VII, 321 pages. 1982.

Vol. 165: N. Mukunda, H. van Dam, L.C. Biedenharn, Relativistic Models of Extended Hadrons Obeying a Mass-Spin Trajectory Constraint. Edited by A. Böhm and J.D. Dollard. VI, 163 pages. 1982.

Vol. 166: Computer Simulation of Solids. Edited by C.R.A. Catlow and W.C. Mackrodt. XII, 320 pages. 1982.

Vol. 167: G. Fieck, Symmetry of Polycentric Systems. VI, 137 pages, 1982.

Vol. 168: Heavy-Ion Collisions. Proceedings, 1982. Edited by G. Madurga and M. Lozano. VI, 429 pages. 1982.

Vol. 169: K. Sundermeyer, Constrained Dynamics. IV, 318 pages. 1982.

Vol. 170: Eighth International Conference on Numerical Methods in Fluid Dynamics. Proceedings, 1982. Edited by E. Krause. X, 569 pages. 1982.

Vol. 171: Time-Dependent Hartree-Fock and Beyond. Proceedings, 1982. Edited by K. Goeke and P.-G. Reinhard. VIII, 426 pages. 1982.

Vol. 172: Ionic Liquids, Molten Salts and Polyelectrolytes. Proceedings, 1982. Edited by K.-H. Bennemann, F. Brouers, and D. Quitmann. VII, 253 pages. 1982.

Vol. 173: Stochastic Processes in Quantum Theory and Statistical Physics. Proceedings, 1981. Edited by S. Albeverio, Ph. Combe, and M. Sirugue-Collin. VIII, 337 pages. 1982.

Vol. 174: A. Kadić, D.G.B. Edelen, A Gauge Theory of Dislocations and Disclinations. VII, 290 pages. 1983.

Vol. 175: Defect Complexes in Semiconductor Structures. Proceedings, 1982. Edited by J. Giber, F. Beleznay, J.C. Szép, and J. László. VI, 308 pages. 1983.

Vol. 176: Gauge Theory and Gravitation. Proceedings, 1982. Edited by K. Kikkawa, N. Nakanishi, and H. Nariai. X, 316 pages. 1983.

Vol. 177: Application of High Magnetic Fields in Semiconductor Physics. Proceedings, 1982. Edited by G. Landwehr. XII, 552 pages. 1983.

Vol. 178: Detectors in Heavy-Ion Reactions. Proceedings, 1982. Edited by W. von Oertzen. VIII, 258 pages. 1983.

Vol. 179: Dynamical Systems and Chaos. Proceedings, 1982. Edited by L. Garrido. XIV, 298 pages. 1983.

Vol. 180: Group Theoretical Methods in Physics. Proceedings, 1982. Edited by M. Serdaroğlu and E. İnönü. XI, 569 pages. 1983.

Vol. 181: Gauge Theories of the Eighties. Proceedings, 1982. Edited by R. Raitio and J. Lindfors. V, 644 pages. 1983.

Vol. 182: Laser Physics. Proceedings, 1983. Edited by J.D. Harvey and D.F. Walls. V, 263 pages. 1983.

Vol. 183: J.D. Gunton, M. Droz, Introduction to the Theory of Metastable and Unstable States. VI, 140 pages. 1983.

Vol. 184: Stochastic Processes – Formalism and Applications. Proceedings, 1982. Edited by G.S. Agarwal and S. Dattagupta. VI, 324 pages. 1983.

Vol. 185: H.N. Shirer, R. Wells, Mathematical Structure of the Singularities at the Transitions between Steady States in Hydrodynamic Systems. XI, 276 pages. 1983.

Vol. 186: Critical Phenomena. Proceedings, 1982. Edited by F.J.W. Hahne. VII, 353 pages. 1983.

Vol. 187: Density Functional Theory. Edited by J. Keller and J.L. Gázquez. V, 301 pages. 1983.

Vol. 188: A.P. Balachandran, G. Marmo, B.-S. Skagerstam, A. Stern, Gauge Symmetries and Fibre Bundles. IV, 140 pages. 1983.

Vol. 189: Nonlinear Phenomena. Proceedings, 1982. Edited by K.B. Wolf. XII, 453 pages. 1983.

Vol. 190: K. Kraus, States, Effects, and Operations. Edited by A. Böhm, J.W. Dollard and W.H. Wootters. IX, 151 pages. 1983.

Vol. 191: Photon Photon Collisions. Proceedings, 1983. Edited by Ch. Berger. V, 417 pages. 1983.

Vol. 192: Heidelberg Colloquium on Spin Glasses. Proceedings, 1983. Edited by J.L. van Hemmen and I. Morgenstern. VII, 356 pages. 1983.

Vol. 193: Cool Stars, Stellar Systems, and the Sun. Proceedings, 1983. Edited by S.L. Balliunas and L. Hartmann. VII, 364 pages. 1984.

Vol. 194: P. Pascual, R. Tarrach, QCD: Renormalization for the Practitioner. V, 277 pages. 1984.

Lecture Notes in Physics

217

Charge Density Waves in Solids

Proceedings of the International Conference
Held in Budapest, Hungary, September 3–7, 1984

Edited by Gy. Hutiray and J. Sólyom

Springer-Verlag Berlin Heidelberg GmbH 1985

Editors

Gyula Hutiray
Jenö Sólyom
Central Research Institute for Physics
H-1525 Budapest, P.O. Box 49, Hungary

ISBN 978-3-540-13913-3 ISBN 978-3-540-39137-1 (eBook)
DOI 10.1007/978-3-540-39137-1

Originally published by Springer-Verlag Berlin Heidelberg New York Tokyo in 1985

THIS VOLUME IS DEDICATED

TO THE MEMORY OF

WILLIAM L. MCMILLAN

The tragic death of William L. McMillan, known to his many friends as Bill, occurred just three days before the opening of the conference. Members of the conference were informed of his death by John Bardeen, who chaired the opening session. His remarks follow:

Last Saturday I received a cable with the shocking news of the death of Bill McMillan. He was killed last Friday, August 30, when struck by a car while riding a bicycle on a country road near his home in Urbana, Illinois. He was a good friend and colleague of many of us here. He was invited to attend this meeting, but declined because of his reluctance to travel.

Bill was very close to his family and left his wife, Joyce, and four children, three boys and a girl. The oldest is just starting graduate studies at Stanford, and three are in college.

He is noted for his basic and unique contributions to many areas of condensed matter physics, including liquid helium, superconductivity, liquid crystals, layer compounds and localization phenomena. His work will remain a permanent part of physics. His achievements have been recognized by the Fritz London award and election to the National Academy of Sciences and to the American Academy of Arts and Sciences.

I feel a deep sense of personal loss of a former student and close associate on the staff of the Physics Department at the University of Illinois.

I suggest that we take a moment of silence in memory of Bill McMillan.

WILLIAM L. McMILLAN

McMillan was born January 13, 1936 in Little Rock, Arkansas, so he died in his forty-eighth year. He held a bachelor's degree in electrical engineering and a masters degree in physics from the University of Arkansas, which also awarded him an honorary degree in 1979. He received a doctorate in physics from the University of Illinois in 1964 with a thesis on the ground state of liquid helium, in which he made an early application of Monte Carlo techniques. After graduation he went to Bell Labs, where, he and John Rowell did their famous work on deriving phonon spectra in superconductors from electron tunneling data. On a year's leave at Cavendish Laboratory and at Orsay, he began a study of liquid crystals, using a Landau-type theory to analyze the various phase transitions and carried out experiments to confirm predictions of the theory. He returned to Illinois as Professor of Physics in 1972 when he was still interested in liquid crystals. An outgrowth of this work was his theory of discommensurations and the commensurate-incommensurate charge density wave transitons in 2D layer compounds. This theory was referred to frequently at this conference and is basic to much of the work reported. More recently, using a novel computer he designed and built himself, he made use of powerful computer techniques to study localization, the metal-insulator transition and other problems of statistical physics. His death is a great loss to his family and friends and to the world of physics.

Preface

This volume contains most of the papers presented at the International Conference on Charge Density Waves in Solids, which took place in Budapest, Hungary, from 3-7 September, 1984. The reader will also find a few papers marked with a * in the contents; these were not delivered at the conference.

This conference grew out of the series of conferences on one-dimensional conductors. The dynamics of charge density wave systems and their non-linear properties have been so extensively studied recently, that the time was ripe for a specialized meeting concentrating on these aspects. That the field of one-dimensional conductors will develop in this direction had been already foreseen by Professor J. Bardeen in 1978, when in his concluding remarks after the International Conference on Quasi-One-Dimensional Conductors in Dubrovnik he said:

> *"It is evident that the field of quasi-one-dimensional conductors continues to be one of great interest and vitality. To me, the greatest interest is how charge transfer and charge transport occur in such systems, including the role of Fröhlich conduction and phonon drag. Although $NbSe_3$ is really a highly anisotropic 3D material with a 3D band structure, further studies should give considerable insight into Fröhlich conduction and pinning."*

Research conducted since that time proved fully that this field is of great vitality and it is our hope that the papers presented at the conference and published in these proceedings attest also to its interest. The volume contains both the invited and contributed papers presented at the conference.

The idea to organize this conference come originally from Professor G. Grüner, to whom we are indebted for his help in shaping the program. We regret very much that he could not attend the conference.

On behalf of the Local Organizing Committee

N. Kroó (chairman)		Gy. Hutiray (secretary)	
A. Jánossy	K. Kamarás	G. Mihály	J. Sólyom
A. Virosztek	A. Zawadowski	G. Zimányi	

we wish to thank the members of the International Organizing Committee

Y. Abe	P. Fulde	K. Maki	J. Rouxel
S. Barišić	B. Horovitz	N.P. Ong	F. di Salvo
W.G. Clark	A. Jánossy	V.L. Pokrovsky	E. Tosatti
I.Dzyaloshinsky	P.A. Lee	M.J. Rice	A. Zawadowski

and the Program Committee

J. Bardeen	V.J. Emery	P. Monceau	J. Sólyom
R. Blinc	R.M. Fleming	L. Mihály	C.M. Varma
S. Brazovsky	H. Fukuyama	T.M. Rice	J.A. Wilson
J. Cooper	G. Grüner	J.R. Schrieffer	

for their help and their valuable suggestions concerning the program.

The conference was sponsored by
IUPAP
European Physical Society
Hungarian Academy of Sciences
Roland Eötvös Physical Society
Central Research Institute for Physics, Budapest.

The participants of the conference learned from Professor J. Bardeen of the tragic death of Professor W.L. McMillan, which occurred a few days before the start of the conference. For his outstanding contribution to the subject, it was decided to dedicate these proceedings to his memory.

Budapest, September 1984

Gy. Hutiray
J. Sólyom

TABLE OF CONTENTS

II. Static Properties of CDW Systems

III. Dynamics of Charge Density Waves, Theory

IV. Charge Density Wave Transport

V. Hysteresis and Metastability

VI. Related Topics

ERRATA

LECTURE NOTES IN PHYSICS 217

Author Index

Charge Density Waves in Solids
Edited by G. Hutiray and J. Sólyom

SUMMARY

T.M. Rice
Theoretische Physik, ETH-Hönggerberg,
CH-8093 Zürich, Switzerland

Previous conferences on charge density waves were mostly concerned with electronic properties and the microscopic theory of the origin of the charge density waves (CDW). This conference focussed on the macroscopic properties, with around half the one hundred papers this week concerned with non-linear conductivity, narrow band noise, anomalous low frequency conductivity, hysteresis and metastability, etc. The start of this trend is easy to identify. It goes back to the discovery by N. Ong and P. Monceau in 1977 of the non-linear conductivity in $NbSe_3$. Since that time interest has built in these effects and the conviction has grown that we are studying an example of Fröhlich conductivity or the macroscopic motion of the charge density wave through the crystalline lattice although to date no completely definitive and irrefutible experiments have been made. The study of such macroscopic motion is clearly an exciting and challenging problem both theoretically and experimentally. Indeed in his new book P.W. Anderson cites this problem as one of the most challenging in many-body physics.

However during the week we also heard some excellent review talks on related topics such as the commensurate-incommensurate transition in insulating ferroelectric crystals by R. Blinc and the Abrikosov flux lattice in superconductors by P. Martinoli. In addition there were often comparisons to other systems such as the spin density wave in Cr, the phenomena of slip and creep in usual crystals and the motion of Bloch walls in ferromagnets. While none of the systems is an exact parallel to the CDW systems we can benefit a lot by the comparison to these analogous systems.

Returning to the main themes of the conference, a series of fascinating experiments were reported on the macroscopic properties and the impressive theoretical progress on these questions was extensively covered. It is impossible for me in this short summary to comment on all, or even most, of the work presented here and for that I request your understanding. I don't need to remind you that during the week, there were many disputed questions in both theory and experiment. Further the differences were not always just over details but often over fundamental points of the theory and experiments such as whether the relevant processes are classical or quantum, bulk or at the contacts. These discussions attest to the vitality of the field and hopefully will lead in time to a consensus on the nature of the underlying phenomena.

In a summary such as this I think I should present my views, or prejudices if you prefer, on some of the disputed points. Thus I favored the classical depinning theory of a CDW from bulk impurity pinning presented in the talks by D. Fisher and L. Sneddon. While the mean field theory is not exact, it seems to give a good description of the dependence of the current on the field and frequency dependence of the conductivity. It also should be extendable to cover such phenomena as onset delays, hysteresis and essential temperature dependences of the threshold field. Indeed a start on these problems was reported by P. Littlewood. The quantum theory based on the tunnelling of solitons and antisolitons in pairs was presented by J. Bardeen and as you saw, the fits to the experimental data on the nonlinear dependence of the current and the frequency dependence of the conductivity are impressive. The extension to account for such phenomena as the time delays in the onset of the nonlinear current and the temperature dependence of threshold fields is a challenge here as it is for the classical theories.

The experiments on selective heating of $NbSe_3$ samples presented by N. Ong were also convincing and to my mind demonstrated that, at least in their samples, the periodic noise is an end, or contact, effect. These experiments clearly support the elegant theories of such effects presented by K. Maki and L. Gorkov.

Nonetheless there are some nagging questions. One is the role of phase slip processes. Do they contribute also to the threshold field (E_{th}) ? In this regard I found the talk by J. Gill on $NbSe_3$ very interesting. He found that he could split E_{th} into a bulk term and a term inversely proportional to the sample length. The former he interpreted as due to bulk pinning. The bulk term was temperature independent. The latter he ascribed to the phase slip processes at the contacts and it was the contact term that was responsible for the rise in E_{th} as the temperature is lowered in $NbSe_3$. This may be an important clue to the understanding of the temperature dependence of E_{th} which up till now has been a puzzle. Clearly more effort should be expended on this question and on the related question of the possible role of bulk phase slip centers. In this regard a search with electron microscopes, etc. for the existence of dislocations in the CDW state would be most useful. Since one cannot easily grow a dislocation - free crystal so one must expect on general grounds that dislocations can exist in the CDW. In the case of 2H-$TaSe_2$ where the electron microscope is particularly effective, such dislocations have been seen in both, the incommensurate and commensurate phases. Unfortunately in 2H-$TaSe_2$ none of the interesting macroscopic effects are observed as I discussed in my own talk.

Another, to my mind, puzzling question is the relation between these macroscopic phenomena in commensurate and incommensurate systems. The discovery of these phenomena in other Nb and also Ta chain compounds, as we heard during the week, is reassuring. It shows that these phenomena are general and not simply confined to $NbSe_3$. For some years the orthorhombic form of TaS_3 has been known as another example for the macroscopic phenomena. However, this form of TaS_3 has an incommensurate--commensurate (IC) transition as the temperature is lowered through 130 K. The surprising experimental result is that the threshold field is not drastically changed in this process. One possible way out of this dilemma was suggested by B. Horovitz, namely that the commensurability threshold field was smaller than expected due to a special cancellation of matrix elements in this case. However thanks to the work of J. Dumas and C. Schlenker at Grenoble we now have another example of non-linear conductivity on both sides of a CI transition, namely the blue bronze $K_{0.3}MoO_3$. While several groups reported somewhat different values of the threshold field, the key result is that the CI transition does not lead to a dramatic increase in E_{th} in $K_{0.3}MoO_3$ either. Further as we heard from G. Travaglini a pinned phase mode can be clearly identified in the far infrared and a simple extrapolation with a cosine form for the commensurate potential gives a value for E_{th} that is much too large. This establishes that the CDW as a whole cannot be moving and leads one to conclude that the non-linear current is carried by charged discommensurations (DC) in this case. The very careful X-ray measurements reported by D. Moncton and K. Tsutsumi showed that the $\vec{Q}$-vector is not perfectly commensurate and so there is a small density of DC remaining in the C-phase. The very low frequency anomalies in the dielectric constants that R. Fleming discussed are presumably due to motion of these DC and leads to the conclusion that the DC are charged as M.J. Rice and coworkers predicted some years ago. Then there are the intriguing, hysteresis, remanence and metastability effects in electric fields that we heard about from the Budapest group in the talks of A. Jånossy, G. and L. Mihåly and from N. Ong. Clearly charged DC can move around in the electric field to maximize their electric dipole parallel to the field. Is this enough or is the CDW state itself a ferroelectric ? I don't see why on general principles it cannot be so but in the blue bronze the X-ray data that R. Fleming reported showed that an electric field made the sample less ordered in the transverse direction which would seem to rule out a ferroelectric ground state. A clear challenge to our experimental colleagues is to establish the charge and mobility of a single DC. All this talk of DC makes me somewhat uneasy about $NbSe_3$ where until now my prejudices were against the DC models of P. Bak and J. Wilson since the X-ray experiments do not show the expected high order satellites.

In this summary the coverage has of necessity been very limited because of time and not because of interest in the other work. Most notable is the omission of the microscopic theories, where there has also been a lot of progress. In this regard I am hopeful that the blue bronze, $K_{0.3}MoO_3$ may prove a simpler test case and that we can progress to the stage of quantitative tests of the microscopic theories. Certainly this will require us to include realistically the Coulomb effects in the DC as L. Gorkov and T. Holstein have been emphasizing this week.

I hope to have conveyed the impression of a lively conference on a lively topic with many fascinating questions, honest differences on experiments and theory and a challenging future to resolve these issues for all of us. Last but not least, I would like to thank the organizers of this conference for a splendid week. I know you have worked hard to achieve this success and I ask you all to join me in expressing our appreciation.

STRUCTURE

NEUTRON AND X-RAY SCATTERING STUDY ON $K_{0.3}MoO_3$ AND OTHER QUASI ONE DIMENSIONAL CONDUCTORS

Masatoshi Sato
Institute for Solid State Physics, University of Tokyo,
7-22-1, Roppongi, Minato-ku, Tokyo 106 Japan

Phase transitions to the charge density wave states in $K_{0.3}MoO_3$, $(TaSe_4)_2I$, $(NbSe_4)_2I$, Mo_8O_{23} and γ-Mo_4O_{11} are studied mainly by neutron and X-ray scattering techniques in order to see the natures of the electron phonon interaction in materials near the metal-insulator boundary. The transition in $K_{0.3}MoO_3$ may be understood by a band picture. The behavior of the transition in $(NbSe_4)_2I$ cannot be considered simply as the one of the band electrons. The magnetic susceptibility of the compound Mo_8O_{23} seems to be a unique example of the behavior predicted by Lee et al. for the fluctuation of one dimensional conductors.

§1 Introduction

Experimental studies mainly by neutron and X-ray scattering on the structural transitions to the charge density wave states in $K_{0.3}MoO_3$, $(TaSe_4)_2I$, $(NbSe_4)_2I$, Mo_8O_{23} and γ-Mo_4O_{11} have been carried out in order to clarify the dynamical natures of the transitions and the electron correlation problem. These compounds seem to be intermediate between the materials with wide band metallic electrons and with localized and strongly correlated electrons. Chakraverty[1] and Rice and Sneddon[2] proposed the similar insulating phases with local singlet electron pairs (bipolarons) as a neighboring phase of a superconducting one; with an increasing electron phonon coupling constant λ, the transition from the superconducting to the insulating phases takes place. While $M_xV_2O_5$ with 3d electrons is proposed to be a bipolaron insulator,[3,4] the so called tungsten bronzes, M_xWO_3 with wide spread 5d electrons are superconductors. The molybdenum bronzes, M_xMoO_3 with 4d electrons are between them. Although the present materials might not be good examples for the direct study of their arguments, they are interesting at least for the study of the electron phonon interaction near the metal-insulator (M-I) boundary, where a new situation may appear; the electron Fermi energy often becomes comparable with the phonon energy. Moreover, in some cases, due to the small polaron-like behavior, the electrons may be confined in the narrow regions of the crystals. Then, new low dimensional conductors can be found with the crystals being kept to have a three dimensional characters. This three dimensionality of the crystals is important for neutron scattering study, by which one can go beyond X-ray studies in some cases. Inelastic scattering has already been carried out on $K_{0.3}MoO_3$ [5,6] and it is being done on $(TaSe_4)_2I$. In the following section, the results on each kind of the compounds are presented.

§2 Results and discussions

2.1 Neutron and X-ray study on $K_{0.3}MoO_3$

$K_{0.3}MoO_3$ has a monoclinic structure. The ten octahedra share edges or corners and form a cluster. By the piling-up of the clusters along $\vec{b}$, four one dimensional chains are formed through a cluster. By the linkage of the chains, the sheets are formed in a plane perpendicular to the [20$\bar{1}$] reciprocal direction (see ref.6 for details). The incommensurate wave vecter q_b of the CDWalong $\vec{b}^*$ approaches 0.75b*. However, it does not seem to undergo a lock-in transition[5] and it has been pointed out that the behavior of q_b does not have a direct correlation with the anomalous T-dependence of the threshold field E_t of the nonlinear conduction observed by Dumas et al.[7] The X-ray work by Tamegai et al.[8] almost reproduced the T-dependence of q_b. The measurement of E_t by Tsutsumi et al.[9] has not found the anomaly observed by Dumas et al. In these two works, the crystals from the same batch as the present ones are used. The arguments on this problem are in ref. 9.

In Fig. 1, the phonon branches which contain the giant Kohn anomaly are shown. The T-dependence of the soft phonon energy is also shown. The softening seems almost complete as T approaches $T_c \simeq 180$ K. In the $S(\vec{q},\omega)$, there can be seen the

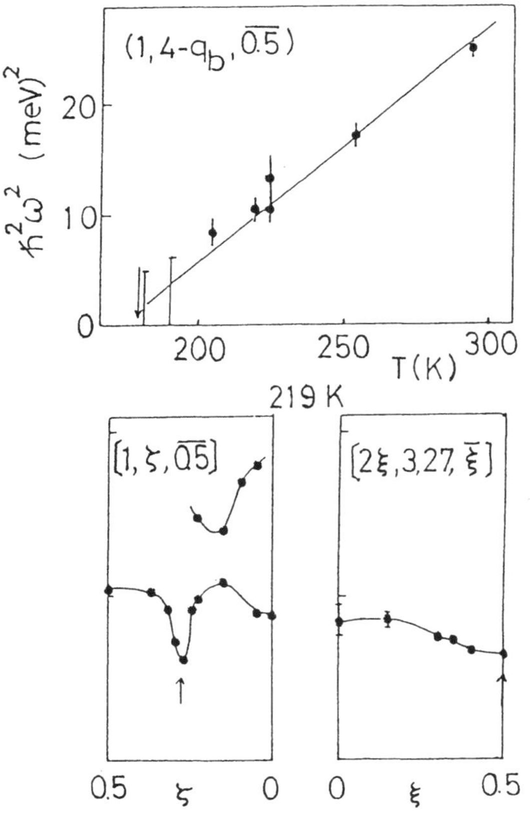

Fig.1 Dispersion curves which involve the giant Kohn anomaly indicated by the arrows. T=219 K. The T-dependence of the soft phonon energy is also shown.

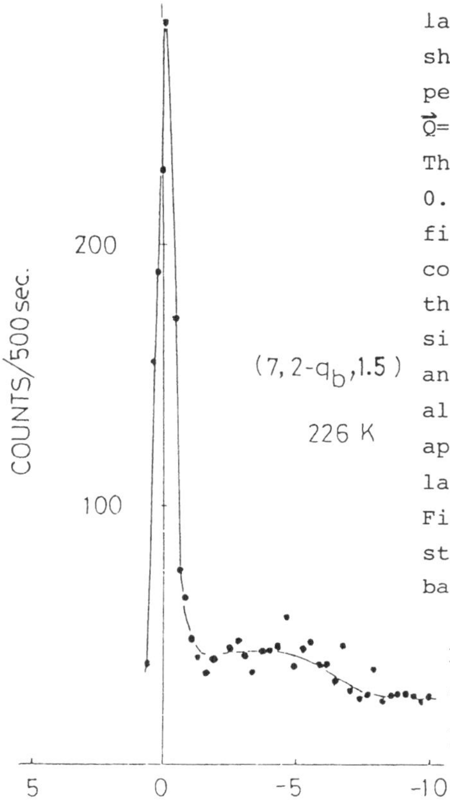

large central peak, an example of which is shown in Fig.2. Fig.3a shows the $\vec{Q}$-dependence of the elastic peak along the line $\vec{Q}=\eta(2\vec{a}^*-\vec{c}^*)+(4-q_b)\vec{b}^*$ with $0\leq\eta\leq1$ above T_c. The superlattice point corresponds to $\eta=0.5$. The significant asymmetry in the figure can be explained by the simple consideration of the structure factor of the atomic displacements. A model of a sinusoidal wave with a uniform amplitude and the wave vector $q_b b^*$ along $\vec{b}^*$ for all atoms within a layer seems a good approximation, where the neighboring layers have the anti-phase modulations. Fig.3b shows the results after the structure factor correction, where the back ground counts are already subtracted.

Fig.2 An example of $S(\vec{q},\omega)$ in $K_{0.3}MoO_3$[6] at the point which corresponds to the superlattice point below T_c. The elastic back ground is already subtracted.

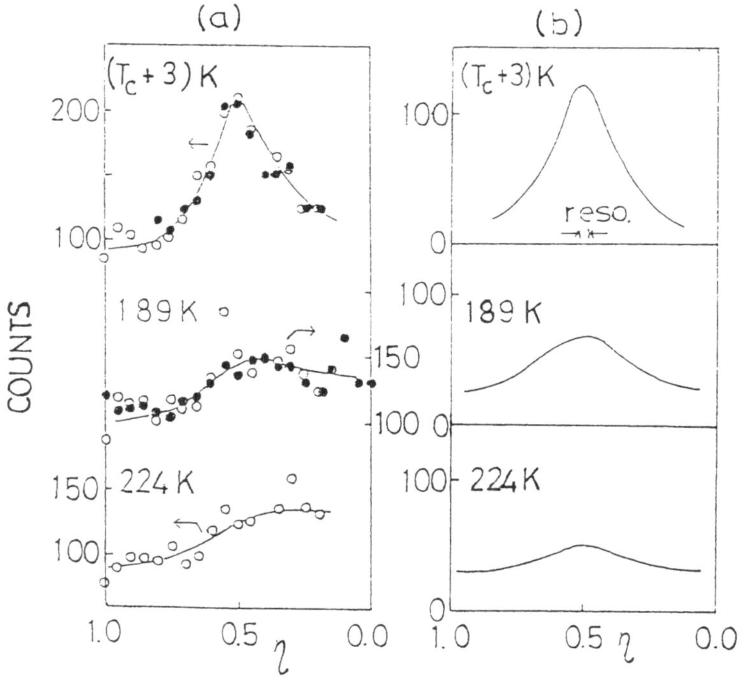

Fig.3
(a) Intensities of the elastic scattering along the line of the scattering vector $\vec{Q}=\eta(2\vec{a}^*-\vec{c}^*)+(4-q_b)\vec{b}^*$ for $0\leq\eta\leq1$. The superlattice point corresponds to $\eta=0.5$. Two kinds of symbols show the results of two different crystals. Asymmetry with respect to $\eta=0.5$ can be seen.
(b) Intensities of the elastic scattering after structure factor correction. The back grounds are subtracted.

If the relaxation time τ of the electron system is long compared with the inverse phonon frequency, the central peak appears with the incomplete phonon softening.[10] A possible mechanism to make the relaxation time longer is the strong electron correlation whatever the microscopic origins are, the Coulomb interaction or a formation of bipolaron through a strong electron phonon interaction. Although a problem remains that the central peak coexists with the almost complete phonon softening, further studies to search the experimental evidence of the correlation effect has been carried out. The $2q_b$ component of the modulation has not been observed by neutrons within an accuracy of 1/300. The X-ray Diffraction study with four circle diffractometer has also been adopted. The 212 superlattice reflections stronger than a significant level were used in the analysis. Gross features of the atomic displacements due to the CDW are as follows.

(i) Strong intensities are observed only for the superlattice points with $k=2n\pm q_b$ (n: integer), which indicates the modulation wave vector along $\vec{b}^*$ is $q_b b^*$.

(ii) The modulation is mainly transverse and the mean atomic displacement of the Mo atoms is almost within a plane formed with three K atoms between the layers (see Fig.4).

(iii) All clusters within a plane have in-phase modulations. The nieghboring layers have anti-phase modulations.

(iv) No $2q_b$ component of the modulation was detected within an accuracy of 1/500.

Although the displacements of the oxygen atoms could not be determined satisfactorily, it was found that the displacements of the Mo atoms were insensitive to those of other atoms. Then, it may be valuable to show schematically, in Fig.4, the (010) projection of the amplitude vectors, $\vec{\delta}_{0j}$ of the sinusoidal modulations of the Mo chains through the j-th crystallographically equivalent sites (j=1,2 and 3). The Mo2 atoms have a component parallel to $\vec{b}$ almost equal to that within the plane. The Mol and Mo2 atoms have only negligibly small components along $\vec{b}$. No sign of the superlattice reflection has been observed which comes from the difference between the $\vec{\delta}_{0j}$ values of the two chains through the crystallographically equivalent sites in a cluster, within the same accuracy as the case of the $2q_b$ component; in the sense that the two Mo chains have equal amplitude sinusoidal modulations, a local correlation which may bring the additional pairing or anti-pairing type displacements of the interchain Mo pairs in a cluster is very small. Then, there has been found no evidence of the correlation suggested by the existence of the central peak in $S(\vec{q},\omega)$.

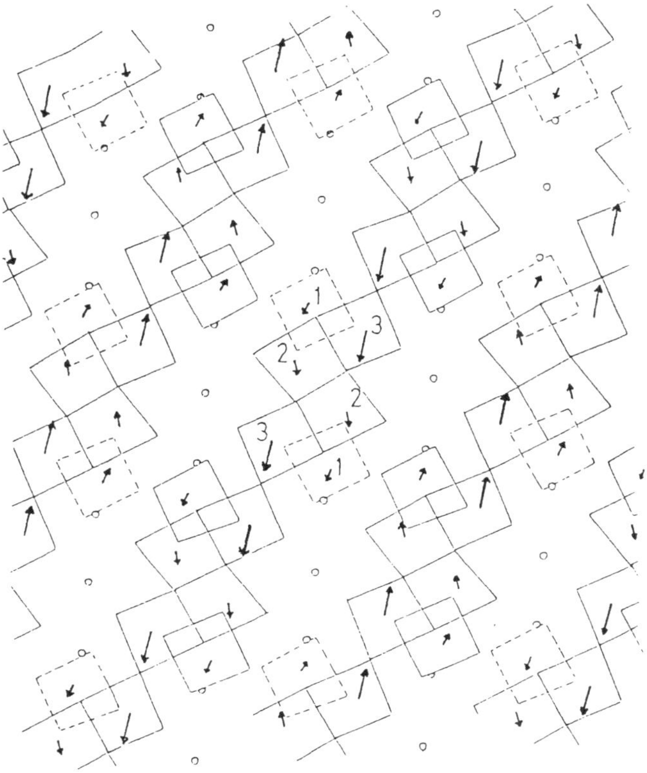

Fig.4 Schematic figure of the amplitude vectors, $\vec{\delta}_{0j}$ of the sinusoidal modulations of the Mo atoms. Only the component within the (010) plane are shown. The Mo1 and Mo3 atoms have negligibly small compnents along $\vec{b}$. The Mo2 atoms have a component parallel to $\vec{b}$ almost equal to that within the plane.
The numbers 1,2 and 3 indicate the Mo1, Mo2 and Mo3 sites, respectively. Open circles show the K atom sites. Oxygens are at the corners of the octahedra.

The CDW is well described by the sinusoidal modulations of the wave vector $q_b b^*$ along $\vec{b}$ and the band picture seems to be applicable rather well. Further study is necessary on the coexistence of the central peak with the almost complete phonon softening.

2.2 $(TaSe_4)_2I$ and $(NbSe_4)_2I$

$(TaSe_4)_2I$ has a tetragonal symmetry.[11] The Ta chains surrounded by the Se atoms indicate a typical one dimensional character. It undergoes the Peierls transition at about $T_c \simeq 260$ K. $(NbSe_4)_2I$ is first synthesized at Tokyo University and found to have the same type of the CDW state as that of $(TaSe_4)_2I$. The transition temperature $T_c \simeq 210$ K. The incommensurate $\vec{q}$ vectors of the CDW modulations are determined by X-ray diffraction[12,13] to be (0.05,0.05,0.085) and (0.065,0.065,0.159) for $(TaSe_4)_2I$ and $(NbSe_4)_2I$, respectively. Fig. 5 shows the examples of the X-ray data for the $(NbSe_4)_2I$. Similar data were also taken on $(TaSe_4)_2I$. Detailed analysis of the intensities of several superlattice reflections of $(TaSe_4)_2I$ has revealed that the transition is due to the condensation of the transverse acoustic phonon. As the

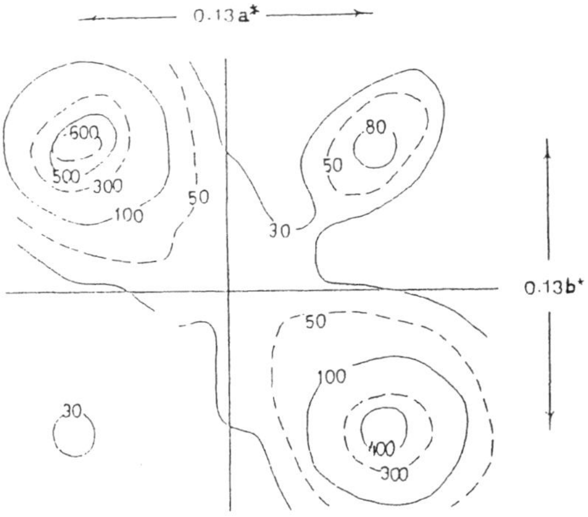

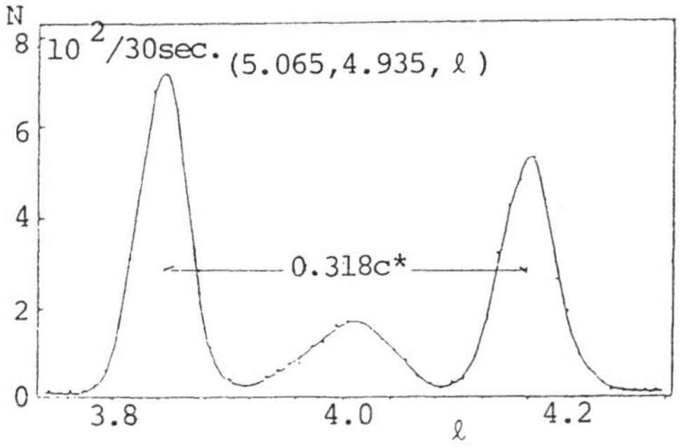

Fig.5 Top:Intensity map of the X-ray diffraction from $(NbSe_4)_2I$ around (554) Bragg point on the plane with ℓ = 4.159. The data were taken at about 110 K.
Bottom:X-ray superlattice reflection profiles of $(NbSe_4)_2I$ along the line (5.065,4.935, ℓ).

Although eight peaks are seen, they are just due to the existence of the different domains. Only two of them at the points symmetric with respect to the fundamental Bragg point should be observed if there exists only one domain.

case of $K_{0.3}MoO_3$, all atoms in a thick chain of Ta and Se are modulated transversely with almost equal amplitudes. Since the electron numbers are very small and roughly approximated to be 0.085 and 0.159 per two Ta and Nb atoms, respectively, the Fermi energies are estimated to be of the order of room temperature. Therefore in these compounds the electrons are easily localized and there is a possibility that the electron correlation has an essential role. Fig.6 shows the T-dependence of the magnetic susceptibility χ and the intensity I of the superlattice reflections of the two compounds.[14] The excess Curie type and the T-independent diamagnetic contributions are subtracted. While the $\chi(T)$ of $(TaSe_4)_2I$ is similar to that of other typical quasi one dimensional conductors, the $\chi(T)$ of $(NbSe_4)_2I$ behaves in rather different way; at T_c determined by the linear extrapolation of the low temperature part of the intensity I, there seems to be negligibly small anomaly. The maximum of $\chi(T)$ seems to take place at a temperature higher than T_c. The I(T) also behaves quite differently from that of $(TaSe_4)_2I$. The Nb compound with 4d electrons is expected to be closer to the M-I boundary than the Ta compound with 5d electrons. This may explain the difference between the behaviors of these compounds.

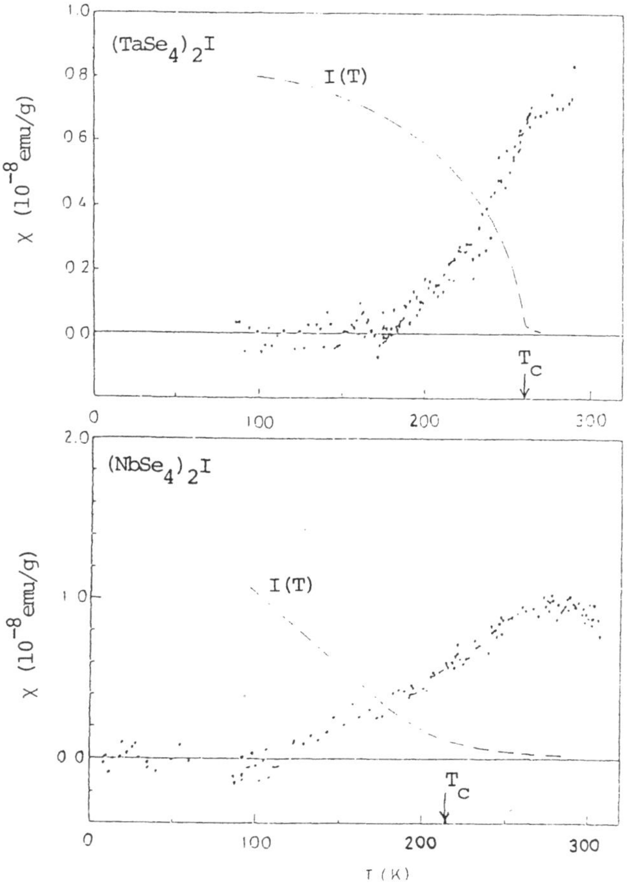

Fig.6
Temperature dependence of the magnetic susceptibilities and the intensities of the superlattice reflections for $(TaSe_4)_2I$ and $(NbSe_4)_2I$ are shown.
For the magnetic susceptibilities, the Curie parts and the T-independent diamagnetic contributions are subtracted.

It can be said at least that a simple band picture cannot be applied to explain the Peierls transition of $(NbSe_4)_2I$. Although it is not clear what microscopic mechanism is relevant in the compounds, the Coulomb interaction or the formation of bipolarons or any others, it is interesting to note certain resemblance of the behaviors of the anomalies in the $(NbSe_4)_2I$ to those of $Na_{0.4}V_2O_5$, which has been proposed to be explained by the formation of bipolarons of the almost localized electrons. Further study may be necessary to understand the behaviors of the lattice and electron systems consistently.

2.3 Mo_8O_{23} and γ-Mo_4O_{11}

In the series of work to search the new materials near the boundary of the M-I transition, several compounds with the CDW transition have been found. Here, let me briefly show the CDW transitions in Mo_8O_{23} and γ-Mo_4O_{11},[15] the crystal structures of which are monoclinic and orthorhombic, respectively. The compound $(MoO_3)_n$ is an insulator and Mo_nO_{3n-1} is conductive due to the presence of the small number of the oxygen deficiency. Fig.7 shows the temperature dependence of the resistivities along the three orthogonal directions. The anomalous maximum is seen in each curve. Especially the anomaly in the

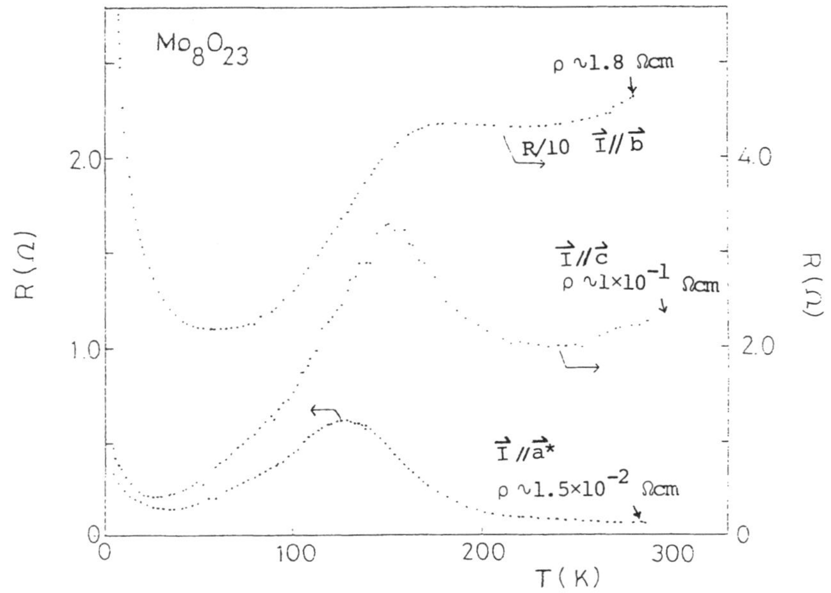

Fig. 7 Electrical resistivities of the compound Mo_8O_{23} along the principal directions. Rough values of the specific resistivities are attached at the corresponding temperatures. Here the c axis is defined in the crystallographical shear plane.

direction parallel to a* is remarkable. Fig.8 shows the temperature dependence of the magnetic susceptibility χ of the Mo_8O_{23} polycrystal. The Curie paramagnetic part probably due to the lattice imperfections and the T-independent diamagnetic contribution are subtracted. It is interesting that the T-dependence of χ is approximately explained by the simple fluctuation theory by Lee et al.[16] for one dimensional conductors. In the figure the broken line shows the calculated results with the mean field transition temperature T_c=455 K. The T-dependence of the ratios of the specific resistivities shown in Fig.7 suggests that Mo_8O_{23} is a quasi one dimensional conductor in high temperature region and a two dimensioanal one at low temperatures.

The γ-Mo_4O_{11} compound has a CDW transition at about 70 K.[15] Although the characteristics of the resistivities seem to be two dimensional-like, the T-dependence of the susceptibility χ suggests that the CDW fluctuation persists up to rather high temperature.

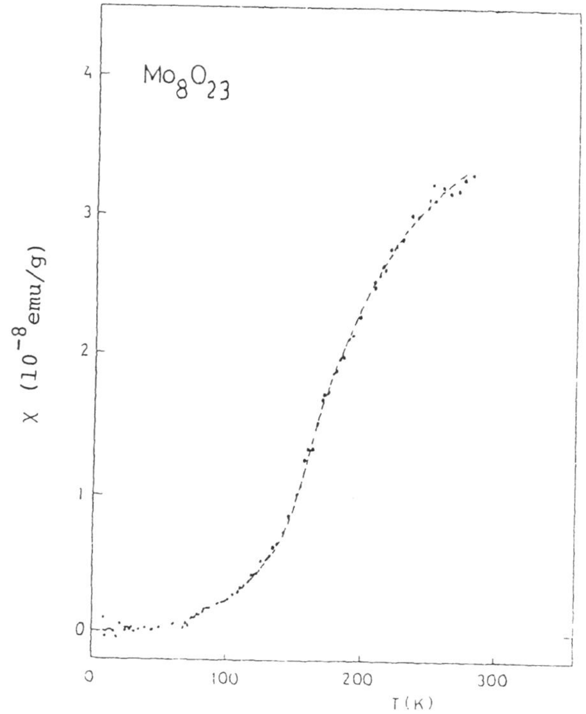

Fig.8 Temperature dependence of the magnetic susceptibilitiy of Mo_8O_{23}.[16] The T-independent diamagnetic part and the Curie paramagnetic part due to the lattice imperfections are already subtracted. The broken line indicates the behavior predicted by the theory of Lee et al. with the mean field T_c=455 K.

§3 Summary

Natures of the CDW transitions in $K_{0.3}MoO_3$, $(TaSe_4)_2I$ and $(NbSe_4)_2I$ are discussed. The transition in $K_{0.3}MoO_3$ seems to be explained by a simple band picture. Further detailed study may be necessary to understand the behaviors of the lattice and the electron systems of $(MSe_4)_2I$ (M=Ta and Nb) in a consistent way. The new materials with the CDW transitions are shown. (Mo_nO_{3n-1}) series is hopeful to study the strong electron-phonon interaction near the metal -insulator boundary.

The author is grateful to Prof. S. Hoshino and Drs. H. Fujishita, H. Nishihara, K. Nakao and S. Sato for their collaborations. He is also grateful to Dr. S. Kurihara for stimulating discussions. He is indebted to Prof. H. Ikeda for allowing his SQUID magnetometer.

References

1. B.K. Chakraverty: J. Physique Lett. 40 L199 (1979)
2. T.M. Rice and L. Sneddon: Phys. Rev. Lett. 47, 689 (1982)

3. B.K. Chakraverty, M.J. Sienko and J. Bonnerot: Phys. Rev. B17 3781 (1978)
4. H. Nagasawa, T. Erata, M. Onoda, H. Suzuki, Y. Kanai and S. Kagoshima: J. Physique C3 1737 (1983)
5. M. Sato, H. Fujishita and S. Hoshino: J. Phys. C16 L877 (1983)
6. M. Sato, H. Fujishita, S. Sato and S. Hoshino: submitted to J. Phys. C
7. J. Dumas, C. Schlenker, J. Marcus and R. Buder: Phys. Rev. Lett. 50 757 (1983)
8. T. Tamegai, K. Tsutsumi, S. Kagoshima, Y. Kanai, K. Tani, H. Tomozawa, M. sato, K. Tsuji, J. Harada, M. sakata and T. Nakajima: Solid State Commun. to be published
9. K. Tsutsumi, T. Tamegai, S. Kagoshima and M. Sato: J. Phys. Soc. Jpn to be published
10. S.K. Sinha: Dynamical Properties of Solids vol.3 edited by G.K. Horton and A.A. Maradudin North Holland (1980)
11. A. Meerschaut, P. Palvadeau and J. Rouxel: J. Solid State Chem. 20 21 (1977)
12. H. Fujishita, M. Sato and S. Hoshino: Solid State Commun. 49 313 (1984)
13. H. Fujishita, M. Sato, S. Sato and S. Hoshino: J. Phys. C to be published
14. M. Sato and H. Nishihara: submitted to J. Phys. C
15. M. Sato, K. Nakao and S. Hoshino: J. Phys. C to be published
16. P.A. Lee, T.M. Rice and P.W. Anderson Phys. Rev. Lett. 31 462 (1973)

X-RAY STUDY OF CHARGE-DENSITY WAVE IN $K_{0.30}MoO_3$ UNDER ELECTRIC FIELDS

K. TSUTSUMI, T. TAMEGAI, S. KAGOSHIMA and M. SATO†

Department of Pure and Applied Siences, University of Tokyo, Komaba 3-8-1, Meguro-ku, Tokyo 153, Japan

†Institute for Solid State Physics, University of Tokyo, Roppongi 7-22-1, Minato-ku, Tokyo 106, Japan

We performed a X-ray diffraction study of the charge-density wave in $K_{0.30}MoO_3$ under electric fields. In this experiment, we observed the change of the CDW wave vector above the threshold electric field E_T. The change occurs mainly in the ($2a^* - c^*$) direction which is perpendicular to the one-dimensional axis (b - axis). That also depends on the direction of the electric field and its magnitude. The time resolved X-ray diffraction study showed that the characteristic time of this change is of the order of one millisecond and depends on the electric field magnitude. We also measured transport properties of the CDW state in $K_{0.30}MoO_3$. They contain a thermal hysterysis in the ohmic resistance, an electrical hysterysis in the differential resistance and the temperature dependence of E_T which is different from that reported already by Dumas et al. On the nonlinear conductivity in $K_{0.30}MoO_3$, there are two types of samples. One is the swithing sample and the other the non-swithing sample.

The so-called molybdenum blue bronze $K_{0.30}MoO_3$ undergoes a metal-insulator transition at 180 K.[1)] The recent X-ray diffraction study revealed that the metal-insulator transition in $K_{0.30}MoO_3$ is accompanied by the growth of the charge-density wave (CDW).[2)] Together with an anisotropy of the electrical conductivity, it is concluded that the phase transition in molybdenum blue bronze is the Peierls one . Anomalous transport properties in the CDW state such as a nonlinear conductivity, a broad band noise, a narrow band noise and a swithing phenomenon are subjects of recent investigations in the physics of quasi one dimensional conductors.[3)]

In this article, we report and discuss our experimental results of "X-ray study of the CDW in $K_{0.30}MoO_3$ under electric fields". There are two kinds of samples on the nonlinear conductivity in $K_{0.30}MoO_3$. One is a swithing sample and the other a non-swithing sample. With the non-swithing samples, we performed two kinds of experiments ; (i) X-ray diffraction under d.c. electric fields, (ii) time resolved X-ray diffraction under pulsed electric fields. We confine ourselves only to the time resolved X-ray diffraction because of the page limitation. Time resolved X-ray diffraction was performed with the wave length of 1.198 A which was taken out by double crystal monochromator from the synchrotron orbital radiation and the diffracted beam was counted by a solid state detector (SSD). Samples are hanged by gold wires to the sample holder made of silicon wafers. We set samples so that the b^*-($2a^*-c^*$) reciprocal lattice plane is parallel to the reflection plane. The resolutions estimated from the width of Bragg reflections are 0.009 $\mathring{A}^{-1}$ along the b-axis and 0.003 $\mathring{A}^{-1}$ along the ($2a^*-c^*$) direction. Pulsed electric fields of positive and negative polarities were Zalternately applied as shown in Fig. 1 (a) by the two probe method. We accumlated the diffracted beam by a multichannel analyzer (MCA) in the multichannel scaling (MCS) mode with channel width of 10 microseconds. The start pulse of MCS mode was synchronized withthe beginning of the pulsed electric field as shown in Fig. 1 (b). The pulse response in the CDW state is shown in Figures 1 (c). Time dependent intensities of the diffracted beam at three positions are also shown in Figures 1 (d), (e) and (f) ; (d) at the satellite peak position, (e) +0.0014 ($2a^*-c^*$) apart from the peak position and (f) -0.0014 ($2a^*-c^*$) apart from the peak position. By measuring the time dependency of the satellite reflection profile, we could obtain the temporal evolution of the satellite reflection along the ($2a^*-c^*$) direction. It is shown in Fig. 2 (a). Figure 2 (b) shows the peak

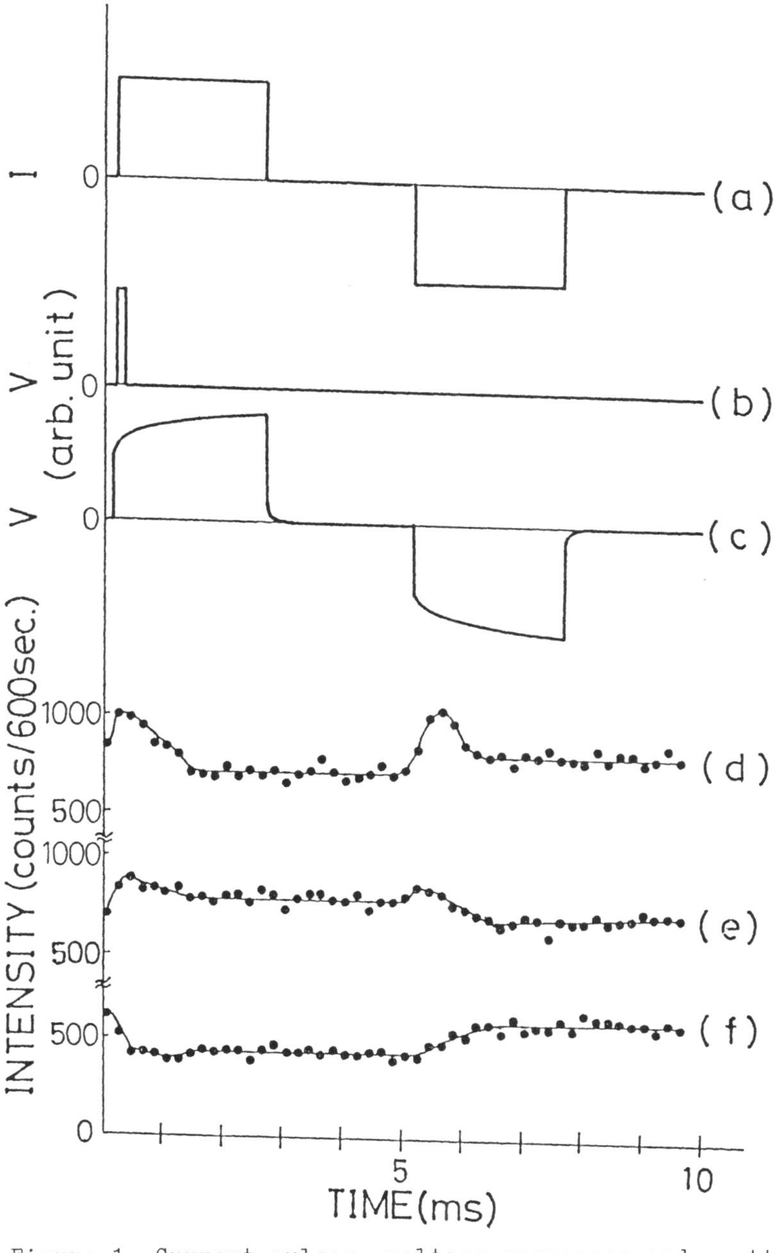

Figure 1. Current pulses, voltage responses and scattering intensities as functions of time. (a) alternating current pulses (b) MCS start pulse is synchronized to the current pulse (c) voltage response of the sample which is time depend with transient time of order of 1ms. Time resolved scattering intensities ; (d) at peak position (e) +0.0014($2a^* - c^*$) apart from peak position (f) -0.0014($2a^* - c^*$) apart from peak position.

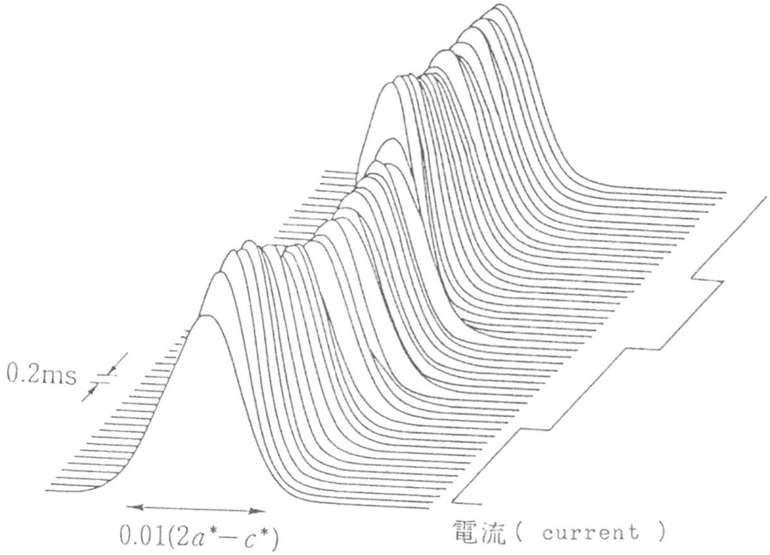

Figure 2(a) . The temporal evolution of the satellite reflection along the (2a* - c*) direction for I=4.0mA at 65 K. Each profile is best fitted to 6 data .

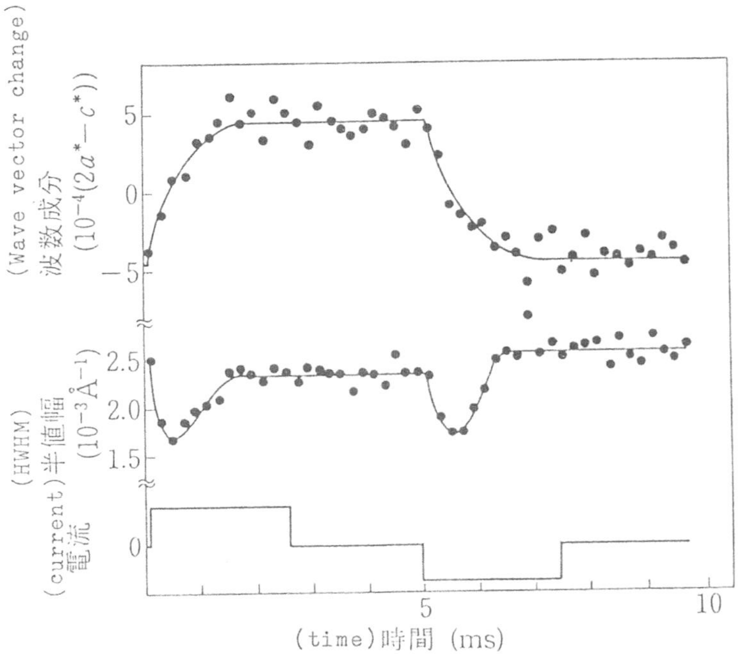

Figure 2(b). The relative position of the satellite reflection along the (2a* - c*) direction and the observed HWHM which are calculated by a Lorenzian best fitting to 6 data. The position 0 corresponds to -0.5 (2a* - c*).

position and the half width at half maximum (HWHM) as functions of time . Although the movement of the satellite reflection is only one-fifth of the full width at half maximum (FWHM) of the satellite reflection, the time dependent intensity is an undoubtful evidence of the wave vector change of the CDW in $K_{0.30}MoO_3$. We can find two stable states for the positive and negative polarities respectively. The transient time from one state to the other is of the order of one millisecond. This corresponds to the electrical transient time and decreases with increasing electric fields. In the transient time , the correlation length along the ($2a^*-c^*$) direction becomes longer and approaches to that of the Bragg reflection (> 1000 Å). However, it is about 500 Å in both steady states. The change of the wave vector is most remarkably seen around 70 K. Above 90 K and below 40 K, we can not observe such a change.

The change of the wave vector along the ($2a^*-c^*$) direction means that the phase difference between CDWs in the neighboring layers of MoO_6 varies as a function of electric fields. We speculate this change in the following way. We must keep in mind that real electrical contacts may be made at a point or a small area although we cover both ends of the sample by the silver paste. In case of such contacts, samples may have an electric field gradient in them as shown in Fig.3 (c). Such a field gradient results in the gradient of the force exerted to the CDW. The CDWs in layers of MoO_6 are expected to be strongly coupled with each other. The CDW in a layer will slide as a whole without deformations even when the field gradient might be present. However, the interlayer coupling is expected to be weak. Therefore, the CDW in a layer where the field is stronger will pull other CDWs in layers with weaker fields. Thus the phase difference between CDWs in neighboring layers of MoO_6 may change its magnitude and sign corresponding to the strength of the electric field and the polarity respectively. This phase change results in that of the ($2a^*-c^*$) component of the CDW wave vector in $K_{0.30}MoO_3$.

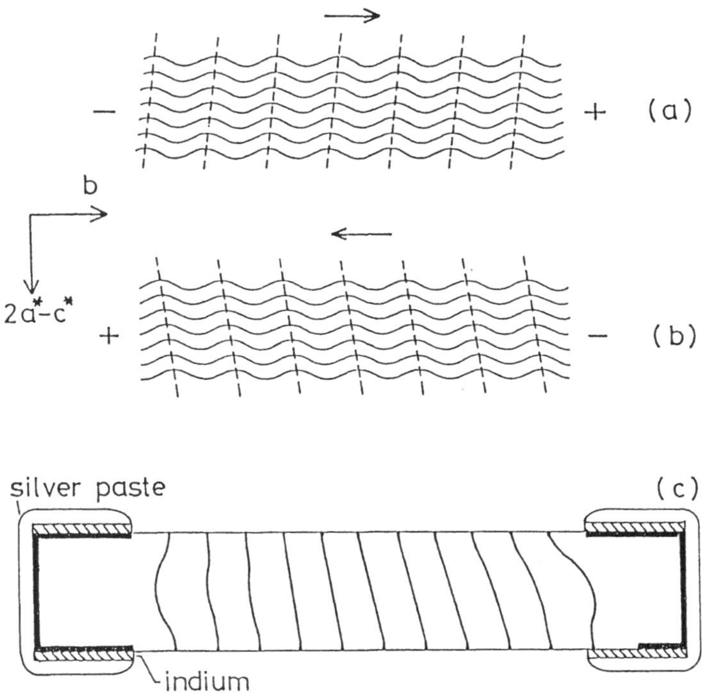

Figure 3. A possible configuration of the CDW in real space. In (a) and (b), sinusoidal curves represents the CDW in each layer of MoO_6. Dotted lines show planes of the same phase. Arrows indicate the sliding direction of the CDW. (c) equi-potential plane in the sample. The potential just under the electrical contact might not be equal. This may cause a field gradient in the sample.

References

1) W. Fogle and J.H. Perlstein : Phys. Rev. B 6 (1972) 1402.
2) J. P. Pouget, S. Kagoshima, J. Marcus and C. Schlenker : J. Physique-Letters 44 (1978) L 113.
3) G. Gruner : Comments on Solid State Phys. 10 (1983) 183 .

TRANSMISSION ELECTRON MICROSCOPY FOR IMAGING AND DIFFRACTION STUDIES OF LOW DIMENSIONAL TRANSITION METAL CHALCOGENIDES

D.M.Bird, D.J.Eaglesham, R.L.Withers, S. McKernan and J.W.Steeds
H.H.Wills Physics Laboratory, Tyndall Avenue, Bristol BS8 1TL, U.K.

Three techniques of transmission electron microscopy (satellite dark field imaging, high resolution imaging and convergent beam diffraction) and their use in the study of charge density wave materials are discussed. Several recent results on 1T-VSe_2, 1T-TaS_2, $TaTe_4$, $NbTe_4$, $(TaSe_4)_2I$ and 2H-$NbSe_2$ are presented.

With the introduction of liquid helium cooled specimen stages transmission electron microscopy has developed into an extremely powerful technique in the study of charge density waves in solids. This is due to its versatility in probing large scale (several μm down to around 5nm) CDW structures by satellite dark field imaging, intermediate structures (5nm to around 0.3nm) by high resolution imaging and the detailed arrangement and symmetry of atoms within CDW supercells by convergent beam electron diffraction. In this paper we review these techniques and present several recent results on quasi one and two dimensional CDW systems.

Satellite Dark Field Microscopy

When a crystalline specimen is illuminated by a parallel electron beam, a pattern of diffracted spots is formed in the back focal plane of the objective lens. By using apertures to reconstruct this pattern in various ways a number of different images can be produced. Bright field and dark field images are obtained by using only the undiffracted beam or one particular diffracted beam respectively. The resulting image is a map of how strongly each part of the crystal scatters into the chosen reflection. Were the specimen a perfect crystalline slab the image would be uniform because all parts diffract equally. Any contrast is therefore due to thickness variation and buckling of the specimen or, more interestingly, real defects such as dislocations and stacking faults. An important aspect of bright or dark field imaging is that images are projections of three dimensional structures. For example, an image of a boundary inclined relative to the incident beam consists of three regions, A and B well away from the fault and C, the area covered by the projection of the fault. A and B might be equivalent regions or differently oriented domains, this would be immediately apparent from the observed contrast. At C the contrast is more difficult to interpret because it depends critically on thickness, the angle of the boundary, the fault vector, the reflection used for imaging etc.However it is usually possible to determine the nature of the fault and the fault vector even if a detailed characterisation of the structure close to the fault is inaccessible.

With a modulated structure, if the aperture is placed around a satellite reflection a satellite dark field image results. Again the parts of the crystal diffracting into that spot are observed, so any contrast can be attributed to, for example, a CDW domain structure or defects in the CDW such as dislocations, stacking faults and discommensurations. The power of this technique in the study of CDW materials is to observe directly the evolution of CDW microstructures as a function of temperature (eg. through phase transitions) or of time at a given temperature. The resolution obtainable is limited to around 5-10nm.

(i) 1T transition metal dichalcogenides. 1T-VSe_2 undergoes two phase transitions at 110K and 85K. In the first modulated state the amplitude of the modulation is weak and has fundamental wavevector $\underline{q} \simeq 0.25\underline{a}^* + 0.314\underline{c}^*$[1]. At 85K the amplitude increases and the wavevector changes to $0.25\underline{a}^* + 0.307\underline{c}^*$. Satellite dark field images have been obtained using the three trigonally related fundamental satellites q_1, q_2, q_3 in both states. Above 85K each reflection produces similar and fairly uniform contrast, indicating that the whole specimen is in a 3q state. This is confirmed by convergent beam diffraction (see later) where none of the mirror planes of the basic 1T structure are broken in the modulated crystal. The lower temperature state is more interesting. Fig. 1a is a dark field image taken in a primary satellite reflection and clearly shows a well defined domain structure. We observe that using q_1, q_2 and q_3 type satellites, each domain is bright in two reflections and dark in the other. The contrast in the bright domains is fairly uniform and the dark domains are almost as dark as holes in the crystal. These results indicate that below 85K, 1T-VSe_2 transforms into a 2q CDW state. This has been confirmed by convergent beam diffraction (Fig. 1b) where the pattern from a single domain shows only two (q_2, q_3) of the three satellites and only one mirror of the parent 1T structure is retained. The observed domains correspond to the three variants of a 2q state with respect to the underlying lattice. Interestingly it can be seen from Fig. 1a that the domain boundaries do not preferentially lie along any crystallographic directions.

On cooling from the incommensurate $1T_1$ state 1T-TaS_2 undergoes two transformations, into the nearly commensurate $1T_2$ state at around 350K and into the commensurate $1T_3$ state around 180K. On warming, a new state, $1T_{2/3}$, is formed at 220K before $1T_2$ reappears at around 280K[2]. Our recent work has concentrated on the $1T_2$ and $1T_{2/3}$ phases. The microstructure of $1T_2$ is complex and satellite dark field images often show a mottled contrast and irregular fringes which are difficult to interpret. However, some features are consistently observed. Figs. 2a and 2b show dark field images of the same area using a primary $1T_2$ satellite at 250K and 200K. The weak linear features in Fig. 2b have sharpened into well defined boundaries at the lower temperature. These boundaries have been determined as being stacking faults in the

a)

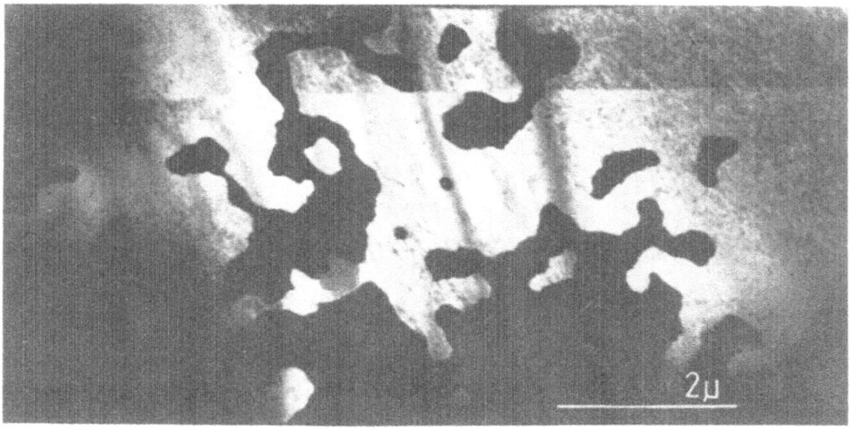

b)

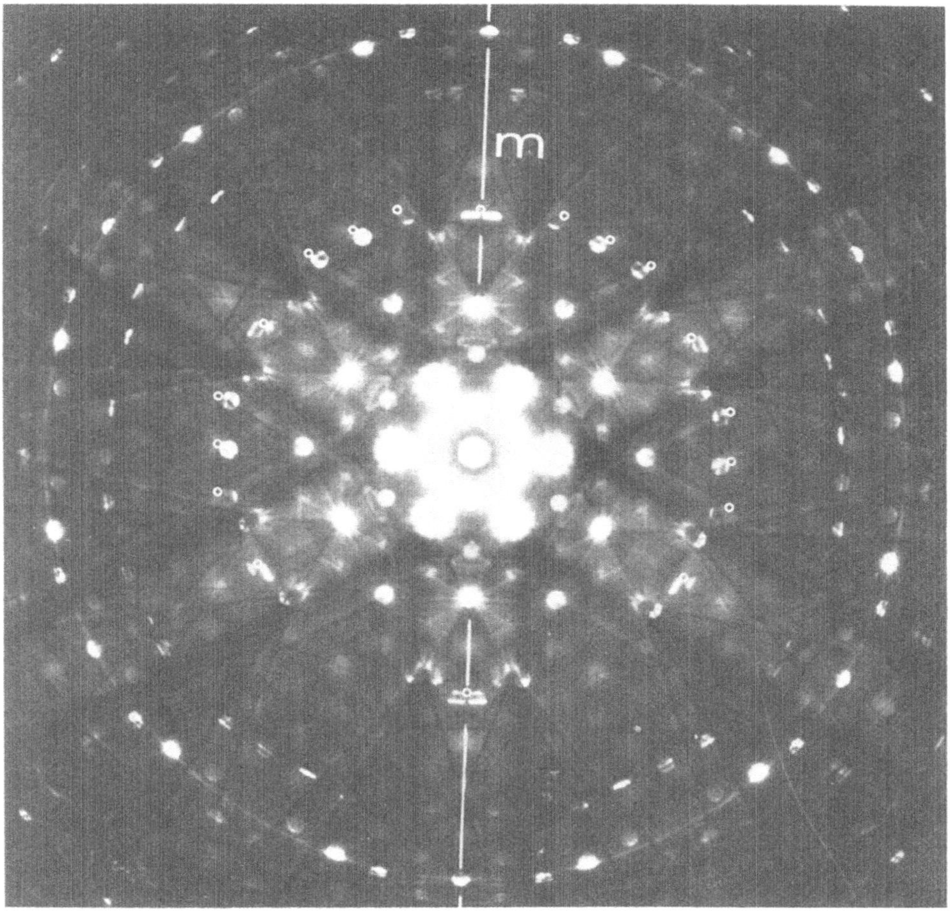

Fig. 1 a) Satellite dark field image showing domains in 1T-VSe_2 below 85K.
b) Convergent beam diffraction pattern from a single domain. The small circles in the first satellite HOLZ ring mark the position of basic lattice reflections.

a)

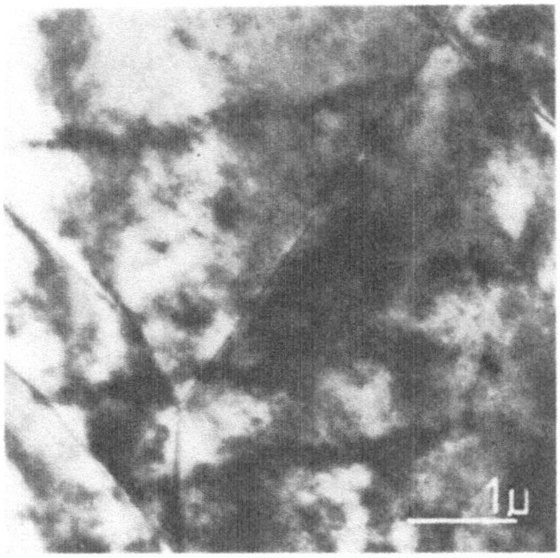

b)

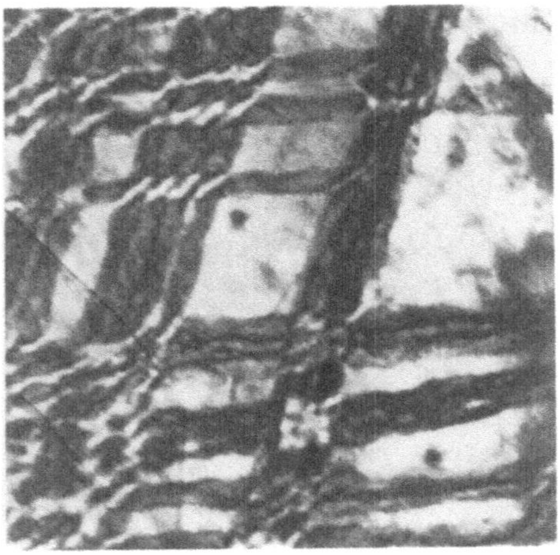

c)

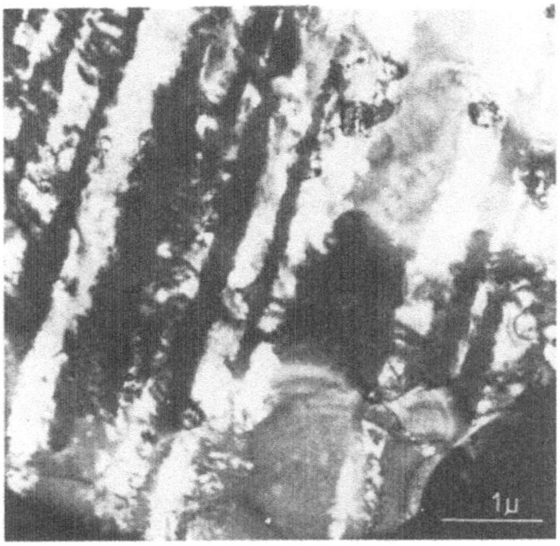

Fig. 2

a), b) Satellite dark field images of 1T-TaS_2 at 250K and 200K.

c) Satellite dark field image using a $1T_{2/3}$ reflection on warming at around 230K, showing intergrowths between $1T_2$ (dark) and $1T_{2/3}$ (bright).

d), e) Convergent beam diffraction patterns from $1T_2$ and $1T_{2/3}$ states.

d)

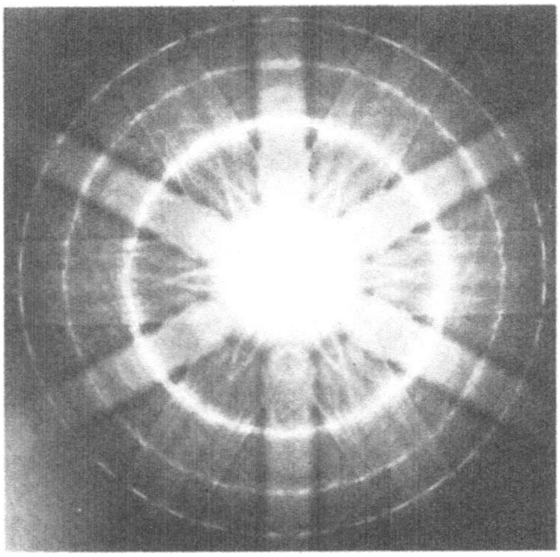

e)

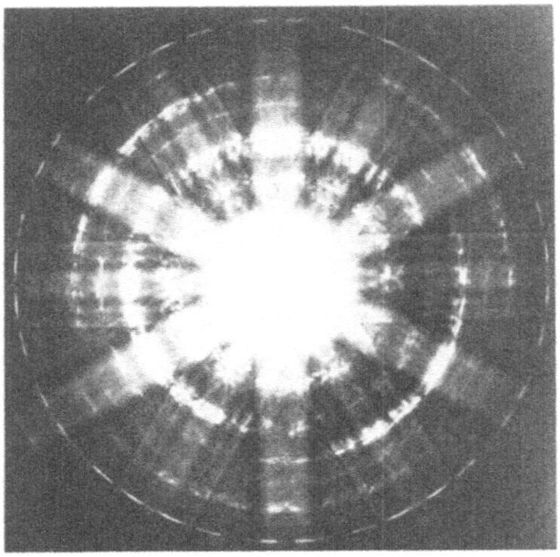

charge density wave, with fault plane close to $(1\bar{1}03)$ and fault vector $\underline{a}+\underline{c}$ [3]. While they are too far apart to be discommensurations, the consistency of their density and temperature dependence seems to indicate that they are equilibrium properties of the CDW. Our results on the $1T_{2/3}$ state partly contradict those of Tanda et al.[2] We observe $1T_{2/3}$ only on warming and satellite dark field images (Fig. 2c) show that it is intergrown with regions of $1T_2$. This is revealed by convergent beam patterns taken in a scan across the area shown in Fig. 2c. As the boundaries in 2c are crossed, the pattern alternates between the trigonal $1T_2$ pattern (Fig. 2d) and the characteristic triclinic $1T_{2/3}$ pattern (Fig. 2e).

(ii) 2H transition metal dichalcogenides. Below 33K 2H-$NbSe_2$ transforms into a triply commensurate CDW state.[4] A satellite dark field image of this phase is shown in Fig. 3a. The contrast is very similar to that previously observed in the triply incommensurate state of 2H-$TaSe_2$[5] (Fig. 4b). In particular, both materials exhibit characteristic 3-fold nodes in the pattern of fringes and beading along the fringes (Fig. 3). The contrast in 2H-$TaSe_2$ was interpreted as coming from a double honeycomb array of three differently oriented orthorhombic domains.[5] We conclude that a similar symmetry breaking occurs in 2H-$NbSe_2$.

(iii) Quasi one dimensional materials. Electron microscopy of the non-linear conductor $(TaSe_4)_2I$ is made difficult because specimens are beam sensitive and damage after only about 30 minutes at an accelerating voltage of 120kV. However, some diffraction patterns and images have been obtained. Satellite spots occur very close to the basic lattice reflections,[6] so dark field images must be formed using a basic reflection and some of its star of 8 satellites rather than from each satellite separately. Fig. 4 shows such an image, taken close to the [110] axis so that 4 satellites are strongly excited. An interesting pattern of fringes arising from the satellites is observed, particularly near the bottom right edge of the image, but a detailed interpretation remains for the future.

As much of our recent work on $TaTe_4$ and $NbTe_4$ has been written up in detail,[7] only a brief summary of the results is presented here. $TaTe_4$ is formed in a 2x2x3 commensurate superlattice and does not transform further. At room temperature $NbTe_4$ has a commensurate $\sqrt{2}x\sqrt{2}$ superlattice in the basal plane but is incommensurate in the chain direction with fundamental wavevector $q=0.344x2c^*$. At $\sim$50K it undergoes a first order lock-in into a 2x2x3 commensurate cell. Satellite dark field imaging of $TaTe_4$ (Fig. 5a) shows that the charge density wave is divided into equivalent domains by CDW stacking faults. These run both parallel and perpendicular to the chain direction and several of the fault vectors have been determined.[7] In $NbTe_4$ the lock-in transition is observed by dark field imaging using a satellite of the commensurate 2x2 basal plane supercell. This spot is present only as a diffuse streak in the incommensurate phase.[7] Fig. 5b shows the transition occuring; the bright

a) b)

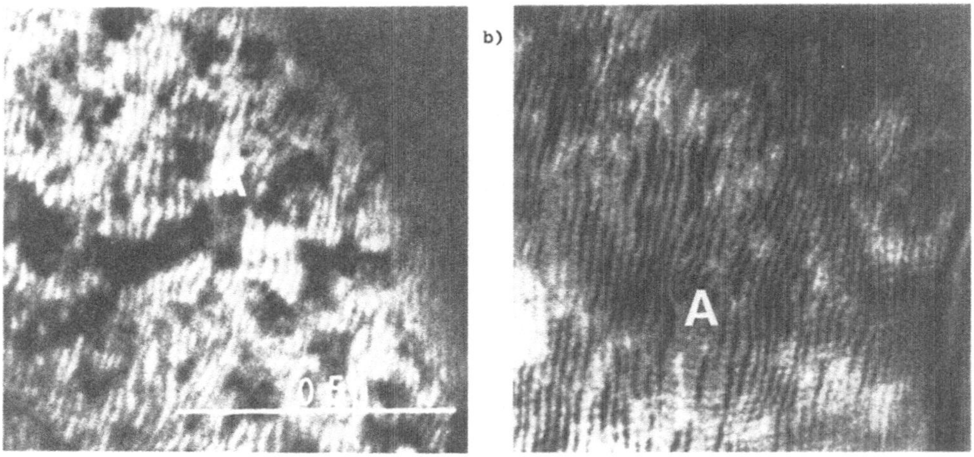

Fig. 3 Satellite dark field images in the triply incommensurate state of a) $2H\text{-}NbSe_2$ and b) $2H\text{-}TaSe_2$. Three fold nodes are marked A.

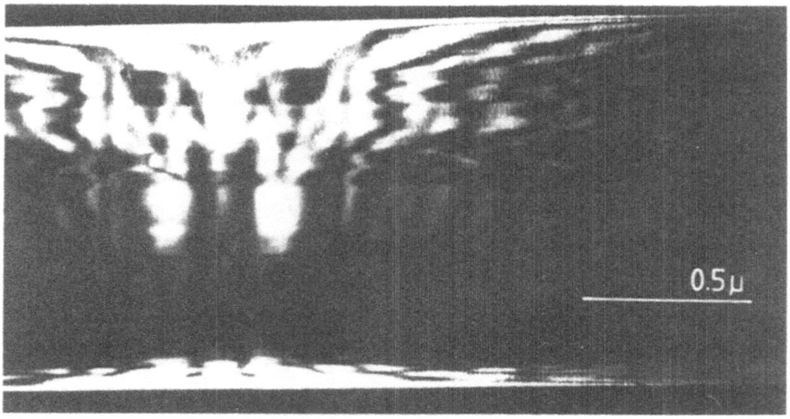

Fig. 4

Dark field image of $(TaSe_4)_2I$, showing fine fringes arising from satellite reflections.

Chain direction is horizontal.

a) b)

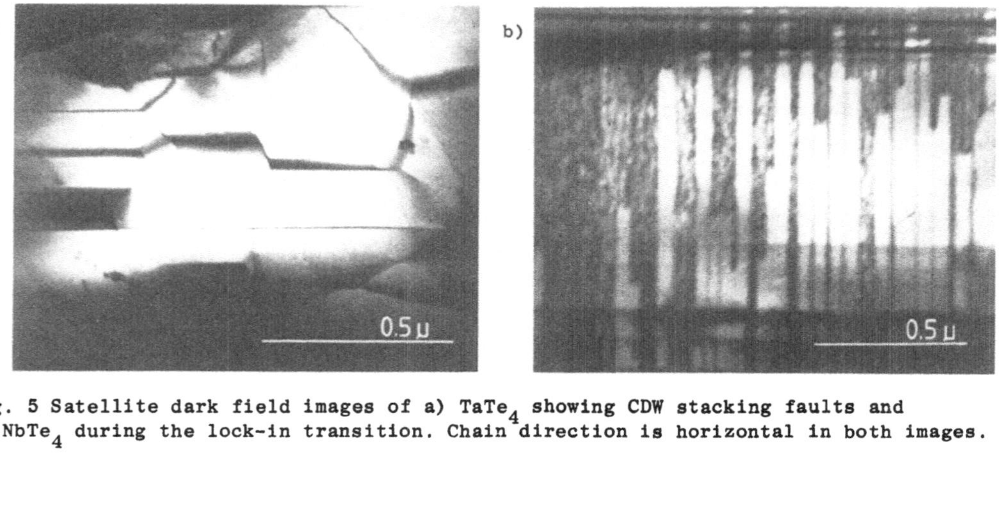

Fig. 5 Satellite dark field images of a) $TaTe_4$ showing CDW stacking faults and b) $NbTe_4$ during the lock-in transition. Chain direction is horizontal in both images.

patches are the commensurate phase, the incommensurate regions are darker. The transformation is very slow and is observed to proceed by the growth of the commensurate phase across the chain direction. When the transition is complete, the commensurate phase is found to have a similar structure to that in $TaTe_4$.

High Resolution Imaging

In order to determine the detailed phasing of an incommensurate CDW onto the underlying lattice (eg. through discommensurations) or the structure near faults in the CDW, imaging with a resolution better than 2-3Å is required. In principle this is done by allowing several reflections in the back focal plane (both basic lattice and superlattice) to interfere and form an image. In practice, the microscope has to have high electrical and mechanical stability to achieve this resolution, which effectively rules out the use of liquid helium and nitrogen cooled specimen stages. Also, quantitative interpretation of high resolution images is not straightforward because electrons are diffracted strongly in crystals thicker than about 50Å, and there are unavoidable aberrations in the imaging system.

$NbTe_4$ is a good test material for high resolution imaging because it has an incommensurate CDW at room temperature and fairly strong higher order satellites[7] which may indicate the presence of discommensurations. It is difficult to see the effects of the CDW in images of the thin crystals required for useful high resolution work. Fig. 6a therefore shows a processed image in projection down [110] from a rather thick crystal with the high and low frequencies filtered to reduce noise. A trebling in the chain direction and a cell doubling (consistent with a $\sqrt{2}x\sqrt{2}$ superlattice) across the chains is clearly visible. Fig. 6b shows line scans in the chain direction from a thin crystal image averaged over several alternate chains. The superlattice has been enhanced by a factor of three for greater clarity. A full quantitative analysis comparing computed images of trial structures with these images has not yet been carried out. However, by inspection of the scans in Fig. 6b it is apparent that the charge density wave is not strongly discommensurate in that there are not extended regions with a constant commensurate structure. Nevertheless, near the parts marked (a) there is some suggestion that the phase is slipping more rapidly than elsewhere.

Convergent Beam Diffraction

Two important features distinguish convergent beam electron diffraction (CBED) from more conventional diffraction techniques. First, electrons with a range of orientations are incident on the specimen. Diffraction patterns therefore consist of a set of discs rather than spots; within each disc a map of diffracted intensity as a function of orientation is observed. It is usual to choose the angle of convergence

a)

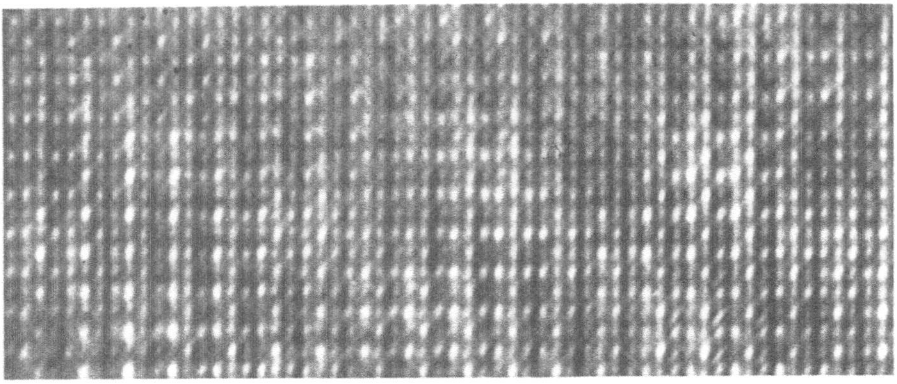

b)

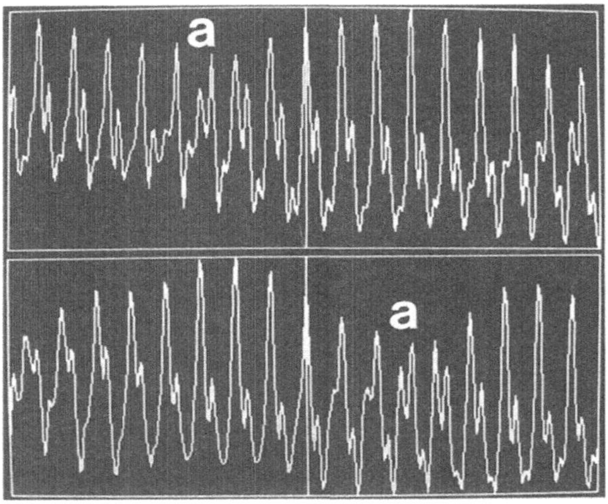

Fig. 6

a) High resolution image of incommensurate $NbTe_4$ at room temperature. The incommensurate modulation is best seen by viewing down the chain direction (horizontal).

b) Line scans in the chain direction averaged over several alternate chains.

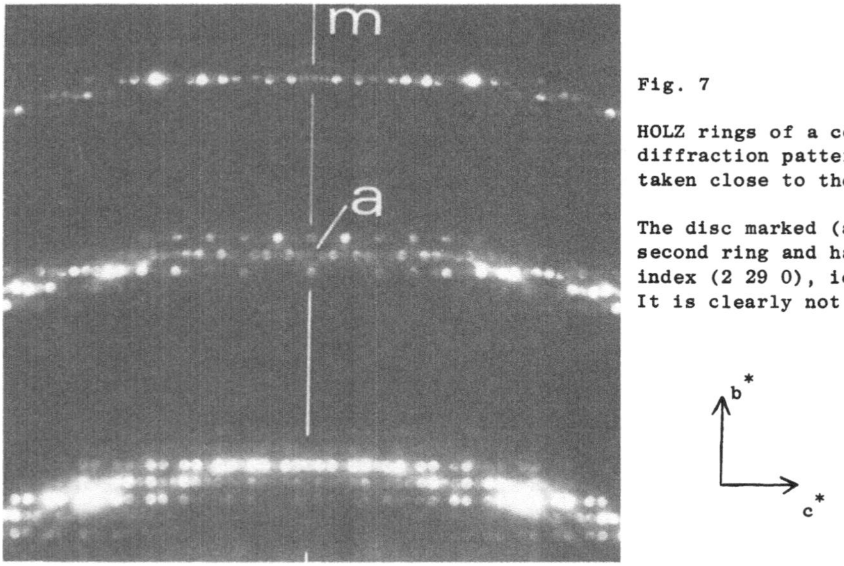

Fig. 7

HOLZ rings of a convergent beam diffraction pattern from $TaTe_4$ taken close to the [100] axis.

The disc marked (a) is in the second ring and has superlattice index (2 29 0), ie. h+k is odd. It is clearly not absent.

so that adjacent discs do not overlap (eg. Figs. 1b,7). Second, a convergent beam can be focussed down to a small area on the specimen, typically the probe size is between 50-500Å. It follows that patterns can be obtained from perfect regions of a crystal and, importantly for CDW studies, from single domains within a domain structure (eg. Figs. 1b, 2d, 2e). Patterns are generally taken at orientations close to zone axes. In this case the centre of the pattern represents diffraction from the structure projected down that axis. Three dimensional information is also present, however, in the concentric higher order Laue zone (HOLZ) rings which surround the central region. This is apparent in Figs. 2d, 2e and in Fig. 1b where, because the fundamental satellite in 1T-VSe_2 has a wavevector component parallel to $\underline{c}^*$, all the satellite reflections occur in two HOLZ rings, corresponding to c^* components q_z and c^*-q_z.

One major use of CBED is in symmetry determination. The symmetry of a pattern is a direct reflection of the point group elements of the crystal structure.[8] For example, the mirror in Fig. 1b reveals the presence of mirror planes, parallel to one of the basic lattice $\{11\bar{2}0\}$ planes, within the modulated crystal. Space group determination is also possible. In kinematic (single scattering) diffraction, glide planes and screw axes lead to forbidden reflections. In CBED these are replaced by lines of zero intensity (dynamic absences) in kinematically forbidden reflections.[8]

There has been some confusion in the literature over the basic lattice space group of $NbTe_4$ and $TaTe_4$. We have confirmed both to be P4/mcc.[7] The symmetry of the 2x2x3 superlattice of $TaTe_4$ and locked-in $NbTe_4$ has also been studied. It is found that the point group elements of the basic cells are retained[7], which leads to possible space groups P4/mcc and P4/ncc. A recent Landau theory investigation[9] has predicted the latter to be the most likely. Were this the case, the n glide would lead to absent reflections for (h+k) odd in the plane l=0. However, in convergent beam patterns from $TaTe_4$ such reflections are observed not to have dynamic absences (Fig. 7, see caption for details). The space group is thus determined to be P4/mcc. The reason for the discrepancy between this result and the Landau theory prediction is under investigation. One possibility is that it is not valid to make the assumption of ref. 9 that the chains, being only weakly coupled, can be treated independently.

A recent advance in CBED has been the development of a technique to fix atomic positions in the unit cell to within better than 0.01Å.[10] While not yet matching the accuracy of X-ray and neutron diffraction structure determination, this method has the usual advantage of convergent beam work that minute areas of a specimen can be probed. Indeed, suitable patterns may be obtained from as few as $\sim 10^5$ atoms. The technique relies on being able to show that in certain circumstances, the intensity of the fine branch structure observed in the HOLZ reflections of zone axis convergent beam patterns (eg. Fig. 1b) can be related to quantities which are similar to the kinematic structure factors for each reflection. By analysing the intensity variation in superlattice HOLZ reflections of patterns from the commensurate CDW state of 2H-$TaSe_2$, it has proved possible to determine the atomic displacements which accompany the CDW.[11] Fig. 8 shows these displacements in one layer of the 2H structure, the Ta and Se motions are exaggerated by factors of 20 and 50 respectively. The interpretation of this structure is discussed by Bird and Withers elsewhere in this conference.

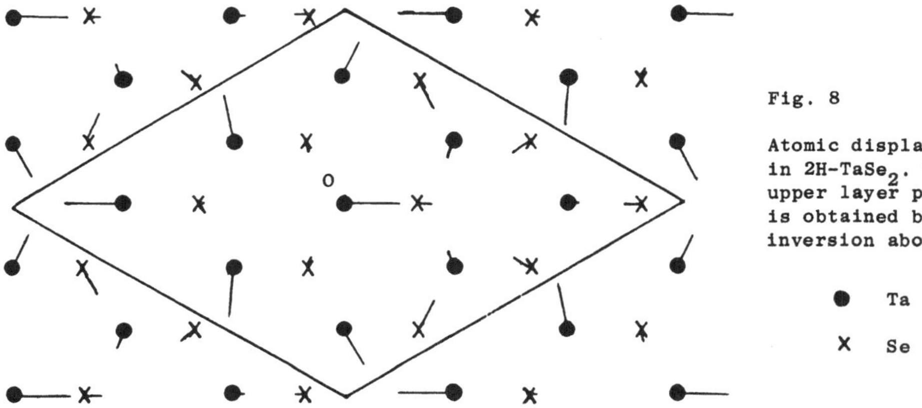

Fig. 8

Atomic displacements in 2H-$TaSe_2$. The upper layer pattern is obtained by inversion about O.

Acknowledgements

Four of the authors (DMB, DJE, RLW, SM) acknowledge financial support from the SERC. Most of the crystals used in these studies were grown by Mr. J. Burrow.

References

1. K. Tsutsumi 1982 Phys. Rev. 26, 5756
2. S. Tanda et al. 1984 J. Phys. Soc. Japan 53, 476
3. R. L. Withers and J. W. Steeds to be published
4. D. E. Moncton, J. D. Axe and F. DiSalvo 1977 Phys. Rev. B16, 801
5. K. K. Fung et al. 1981 J. Phys. C 14, 5417
6. H. Fujishita, M. Sato and S. Hoshino 1984 Solid State Comm. 49, 313
7. D. J. Eaglesham et al. 1984 J. Phys. C in press
8. R. Vincent and J. W. Steeds 1983 J. Appl. Cryst. 16, 317 and refs. therein
9. M. B. Walker preprint and this conference
10. R. Vincent, D. M. Bird and J. W. Steeds 1984 Phil. Mag. in press
11. D. M. Bird, S. McKernan and J. W. Steeds 1984 J. Phys. C in press

ASPECTS OF CHARGE-DENSITY WAVES IN THE $TaTe_4$-$NbTe_4$ STRUCTURES AND IN 2H-$TaSe_2$

M.B. Walker
Department of Physics and Scarborough College
University of Toronto, Toronto, Ontario M5S 1A7, Canada

Theoretical predictions of the charge-density-wave structure of the commensurate and incommensurate phases of $TaTe_4$ and $NbTe_4$ are reviewed and assessed in the light of recent experimental results. Some properties of domain walls in 2H-$TaSe_2$ are reviewed and the fact the even-layer and old-layer domain walls have different basal-plane projections is deduced from a symmetry argument.

In this talk I will describe some new work on a theoretical model of the commensurate and incommensurate distorted structures of the $TaTe_4$ - $NbTe_4$ series of compounds, and review in a new way, some older work on the interpretation of the charge-density-wave (CDW) state of 2H-$TaSe_2$.

Recently Boswell, Prodan and Brandon[2] have discovered an incommensurate phase in $NbTe_4$ and have suggested that both this phase and the closely related commensurate[2,3] phase of $TaTe_4$ might be interpretable in terms of CDW's (see also Ref. 4). They also pointed out that these materials should be of special interest since not only do they display interesting phases, but they are among the simplest materials which allow a study of CDW's by diffraction techniques. This simplicity has allowed a detailed theoretical prediction[1,5] of the nature of the distorted state to be made on the basis of a phenomenological model. I would like to begin by describing these results, and assessing their validity by comparing them with recent experimental work[6,7]. A point of special interest is that the model for CDW's on a single chain of the $NbTe_4$ structure is a prototypical model for a commensurate to incommensurate phase transition which has been much studied theoretically; to my knowledge, the $TaTe_4$ - $NbTe_4$ series provides the best physical realization of this model to date.

The undistorted structure of $TaTe_4$ and $NbTe_4$ is assumed to consist of a series of equivalent columns arranged as shown in Fig. 1. The assumed CDW distortion increases the size of the unit cell from the $a \times a \times c$ unit cell of Fig. 1 to a commensurate[2,3] $2a \times 2a \times 3c$ unit cell for $TaTe_4$ and gives both incommensurate[2,4,6] and commensurate[6] structures in $NbTe_4$.

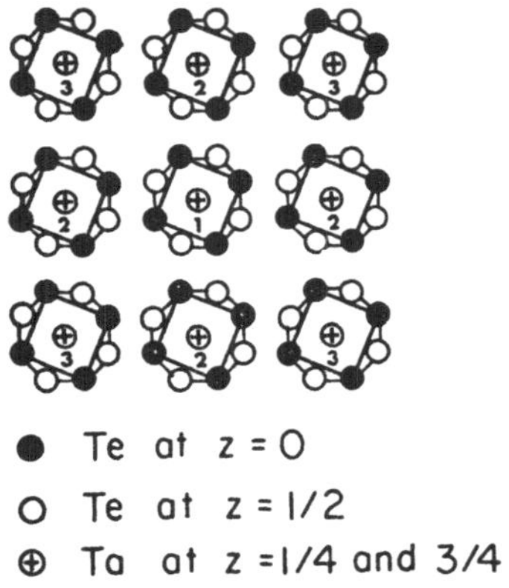

Fig. 1 The undistorted structure of $TaTe_4$ (Refs. 2, 3, 5, 6).

The approach of Ref. 1 assumes that the properties of a CDW on a single column can be discussed first and that the interaction between the columns can be added later. The CDW on a single column is characterized in terms of the additional charge per unit length, $\delta\rho(z) = \mathrm{Re}\,[\psi(z)\exp(iQz)]$, associated with its formation (here $Q = 2c^*/3$ and not $c^*/3$ as explained in Refs. 1 and 5). The complex order parameter ψ is determined by minimizing the free energy functional

$$F = \int\left[A|\psi|^2 + B\left|\left(i\frac{d}{dz} + \delta c^*\right)\psi\right|^2 - C\mathrm{Re}(\psi^3) + D|\psi|^4\right]dz \tag{1}$$

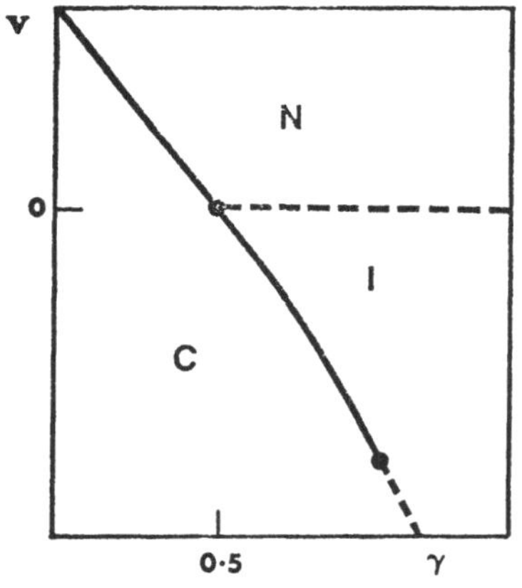

Fig. 2 Mean field theory phase diagram for a single chain; $v=AD/(2C^2)$ and $\gamma=B\delta^2 c^{*2} D/(2C^2)$. The phases are N, normal; C, commensurate; and I, incommensurate. Solid and dashed lines indicate first and second order phase transitions, respectively.

(F is determined by symmetry arguments). The minimization has been previously carried out by Jacobs and Walker[8] and yields the phase diagram shown in Fig. 2. The competition between the terms in C, which stabilizes the commensurate phase, and B, which stabilizes the incommensurate phase, is evident in the phase diagram. If one writes $\psi(z)=a \times \exp[i\theta(z)]$ and then makes the approximation that a is independent of z the Frenkel-Kontorova (or sine-Gordon) model analyzed for example by Frank and van der Merwe[9] and by McMillan[10] is obtained. It is worth emphasizing that the predictions of the two models differ significantly in that the interaction between two discommensurations, i.e., domain walls, is purely repulsive for the sine-Gordon model[11] whereas for the model of Eq. (1), the interaction between two domain walls can be attractive[8] as shown in Fig. 3.

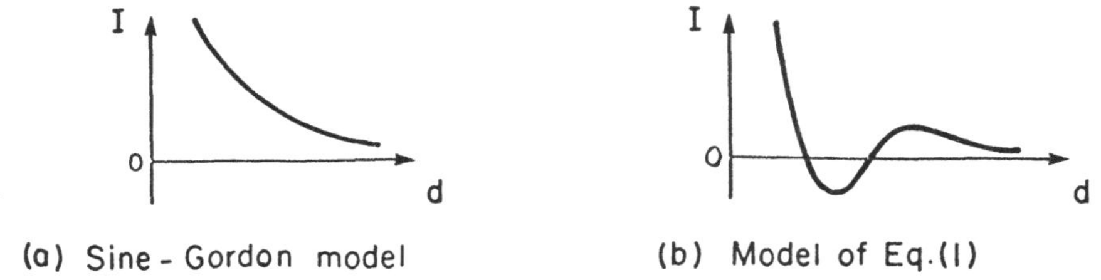

Fig. 3 Interaction energy of two domain walls versus their separation d.

The attractive component of the interaction between walls for the model of Eq. (1) causes the commensurate to incommensurate phase transition to be first order, as shown in Fig. 2. The strong first order commensurate to incommensurate transition in $NbTe_4$ recently described by Eaglesham et al[6] may be a partial reflection of this effect, but interchain couplings undoubtedly also play an essential role in causing the transition to be first order since the basal plane unit cell changes size in this transition.

A good test of the model[1] for CDW's in the $TaTe_4$ - $NbTe_4$ system is its ability to predict the detailed structure of the commensurate phase; detailed predictions have already been given by Walker [1]. For a single column the ground state is three-fold degenerate, the phase θ of the order parameter $\psi = a\exp[i\theta]$ (assumed constant in the commensurate phase) being found to be $\theta=0, +2\pi/3$ or $-2\pi/3$ by minimizing Eq. (1) for $C>0$. The three-fold degeneracy corresponds to the fact that different states of the same energy are produced by translating the CDW of wavelength $3c/2$ by distances of $+c/2$ or $-c/2$ along the c-axis. The problem of describing the commensurate state of the whole crystal is essentially a problem of determining the relative phases θ_j of the charge-density waves on the different columns; the index j = 1,2 or 3 refers to the three different columns shown in the $2a\times 2a$ unit cell of Fig. 1. A $2a\times 2a$ basal plane unit cell can only be obtained if both nearest-neighbor and next-nearest neighbor interactions are assumed to be such that the relative phases on the different columns like to be different; this is achieved by taking $\theta_2=0$, and $\theta_1=-\theta_3=2\pi/3$ assuming very weak intercolumn interactions. For stronger intercolumn interactions, $\theta_2=0$ and $\theta_1=-\theta_3=\theta_0$ where $\theta_0>2\pi/3$. This solution gives a commensurate phase with the space group No. 130, P4/ncc (D_{4h}^{8}).

It is of interest to compare the theory of the commensurate phase of a single column of $TaTe_4$ with that for the commensurate phase of a single layer of $2H-TaSe_2$. As was seen above the phase of the order parameter in $TaTe_4$ is determined by the sign of one real parameter, namely C, in the phenomenological model, and only one (degenerate) type of ground state is found. In $2H-TaSe_2$ on the other hand, the phase-dependent terms in the free energy contain three complex, i.e. six real parameters[8], and three distinct ground states are possible[12] depending on the values of these parameters. These considerations forcefully illustrate the relative simplicity of the $TaTe_4$-$NbTe_4$ system.

Independently of the theoretical study[1], Eaglesham et al[6] had carried out a convergent beam electron diffraction study directed at determining the space group symmetry of the commensurate phases of $TaTe_4$ and $NbTe_4$. They found the c-axis glide planes which the theoretical model predicts, but had suggested that the appropriate space group was No. 124 P4/mcc (D_{4h}^{2}). However, it was quickly realized after an exchange of preprints (Ref. 1 <-> Ref. 6) during a visit by the author to Bristol that the convergent beam results were consistent with either P4/mcc or P4/ncc and that further work needed to be done to experimentally distinguish the two possibilities. Recently, J. Brandon[7] has examined his electron diffraction data on $TaTe_4$ with a view to resolving this problem and has found the systematic absences characteristic of the basal plane being a diagonal glide plane, thus determining that the appropriate space group is P4/ncc and not P4/mcc, and confirming, to within the accuracy of his measurements, the theoretical prediction.

As described in Ref. 1, the predictions of the theoretical model are also consistent with the known properties of the incommensurate phase of $NbTe_4$.

I would now like to make a few remarks about discommensurations, i.e. domain walls, in $2H-TaSe_2$. In the time available I can focus on only one or two aspects of this fascinating subject; those interested in further information are referred to Refs. 12 to 18. In order to understand the structure of a domain wall, it is necessary first to understand the structure of the commensurate phase of $2H-TaSe_2$. In

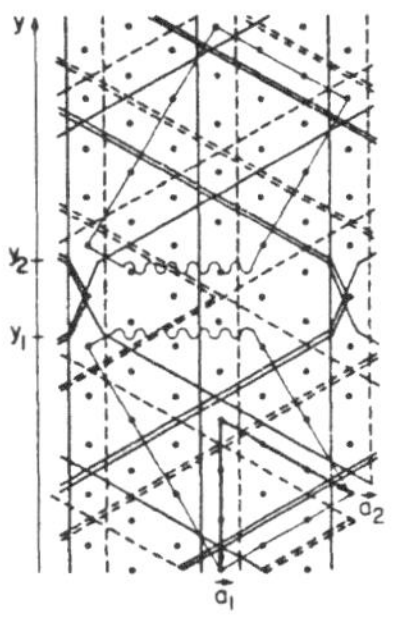

Fig. 4 Basal plane projection of an even-layer domain wall in the commensurate charge-density-wave state of 2H-$TaSe_2$. The solid circles represent Ta ions. The CDW state in a given layer is a superposition of three sinusoidal CDW's having their wave vectors in the basal plane and oriented at 120° relative to one another. Solid (dashed) lines represent the maxima of the charge-density of the individual sinusoidal waves in the even (odd) layers; the waves whose maxima are represented by double solid or dashed lines have different amplitudes and phases from those represented by single lines.

the absence of CDW's 2H-$TaSe_2$ has a hexagonal layered structure, the layers being parallel to the basal plane. There are two different types of layers (even and odd) which are stacked alternately as one goes up the c-axis. In the commensurate state, each layer contains three charge-density waves whose $\vec{Q}$-vectors make angles of 120° relative to each other, as illustrated in the bottom half of Fig. 4.

A domain wall is formed by starting in Fig. 4, for example, with the entire region of Fig. 4 having the commensurate structure of the region $y < y_1$; the charge density in the region $y < y_1$ is then held constant while the charge density in the even layers only in the region $y > y_1$ is translated by a normal state lattice constant (i.e. the distance between two Ta nearest neighbours). This model of a domain wall has been successfully used[13,17] to interpret the observed domain structures[15,16] associated with the stripe and double honeycomb phases of 2H-$TaSe_2$, and their associated CDW dislocations.

An important aspect of this interpretation is that the even-layer and odd-layer domain walls do not have the same basal plane projection[13]. This result has been previously shown to follow from a detailed analysis[14] of a phenomenological model. What I will try to do here is to show that this result is expected on the basis of a simple symmetry argument.

The essential property of an even-layer (or odd-layer) domain wall is that it displaces the even-layer (or odd-layer) CDW's by a lattice constant (see Fig. 4). This relative change in stacking of the CDW's gives a different overall crystal structure on either side of the wall (see Fig. 4).

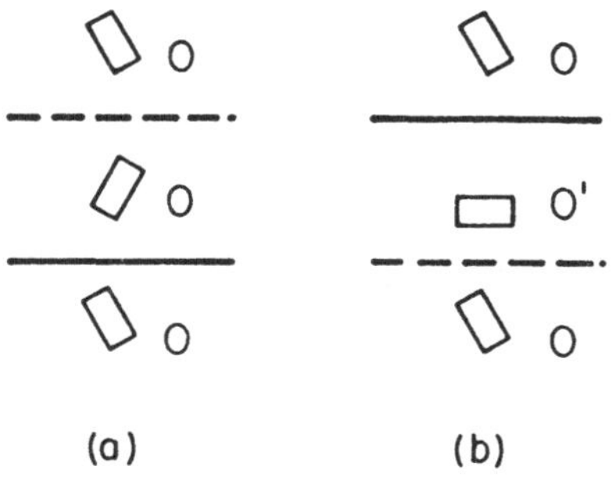

Fig. 5 Basal plane projections showing the changes in crystal structure caused by differing relative positions of even-layer (solid line) and odd-layer (dashed line) domain walls. O and O' label orthorhombic unit cells with different internal structures.

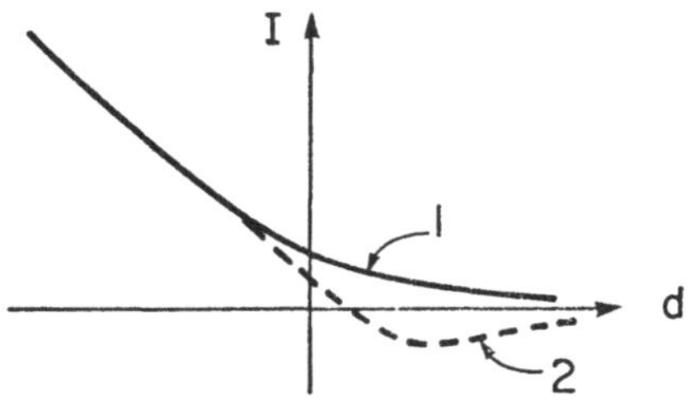

Fig. 6 The interaction energy I of even-layer and odd-layer domain walls of separation d. Curves 1 and 2 give two different possibilities.

Fig. 5 shows the change in overall crystal structure which results when even-layer and odd-layer walls are inserted in two different relative positions. In (a) starting at the bottom one finds that the unit cell is rotated by 120° when the even layer wall is crossed and then rotated back to its original position when the odd-layer wall is crossed. In (b) however, when the odd-layer wall is crossed on starting from the bottom, not only is the unit cell rotated, but the internal structure of the unit cell is changed as indicated by the prime attached to the symbol O'. Assuming the structure O to be the lowest energy structure, the structure O' must have a higher energy, and the overall energy of the system then increases linearly with wall separation in case (b); such is not the case in (a) where the states in the different regions differ only in unit cell orientation and not in internal structure. Thus the interaction energy of the two walls as a function of their relative separation varies schematically as shown in Fig. 6, where the wall separation is by definition positive for (a) and negative for (b) of Fig. 5. Note that the asymmetry with respect to positive and negative wall separations indicates that the two types of walls will in general have a positive and non-zero separation.

The model of a domain wall shown in Fig. 4 also enables one to deduce the different possible ways in which the domain walls can be joined together and thus allows an interpretation[13,17] of the microstructure observed[15,16] by electron microscopy. Fig. 7 shows the most commonly observed domain wall vertices; the interpretation of the electron microscope observations in terms of these vertices has been very successful.

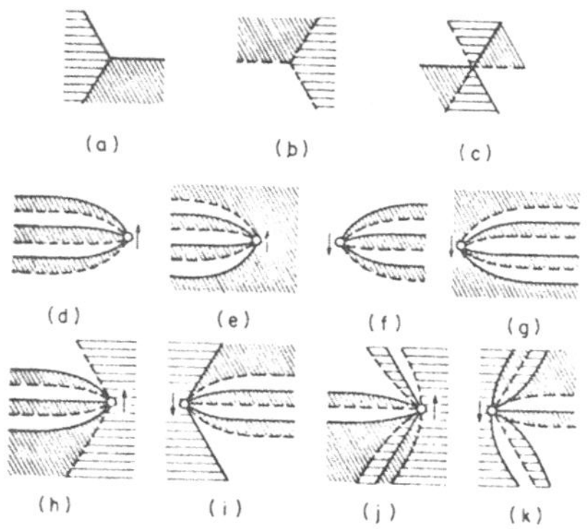

Fig. 7 The domain wall vertices most commonly observed in 2H-$TaSe_2$. (a) and (b) are the two three fold intersections and (c) is the sixfold intersection. (d) - (g) are six-line dislocations, (h) and (i) are seven-line dislocations, and (j) and (k) are nine-line dislocations. The arrow by each dislocation gives its Burgers vector; the intersections have Burgers vector zero.

References

1. M.B. Walker, Can. J. Phys. (expected issue, January, 1985).
2. F.W. Boswell, A. Prodan and J.K. Brandon, J. Phys. C 16, 1067 (1983).
3. E. Bjerkelund and A. Kjekshus, J. Less-Common Metals 7, 231 (1964).
4. J. Mahy, J. Van Landuyt and S. Amelinckx, Phys. Stat. Sol. (9) 77, K1 (1983).
5. D. Sahu and M.B. Walker, unpublished.
6. D.J. Eaglesham, D. Bird, R.L. Withers and J.W. Steeds, J. Phys. C (in press).
7. J. Brandon, private communication.
8. A.E. Jacobs and M.B. Walker, Phys. Rev. B21, 4132 (1980).

9. F.C. Frank and J.H. van der Merwe, Proc. Roy. Soc. A198, 205 (1949).

10. W.L. McMillan, Phys. Rev. B14, 1496 (1976).

11. J.K. Perring and T.H.R. Skyrme, Nucl. Phys. 31, 550 (1962).

12. M.B. Walker and A.E. Jacobs, Phys. Rev. B24, 6770 (1981).

13. M.B. Walker and A.E. Jacobs, Phys. Rev. B25, 3424 (1982).

14. A.E. Jacobs and M.B. Walker, Phys. Rev. B26, 206 (1982).

15. K.K. Fung, S. McKernan, J.W. Steeds and J.A. Wilson, J. Phys. C14, 5417 (1981).

16. C.H. Chen, J.M. Gibson and R.M. Fleming, Phys. Rev. B26, 184 (1982).

17. M.B. Walker, Phys. Rev. B26, 6208 (1982).

18. P. Prelovsek and T.M. Rice, J. Phys. C16, 6513 (1983).

CHARGE DENSITY WAVES, PHASING, SLIDING AND RELATED PHENOMENA IN $NbSe_3$ AND OTHER TRANSITION METAL CHALCOGENIDES

J.A. Wilson
H.H. Wills Physics Laboratory University of Bristol, BRISTOL BS8 ITL. U.K.

The material presented is largely drawn from a rewiev paper given at the Royal Society Symposium on Low Dimensional Solids held in May 1984 and to be published in January 1985 in The Philosophical Transactions of the Royal Society (Phil. Trans. A). The following is the abstract for that paper.

"The establishment of CDWs in $NbSe_3$ and the phenomena associated with their sliding are examined against the background of behaviour presented for a variety of other transition metal chalcogenides. The work has been set in terms of discommensuration arrays, but with these not so strongly defined as in the authors previous work (see J.Phys.F.Met. 12, 2469 1982 and Phys.Rev. B19, 6456, 1979). There the discommensuration array was used as generator of regular soliton shot noise. The ac noise is now related rather to creation and destruction of the spatial modulation of the CDW itself, occuring in close proximity to the contacts, and with the characteristic length $\lambda/2$. The discommensuration array governs the depinning processes from the lattice and impurities. Unlike in 2H-$TaSe_2$ the discommensurations carry appreciable charge ($\sim e/2$) because of the charge transfer between the inequivalent chains in the structure.

The differences between $NbSe_3$ and m-and o-TaS_3 (where localization effects are more important) have been examined. Related results on $(TaSe_4)_2I$ and other seleno-halides are discussed in conjun tion with recent results on the tellurides $TaTe_4$, $ZrTe_3$, $ZrTe_5$ to provide an overview of the materials aspect of the field. Suggestions are made for further-experiments".

Electron microscopy was undertaken in the laboratory by D. Eaglesham on the tellurides and will be reported shortly in J.Phys.C., Solid State.

Following the analysis made by D. Bird and coworkers of the structure of the commensurate phase of 2H-$TaSe_2$ (see this conference), a further discussion of the problems raised by the paper J.Phys.F. Metals 14, 123

(1984) has been given, which will appear shortly in J.Phys.F. The origin of the orthorhombicity would seen to be in line with the Γ K saddle points of the band structure finally playing a significant part in the broadly based nesting going on in 2H-$TaSe_2$. It should by emphasized with regard to material in the present conference that each discommensuration in 2H-$TaSe_2$ effects phase slip in each sandwich: double honeycomb geometry has a very general parantage.

We have unfortunately not been able to demonstrate any effect of a magnetic field on the phasing of the CDW in 2H-$TaSe_2$ (see Herrenden Harker this conference), but for reasons given earlier I would encourage those with very high field facilities to make examination of this possibility. All the various phasing possibilities are very close in energy ad demonstrated by the work of Doran and Woolley.

STRUCTURAL AND ELECTRICAL PROPERTIES INTERPRETATION THROUGH BAND STRUCTURE CALCULATIONS ON THE $(MSe_4)_nI$ SERIES (M = Nb, Ta).

P. Gressier, A. Meerschaut, J. Rouxel and M.H. Whangbo*

Laboratoire de Physicochimie des Solides, LA 279
2, rue de la Houssinière - 44072 NANTES Cédex, France.

* permanent address: Department of Chemistry, North Carolina State University, Raleigh, NC 27695-8204, U.S.A.

Halogened transition metal tetrachalcogenides $(MX_4)_nY$ [M = Nb, Ta ; X = S, Se ; Y = I, Br, Cl , n = 2, 3, 10/3, 4] provide us with a new series of pseudo-1D compounds with exciting properties. In these compounds infinite (MX_4) chains are well separated from each other by halogen chains. Metal ions are sandwiched by two rectangular chalcogen units with a dihedral angle of approximately 45°. According to n, different sequences of metal-metal distances can be observed in unit cells. There are also various possible arrangements of rectangular units and different halogen ions environments. Differences in structures are reflected in drastically different electronic properties such as semiconducting properties or occurrence of charge-density-waves (CDW) with related non-linear effects. A systematic study of the structure of $(MSe_4)_nI$ compounds is presented along with tight binding band calculations. It is shown that Peierls distortion and CDW phenomena are strongly affected by dz^2 band filling. As n decreases, dz^2 band filling decreases from 1/2 leading to a decrease in the tendency to metal chain distortion. 3D interactions are also discussed in connection with CDW occurrence.

I - INTRODUCTION

Transition metal trichalcogenides, especially $NbSe_3$ and TaS_3, have been largely studied for their one-dimensional properties **(1)**. They consist of MX_6 trigonal prisms (M = Nb, Ta ; X = chalcogen) which share their X_3 triangular faces to form MX_3 chains parallel to the growing axis of the crystals (fig. 1). Transition metal atoms located at the center of each MX_6 prism interact along the chain axis through dz^2 orbitals. There is also some interchain interaction through interchain M-X bonding. What is remarkable in those compounds is the occurrence of X_2^{2-} pairs as well as X^{2-} ions.

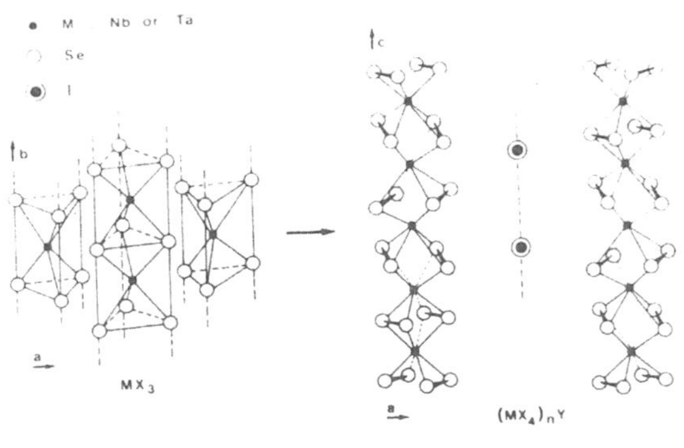

FIG. 1

These anions behave as "electrons reservoirs", which governs the amount of electron density available along the metallic chains and therefore has drastic effects upon their physical properties.

Transition metal tetrachalcogenides $(MX_4)_nY$ (M = Nb, Ta ; X = S, Se ; Y = halogen ; n = 2, 3, 10/3, 4) are also one-dimensional in structure **(2)**. In the following we will focus on the iodine derivatives $(MSe_4)_nI$. These compounds crystallize with tetragonal symmetry and consist of MSe_4 chains (fig. 1) which are parallel to c axis and separated by iodine atoms. In an MSe_4 infinite chain, each metal is sandwiched by two rectangular selenium units. The dihedral angle between adjacent rectangles is 45°, so that the stacking unit is an MSe_8 rectangular antiprism. The interaction between metal atoms is only through dz^2 overlap along the chain. The shortest interchain metal-metal distance is about 6.7 Å, significantly larger than the intrachain average value d of about 3.2 Å. The shorter Se-Se side of rectangles is about 2.35 - 2.40 Å while the longer one is about 3.50 - 3.60 Å, which is comparable to the shortest interchain Se-Se distance. The former value is typical of a Se_2^{2-} pair so that, if there were no iodine in the structure, we have the formal oxidation state $M^{4+}(Se_2^{2-})_2$, (i.e. a metal d^1 configuration). Iodine atoms are well separated from one another and can be considered as I^- ions. This leads to a decrease in the number of available d electrons : for $(MSe_4)_nI$, the average number of d electrons on each metal ion is (n-1)/n.

Every two adjacent Se_4^{4-} units have a dihedral angle θ of about 45° (fig. 2 a,b). This is mainly due to their occupied π and π^* orbitals. First, the overlap between such adjacent orbitals is minimized, therefore minimizing repulsive interactions. Also, at θ =45°, π and π^* orbitals provide with symmetry-adapted orbitals which interact with all metal d orbitals. except for dz^2. Fig. 3 represents the orbitals levels of a MSe_8^{4-} unit as a function of θ **(3)**.

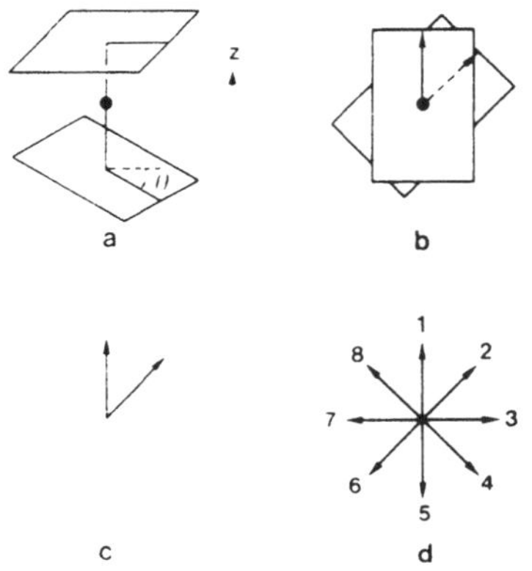

FIG. 2

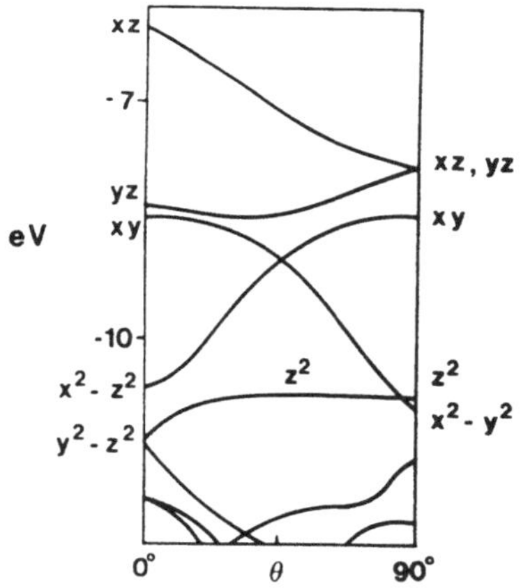

FIG. 3

The p-block orbitals are lowered in energy at $\theta = 45°$, while all the d-block levels except for dz^2, labelled here as xy for d_{xy}, are raised in energy. With a d^1 ion, the dz^2 level is singly occupied and well separated from both the p-block and the other d-block levels. Thus the dz^2 band of a MSe_4 chain is rather well separated from other bands and the electronic band structures of MSe_4 chains can be approximated by considering only their dz^2 bands. Consequently, the band filling f of a dz^2 band for $(MSe_4)_nI$ is $f = (n-1)/2n$. For instance, $(NbSe_4)_3I$, $(TaSe_4)_2I$ and $(NbSe_4)_{10/3}I$ would have 1/3-, 1/4-, and 7/20-filled bands, respectively. As n increases, f becomes closer to 1/2, which is the limit for $n \to \infty$.

A linear chain with incomplete band filling is susceptible to a Peierls distortion, which opens up a band gap at the Fermi level **(4)**. For $f \lesssim 1/2$, the distortion increases the repeat distance along the chain by a factor 1/f. Therefore 1/2-, 1/3- and 1/4-filled bands induce chain dimerization, trimerization and tetramerization, respectively. This can be accompanied by a charge density wave (CDW) state. When a CDW state arises from a f-filled band, it is expected to be commensurate if f is a simple fraction of unity (i.e., if $2k_F$ is a simple fraction of the reciprocal vector d*, where k_F is the Fermi wavevector and $d^* = 2\pi/d$ with d being the metal-metal distance). The occurrence of a CDW is detected either by non-linear effects in resistivity measurements or by the satellites spots on electron diffraction patterns and diffuse lines in X-ray diffraction patterns which give the components of the CDW wavevector.

For rectangular Se_4 units in $(MSe_4)_nI$ various orientations are possible even with the dihedral angle of 45° imposed between any two adjacent Se_4 units. We may represent a rectangle by an arrow from its center to the midpoint of a short side. For example, the representation of fig. 2b is given by fig. 2c. Fig. 2d shows the various possible orientations of a rectangle, which are numbered starting from an arbitrary position. The relative arrangement of m Se_4^{4-} units may be given by a set of m integers $(\alpha_1\ \alpha_2\ \ldots\ \alpha_m)$, α_i referring to the i-th unit. For instance, fig. 2c is represented by (12). Then the dihedral angle between i-th and j-th units will be given by $\theta = (\alpha_j - \alpha_i) \times 45°$, and we can arbitrarily take $\alpha_1 = 1$ without loss of generality. As Se_4 units are rectangular, α_i is equivalent to $\alpha_i \pm 4$. To avoid complications arising from this equivalence we will choose α_i's such that $\alpha_{i+1} - \alpha_i = \pm 1$.

In the following, we will first describe the structural and electrical properties of $(NbSe_4)_3I$, $(TaSe_4)_2I$ and $(NbSe_4)_{10/3}I$. The occurrence of various band fillings and that of different metal-metal distances, and electron transport properties led us to perform band calculations on these compounds, which will be discussed as well. Band and molecular electronic structures were calculated based upon the extended Hückel method **(5)**.

II - STRUCTURE AND PROPERTIES OF $(MSe_4)_nI$ COMPOUNDS

A - $(NbSe_4)_3I$

Table I shows the unit cell parameters of $(NbSe_4)_3I$, which crystallizes with tetragonal symmetry, space group P4/mnc **(6)**.

TABLE I
Iodine tetraselenides unit cell parameters

$(NbSe_4)_3I$	$(NbSe_4)_{10/3}I$	$(TaSe_4)_3I$	$(TaSe_4)_2I$
a = 9.489(1) Å	a = 9.464(1) Å	a = 9.4767(7) Å	a = 9.5317(9) Å
c*= 19.13 (3) Å	c*= 31.910(8) Å	c*= 19.071 (3) Å	c*= 12.761 (3) Å
V = 1722.48 Å^3	V = 2858.09 Å^3	V = 1712.72 Å^3	V = 1159.38 Å^3
tetragonal P4/mnc	tetragonal P4/mcc	tetragonal P4/mnc	tetragonal I422
Z = 4	Z = 6	Z = 4	Z = 4

*NB * growing axis*

Along the two chains of a unit cell, six selenium rectangles and six niobium atoms are located within one c parameter (fig. 4) because c = 6d. Each chain has the (123432)-arrangement of Se_4 units (fig. 5b). The shortest I-I distance is 4.96 Å, to be compared with 2.68 Å in solid I_2 or 3.07 Å in I_3^-.

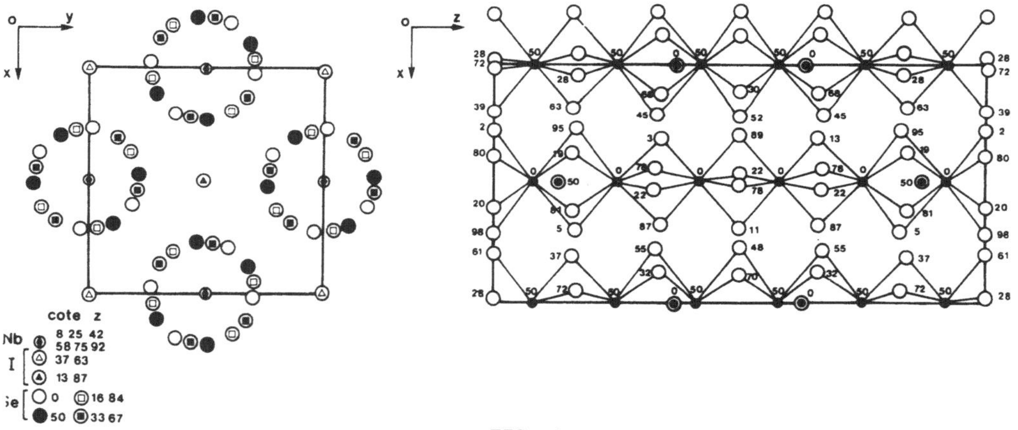

FIG. 4

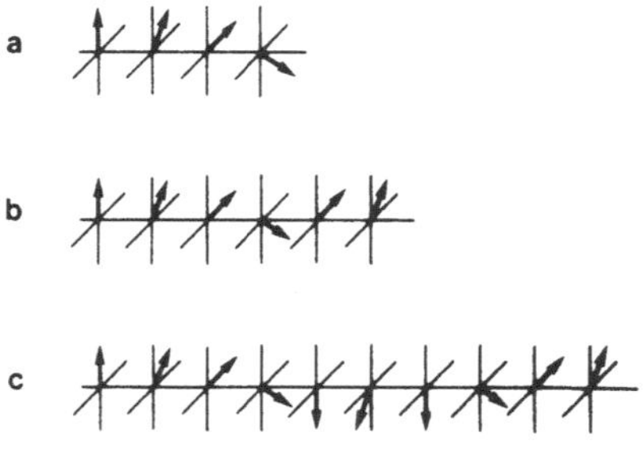

FIG. 5

For $(NbSe_4)_3I$, f is equal to 1/3, so that we expect a metal chain distortion to be one of the two shown in fig. 6 b-c. The latter is observed at room temperature, that is, two different Nb-Nb distances (3.06 Å and 3.25 A, i.e. $\delta \simeq 0.06$ Å) occur along c axis. Each short Nb-Nb bond is followed by two longer bonds as depicted in Table II.

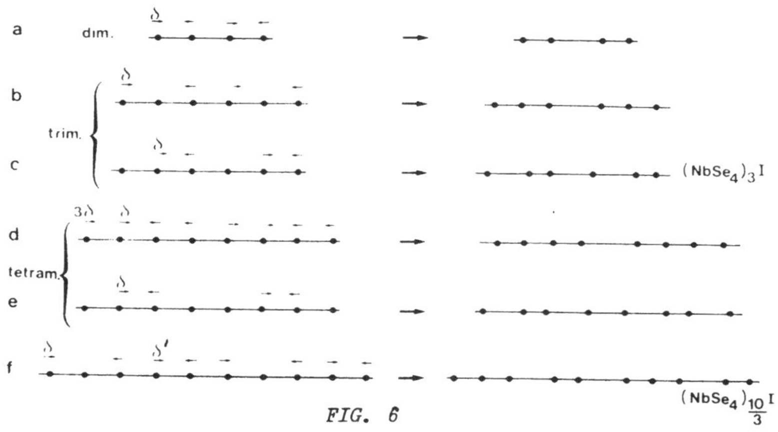

FIG. 6

TABLE II
Various metal-metal sequences in the $(MX_4)_nY$ compounds

Compound	M-M sequence	ρ_{RT}(Ω .cm)	E(eV)
$(NbSe_4)_3I$	— Nb —3.25— Nb —3.25— Nb —3.06— Nb—	1	0.19 at RT
$(NbSe_4)_{10/3}I$	— Nb —3.17— Nb —3.17— Nb —3.23— Nb —3.15— Nb —3.23— Nb —	10^{-2}	0.13
$(TaSe_4)_2I$	— Ta —3.206— Ta —3.206— Ta —	1.5×10^{-3}	0.25
$(TaSe_4)_3I$	(same sequence as for $(NbSe_4)_3I$)	0.2	?

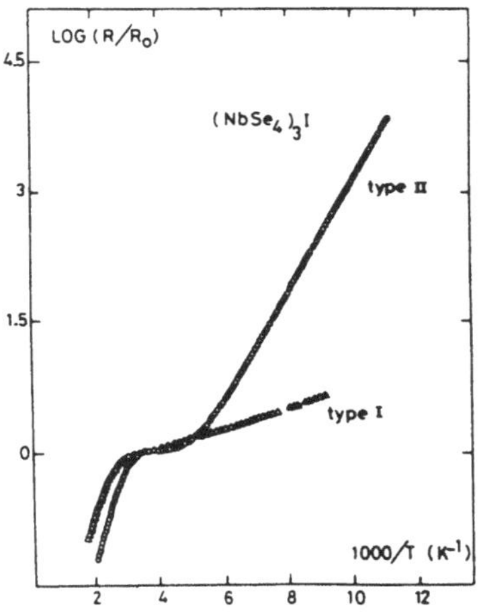

FIG. 7

Room temperature resistivity is about 1 Ω.cm **(2)**. Resistivity measurements indicate a semiconducting behavior above room temperature with a gap of 0.19 eV. There is a decrease of the activation energy around 275 K (fig. 7). Below this temperature, it appears that there are two types of curves, corresponding to two different crystals which can not be distinguished by X-ray techniques. Type I crystals exhibit a very small semiconducting gap while that of type II crystals increases back at lower temperatures (E = 0.11 eV).

Roucau et al. **(7)** observed that one condition limiting possible reflections (viz., h+l must be even for h0l spots) on diffraction patterns of $(NbSe_4)_3I$ vanishes by lowering temperature. This is due to a change of the space group corresponding to a disappearing of the glide plane n. $(TaSe_4)_3I$, not completely studied, yet, is very similar to $(NbSe_4)_3I$ (table I). Its room temperature resistivity is 0.2 Ω.cm, which is not far from that of $(NbSe_4)_3I$ **(2)**.

B – $(TaSe_4)_2I$

Table I shows the unit cell parameters of $(TaSe_4)_2I$ which crystallizes with tetragonal symmetry, space group I422 **(8)**. Each chain has the (1234)- arrangement of Se_4 rectangles so that c = 4d, i.e., there are four selenium rectangles and four tantalum atoms within one c parameter along each chain (fig. 8). The shortest I-I distance is 3.96 Å. As f is equal to 1/4, we expect a chain tetramerization as in fig. 6d-e. Nevertheless, all Ta-Ta distances are found to be equivalent (3.206 Å).

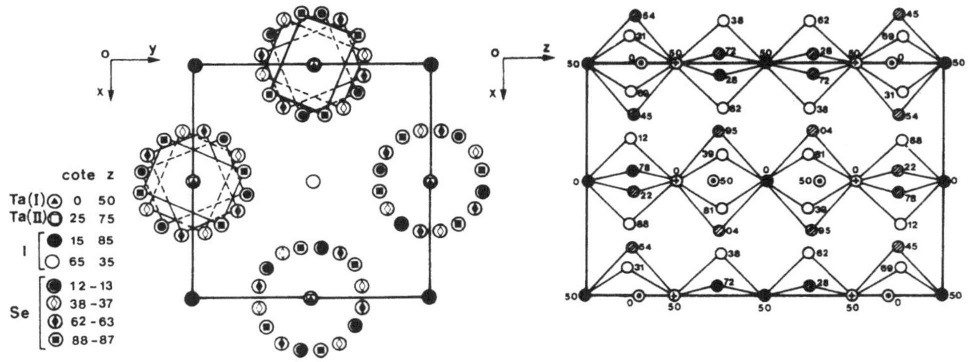

FIG. 8

The room temperature resistivity (Table II) is 1.5×10^{-3} Ω.cm, which is much lower than that of $(NbSe_4)_3I$. Fig. 9 is a logarithmic plot of the relative resistivity R/R_0 as a function of 1/T, together with its derivative **(9)**. There is a transition at T = 263 K to a semi-conducting state (E = 0.25 eV). Non linear transport properties were observed below the transition temperature **(9,10)**, which shows that this transition is a CDW-induced one. This was confirmed by electron **(7)** and X-ray **(11)** diffraction measurements.

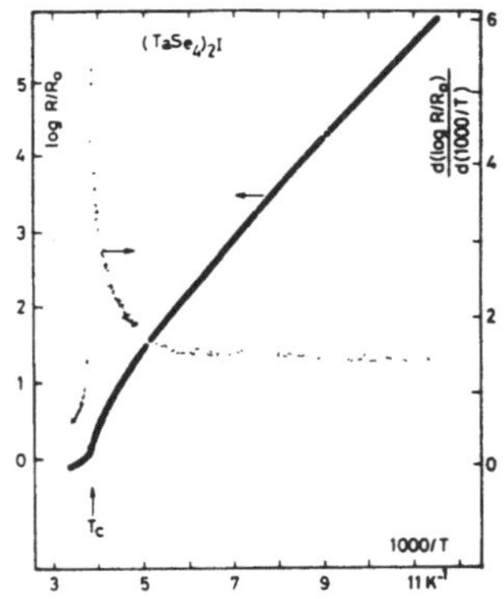

FIG. 9

C - $(NbSe_4)_{10/3}I$

Table I shows the unit cell parameters of $(NbSe_4)_{10/3}I$, which crystallizes with tetragonal symmetry, space group P4/mcc **(12)**. The a parameter is the same as before but c parameter is 10d : there are 10 selenium rectangles and 10 niobium atoms within one c parameter along each chain (fig. 10), which has the (1234565432)-arrangement of rectangular Se_4 units (fig. 5c). For $(NbSe_4)_3I$ and $(TaSe_4)_2I$, iodine atoms are similarly distributed in all their channels. For $(NbSe_4)_{10/3}I$ there is a difference between 00z and ½,½,z channels, which are occupied by four and two iodine atoms, respectively : in the first type, iodine is rather closely bonded to four selenium as in $(NbSe_4)_3I$ and $(TaSe_4)_2I$. The shortest iodine-iodine distance is 5.07 Å. Each iodine atom of the second type is weakly bonded to eight selenium atoms in a square antiprismatic arrangement.

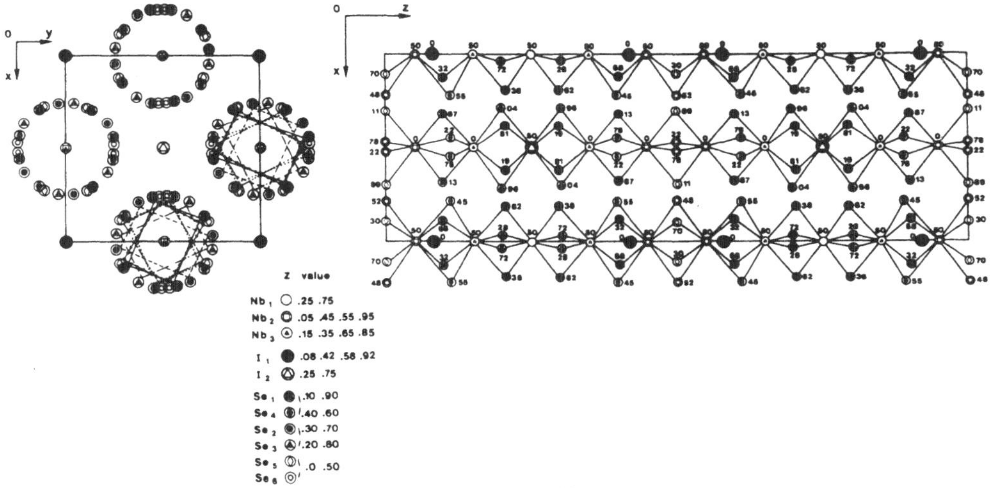

FIG. 10

In agreement with a f value (f = 7/20), smaller than 1/2, the metal chain is distorted (fig. 6f) with 3 different Nb-Nb distances, which are given in table II. This shows that $\delta \simeq \delta' \simeq 0.02$ Å.

Room remperature resistivity is about 1.5×10^{-2} Ω.cm, which is two orders of magnitude lower than that of $(NbSe_4)_3I$. Fig. 11 shows an electrical behavior that is very similar to that of $(TaSe_4)_2I$ **(13)**. There is a transition at T = 285 K, and at lower temperatures $(NbSe_4)_{10/3}I$ has a semiconducting gap E = 0.13 eV. Once again, non-linear electrical properties **(13)** and electron diffraction **(7)** show that we are dealing with a CDW transition.

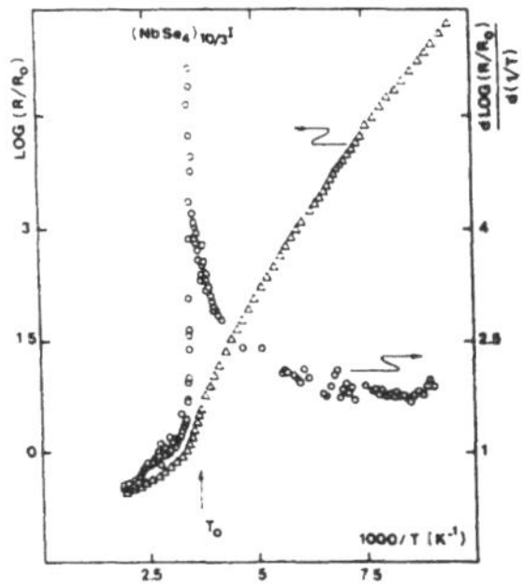

FIG. 11

III - BAND STRUCTURE CALCULATIONS

The $(MSe_4)_nI$ compounds discussed above provide us with many challenging problems. First, why do $(NbSe_4)_3I$ (f = 0.33) and $(NbSe_4)_{10/3}I$ (f = 0.35) prefer the particular metal chain distortions and why are they so different ? $(TaSe_4)_2I$ is surprising, too. Despite a band filling f = 1/4, its structure has no permanent metal chain distortion. In addition, it undergoes a CDW transition with a k_F vector which is incommensurate.

Semiconducting properties of $(NbSe_4)_3I$ above room temperature can be interpreted by considering the electron counting $Nb^{5+}(Nb^{4+})_2(Se_2^{\,2-})_6(I^-)$. Both d electrons are involved in a Nb-Nb bond, leading to short distances separated by two long distances. This kind of reasoning is much more difficult for $(NbSe_4)_{10/3}I$. With f = 0.35, its metal chain is slightly distorted at room temperature. Nevertheless, its resistivity is 100 times smaller than that of $(NbSe_4)_3I$, which is primarily related to a smaller magnitude of the distortion. Moreover, it undergoes an incommensurate CDW transition at T = 285 K.

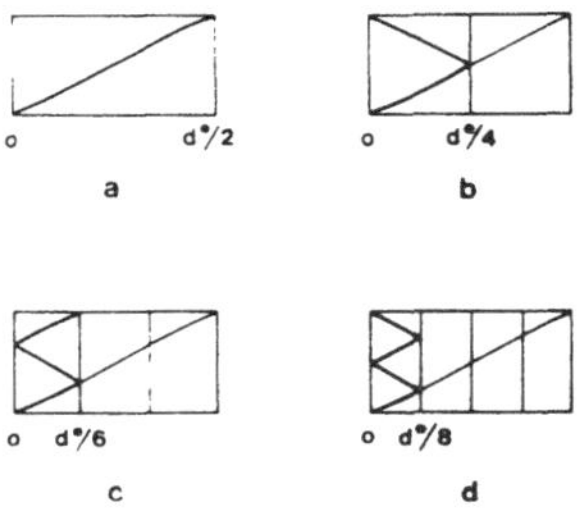

FIG. 12

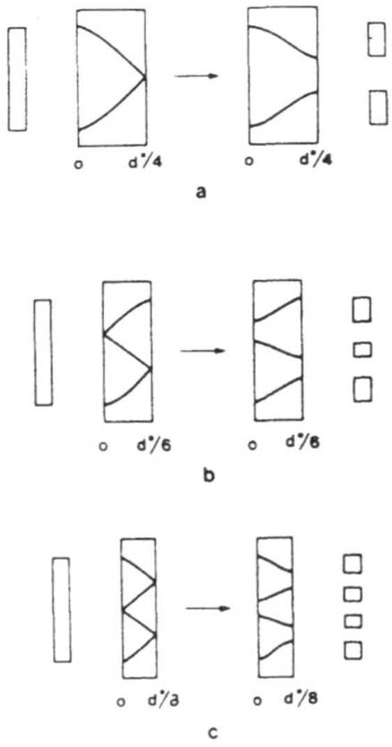

FIG. 13

In order to study how band filling is related to metal chain distortions and CDW phenomena in $(MSe_4)_nI$ compounds we carried out band structure calculations on those compounds (3). As shown by the $NbSe_8^{\,4-}$ study, the band of interest for our compounds will be dz^2 band. For a metal chain with a repeat distance d, its shape is depicted in fig. 12a. The first Brillouin zone is made of wavevectors k with $-d^*/2 \leqslant k \leqslant d^*/2$. When the repeat distance becomes md, because of distortion and/or Se_4 units arrangement, the first Brillouin zone is given by $-d^*/2m \leqslant k \leqslant d^*/2m$ and there occur m bands since a unit cell contains m metal atoms. In the case where there is no distortion, the m bands are derived by subdividing the band structure of fig. 12a into m parts of equal zone length and then folding those sections on the first one. That was done in fig. 12b-d for m = 2-4, respectively. At zone center and edge, bands are degenerate, due to the fact that the increase of the unit cell size is artificial. When it is real, those degeneracies are lifted. For example, fig. 13 a-c show the band splitting that occurs from fig. 12b-d due to dimerization (fig. 6a), trimerization (fig. 6b-c) and tetramerization (fig. 6d-e), respectively. In the following, results of band calculations on (12)-$NbSe_4$, (123432)-$NbSe_4$

and (1234)-$TaSe_4$ chains are discussed. In all cases, we will compare distorted and undistorted structures and take interchain interactions into account. In addition, we will comment on $(NbSe_4)_{10/3}I$.

1. 3D interactions of a (12)-$NbSe_4$ chain

With the condition $\theta = 45°$, the smallest unit cell of a MSe_4 chain will have the (12)-arrangement of Se_4^{4-} units. The band structure of a (12)-$NbSe_4$ chain is shown in fig. 14a. The width of the folded dz^2 band (a,b) is very large. A flat band of d_{xy} and $d_{x^2-y^2}$ character overlaps slightly with b. The dz^2 band is half-filled by the two d electrons of niobium ions, so that it is assumed to undergo a dimerization (fig. 6a). We calculated band structure and total energy as a function of δ. Table III summarizes the results. Fig. 14b is the band structure for δ = 0.1 Å. The minimum energy occurs for δ =0.17 Å with a relative stability of 2.4 k.cal/mol per $NbSe_4$.

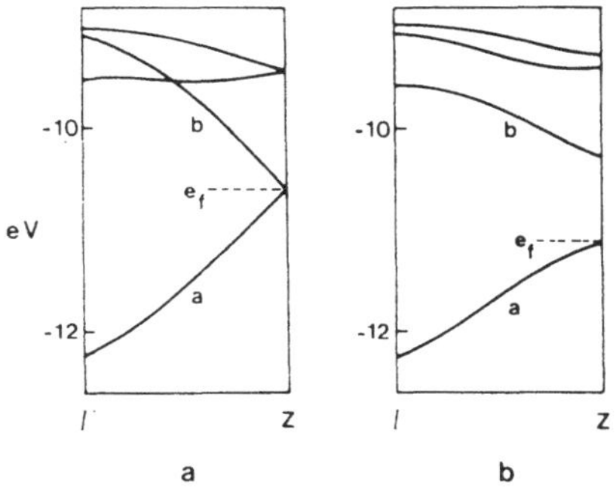

FIG. 14

TABLE III - The band fillings f, optimum displacements δ_{opt}, and relative energies ΔE of the distorted MSe_4 chains 6a-e.

Chain	f	δ_{opt} (Å)	ΔE (kcal/mol)
6a	0.5	0.17	-2.4
6b	0.333	0.16	-0.6
6c	0.333	0.11	-0.4
6d	0.25	0.04	-0.2
6e	0.25	0.00	0.0

Before studying real $(MSe_4)_nI$ systems, we must estimate how the interchain interaction affects band structures. Fig. 15a shows the unit cell of 3D-(12)-$NbSe_4$ on which calculations were performed. The relative positions of chains is that found in $(TaSe_4)_2I$. The results are shown in fig. 16 where special points are those of the Brillouin zone (fig. 15b). The number of bands is doubled, since there are two chains in a unit cell. The dz^2 bands show a large dispersion only along chain direction, (i.e. Γ → Z and R → X), and are split along

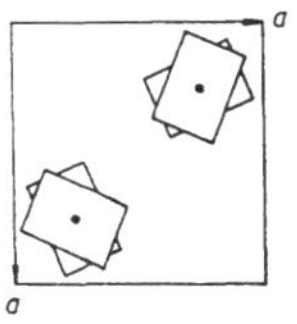

FIG. 15a

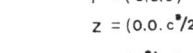

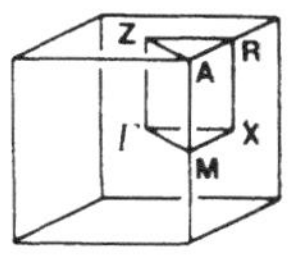

Γ = (0,0,0)
Z = (0,0,c*/2)
R = (a*/2,0,c*/2)
A = (a*/2,a*/2,c*/2)

FIG. 15b

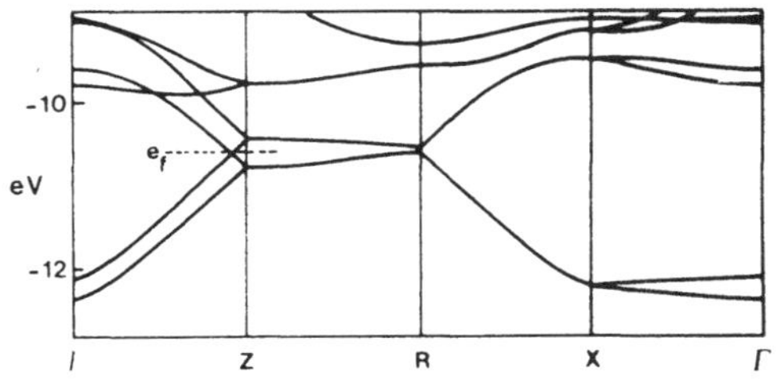

FIG. 16

Γ → Z but not at all along R → X (or M → A not shown here). One can show that this is the case for all $(MSe_4)_nI$ compounds due to tetragonal symmetry and to the fact that dz^2 orbitals are directed along chain direction. Consequently, the magnitude of interchain Se-Se interaction, reflected by the dz^2 band splitting, is significant only along Γ → Z.

The calculations on 3D-(12)-$NbSe_4$ show that the width of the dz^2 band along the chain direction is about 3 eV while that along the interchain direction is about 0.25 eV. In the following, our discussion will be based upon calculations on single MSe_4 chains. Fig. 17 shows how the effect of the interchain Se-Se interactions can be taken into account by doubling dz^2 bands of a single chain along Γ → Z with a separation of 0.25 eV. By "folding" the bands as in fig. 12, the folded bands cross each other because of different symmetry character.

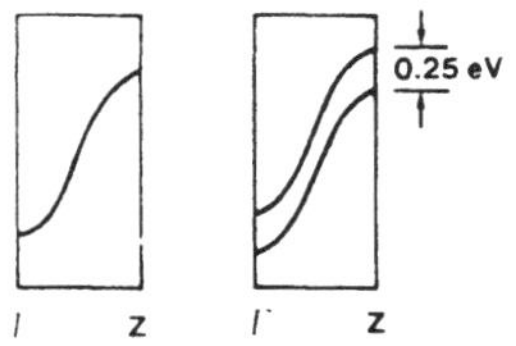

FIG. 17

2. $\underline{(NbSe_4)_3I}$

We expect a band structure with six subbands obtained by folding fig. 12c once again. The bottom four subbands of a undistorted (123432)-$NbSe_4$ chain are shown in fig. 18. There is a small gap between b and c due to a lack of a screw symmetry in the chain. Distortions of fig. 6b and 6c were also examined. Table III gives the optimum value δ_{opt} obtained for both distortions. Both appear to be more stable than undistorted structure but the difference between them is very small. Distortion 6b, which is not observed, looks more stable but is expected to be less preferred in terms of lattice strain. The band gap at the optimum structures are 0.63 eV and 0.95 eV, respectively, which explains semiconducting properties of $(NbSe_4)_3I$. This is so even if we take the doubling of dz^2 bands by interchain interaction into account.

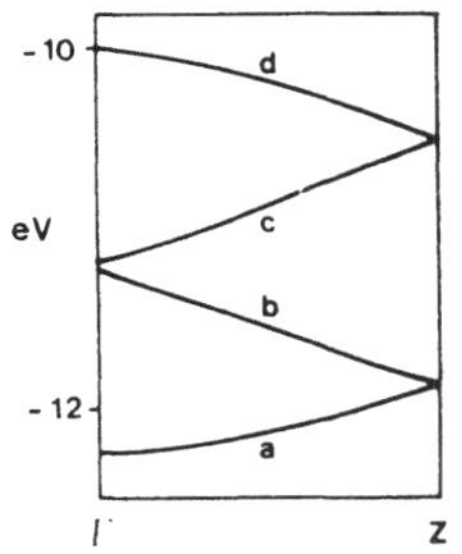

FIG. 18

3. $(TaSe_4)_2I$

The band structure of an undistorted (1234)-$TaSe_4$ chain is given by fig. 19. The a,b,c and d bands correspond to those of fig. 12d. With two d-electrons (f = 1/4), only the band a is filled, which gives Fermi level at the zone edge.

We examined the distortion 6d and 6e. A band gap opens up between a and b and increases with δ. Table III shows that very little energy is gained by distortion 6d and nothing by 6e. This table gives evidence that the tendency towards metal chain distortion decreases sharply as the band filling decreases from 1/2.

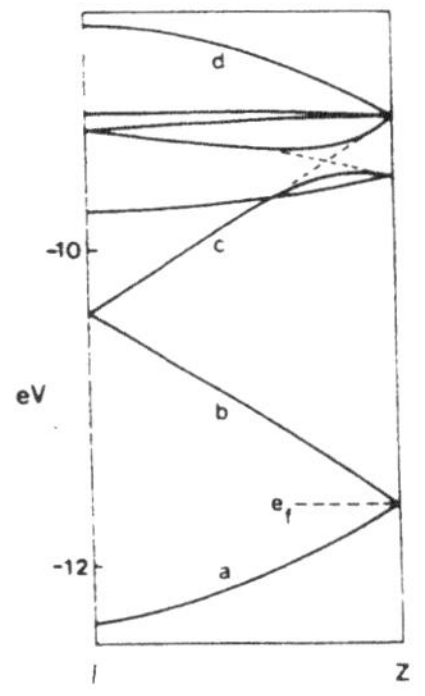

FIG. 19

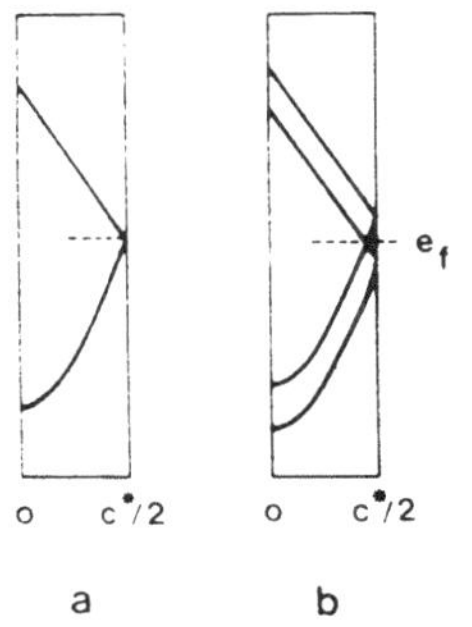

FIG. 20

Fig. 20 shows the effect of interchain interaction on the dz^2 bands a and b around the Fermi level. The two overlapping bands cross each other. This leads to a Fermi wavevector $k_F \approx 0.44$ c* which is in good agreement with experimental values. Electron **(7)** and X-ray **(11)** diffraction measurements give $k_F = 0.47$ c* and $k_F = 0.458$ c*, respectively.

4. $(NbSe_4)_{10/3}I$

Actual band calculations were not performed on (1234565432)-$NbSe_4$ chain of $(NbSe_4)_{10/3}I$ because the unit cell is too large. Nevertheless, we can derive dz^2 band structure of an undistorted chain in the same manner as in fig. 12, thus obtaining fig. 21. Seven d-electrons are available per unit cell, so that the a-c bands are completely filled while d band becomes half-filled. Together with the effect of interchain Se-Se interaction, this gives a Fermi wavevector slightly different from c*/4. The metal chain distortion does not open up any gap in the middle of subband d, so that we expect a Fermi wavevector which is incommensurate and close to c*/4. This is in reasonable agreement with the experimental value of 0.2435 c* **(7)**.

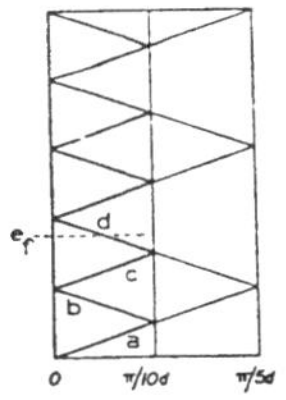

FIG. 21

IV - CONCLUSION

$(MSe_4)_nI$ compounds contain MSe_4 chains with various degrees of dz^2 band fillings. This enables us to examine how structural and electronic properties are affected by small differences in band fillings. As the band filling decreases from 1/2, the tendency toward metal chain distortion decreases, which is the cause of the undistorted structure of $(TaSe_4)_2I$. We showed that the dz^2 band of an MSe_4 chain is well separated from other bands. The effect of interchain Se-Se interaction on dz^2

band is appreciable only along $\Gamma \rightarrow Z$ direction. This explains the pseudo one-dimensionality of $(MSe_4)_nI$ compounds and the incommensurate values of k_F vectors for $(TaSe_4)_2I$ and $(NbSe_4)_{10/3}I$. A good knowledge of band structures is expected to be of great help in the interpretation of the growing number of experimental works on these compounds.

Acknowledgements

All compounds were prepared at Nantes by L. Guémas. Transport properties and electron diffraction measurements were performed by P. Monceau's group at Grenoble and by R. Ayroles's group at Toulouse. M.H.W is thankful to the Camille and Henry Dreyfus Foundation for a Teacher-Scholar Award (1980-1985).

References

(1) a) P. Monceau, Statics and dynamics of non-linear systems, Solid State Science, 47, G. Benerek, H. Biltz and R. Zeyer Ed, Springer Verlag 1983, p. 144.
b) N.P. Ong, Can. J. Phys. 60, 757 (1982).
c) G. Grüner, Physica 8D, 1 (1983).
(2) P. Gressier, A. Meerschaut, L. Guémas, J. Rouxel, P. Monceau, J. Solid State Chem. 51, 141 (1984).
(3) P. Gressier, M.H. Whangbo, A. Meerschaut, J. Rouxel, Inorg. Chem. 23, 1221 (1984).
(4) a) R.E. Peierls, "Quantum Theory of Solids", Oxford University Press, London, 1955, p. 108.
b) M.H. Whangbo, Acc. Chem. Res. 16, 95 (1983).
(5) a) R. Hoffmann, J. Chem. Phys. 39, 1397 (1963).
b) M.H. Whangbo, R. Hoffmann, J. Am. Chem. Soc. 100, 6093 (1978).
c) M.H. Whangbo, R. Hoffmann, R.B. Woodward, Proc. R. Soc. London, Ser. A 366, 23 (1979).
(6) A. Meerschaut, P. Palvadeau, J. Rouxel, J. Solid State Chem. 20, 21 (1977).
(7) C. Roucau, R. Ayroles, P. Gressier, A. Meerschaut, J. Phys. C : Solid State Phys. 17, 2993 (1984).
(8) P. Gressier, L. Guémas, A. Meerschaut, Acta Cryst. B38, 2877 (1982).
(9) Z.Z. Wang, M.C. Saint Lager, P. Monceau, M. Renard, P. Gressier, A. Meerschaut, L. Guémas, J. Rouxel, Solid State Commun 46, 325 (1983).
(10) M. Maki, M. Kaiser, A. Zettl, G. Grüner, Solid State Commun 46, 497 (1983).
(11) H. Fujishita, M. Sato, S. Hoshino, Solid State Commun 49, 313 (1984).
(12) A. Meerschaut, P. Gressier, L. Guémas, J. Rouxel, J. Solid State Chem. 51, 307 (1984).
(13) Z.Z. Wang, P. Monceau, M. Renard, P. Gressier, L. Guémas, A. Meerschaut, Solid State Commun. 47, 439 (1983).

DEFECTS AND CHARGE DENSITY WAVES IN IRRADIATED LAYER AND CHAIN COMPOUNDS

H. Mutka+, S. Bouffard and L. Zuppiroli
DTech. SESI, Centre d'Etudes Nucléaires
B.P. 6, F-92260 Fontenay-aux-Roses, France

Controlling the defect concentration by irradiation permits detailed investigations of the effects of defects on the charge density wave phenomena. With experiments on irradiated layer and chain compounds we have characterised these effects. The electronic transport properties, the phase transition temperatures and the structural coherence of the charge density waves have been followed as a function of the concentration of irradiation induced defects from the ppm level up to 10^{-2} atomic fraction. The pinning by defects affects the dynamic properties of the charge density wave already at defect concentrations of the ppm level. At higher defect concentrations of the level of 10^{-3} atomic fraction, the pinned charge density wave becomes strongly strained. Its structural coherence breaks down and a concomitant decrease of the critical temperature of the charge density wave transition is observed.

1. Introduction

The importance of defects and disorder in the charge density wave (CDW) materials has been demonstrated both in theory and in experiments. However, it is not always easy to extract the specific action of defects in decreasing the critical temperatures of the CDW transitions or in attenuating the collective dynamics by pinning the CDW-condensate. The practical investigation of these problems necessitates a good control of sample quality, the concentration and distribution of defects must be correlated with the observed properties. This is not easy in case of impurities or non-stoichiometry at the level below 10^{-2} atomic fraction but can be realised in careful irradiation experiments[1,2]. Using energetic particle irradiation, for example fast (MeV) electrons, it is possible to produce point defects at random positions, by knocking off atoms from their normal lattice sites. In addition, in-situ experiments permit to follow the interesting properties on the same sample without changing anything else but the concentration of

+permanent address: Technical Research Centre of Finland, SF-02150 Espoo 15, Finland defects. Even at the lowest irradiation doses we have a uniform damage and the defect concentration is well proportional to the dose. This gives a good relative accuracy when studying the variations of different phenomena. The determination of the absolute defect concentration scale is a problem specific to each compound[2] which will not be discussed here. It has been done for the electron irradiated compounds discussed in this paper and the numbers given below should be accurate within a factor of three.

Several low-dimensional metals and CDW compounds have been and are irradiated in Fontenay-aux-Roses[1,2]. This report is intended to give account on the observations made on the layered transition metal dichalcogenides and on some transition metal chain compounds (trichalcogenides, bronzes). In the first part we shall concentrate on the effects of defects on the static CDW, results concerning the stability of the CDW phase and its structural coherence will be discussed. The second part is devoted to the depinning of the CDW-condensate under electric field that is observed in the chain compounds.

2. Stability and coherence of CDW in presence of defects

There are two basic theoretical predictions concerning the stability and the coherence of the CDW phase in presence of defects. The amplitude of the CDW should be rapidly suppressed with increasing defect scattering rate, the critical temperature first decreases and then falls down to zero at a critical scattering rate[3-5]. On the other hand, charged defects are expected to enhance CDW fluctuations above the critical temperature: the Friedel oscillations have the same $2k_F$ periodicity, thus the CDW-instability is triggered near defects. Accordingly the CDW becomes pinned by the defects and a random defect distribution should then break the long-range coherence of the CDW[6-9].

2.1 Layer compounds

In the layered dichalcogenides the strength of the CDW distortion varies from quite weak (2H-$NbSe_2$, onset temperature T_0= 35 K) to very strong (1T-TaS_2 and 1T-$TaSe_2$, $T_0 \approx$ 600 K)[10]. This variation of transition temperature seems to correlate with the electronic anisotropy: in

$2H-NbSe_2$ the conductivity parallel to the layers is about 30 times the one perpendicular to them[11], in $1T-TaS_2$ this ratio is about 500(ref.12). This interesting feature of the dichalcogenides permits us to examine the defect-CDW interaction as a function of the strength of the CDW instability (or dimensionality).

The effect of defects on the CDW was studied[13] within an approach already used by other authors[14,15]: measuring simultaneously the normal state resistivity increase and the phase transition temperature we can correlate the scattering rate of the conduction electrons and the critical temperature of the CDW onset. In $2H-NbSe_2$ this correlation is in agreement with the supposition that the scattering time of conduction electrons controls the CDW formation, as it was already concluded by Stiles et al.[14] and by Huntley et al.[15]. In $2H-TaS_2$ and $1T-VSe_2$ with stronger CDW distortions the effect of defects on the CDW onset is weaker, in spite of the fact that the resistivity increases more rapidly with defect concentration, see Fig.1. In fact we do not observe a decrease of the critical temperature towards zero at some critical concentration of defects . Instead, at concentration of a few 10^{-3} atomic fraction, the well defined phase transition disappears at a relatively high temperature. Electron diffraction has revealed that in this state without phase transition incoherent CDW exist, the amplitude of the distortion does not vanish[13]. The incoherence of the remaining CDW distortion is evidenced by the broadening of the diffraction satellites, suggesting that the distortion is composed of domains whose size is decreasing when the defect concentration increases. Indeed,

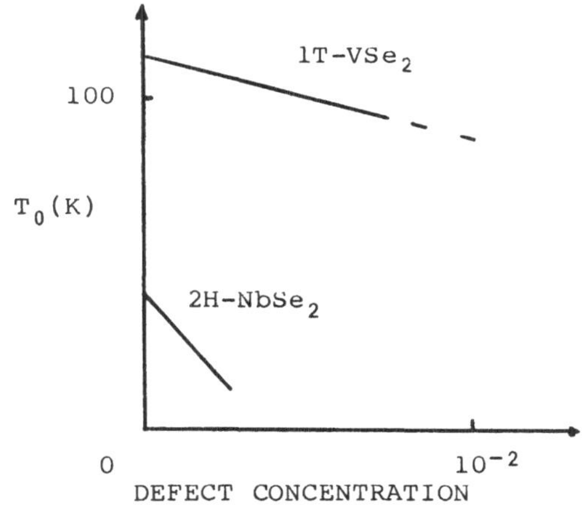

Fig.1 The CDW onset temperature decreases less rapidly when the initial value is higher. The transition disappears before reaching zero temperature.

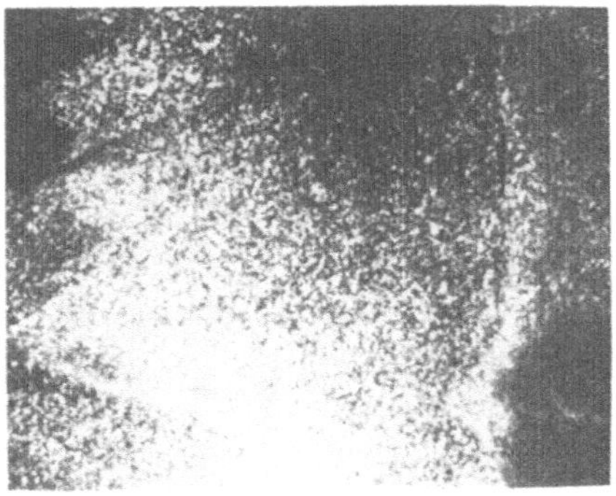
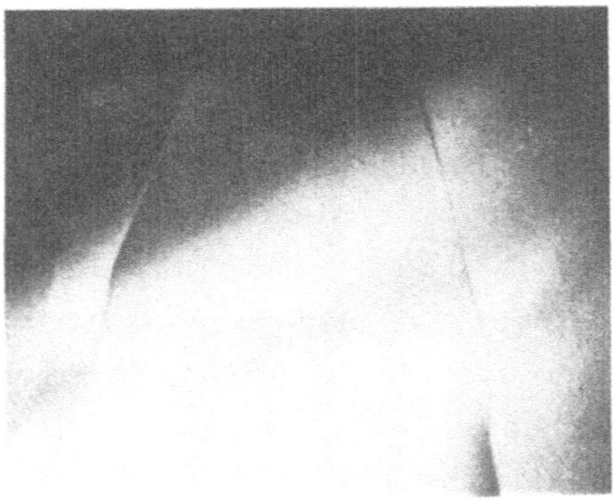

defect concentration 10^{-2}
domain size ≈ 100 A

3×10^{-2}
"glassy" CDW

Fig. 2 Satellite reflection dark-field micrographs of the defect controlled CDW domains in 1T-TaS_2(T=100 K, magnification 15000).

this happens as it has been demonstrated by the satellite reflection electron microscopy on 1T-TaS_2[16]. Some results are shown in the Fig. 2 of the incommensurate CDW phase of this material. It can be seen that even a "glassy" CDW, characterised by very broad diffraction satellites and vanishingly small domain size, can be obtained. A more detailed characterisation of the effect of defects on the domains and boundaries of the CDW is going on and reported also in this conference[17].

The correlation between the onset temperature and its rate of decrease reflects the importance of the normal conducting electrons in screening the defect potentials. In materials with higher onset temperature the defect induced CDW fluctuations are enhanced and the amplitude of the CDW is less rapidly suppressed. However, this CDW pinned by defects cannot gain any long-range coherence, it breaks into domains whose size decreases with increasing defect concentration.

2.2 Chain compounds

The decrease of the critical temperature of the CDW transition by defects has been well demonstrated for the materials of the $NbSe_3$ family[18-20] and the blue bronzes[21]. In general the observations are similar to those made on the layer compounds, the initial decrease is followed by a smearing of the phase transition far before it has fallen down to zero temperature. Here, too, electron diffraction has been used to follow the coherence of the CDW as a function of defect concentration[19,20]. Once again, it has been observed that when the phase transition becomes smeared the coherence of the distortion is lost, but its amplitude does not vanish. Continuing the reasoning of the precedent section we can expect that in these materials, with complete destruction of the free Fermi surface in the CDW phase(except in $NbSe_3$), the defect driven CDW amplitude is even more important.

The existence of CDW domains in the chain compounds is an important question in what concerns the collective CDW transport properties. Observations by dark-field electron microscopy have shown contrasts that have been interpreted as CDW domains[22-24]. It has also been suggested that these domains can be influenced by the radiation damage induced during observation under electron beam. We have not studied this problem in detail but we want to show a result that concerns the effect of defects on the domain contrasts in the orthorhombic TaS_3, Fig. 3. There cannot be seen any spectacular effect at a defect concentration that can dramatically affect the CDW transport, for example increase more than tenfold the threshold field and smear strongly the periodic noise peaks (see section 3.).

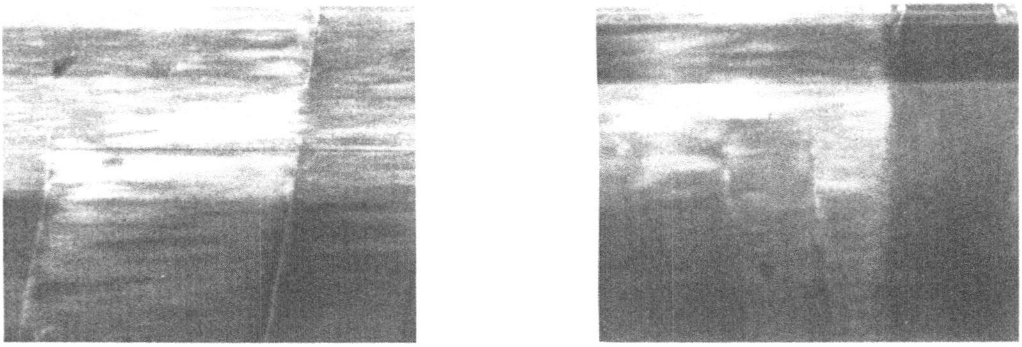

Fig. 3 Satellite reflection dark-field micrographs of the orthorhombic TaS_3. The domain contrasts are quite similar in the pure sample (left) and in the one containing 3×10^{-4} atomic fraction of defects(right). The temperature of observation is about 130 K (magnification 30000).

2.3 Remarks on the pinned static CDW

Considering the static structural properties of the CDW the manifestations of the pinning by defects become experimentally evident at defect concentrations exceeding 10^{-3} atomic fraction, both in the layer and the chain compounds. At such a concentration the strain induced by the pinning in the CDW exceeds the limits of accommodation by long-ranged deformations. It is then possible to observe the formation of localised defects in the CDW, such as dislocations and domain boundaries[17]. In diffraction experiments one begins to observe the broadening of the CDW satellites due to this loss of coherence.

The CDW amplitude is affected exactly at the same defect concentration range (10^{-3} atomic fraction) as it can be concluded from the behaviour of the transition temperature. Most of the theoretical considerations associate the decrease of the amplitude to scattering of conduction electrons by defects but experimentally this hypothesis is hard to verify. It is clear that the models taking into account only the scattering time are not enough to explain the stability of the CDW in presence of defects. The phenomenological ideas of defect driven CDW amplitude and elastic energy of the pinned CDW should surely be included in the adequate treatment of the CDW amplitude.

It is usual to speak about strong or weak pinning of the CDW depending on whether the pinning energy gained by fixing the phase of the CDW or the elastic energy lost in accommodating the CDW between defects is greater[25,26]. The weak pinning is associated with defect controlled domain structures that are formed to take advantage of the fluctuations of the defect distribution. Experimentally we have not seen any important effects on the CDW domains at defect concentrations below a few 10^{-4} atomic fraction. At such concentrations the average distance between randomly distributed defects is only about ten CDW wavelengths (for three dimensions, because the accommodation of the CDW phase is 3D problem), which surely should result important elastic accommodations. However, in real systems the defects may be mobile and even a short-range migration could lead to the formation of a defect density wave that is coherent with the CDW[27]. In such case, the elastic energy contribution could be greatly reduced and the coherence of the CDW restored. We shall return to this question of strong/weak or coherent/incoherent pinning in the next section.

3. Depinning and dynamics of the CDW

The interesting dynamical properties of the CDW condensate in the inorganic quasi-1D metals form a major part of this conference. Ever since the early studies on substitutionally disordered or irradiated $NbSe_3$[28-30] it has been clear that pinning by defects controls these phenomena, as it had been predicted in theory[25,26,31]. With the increase of the number of compounds presenting collective CDW transport more systematic experimental examination of the pinning has now become possible.

In the foregoing section we have seen that the static CDW becomes visibly affected by the pinning at defect concentrations of the order of 10^{-3} atomic fraction. However, there is no doubt that the strong local electronic perturbations of the irradiation induced defects (vacancies and interstitial atoms) pin the CDW effectively already before it will be deformed so evidently. This has been clearly shown in the in-situ irradiation experiments on orthorhombic TaS_3 and on the blue bronzes: the introduction of $10^{-5}...10^{-4}$ atomic fraction of defects can increase the threshold field by one order of magnitude[20,32]. Moreover, the increase of the threshold field is in linear proportion with the defect concentration suggesting the strong pinning regime where the pinning energy dominates[25,26].

In fact the entire depinning characteristics are scaled linearly with the defect concentration as it can be seen in Fig. 4. The results are for a sample of orthorhombic TaS_3 that was irradiated and observed in-situ at 150 K. The left part of the figure shows the dynamic resistance curves obtained at different irradiation doses up to 56×10^{-6} atomic fraction of defects. In the right part the electric field has been scaled with the threshold value (or the defect concentration, because the dependence is linear) and a single depinning curve results. It seems that the pinning force of defects and the electric field find a similar equilibrium state independently of the defect concentration.

In orthorhombic TaS_3 at defect concentrations approaching and exceeding 10^{-4} atomic fraction the depinning becomes smeared and the determination of the threshold field ambiguous[20]. The measurements of the noise associated to the sliding of the CDW have shown that the narrow peaks broaden at the same dose scale[20]. This may reflect the beginning of the loss of coherence that is observed in diffraction experiment at somewhat higher defect concentration. However, Fig. 3 shows that the domain contrasts remain practically intact up to 3×10^{-4} defects.

In the light of these observations the pinning in the orthorhombic TaS_3 below 10^{-4} atomic fraction of irradiation induced defects is strong (linear increase of threshold field) and coherent (the defects do not control the structural coherence). It might be tempting to describe the subsequent loss of structural coherence and smearing of depinning characteristics as a passage to the weak pinning regime. But we cannot ascertain that the elastic energy contribution is indeed dominating even though it is important enough to break the coherence of the CDW.

The strength and the coherence of the pinning have been often evoked in discussions concerning the metastability of the pinned CDW and of the depinning. The recent observations on the blue bronzes have shown that metastability can be induced by introducing extremely low defect con-

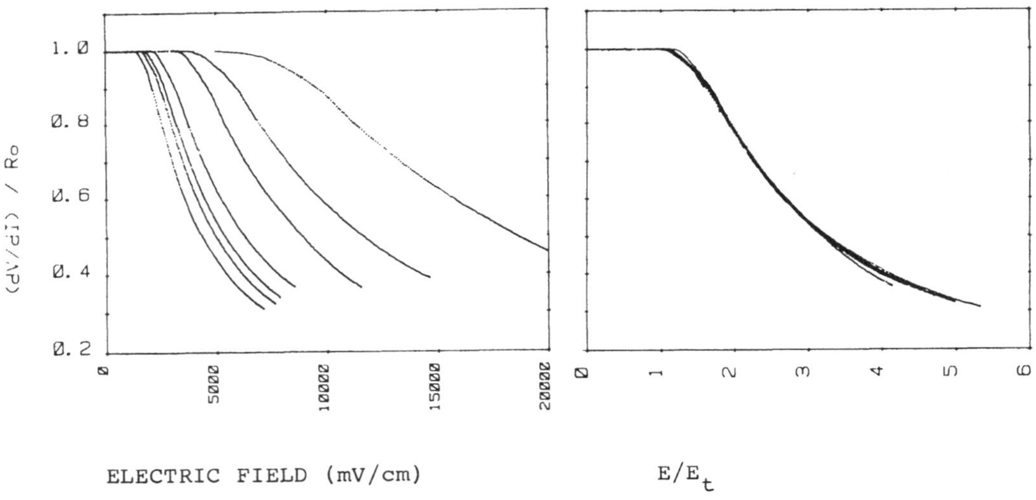

Fig. 4 The depinning curves of orthorhombic TaS_3 obtained at different defect concentrations for the same sample (T=150 K) can be brought on a single curve by scaling the electric field with the threshold value, which is equivalent to scaling by defect concentration. The defect concentrations are 0, 2, 4, 8, 16, 32 and 56 times 10^{-6} atomic fraction.

centrations of the ppm level[32,33]. Considering the above presented results one would not expect any variation of the coherence at such low defect concentration and, accordingly, the metastability might be related to the strong pinning of the irradiation induced defects[33].

4. Conclusion

The experiments on irradiated layer and chain compounds have revealed nicely how the pinning of the CDW by defects develops. At defect concentrations of the order of $10^{-6}...10^{-4}$ atomic fraction the CDW is able to accommodate the strain associated with the pinning by long ranged deformations. The static configuration is hardly affected but the dynamic properties show important variations: the threshold field of the non-linear conductivity can increase tenfold. At concentrations approaching 10^{-3} atomic fraction the static coherence of the CDW begins to be perturbed: broadening of the diffraction satellites and formation of localised defects in the CDW are observed.

References

1. L. Zuppiroli, Rad. Effects 62, 53 (1982)
2. L. Zuppiroli, H. Mutka, S. Bouffard, Mol. Cryst. Liq. Cryst. 85,1 (1982)
3. B.R. Patton, L.J. Sham, Phys. Rev. Lett. 31, 631 (1973)
4. H. Schuster, Solid St. Commun. 14, 127 (1974)
5. L.N. Bulaevskii, M.V. Sadovskii, Sov. Phys. Solid State 16, 743 (1974)
6. L.J. Sham, B.R. Patton,Phys. Rev. Lett. 36, 733 (1976)
7. L.J. Sham, B.R. Patton Phys. Rev. B 13, 3151 (1976)
8. T. Tsuzuki, K. Sasaki, Progr. Theor. Phys. 65, 19 (1981)
9. Y. Imry, S.-K. Ma, Phys. Rev. Lett. 35, 1399 (1975)
10. J.A. Wilson, F.J. Di Salvo, A. Mahajan, Adv. Phys. 24, 117 (1975)
11. J. Edwards, R.F. Frindt, J. Phys. Chem. Solids, 32, 2217 (1971)
12. P.D. Hambourger, F.J. Di Salvo, Physica 99B, 173 (1980)
13. H. Mutka, N. Housseau, J. Pelissier, R. Ayroles, C. Roucau, Solid St. Commun. 50, 161 (1984)
14. J.A.R. Stiles, D.L. Williams, M.J. Zuckermann, J. Phys.C: Solid State Phys. 9, L489 (1976)

15. D.J. Huntley, F.J. Di Salvo, T.M. Rice, J. Phys.C: Solid State Phys. 11, L767 (1978)
16. H. Mutka, N. Housseau, Phil Mag. A47, 797 (1983)
17. G. Salvetti, Thesis, Université de Toulouse, France (1984)
G. Salvetti, R. Ayroles, C. Roucau, H. Mutka, P. Molinié, in this conference
18. W.W. Fuller, P.M. Chaikin, N.P. Ong, Solid St. Commun. 39, 547 (1981)
19. G. Mihaly, N. Housseau, H. Mutka, L. Zuppiroli, J. Pelissier, P. Gressier, A. Meerschaut, J. Rouxel, J. Physique Lett. 42, L263 (1981)
20. H. Mutka, S. Bouffard, G. Mihaly, L. Mihaly, J. Physique Lett. 45 L113 (1984)
21. H. Mutka, S. Bouffard, M. Sanquer, J. Dumas, C. Schlenker, International Conference on Low-Dimensional Synthetic Metals, Abano Terme, Italy (1984), to be published in Mol. Cryst. Liq. Cryst.
22. K.K. Fung, J.W. Steeds, Phys. Rev. Lett. 45, 1696 (1980)
23. C.H. Chen, R.M. Fleming, P.M. Petroff, Phys. Rev. B 27, 4459 (1983)
24. C.H. Chen, R.M. Fleming, Solid St. Commun. 48, 777 (1983)
25. H. Fukuyama, P.A. Lee, Phys. Rev. B 17, 535 (1978)
26. P.A. Lee, T.M. Rice, Phys. Rev. B 19, 3970 (1979)
27. P. Lederer, G.Montambaux, J.P. Jamet, International Conference on Low-Dimensional Synthetic Metals, Abano Terme, Italy (1984)
28. J.W. Brill, N.P. Ong, J.C. Eckert, J.W. Savage, S.K. Khanna, R.B. Somoano, Phys. Rev. B 23, 1517 (1981)
29. P. Monceau, J. Richard, R. Lagnier, J. Phys.C: Solid State Phys., 14, 2995 (1981)
30. W.W. Fuller, G.Grüner, P.M. Chaikin, N.P. Ong, Phys. Rev. B 23, 6259 (1981)
31. for the more recent work see contributions to this conference and references therein
32. H. Mutka, S. Bouffard, J. Dumas, C. Schlenker, J. Physique Lett. 45 L729 (1984)
33. S. Bouffard et al., in this conference

ELECTRON DIFFRACTION CHARGE DENSITY WAVE STUDIES IN THE CHALCOGENIDE COMPOUNDS $(MX_4)_nI$

C. Roucau and R. Ayroles

Laboratoire d'Optique Electronique du C.N.R.S.

29, rue Jeanne Marvig, B.P. 4347, 31055 Toulouse Cedex, France

We report the observations by electron diffraction of two different structural behaviours in chalcogenide compounds $(MX_4)_nI$. The first, in $(TaSe_4)_2I$ and $(NbSe_4)_{10/3}I$ corresponds to a one-dimensional conductor inducing a Peierls transition associated with a resistivity anomaly appearing at low temperature. The second behaviour observed concerns $(TaSe_4)_3I$ and $(NbSe_4)_3I$ which show a space group change determined by the convergent beam method at about 200 K and 280 K respectively.

INTRODUCTION - Numerous transition metal chalcogenides exhibit anomalous electrical resistivity. Various results obtained by X ray, neutron and electron diffraction have permitted the correlation between the physical and structural properties to be determined. From the structural point of view we can observe either of two types of behaviour charge density waves (CDW) appearing coupled to the periodical distortions of the lattice and space group changes due to the atomic displacements.

For our study we have used a series of one-dimensional iodine compounds with chemical formula $(MSe_4)_nI$ where M is a transition metal (M = Ta, Nb). More exactly we have examined $(TaSe_4)_2I$, $(NbSe_4)_{10/3}I$, $(NbSe_4)_3I$ and $(TaSe_4)_3I$ prepared and characterized by Gressier et al.[1]. Their transport properties have been studied by Wang et al.[2,3,4]. These authors have found that the first two compounds have a metallic character at room temperature whereas the other two are semiconducting or weakly metallic.

In this paper we show that these different properties are found again in the variation of the structural behaviour with the temperature. In the first case the resistivity anomaly corresponds to a Peierls transition whereas in the second a space group change appears. Our purpose is to provide a comparative study, by electron microscopy, of the structural properties of the different compounds indicated above as a function of the temperature.

CHARGES DENSITY WAVES IN $(TaSe_4)_2I$ and $(NbSe_4)_{10/3}I$ - The micrograph of fig. 1 shows a diffraction pattern obtained with $(TaSe_4)_2I$ at 90 K, which is far below the transition temperature T_c = 263 K. If we consider, on this pattern, the distortion vector generally defined as the vector joining a satellite spot to the neighbouring Bragg spot we obtain, in these conditions, the Fermi wave vector $\vec{k}_F = 0.028\,\vec{c}^*$ which is not consistent with the metallic properties of the material. In consequence we have tried to find a solution in better agreement with the physical propertie observations. Fig. 2 illustrates the situation corresponding to fig. 1 in the reciprocal lattice. In a recent paper [5] we have reported the components of the lattice distortion vector

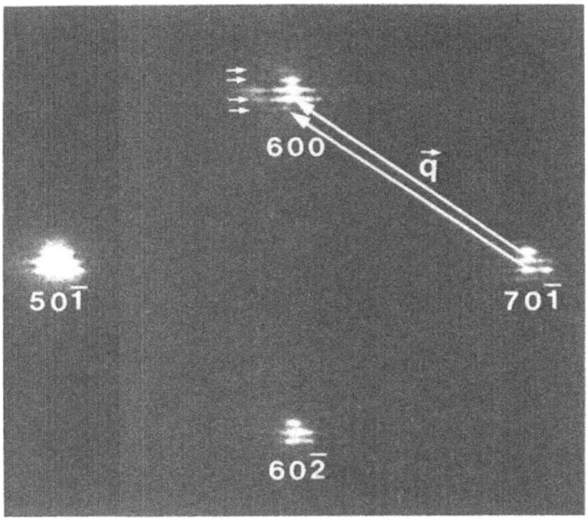

Fig. 1

$(TaSe_4)_2I$ electron diffraction pattern obtained at 90 K and containing the (a^*, c^*) plane. Note the satellite spots defined by the distortion vector $\vec{q}$.

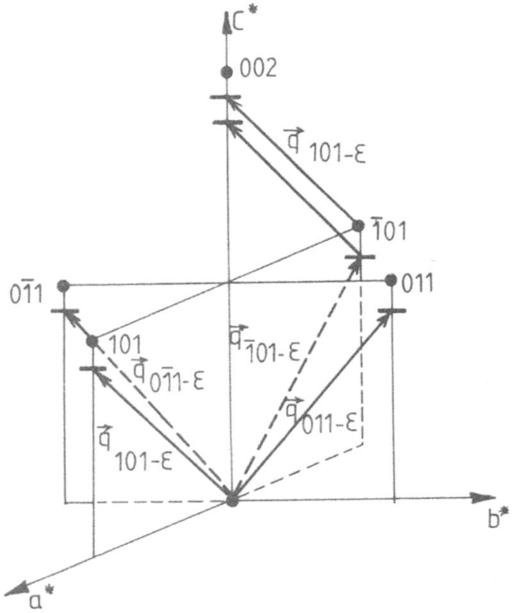

Fig. 2

Representation of the different distortion vectors $\vec{q}$ in the (a^*, c^*) plane of the reciprocal lattice of $(TaSe_4)_2I$.

due to the CDW below the transition temperature. These components have the form $(a^* ; 0 ; (1 - \varepsilon) c^*)$ and taking account of the fourfold $\vec{c}^*$ axis in the tetragonal lattice space group I_{422}, we have 4 possible distortion vectors in the (a^*, c^*) plane as shown in the figure. This representation permits us to explain the presence of the higher order harmonic. For example the second harmonic situated near the (002) spot is obtained from the satellite spot of first order situated near the main spot $(\bar{1}01)$ by using the distortion vector $\vec{q}_{101-\varepsilon}$. There is thus a "double reflection" effect, which explains the successive harmonic spots along the $\vec{c}^*$ direction and shows that the periodic distortion is not sinusoïdal. We can note that Fujishita et al.[6] have determined, by X ray diffraction, the vector $\vec{q} = (\pm 0.05\ a^* ; \pm 0.05\ b^* ; \pm 0.085\ c^*)$ which corresponds in fact to the vector joining the satellite spot to the closest Bragg spot. In our diffraction patterns the spots are elongated in the $\vec{a}^*$ direction and it is not possible to reveal a weakly component along it.

In the $(NbSe_4)_{10/3}I$ case we have observed a similar transition by lowering the temperature below the transition point (T_c = 285 K). Here, the distortion vector has only one component $\vec{q} = 0.487\ \vec{c}^*$. The intensity of the corresponding satellite spots is weak (see arrows on fig. 3). In these conditions it is difficult to detect the possible presence of harmonics. However weak spots appear below the transition temperature. They may correspond either to the second harmonic or to forbidden spots (h0l) with l odd, for which the structure factor is not exactly equal to zero. Our experiments have not permitted us to determine the origin of these spots.

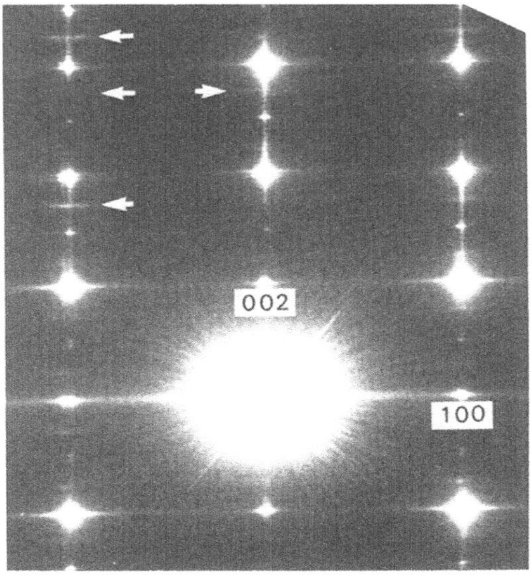

Fig. 3

Satellite spots appearing at low temperature in the $(NbSe_4)_{10/3}I$ electron diffraction pattern (see arrows). Note the additional spots along the direction parallel to c^*.

On the diffraction pattern reproduced in fig. 3 we also note a series of additional spots very near to the rows of Bragg spots parallel to the $\vec{c}^*$ direction. Electron microscope observation permits us to explain these spots. By the high resolution technique we observe the interference fringes corresponding to the atomic planes perpendicular to the $\vec{c}$ axis with periodicity c = 31.9 Å (Fig. 4). The periodicity of these fringes is disturbed by the "fault planes" which occasionally modify the distance between two consecutive lattice planes.

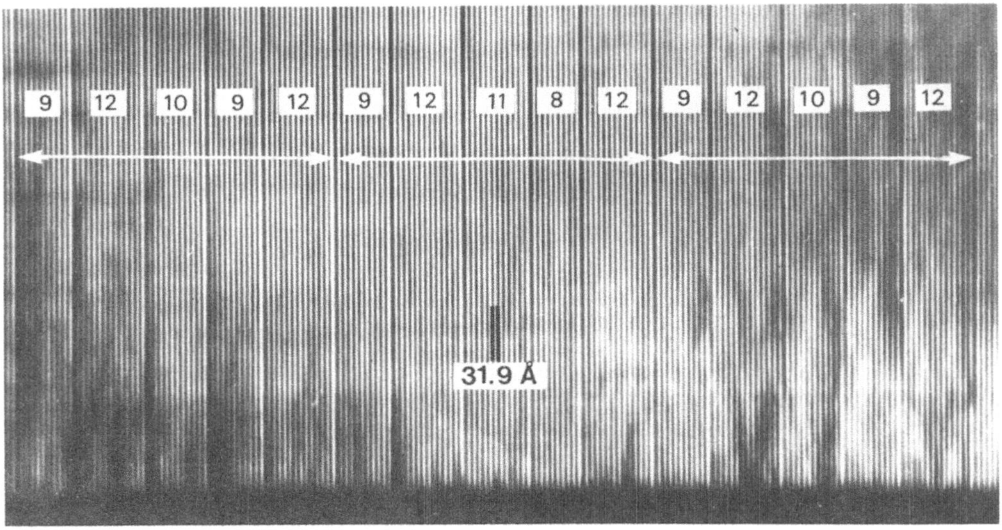

Fig. 4

Interference fringes and "fault planes" in $(NbSe_4)_{10/3}I$ showing a complex sequence (9c + 12c + 10c + 9c + 12c).

This periodicity is about 11 c and corresponds to the additional spots. In fact the sequence is more complex and reappears periodically. In the present case we have the sequence (9c + 12c + 10c + 9c + 12c), namely 52c = 1660 Å . Experiments at various temperatures between 100 and 400 K have shown that these plane defects do not appear to change. We can therefore conclude that they have no influence on the conducting properties or on the hysteresis phenomena observed in the non-linear effect studies[2].

<u>DISPLACIVE TRANSITIONS IN $(NbSe_4)_3I$ AND $(TaSe_4)_3I$</u> - These compounds also exhibit resistivity anomalies at low temperature[1,4]. Electron diffraction experiments have revealed a space group change as shown by the series of patterns (fig. 5) obtained for example on $(NbSe_4)_3I$ at different temperatures in the (a^*, c^*) reciprocal lattice plane. We note the progressive formation of new spots (see arrows) when the temperature decreases below the transition temperature T_c = 275 K. The (h0l) spots, with h+l odd, become more and more intense over a temperature interval of about 50 K. This observation indicates that a second order transition occurs according to Izumi et al.[7]. Taking account of experimental results on the permitted reflections we can determine the new space group which is a sub-group of $P_{4/mnc}$ found at room temperature. We have listed in the following table the different permitted space sub-groups and the imposed conditions on the hkl indices.

Comparing this with the experimental results, it appears that the only compatible sub-group is $P\bar{4}2_1c$. These results are in contradiction with those published by Izumi

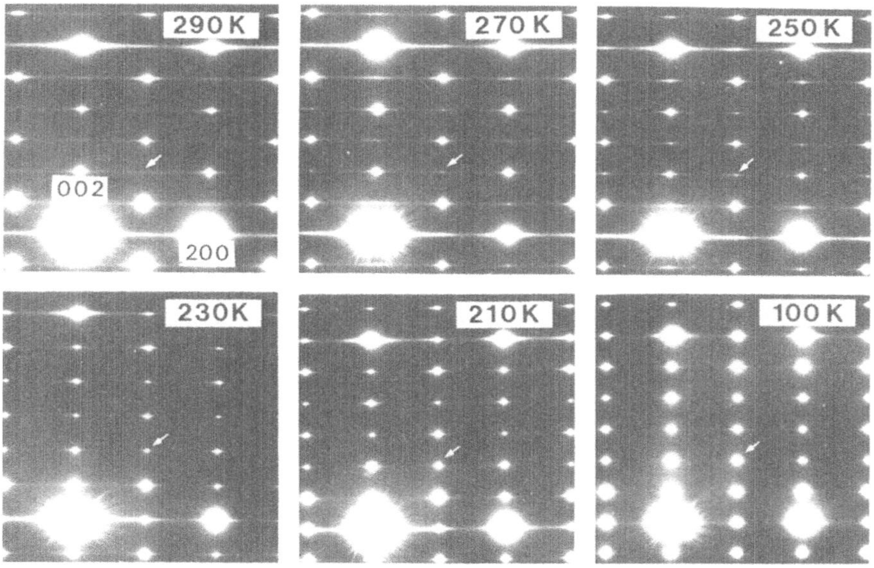

Fig. 5

Electron diffraction patterns obtained with $(NbSe_4)_3I$ at different temperatures.
Note the progressive formation of the new spots (see arrows) indicating a space
groupe change.

$(NbSe_4)_3I$	Space group	h 0 l	h h l	h 0 0	0 0 l
T > 275 K	Exp. Res.	h+l = 2n	l = 2n	h = 2n	l = 2n
	$P4/mnc$	h+l = 2n	l = 2n	h = 2n	l = 2n
T < 275 K	Exp. Res.	-	l = 2n	h = 2n	l = 2n
	$P4/m$	-	-	-	-
	$P4212$	-	-	-	-
	$P4nc$	h+l = 2n	l = 2n	h = 2n	l = 2n
	$P\bar{4}n2$	h+l = 2n	-	h = 2n	l = 2n
	$P\bar{4}21c$	-	l = 2n	h = 2n	l = 2n

et al.[7] who have observed the (hkl) spots with l = 2n + 1 (n integer) at 30 K by X ray diffraction. Our electron diffraction patterns do not confirm this[5]. However we note that our experiments have not permitted us to distinguish between the behaviour of type I and type II compounds as characterized by the resistivity variation with the temperature[1].

We also have used the $(TaSe_4)_3I$ specimen of which the transition temperature T_c is about 200 K. The results obtained concerning the structure are exactly the same. Compared with the preceding compound it is more difficult to observe the specimens by electron microscopy because they are more thick.

CONCLUSION - The chalcogenide compounds $(MX_4)_nI$ that we have studied exhibit quite similar structures formed of MSe_4 chains. Furthermore they all present resistivity anomalies at low temperature between 200 and 300 K though the values of resistivity are different from one material to another. Our study shows that they are very different from the point of view of structural behaviour at the transition. Taking account of this unexpected result, it appears necessary to continue the investigations by using other compounds of the $(MX_4)_nI$ family to establish the precise relation between electronic and structural properties.

1 - P. Gressier, A. Meerschaut, L. Guemas, J. Rouxel and P. Monceau, 1984a, J. Solid State Chem., 51, 141.

2 - ZZ. Wang, P. Monceau, M. Renard, P. Gressier, L. Guemas and A. Meerschaut, 1983a, Solid State Commun., 47, 439.

3 - ZZ. Wang, M.C. Saint Lager, P. Monceau, M. Renard, P. Gressier, A. Meerschaut, L. Guemas and J. Rouxel, 1983b, Solid State Commun., 46, 325.

4 - ZZ. Wang and P. Monceau, 1983, Private Communication.

5 - C. Roucau, R. Ayroles, P. Gressier and A. Meerschaut, 1984, J. Phys. C : Solid State Phys., 17, 2993.

6 - H. Fujishita, M. Sato and S. Hoshimo, 1984, Solid Commun., 49, 313.

7 - M. Izumi, T. Iwazumi, K. Uchinokura, R. Yoshizaki and E. Matsuura, Solid State Commun., 1984, 51, 191.

NEUTRON STUDIES OF THE BLUE BRONZES $K_{0.3}MoO_3$ AND $Rb_{0.3}MoO_3$

C.Escribe-Filippini, J.P.Pouget*, R.Currat**, B.Hennion*** and J.Marcus

L.E.P.E.S. C.N.R.S., 166X, 38042 Grenoble Cedex, France

*L.P.S. Université Paris-Sud, 91405 Orsay, France

**I.L.L., 156X, 38042 Grenoble Cedex, France

***L.L.B., C.E.N. Saclay, 91191 Gif sur Yvette, France

The quasi one-dimensional conductors $K_{0.3}MoO_3$ and $Rb_{0.3}MoO_3$ exhibit a CDW driven Peierls transition at T_c = 180 K. We report on recent inelastic and elastic neutron scattering measurements on these compounds. Some low lying phonon dispersion branches have been measured in the metallic phase. The behaviour of the Kohn anomaly has been studied between room temperature and T_c. Below T_c the value of the CDW wavevector has been determined as a function of temperature.

INTRODUCTION

It is now well established[1] that the molybdenum blue bronzes $K_{0.3}MoO_3$ and $Rb_{0.3}MoO_3$ are quasi one-dimensional metals at room temperature and exhibit a metal-semiconductor phase transition at 180 K.

$K_{0.3}MoO_3$ and $Rb_{0.3}MoO_3$ have space group C 2/m with twenty formulae per unit cell. The crystallographic structure is built from infinite sheets of MoO_6 octahedra separated by alkali ions and can be viewed as containing infinite chains of MoO_6 octahedra sharing corners along the monoclinic b direction (direction of highest conductivity)[2].

X-ray studies have shown that the metal to semiconductor transition is a Peierls transition towards an incommensurate charge density wave (CDW) state[3]: Diffuse X-ray scattering patterns showing characteristic precursor effects have been obtained at room temperature. The diffuse scattering maxima are centered on incommensurate positions in reciprocal space ($q_b \sim 0.28b^*$), with a platelet-shaped intensity distribution in a plane perpendicular to the [010] direction. Below 180 K the diffuse intensity condenses into satellite reflections at $\vec{q}_s = 0.26\vec{b}^* + \frac{1}{2}\vec{c}^*$.

In the semi-conducting phase the blue bronzes show a non linear conductivity, above a sharp threshold electric field, due to CDW transport. Other properties characteristic of CDW transport such as quasi-periodic noise voltage in the frequency range of 10 to 100 kHz have also been reported[4].

In what follows we present results on the temperature dependence of the CDW wavevector as well as on some low lying phonon dispersion branches measured in the metallic phase. Finally the T dependence of the 2 k_F anomaly will be discussed.

The measurements were performed on thermal neutron 3-axis spectrometers at LLB (Saclay) and ILL (Grenoble) and on a cold neutron 3-axis spectrometer (ILL). Different scattering zones containing strong satellite reflections were used.

Temperature dependence of the CDW wavevector

Fig. 1. shows the temperature dependence of the b-component of the CDW wavevector q_b. The error bars estimated at $\pm$ 0.001b* take into account the uncertainties on the lattice parameters and the misalignment of the spectrometer. We notice no appreciable variation below 110K and although the maximum value of q_b is near the commensurate value 1/4 it is, however, unambigously incommensurate as reported by Sato et al[5] and by Pouget et al[6]. The well-defined commensurate-incommensurate phase transition reported in ref. 7 is not confirmed.

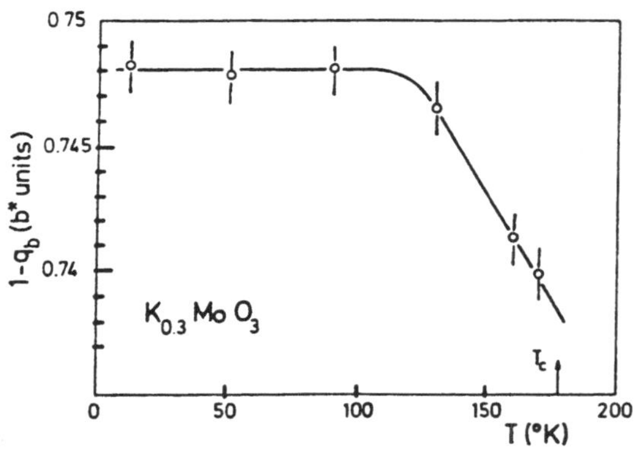

Fig. 1.

Temperature dependence of the CDW wavevector measured by neutron scattering below T_c.

Phonon dispersion curves

A few low-frequency phonon dispersion curves measured in the (0, k, l) and (2h, k, $\pm$h) scattering zones are shown in Fig. 2a and 2b. Sections of the face-centered monoclinic Brillouin zone are shown in Fig. 3a and 3b.

The anisotropy of the LA slopes, as seen in Fig. 2a, reflects the anisotropy in the crystallographic structure with strong metallic binding along the b-axis. The Kohn anomaly is observed not on a TA branch as suggested in ref. 5 but on a branch with optic character and b-polarisation.

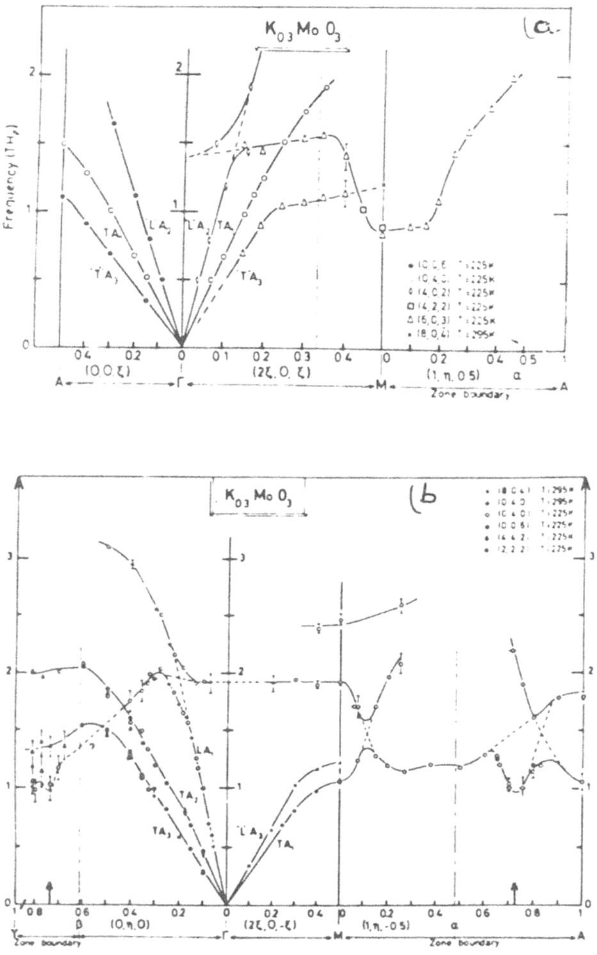

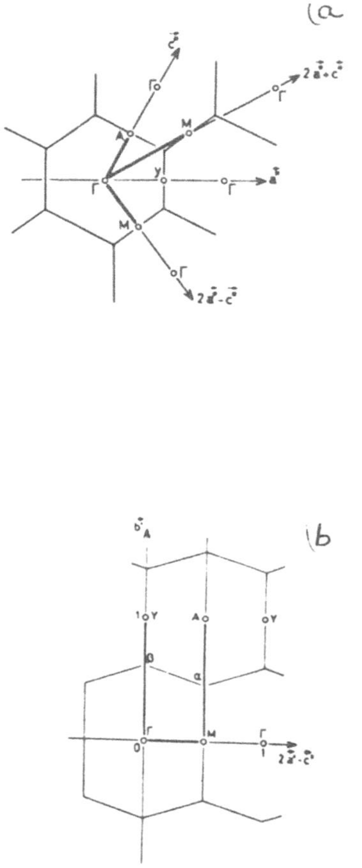

Fig. 2.
Low frequency phonon branches in the blue bronze $K_{0.3}MoO_3$*. Arrows point toward the position of this Kohn anomaly.*

Fig. 3.
Brillouin zone sections by
(a) the $(\vec{a}^*, \vec{c}^*)$ *plane;*
(b) the $(2\vec{a}^* - \vec{c}^*, \vec{b}^*)$ *plane.*

<u>T-behaviour of Kohn anomaly</u>

The behaviour of the soft-branch between T_c and room temperature was examined near the (5, $1-q_b$, 2.5) satellite position using the IN12 cold neutron 3-axis spectrometer. The frequency and wavevector resolution of the instrument was of the order of 60 GHz and $0.01b^*$, respectively.

Intensity limitations inherent to high-resolution inelastic studies require large single-crystal samples. Since Rb-blue bronze single crystals were available to us in larger sizes than for the K-blue bronze, this part of the work was carried out on the Rb compound. The two compounds are known to have the same room temperature structure[2] and the same T_c[1] and display identical X-ray patterns[3].

Typical constant frequency scans along b^* across the Kohn anomaly are shown in Fig. 4. (T = 187K). The double-hump structure observed in Fig. 4

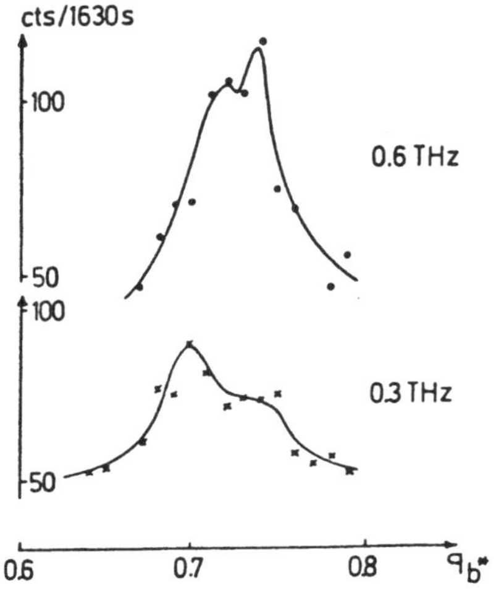

Fig. 4.
Constant frequency scans along (5, ξ, 2.5) at 187K

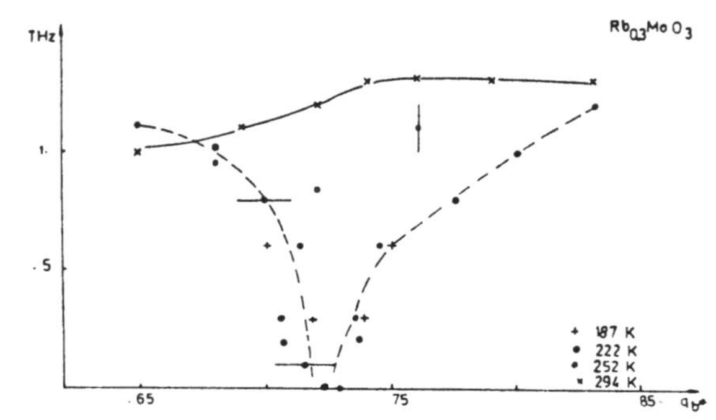

Fig. 5.
Temperature dependence of the Kohn anomaly

indicates a dispersion minimum at the expected position of the anomaly. However, since the two intensity maxima are not fully resolved, the actual shape of the anomaly cannot be specified without separating finite resolution effects from true anharmonic broadening. The position of the intensity maxima obtained from constant-Q and constant-frequency scans at various temperatures are summarized in Fig. 5. Considerable softening is observed between room temperature and T_c.

Concluding remarks

The inelastic results presented above indicate that the Peierls transition in the blue bronzes is associated with a strongly temperature dependent Kohn anomaly on a low-frequency phonon branch with optic character and polarised along the b-direction. This behaviour is consistent with the known anisotropy of the crystal structure and the quasi 1-dimensional character of the electrical conductivity. A more detailed analysis of the present inelastic data in the Kohn anomaly region is in progress. A quantitative comparison with the dynamical behaviour observed in other 1-dimensional conductors such as KCP[8] is, however, difficult due to the limited size of currently available single crystals.

REFERENCES

1. W. Fogle and J.H. Perlstein, Phys. Rev. B6, 1402 (1972).
   R. Brusetti et al in "Recent Developments in Condensed Matter Physics", Vol. 2, Ed. J.T. de Vreese et al (Plenum, 1981) p. 181.
2. M. Ghedira, J. Chenavas, M. Marezio and J. Marcus, submitted for publication.
3. J.P. Pouget, S. Kagoshima, C. Schlenker and J. Marcus, J. Physique--Lettres 44, L113 (1983).
4. J. Dumas and C. Schlenker, Proc. ICSM (1984); This conference and references therein.
5. M. Sato, H. Fujishita and S. Hoshino, J. Phys. C Solid State 16, L877 (1983).
6. J.P. Pouget, C. Escribe-Filippini, B. Hennion, R. Currat, A.H. Moudden, R. Moret, J. Marcus and C. Schlenker, in Proceeding ICSM (1984).
7. C.H. Chen, L.F. Schneemeyer and R.M. Fleming, Phys. Rev. B29, 3765 (1983)
8. K. Carneiro, G. Shirane, S.A. Werner and S. Kaiser, Phys. Rev. B13, 4258 (1976).

THE EFFECT OF A MAGNETIC FIELD ON THE DISCOMMENSURATE TO COMMENSURATE TRANSITION IN 2H $TaSe_2$

W. G. HERRENDEN HARKER

H H Wills Physics Laboratory
University of Bristol, Tyndall Avenue, Bristol BS8 1TL, England.

Abstract Wilson and Vincent have proposed, in a recent theoretical paper, that the discommensurate to commensurate (lock-in) transition in 2H $TaSe_2$ is strongly affected by an applied magnetic field. The proposed phase diagram resembles that found as a function of pressure (0-4 GPa) but with magnetic field in the region 0-2 Tesla as the variable.

The onset($T_O \sim 123K$), lock-in ($T_L=88K$), stripe ($T_S=92K$), and reversion ($T_R=113K$) transitions were monitored by measuring the resultant thermal expansion anomalies along the $\underline{c}$ axis (perpendicular to the layers) using a highly sensitive capacitance dilatometer. Detailed measurements were made of the lock-in and stripe transitions as a function of magnetic field at constant temperatures and constant fields while the temperature was being swept in both directions (the transition shows thermal hysteresis). These measurements clearly demonstrate that the proposed phase diagram is incorrect. As a result the phasings of the C.D.W. discussed by Wilson and Vincent must be questioned.

1. Introduction

The transitions that the layered material 2H $TaSe_2$ displays as a result of the formation of the three charge density waves (C.D.W.) within the layers are summarized in Fig.1. The choice of the phasing of the C.D.W. onto the crystal lattice has important implications for the interpretation of physical measurements that sample the local environment of the ions in the solid, and has therefore been the subject of a number of theoretical studies (e.g. Doran and Woolley[1], and Wilson[2] amongst others). The electron microscope work has shown that in the commensurate C.D.W. state the lattice is orthorhombic; the sample is actually composed of domains of the order of 1μm across arranged in a characteristic double honeycomb geometry. Recently the phasing problem has been considered by Wilson and Vincent[3] in an attempt to interpret a wide variety of experimental data in a consistent way. They concluded that the N.M.R.[4] and Hall coefficient measurements[5] made at 7 and 1-2 Tesla respectively could not be interpreted on the basis of the same phasing of the C.D.W. as the interpretation of the Mossbauer[6], Raman[7], and neutron diffraction[8] measurements (i.e. zero field measurements) would suggest. The effect of pressure on the commensurate C.D.W. state is to significantly alter the interlayer coupling and hence the phasing of the C.D.W. The phase diagram has been measured experimentally by McWhan et al.[9] using X-ray diffraction. By analogy with this phase diagram Wilson and Vincent have proposed a similar phase diagram, Fig. 2, but with magnetic field as the variable. They conclude that the low field/ low pressure structure is the $\gamma\gamma 1^{\triangleright}$ phasing which is converted to the $\alpha\alpha 3^{\oplus}$ structure at high fields/ high pressure (for a discussion of this nomenclature see Wilson and Vincent). This choice of phasing enables the low field phasing to alter smoothly, due to the movement of discommensurations, to the high field structure. It has the added advantage that it

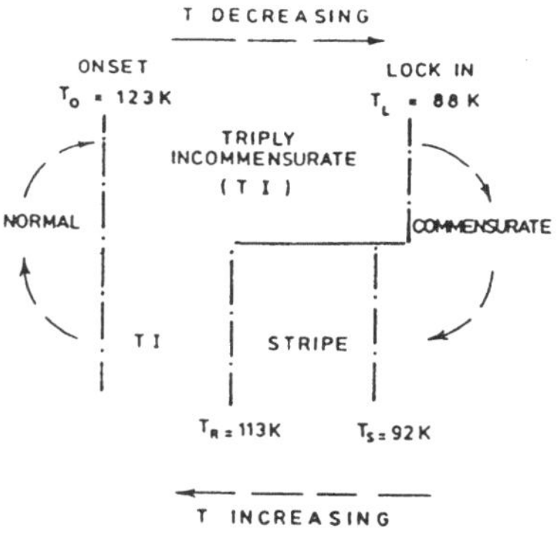

Fig.1. The zero pressure diagram of 2H $TaSe_2$.

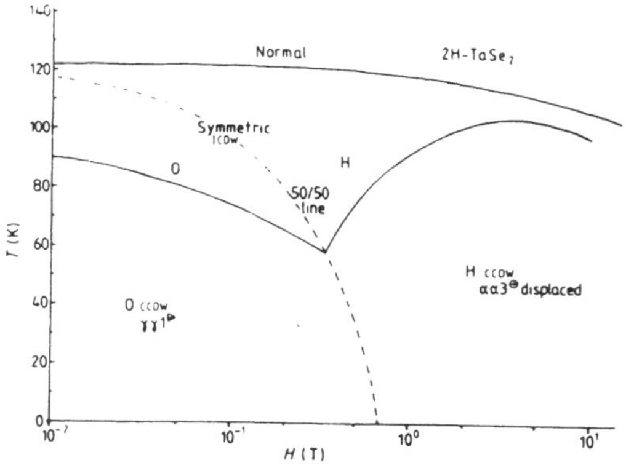

Fig.2. The proposed phase diagram for 2H $TaSe_2$ in the presence of a magnetic field (Wilson and Vincent).

offers an explanation for (1) the lack of any evidence of the lock-in transition in the susceptibility data[10], since at the fields used, 1 Tesla, no lock-in transition would be expected and (2) the anomalous behaviour of the Hall coefficient which again shows no anomaly at T_L but does show some structure at much lower temperatures (~35K). Experimentally the most convenient and sensitive method of investigating this phase diagram is to observe the magnetic field dependence of the c axis thermal expansion associated with the lock-in transition.

2. Experimental method

The measurements were made using a highly sensitive capacitance dilatometer the design of which will be published elsewhere. The capacitance was monitored using a General Radio 1621 transformer ratio arm bridge. The dilatometer was mounted in a continuous flow gas cryostat which enabled temperatures in the range 4.2 - 300K to be achieved. The temperature was monitored using a rhodium/iron resistance thermometer (manufactured by Oxford Instruments Ltd.) with a quoted accuracy of ±0.5K over the entire temperature range and a resolution of ± 0.1K. Slow temperature sweeps (~ 0.1 - 0.5K/min.) with a copper shim sample and of the empty dilatometer in the region 20 - 300K gave smooth chart recorder traces with no discernible abrupt capacitance changes. From these measurements it was concluded that it is possible to detect thermal expansion anomalies of the sample corresponding to a change in sample thickness of $\gtrsim$ 2Å.

3. Results and discussion

The onset, lock-in (see Fig. 3) stripe and reversion transitions were all observed as anomalies in the c axis thermal expansion. The transition temperatures, defined as the centre point of the transition, and the strain at the lock-in and stripe transitions measured during a single low temperature experimental run are summarized in Table I. Provided that the sample is only cycled in the region 60-140K, i.e. over the region in which the C.D.W. is formed and locks onto the lattice, the magnitude of the associated strain is reproducible. However, once the sample is cycled to room temperature the magnitude of the strains associated with these transitions measured in a subsequent experiment is reduced (four such cyclings roughly halve the observed strain). The transition temperatures are unaffected by thermal cycling. The magnitudes of the strains at T_L and T_S show no significant systematic variation and agree quite well with the values quoted by Steinitz et al.[11], (1.63×10^{-5}) and Simpson et al.[12] (2.7×10^{-5}) but no close agreement is to be expected in view of sample degradation on thermal cycling.

There are two possible ways in which to investigate the proposed phase diagram (a) cool (or warm) the sample in a constant magnetic field and measure T_L (or T_S) or (b) hold the temperature constant close to the transitions and sweep the magnetic field to investigate the possibility of any magnetostriction anomaly as a result of a change in phasing of the C.D.W. It is unclear from the paper of Wilson and Vincent whether the direction of the magnetic field with respect to the c axis should have any effect on the phase diagram. Consequently measurements were made with the field both parallel and perpendicular to the layers.

Table I. Sample A thickness = $3.0_6 \times 10^{-4}$ m. Run 4.

Sweep direction	T_oK	T_RK	T_LK	T_SK	$\frac{\Delta c}{c} \times 10^5$ at T_L or T_S
T decreasing	125.5	-	88	-	3.7_4
T increasing	-	113	-	92	4.2_0
T decreasing	121	-	88	-	4.0_7
T increasing	-	113	-	92	4.0_7

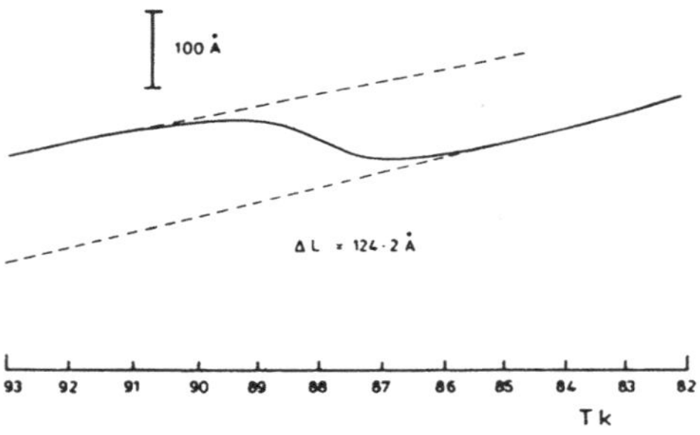

Fig.3. The thermal expansion anomaly, relative to the dilatometer, associated with the lock-in transition T_L=88K. Sample thickness $3.0_5 \times 10^{-4}$m.

Constant-Field results:- The commensurate to stripe phase transition temperature T_S and strain were measured at various fixed magnetic fields both parallel and perpendicular to the c axis. The experimental points are shown in Fig.4. Each fixed field point is the composite of several independent warming cycles. At any particular field the measured values of T_S differed by less than 1K. The values of T_S were independent of increasing the field from zero when the sample was in the commensurate, the stripe or the triply incommensurate phase. Representative values of T_L were also measured at various fields and again no magnetic field dependence was seen. The strain associated with the transition, within any particular low temperature run was independent of the field to within ~10% (- the same reproducibility as seen in zero field measurements) - and showed no correlation with the presence of the field.

Constant-Temperature results:- The temperature of the dilatometer was held as constant as possible (better than 0.1K drifts per min.) and the magnetic field

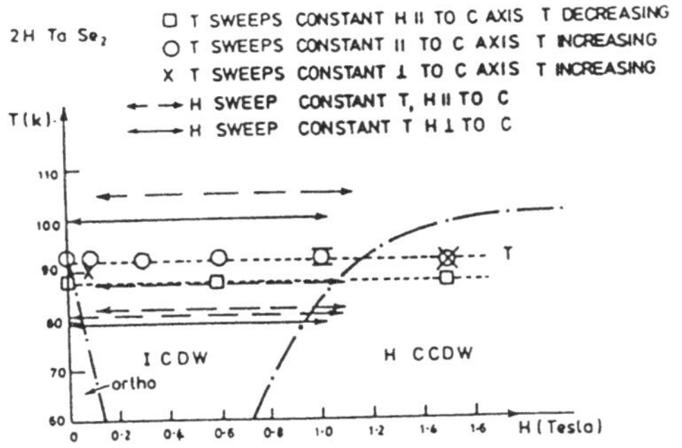

Fig.4. The measured and proposed phase diagram for 2H $TaSe_2$ in the presence of a magnetic field (0-1.6 Teslà).

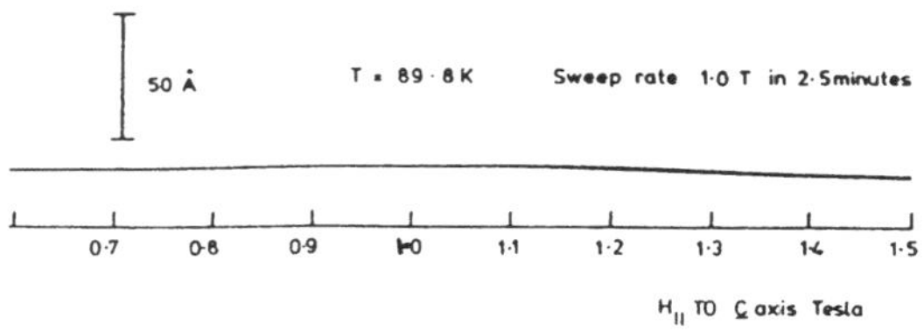

Fig.5. The magnetostriction of 2H $TaSe_2$.

swept; a typical trace is shown in Fig.5. During this experimental run the step due to the commensurate - to - stripe transition would have been ~70Å. When compared with traces obtained in constant field as a function of time, in order to measure changes in capacitance due to small temperature drifts and measurements made with the empty dilatometer, it was clear that there was no significant field effect on the sample. These measurements allowed a range of fields to be investigated above and below the lock-in or stripe transitions,see Fig.4.

The results clearly demonstrate that there is no magnetic field dependence of the lock-in transition and that the phase diagram proposed by Wilson and Vincent cannot be correct. Their identification of the phasings must therefore be in doubt. More recent work by Bird et al.[13] using convergent beam electron diffraction has demonstrated a phasing that enables all of the experimental data to be interpreted without the necessity of invoking a magnetic field dependence.

Acknowledgements I should like to thank Dr. J.A.Wilson for many stimulating discussions and Mr. P.S.A.Field for his help with many technical problems, particularly for his skill in constructing the dilatometer. The financial support of the Royal Society is also gratefully acknowledged.

References

1. Doran N.J. and Woolley A. J.Phys. C 14, 4257, 1981.
2. Wilson J.A. Phys. Rev. B 17, 3880, 1978.
3. Wilson J.A. and Vincent R. J. Phys. C 17, 123, 1984.
4. Pfeiffer L., Walstedt R.E., Bell R.F. and Kovacs T. Phys. Rev. Lett. 44, 1455, 1982.
5. Naito M. and Tanaka S. J. Phys. Soc. Japan 51, 228, 1982.
6. Pfeiffer L., Kovacs T. and Saloman D. Bull. Am. Phys. Soc. 28, 506, 1983.
7. Scott G.K., Bardhan K.K. and Irwin J.C. Phys. Rev. Lett. 50, 771, 1983.
8. Moncton D.E., Axe J.D. and DiSalvo F.J. Phys. Rev. B 16, 801, 1977.
9. McWhan D.B., Axe J.D. and Youngblood R. Phys. Rev. B 24, 5391, 1981
10. Hillenius S.J. and Coleman R.V. Phys. Rev. B 18, 3790, 1982.
11. Steinitz M.O. and Grunzweig-Genossar J. Solid State Commun. 29, 519, 1979.
12. Simpson A.M., Jericho M.M. and DiSalvo F.J. Solid State Commun. 44, 1543, 1982.
13. Bird D.M., McKernan S., Steeds J.W. to be published in J. Phys. C.

HIGH PRESSURE INVESTIGATION OF THE CDW PHASE DIAGRAM OF $1T-TaS_2$

D.R.P. Guy, A.M. Ghorayeb, S.C. Bayliss and R.H. Friend
Cavendish Laboratory, Madingley Road,
Cambridge, U.K.

We have investigated the charge density wave phase diagram of $1T-TaS_2$, including the transition from the triclinic (T-) phase to the nearly-commensurate (NC-) phase found at about 283K on warming, by means of resistivity measurements under pressure. These data are combined with transition temperature and latent heat data from differential scanning calorimetry measurements to give estimates of the volume changes at the transitions using the Clausius-Clapeyron equation. The discrepancy between estimated and measured volume changes at the transition between the NC-phase and the commensurate (C-) phase on cooling is explained by considering the time-dependence of this transition under pressure.

The charge density wave (CDW) phase diagram of $1T-TaS_2$ is now known to be more complex than had been thought[1-7]. The existence of a phase transition observed at about 283K (on warming only) is now well established. Since it was first observed in thermal expansion (capacitance dilatometry) measurements[1], it has been reported in measurements of backscattering yields of 1.00MeV He^+ ions[2], X-ray diffraction[3], van der Pauw resistance[4], resonant flexural vibration period[5] and thermopower and resistivity[6]. In addition we have recently reported evidence for the transition in thermal (differential scanning calorimetry), resistivity and Hall effect results[7] which indicate some in-plane anisotropy.

The X-ray diffraction study by Tanda et al[3] established that there is a nearly-commensurate triclinic (T-) phase in $1T-TaS_2$ observed between 223K (where a transition takes place from the commensurate (C-) phase) and 283K on warming, and that between 283K and the onset of the incommensurate (I-) phase at 355K the phase is nearly-commensurate (NC-phase). Here we adopt the nomenclature of Tanda et al: the T-phase and NC-phase correspond respectively to the phases previously termed $1T_2.2$ and $1T_2.1$[7]. On cooling $1T-TaS_2$ the phases are as previously determined: I-phase ($1T_1$) at T>352K, NC-phase ($1T_2$) at 352K>T>183K and C-phase ($1T_3$) at T<183K. Of interest to this work, we note that the charge density waves are disordered along the c-axis in the C-phase[3,8,9]. Tanda et al[3] reported T-satellites coexisting with NC-satellites between 250K and 200K on cooling. We observe no anomaly at or near 250K (within experimental error) either in the thermal measurements or in the resistivity at any pressure.

In the light of the recent improvement in understanding of the phase diagram of $1T-TaS_2$, we have examined the pressure-dependence of the phase diagram by resistivity measurements extending down to 4K. In-plane electrical resistivity measurements were made on single crystals using the four-contact van der Pauw technique[10]. The resistivity and both van der Pauw resistances were recorded at temperature intervals no greater than 1K. High pressure resistivity measurements were made in the pressure range from ambient pressure to 9.2kbar and in the temperature range from 4K to 300K under isotropic conditions in a beryllium-copper pressure cell with a 1:1 mixture of n-pentane and iso-pentane as the pressure medium. A Mettler Thermal Analysis System DSC was used to determine the transition temperatures and latent heats of the transitions. Crystals for thermal measurements were taken from several growth batches and the results were found to be independent of the batch and of the heating and cooling rate. Crystals for the electrical measurements were taken from one of these batches.

The electrical and thermal results are summarised in Table 1. In the pressure regime up to 4kbar we find excellent agreement with Tani et al[11] who reported dT/dP from resistivity measurements for all the transitions except that between the T- and NC-phases using a helium gas pressure system. This agreement shows that the phase diagram is not dependent upon the nature of the pressure medium used. The electrical and thermal results obtained in this work have been combined through the Clausius-Clapeyron equation for a 1st order phase transition, $dP/dT = L/T\,\Delta V$, to give calculated volume changes at the transitions. Our thermal data for the high temperature transitions between the I- and NC-phases show, when combined with dT/dP measured by Tani et al[11], that the volume change at the NC- to I-phase transition is about 10% greater than that at the I- to NC-phase transition.

On heating, good agreement is obtained between our calculated ΔV and that measured by Sezerman et al[1] for the C- to T-phase transition. The volume change at the T- to NC-phase transition is very small. We calculate a change of only $0.97\times10^{-2}\,cm^3\,mol^{-1}$, half that measured by capacitance dilatometry[1]. Even with this discrepancy the data

TABLE 1 Summary of Electrical and Thermal Results.

Transition	T † /K	L /cal mol^{-1}	dT/dP /K $kbar^{-1}$	V(Calc)* /$10^{-2}cm^3\,mol^{-1}$	V(Meas)* /$10^{-2}cm^3\,mol^{-1}$
HEATING					
C→T	221±1	72±3	-6±.5	8.5±.8	7.5[1]
T→NC	283±3¶	13±1	-5±.5	0.97±.12	1.9[1]
NC→I	355±1	144±2	-3.48[11]	5.9±.1	3.4[1], 4.5[12]
COOLING					
I→NC	352±1	134±3	-3.48[11]	5.5±.1	2.9[1], 4.5[12]
NC→C	183±4	95±4	-12.5±.5	27±2	13[1]

All data are from this work except where indicated by a bracketed superscript

* In all cases the volume is smaller on the high temperature side of the transition

† From thermal measurements at ambient pressure

¶ Tani and Tanaka[6] have reported a spread in transition temperatures from 283K to 275K support the assumption that the resistivity anomaly is indeed caused by the transition between the T- and NC-phases.

On cooling through the NC- to C-phase transition, the measured volume change[1] is large, but consistent with the other volume changes. The calculated volume change is, however, twice that measured and cannot be reconciled with the other volume changes. A time-dependence (possibly connected with the previously mentioned disorder along the c-axis in the C-phase) has been found in the NC- to C-phase transition under pressure[11] and our results confirm this. The time-dependence increases with pressure and is very marked by 5kbar at which pressure the transition takes place over a period of about 2hrs. if the temperature is held constant in the region of the expected transition temperature. Fast cooling causes the transition to take place over a drop of many degrees in temperature (we found a width of 40K centered at about 90K at 5.2kbar and about 2.5K min^{-1}) and the transition can be suppressed completely at this cooling rate by the application of a pressure of 5.9kbar. Consequently very slow cooling is required in the region of the transition to enable an accurate estimate of the transition temperature to be made. We now consider that the cooling rates used in this work in the pressure range up to 4kbar (between 1 and 2K min^{-1}) may not be sufficienly slow to allow for the time-dependence of the transition, causing the transition to occur out of thermodynamic equilibrium and dT/dP to be artificially large. Experiments to test this are planned.

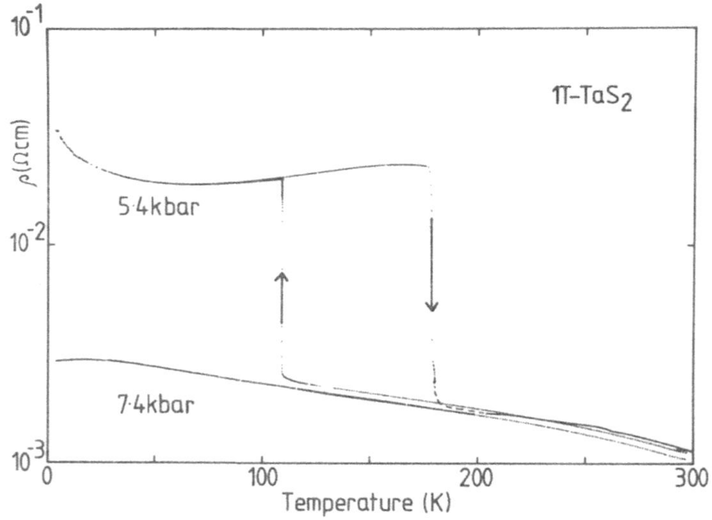

Figure 1: Temperature dependence of the resistivity of 1T-TaS_2 at 5.4kbar and 7.4kbar. Data points are shown except between 182K and 205K on warming at 5.4kbar where the data were affected by a slight inhomogeneity within the pressure medium and a dashed line has been drawn as a guide to the eye. The average cooling rates were 2.0K min^{-1} at 5.4kbar and 1.7K min^{-1} at 7.4kbar. At 5.4kbar the temperature was held at 109.0±.2K for two hours whilst the NC- to C-phase transition occurred. The transition between the T-phase and the NC-phase is at 260±2K. At 7.4kbar the resistivity reached a maximum at 15K and had fallen by 1.5% at 4K.

No time-dependence has been found in the C- to T-phase transition. We note that this transition is exothermic whereas that from the NC-phase to the C-phase is endothermic. The exothermic nature of the the C- to T-phase transition may help the transition, once started, to continue.

Figure 1 shows the temperature-dependence of the resistivity at pressures of 5.4 and 7.4kbar, the latter shows the suppression of the C-phase. The 5.4kbar plot was obtained by holding the temperature at 109.0±.2K for two hours whilst the transition between the NC-phase and the C-phase occurred. On recommencing cooling the resistivity immediately began to fall. A plot similar to that at 7.4kbar was obtained at a uniform cooling rate of 1.3K min^{-1} at 9.2kbar. At 7.4kbar the C-phase has been suppressed and the low temperature resistivity indicates that the sample is still in the NC-phase. This behaviour is very similar to that recently reported for $1T\text{-}TaS_2$ substitutionally doped with 0.1% tungsten, at ambient pressure[4].

Evidence to show that the C-phase is indeed suppressed in resistivity plots such as that obtained at 7.4kbar is provided by a fast-cooling run (2.5K min^{-1}) at 5.9kbar in which, following a resistivity behaviour similar throughout the temperature range 300K to 4K to that found at 7.4kbar, the characteristic NC- to C-phase transition began to occur at 80K on warming and was centered quite sharply on 110K. The C-phase resistivity was about 75% of its expected value and the C- to T-phase and T- to NC-phase transitions occurred at the expected temperatures. Both in this case and in all other cases, the ratio of the van der Pauw resistances exhibited its characteristic oscillations[4], the exact nature of which is dependent upon the positions of the electrical contacts relative to the crystal axes, on passing to and from the C-phase.

References:

1. O.Sezerman, A.M.Simpson and M.H.Jericho, Sol.St.Commun. 36 (1980) 737
2. T.Haga, Y.Abe and Y.Okwamoto, Phys.Rev.Lett. 51 (1983) 678
3. S.Tanda, T.Sambongi, T.Tani and S.Tanaka, J.Phys.Soc.Japan 53 (1984) 476
4. H.Fujimoto and H.Ozaki, Sol.St.Commun. 49 (1984) 1117
5. A.Suzuki, R.Yamamoto, M.Doyama, H.Mizubayashi, S.Okuda, K.Endo and S.Gonda, Sol.St.Commun. 49 (1984) 1173
6. T.Tani and S.Tanaka, J.Phys.Soc.Japan 53 (1984) 1790
7. S.C.Bayliss, A.M.Ghorayeb and D.R.P.Guy, J.Phys.C 17 (1984) L533
8. C.B.Scruby, P.M.Williams and G.S.Parry, Phil.Mag. 31 (1975) 255
9. K.K.Fung, J.W.Steeds and J.A.Eades, Physica 99B (1980) 47
10. L.J.van der Pauw, Philips Research Reports 13 (1958) 1
11. T.Tani, T.Osada and S.Tanaka, Sol.St.Commun. 22 (1977) 269
12. M.O.Robbins and E.A.Marseglia, Phil.Mag.B 42 (1980) 705

LANDAU THEORY OF 2H-$TaSe_2$

D.M.Bird and R.L.Withers

H.H.Wills Physics Laboratory, Tyndall Avenue, Bristol BS8 1TL, U.K.

The recently determined structure of the commensurate superlattice state of 2H-$TaSe_2$ is analysed in terms of the phenomenological free energy expansion of Jacobs and Walker. It is shown that the magnitude of the inter-layer term is considerably larger than has been previously assumed. Some consequences of this are discussed.

Elsewhere in this conference, Bird et al. briefly discuss the technique of convergent beam electron diffraction (CBED) and give some examples of its use in crystal symmetry determination. Recent work in this field has shown that by analysis of the fine structure in higher order Laue zone (HOLZ) reflections it is possible to fix atomic positions within the unit cell to a high degree of accuracy (~0.005Å).[1,2] This technique has now been applied to several materials, including the commensurate superlattice state of 2H-$TaSe_2$.[3] In this paper we discuss the nature of the determined structure in terms of the phenomenological theory of Jacobs and Walker.[4]

Fung et al.[5] found that the commensurate CDW state of 2H-$TaSe_2$ has a microstructure of ∿1μm orthorhombic domains, oriented in three distinct ways relative to the underlying hexagonal lattice. This explained why the previous neutron diffraction work[6] had found overall hexagonal symmetry. It is straightforward to obtain CBED patterns from a single domain; using these Bird, McKernan and Steeds[3] determined the stucture outlined in Table 1. The in-the-layer displacements are written in terms of PSD waves as

$$\underline{\delta}_{\underline{L}}^{Ta} = \sum_{j=1}^{3} \hat{\underline{G}}_j^s \, a_j^{Ta} \cos(\underline{G}_j^s . \underline{L} + \Theta_j^{Ta})$$
$$\underline{\delta}_{\underline{L}}^{Se} = \sum_{j=1}^{3} \hat{\underline{G}}_j^s \, a_j^{Se} \cos(\underline{G}_j^s . \underline{L} + \underline{G}_j^s . \underline{R}_{Se} + \Theta_j^{Se}) \qquad (1)$$

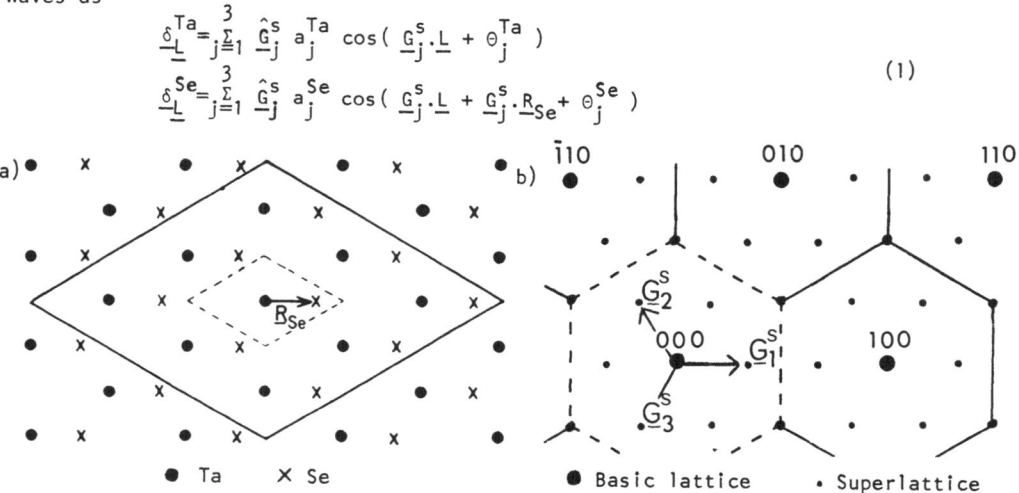

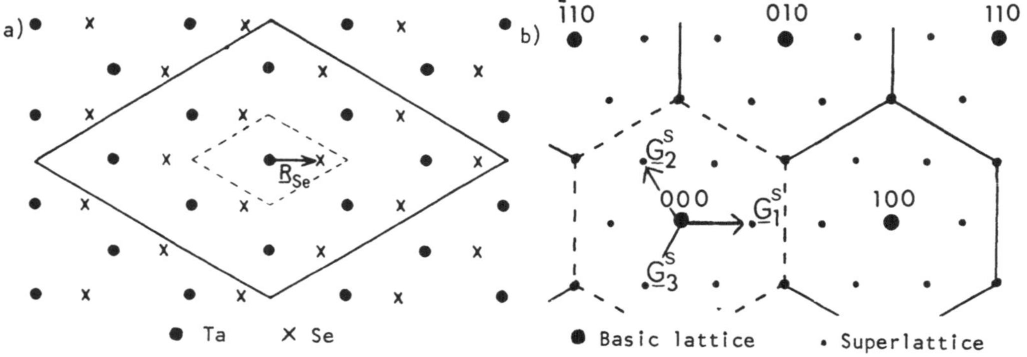

Fig.1 a),b) Real and reciprocal space lattices of 2H-$TaSe_2$

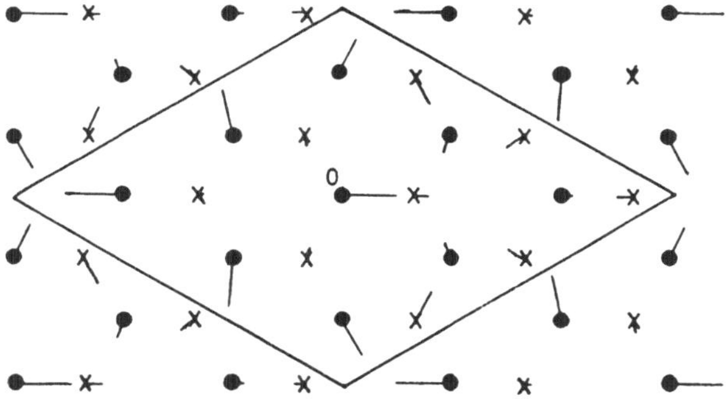

Fig.2 Atomic displacements in one layer of the 2H structure. Upper layer is obtained by inversion at 0.

	Θ_1	$\Theta_2(=\Theta_3)$	$\frac{a_1}{a_2}(=a_3)$
Ta	22°	222°	0.9
Se	303°	132°	1.0

Table 1 PSD amplitudes and phases (from ref.3)

where $\underline{L}$ labels the basic cells, $\underline{R}_{Se}$ the basic selenium position (Fig.1a) and the $\underline{G}_j^s$ are the fundamental superlattice wavevectors (Fig.1b). An inversion centre links the two layers, the displacement pattern of one layer is shown in Fig.2.

Jacobs and Walker[4] wrote down a phenomenological free energy expansion for 2H-$TaSe_2$ in terms of three order parameters for the reciprocal superlattice vectors $\underline{G}_j^s$. In the commensurate state these become complex numbers $a_j e^{i\Theta_j}$. In order to discuss the determined structure we identify the order parameter as being related to the Ta PSD, so $a_j=a_j^{Ta}$ and $\Theta_j=\Theta_j^{Ta}$. Note that while the absolute phase of the Se PSD is different, the relative amplitudes and phases are similar for the two species ie. $\Theta_1^{Ta}-\Theta_2^{Ta}=200^\circ$, $\Theta_1^{Se}-\Theta_2^{Se}=189^\circ$ and $a_1^{Ta}/a_2^{Ta}=0.9$, $a_1^{Se}/a_2^{Se}=1.0$. This identification of the order parameter is discussed further in ref.7. We shall primarily be interested in the phasing of the PSD pattern onto the underlying lattice and not in the magnitude of the modulation. The phase dependent parts of the Jacobs and Walker free energy can be written to cubic order as

$$F_{\substack{phase\\dep.}} \propto |D|\, a_1\, a_2^2 \cos(\Theta_1 + 2\Theta_2 + \Theta_D) + |E|\,(a_1^3 \cos(3\Theta_1 + \Theta_E) + 2a_2^3 \cos(3\Theta_2 + \Theta_E)) + |F|\,(a_1^2 \cos(2\Theta_1 + \Theta_F) + 2a_2^2 \cos(2\Theta_2 + \Theta_F)) \quad (2)$$

where the Landau theory coefficients are written $D=|D|e^{i\Theta_D}$ etc. and orthorhombic symmetry is imposed by putting $a_2=a_3$, $\Theta_2=\Theta_3$. The free energy is the same for each layer in the crystal. In previous treatments of this expansion it has been assumed that the two dimensional character of 2H-$TaSe_2$ will mean that the interlayer term F is small compared to the cubic intralayer terms D and E, even though it is harmonic in the order parameter.[8-11] We will show that this assumption is inconsistent with the determined structure.

Putting $a_1=a_2$ in (2), the free energy is minimised when Θ_1 and Θ_2 satisfy

$$\begin{aligned} &|D|\, a \sin(\Theta_1 + 2\Theta_2 + \Theta_D) + 3|E|\, a \sin(3\Theta_1 + \Theta_E) + 2|F| \sin(2\Theta_1 + \Theta_F) = 0 \\ &|D|\, a \sin(\Theta_1 + 2\Theta_2 + \Theta_D) + 3|E|\, a \sin(3\Theta_2 + \Theta_E) + 2|F| \sin(2\Theta_2 + \Theta_F) = 0 \end{aligned} \quad (3)$$

For general coefficients these equations can not be solved analytically. We will therefore concentrate on the limits where $(|D|,3|E|) \gg |F|$ or $(|D|,3|E|) \ll |F|$. Instead of Θ_1 and Θ_2 it is convenient to use the angles $\phi=(\Theta_2-\Theta_1)$ and $\psi=(\Theta_1+2\Theta_2)/3$ to describe the PSD phase pattern. ϕ represents the symmetry position (point of common phasing[12]) along the mirror line of the orthorhombic supercell and ψ defines its detailed shape. ϕ has special values $2m\pi$, $2\pi/3+2m\pi$ and $4\pi/3+2m\pi$ (labelled α,β and γ respectively[10,12]) where the displacements do not break the trigonal symmetry of individual layers (Fig.3). In Fig.3 we also label the positions δ, where $\phi=\pi+2m\pi$ and the displacement pattern is orthorhombic within a single layer. Only at the α position does the two layer displacement pattern retain the high temperature space group $P6_3/mmc$. At all other values of ϕ the space group becomes Cmcm, the one determined by Fung et al.[5]

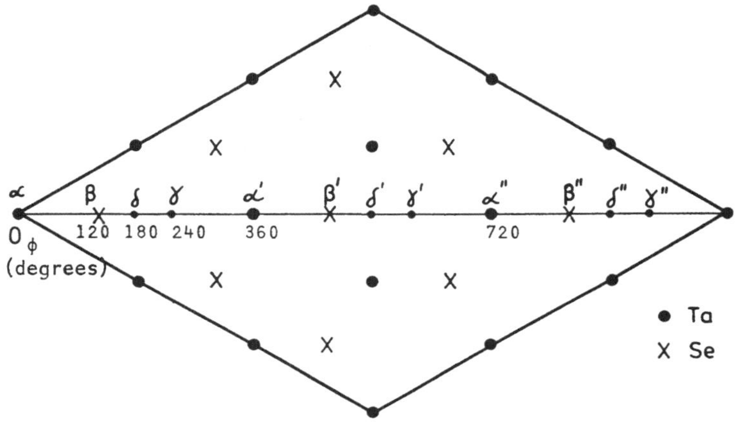

Fig.3

Special positions of the phase origin along the mirror line of the 3x3 supercell.

If $(|D|,3|E|) \gg |F|$ the zeroth order solution of (3) is $\Theta_2^o-\Theta_1^o=2n\pi/3+2m\pi$ [8] where $|D|\sin(3\Theta_1^o+4n\pi/3+\Theta_D)+3|E|\sin(3\Theta_1^o+\Theta_E)=0$. n=0,1,2 corresponds to the α,β,γ solutions, the most favourable is that which minimises the F term in (2). Including the F term as a perturbation pulls the orthorhombic solutions off the exact β and γ positions, but the α solution is not moved. If $(|D|,3|E|) \ll |F|$ the zeroth solution becomes $\Theta_1^o=\frac{1}{2}(\pi-\Theta_F)+n_1\pi$, $\Theta_1^o=\frac{1}{2}(\pi-\Theta_F)+n_2\pi$ ie. $\phi=n\pi+2m\pi$. n can equal 0 or 1, which correspond to the phase pattern being centred at positions α and δ in Fig.3. The cubic correction to the free energy in this limit is given by

$$\delta F \;\alpha\; \pm\left[\,|D|a\cos\left(\frac{3\pi}{2}-\frac{3\Theta_F}{2}+\Theta_D\right)\;\begin{matrix} +\;3|E|a\cos\left(\frac{3\pi}{2}-\frac{3\Theta_F}{2}+\Theta_E\right) & \text{for } \alpha \\ -\;|E|a\cos\left(\frac{3\pi}{2}-\frac{3\Theta_F}{2}+\Theta_E\right) & \text{for } \delta \end{matrix}\right]. \tag{4}$$

It can be seen that δ is favoured over α only if

$$|D\cos\left(\frac{3\pi}{2}-\frac{3\Theta_F}{2}+\Theta_D\right)| > |E\cos\left(\frac{3\pi}{2}-\frac{3\Theta_F}{2}+\Theta_E\right)| \tag{5}$$

and these two contributions have opposite signs. Again, the solution is shifted off δ but not off α if D and E are included as perturbations.

We now compare these findings with the determined structure. From Table 1, $\Theta_2^{Ta}-\Theta_1^{Ta}=$

200°, which lies between the values of 180° and 240° for γ and δ respectively. It is clear that to obtain this solution the interlayer and cubic intralayer coefficients must be of similar orders with the F term slightly dominant. It also follows from the condition (5) that D is likely to be larger than E. We conclude that it is invalid to treat the interlayer term as a perturbation in theoretical treatments of 2H-$TaSe_2$. This conclusion is strengththened in incommensurate phases where, as the amplitude of the modulation falls as the onset transition is approached, the cubic terms are further reduced relative to harmonic terms.

These findings have interesting consequences for the nature of discommensurations in 2H-$TaSe_2$. Discommensurations were found to be twin boundaries, lying preferentially along basic lattice $\{11\bar{2}0\}$ planes, between differently oriented orthorhombic domains[5]. Their presence tends to increase the fundamental wavelength of the CDW. Provided the amplitudes a_1 and a_2 are equal, the CDW/PSD pattern can be described by the position of its common phasing points. Thus the overall effect of a discommensuration can be analysed in terms of the total shift of these points from one side of the boundary to the other. It is straightforward to show that these shifts are $(2-\phi/120^{o})a_o$ and $(-1+\phi/120^{o})a_o$ on the two layers, and that they are in the direction normal to the discommensuration[7]. For the β and γ solutions (where $\phi=120^{o}$ and 240°) one shift is a_o and the other 0[9]. Thus the two layers are effectively uncoupled. This is consistent with the weak interlayer term necessary to find β or γ, but not with the observed rigidity of discommensurations parallel to the c axis[5]. In the determined structure, $\phi=200^{o}$, making the two shifts $0.66a_o$ and $0.33a_o$. Thus the effect of a larger interlayer term is also seen in a stronger coupling of the layers through a discommensuration.

1. R.Vincent, D.M.Bird and J.W.Steeds 1984 Phil. Mag. in press
2. D.M.Bird 1984 J.Phys.C in press
3. D.M.Bird, S.McKernan and J.W.Steeds 1984 J.Phys.C in press
4. A.E.Jacobs and M.B.Walker 1980 Phys. Rev. B21, 4132
5. K.K.Fung et al. 1981 J.Phys.C 14, 5417
6. D.E.Moncton, J.D.Axe and F.J.Disalvo 1977 Phys. Rev B16, 801
7. D.M.Bird and R.L.Withers 1984 to be published
8. M.B.Walker and A.E.Jacobs 1981 Phys. Rev. B24, 6770
9. M.B.Walker and A.E.Jacobs 1982 Phys. Rev. B25, 4856
10. W.L.McMillan 1982 preprint
11. P.B.Littlewood and T.M.Rice 1982 Phys. Rev. Lett. 48, 27
12. J.A.Wilson and R.Vincent 1984 J.Phys.F 14, 123

Both authors acknowledge financial support from the SERC.

MULTIDOMAIN STRUCTURES OF INCOMMENSURATE PHASES IN CDW STATES OF 2H-$TaSe_2$

V. Janovec and V. Dvořák
Institute of Physics, Czech. Acad. Sci.,
Na Slovance 2, 180 40 Prague 8, Czechoslovakia

Incommensurate phases in $TaSe_2$ are described as special types of domain structure associated with the structural phase transition from $P6_3/mmc$ to Cmcm phase. Domain walls of lowest energy forming these structures are deduced from the Landau free energy expansion. Possible stripe phases, double honeycomb phase and coexistence of different phases are discussed within this approach which is not associated with any specific model of $TaSe_2$.

Electron microscopic observations disclosed spectacular mictrostructures of the incommensurate phases in 2H-$TaSe_2$[1,2]. Discussions of these structures[3-5] have been based on the two-layer model[6,7]. Here we describe an alternative approach not relying on the layer structure of 2H-$TaSe_2$.

Within a standard Landau-type treatment the commensurate (C) phase of Cmcm symmetry can be described as a result of a phase transition from the parent (P) phase of $P6_3/mmc$ symmetry. The order parameter has three complex components $Q_j = r_j \exp(i\Theta_j)$, $j = 1,2,3$, with two r_j equal and $\Theta_j = (2\pi/3)m_j$, where $\sum_j m_j = 0 \pmod 3$, $m_j = 0,1,2$[8]. Amplitudes r_j determine the orientation and phases Θ_j the position of the orthorhombic C unit cells in space. Three possible orientations are related by lost $2\pi/3$ and $4\pi/3$ rotations and each orientation can be realized in nine different positions related by translations lost at the P-C transition. Thus there are 27 equivalent C domain states that can be labelled by two numbers, one specifying the orientation and the other the position of the C structure (Fig. 1).

In the domain approximation an ideal incommensurate phase is represented by a C domain structure containing only equivalent domain walls (W's) of the lowest negative energy (such W's are usually called discommensurations[9]). Analysing Lifshitz invariants and anisotropic terms in Q_j and utilizing symmetry arguments it can be deduced that there are 54 equivalent W's of minimum energy (Fig. 2). These W's are perpendicular to primitive translations a_h^j of the P phase and join domains with different orientation of orthorhombic structures which are, in addition, shifted by a_h^j perpendicularly to W (cf., Fig. 2 and Fig. 1) in agreement with the observation[1]. Domains adhere to W's of Fig. 2 in a defined sequence, e.g., the wall $1_1/2_5$ has minimum energy but the wall $2_5/1_1$ with a reversed domain sequence has not. As seen from Fig. 2 parallel W's form regular wall lattices that represent 9 different stripe phases of 3 orientations. Each stripe phase is a repetition of an elementary wall sequence (W unit cell) comprising 6 W's. Different stripe phases are related by symmetry operations lost at the P-C transition and are, therefore, equivalent in the same sense as are the C domain states.

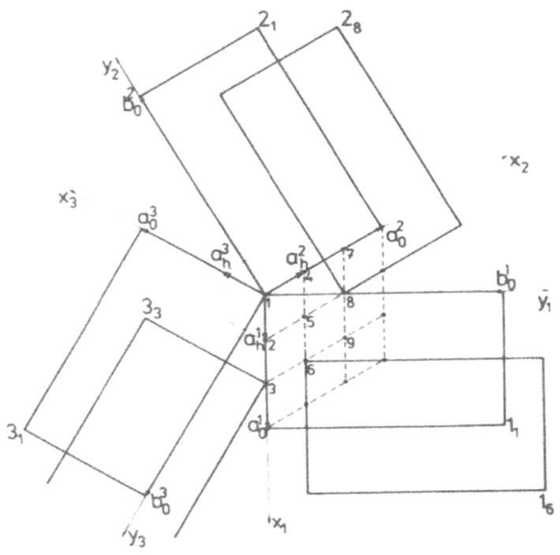

Fig. 1 Designation of 27 C domain states represented by conventional orthorhombic unit cells (solid rectangles). The first number in domain state symbol specifies the orientation and the subscript the position of the C structure with respect to a common coordinate frame. a_h^j and a_o^j, j = 1,2,3, are primitive translations of P and C phases, resp.

Another regular domain pattern that can be formed from W's of minimum energy is the double honeycomb structure (Fig. 3). It is a diperiodic repetition of a primitive rhombic cell containing just once each of 27 domain states, once each of 54 W's of minimum energy and, in addition, 9 different but equivalent left-handed (with respect to translations a_h^j in the W's), 9 right-handed 3-fold wall intersections and 9 equivalent 6-fold wall intersections. This domain pattern is identical with the double honeycomb discommensuration mesh proposed as a possible interpretation of electron diffraction patterns of the triply incommensurate phase[1]. There is only one variant of the double honeycomb structure.

Domain states of the C phase, different stripe phases and double honeycomb phase are topologically different structures. Boundaries between them are formed by topological defects. The changing periodicity at stripe phase boundaries is accommodated by linear defects (called CDW dislocations[9], six-line dislocations[4], deperiodization lines[10]). These are lines on which terminate 6 W's of elementary wall sequences. Examples of coexisting stripe phases are given in

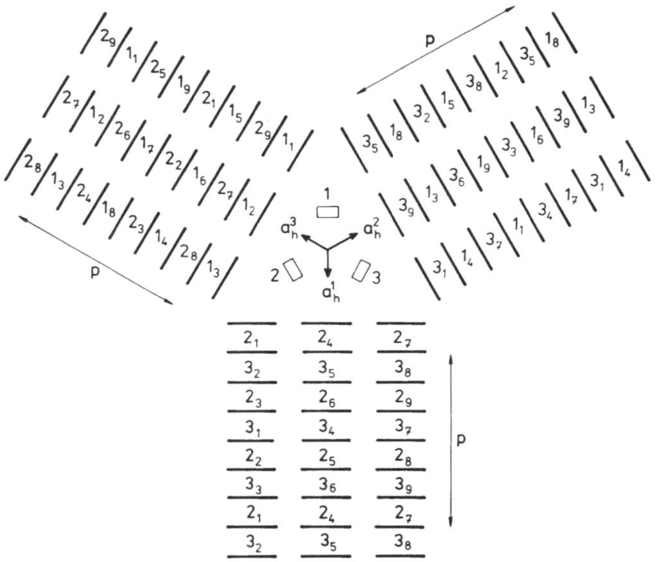

Fig. 2 54 equivalent domain walls with lowest negative energy arranged in wall lattices (periodicity p) of 9 different stripe phases. Rectangles indicate orientation of conventional C unit cells.

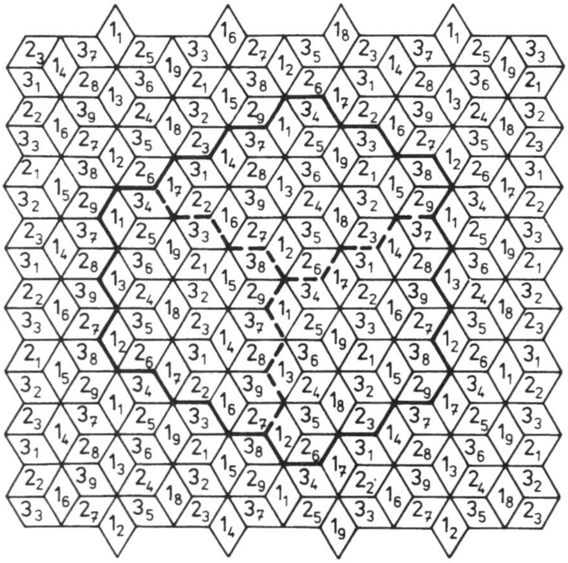

Fig. 3 Double honeycomb structure formed from walls of Fig. 2. Dashed (solid) frames mark primitive (conventional) unit cells of the periodic wall pattern.

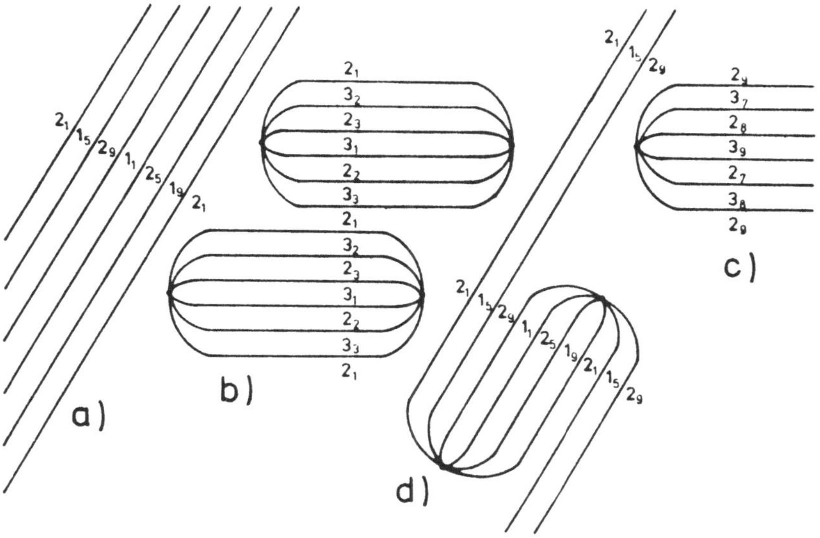

Fig. 4. Boundary between stripe phases a) b) with a common domain state 2_1 is formed by linear defects whereas stripe phases b) c) with no common domain state join along a narrow domain 1_5 bordered on both sides by linear defects. Nucleation of a stripe phase from a narrow domain is sketched in d). All these situations can be found in observed microstructures[1,2]. Linear defects mediating double honeycomb structure with other structures are formed by wall intersections different from that appearing in Fig. 3. Homotopy classification of possible line defects can be performed by mapping closed loops encircling defects into the order parameter space.

1) K.K. Fung, S.McKernan, J.W.Steeds and J.A.Wilson: J.Phys. C: Solid State Phys. 14, 5417 (1981).
2) C.H.Chen, J.M.Gibson and R.M.Fleming: Phys.Rev. B 26, 184 (1982).
3) W.L.McMillan: preprint.
4) M.B.Walker: Phys.Rev. B 26, 6208 (1982).
5) K.Nakanishi and H.Shiba: J.Phys.Soc.Jpn 52, 1278 (1983).
6) P.B.Littlewood and T.M.Rice: Phys.Rev.Lett. 48, 27 (1982).
7) M.B.Walker and A.E.Jacobs: Phys.Rev. B 25, 4856 (1982).
8) V.Dvořák and V.Janovec: J.Phys. C: Solid State Phys. (in print).
9) W.L.McMillan: Phys.Rev. B 14, 1496 (1976).
10) V.Janovec: Phys.Lett. 99A, 384 (1983).

ELECTRON MICROSCOPY OF CHARGE DENSITY WAVE DEFECTS IN 1T-TaS_2 AND 1T-$TaSe_2$

G. Salvetti[1], R. Ayroles[1], C. Roucau[1], H. Mutka[2], P. Molinié[3]

1. Laboratoire d'Optique Electronique du CNRS, BP 4347, 31055 Toulouse Cedex, France
2. Technical Research Center of Finland, 02150 Espoo, Finland
3. Laboratoire de Chimie des Solides, 44072 Nantes Cedex, France

We have studied the charge density wave pinned by defects in the layer compounds 1T-TaS_2 and 1T-$TaSe_2$. The number of pinning defects is controlled by irradiating the samples with high energy electrons (2 MeV). Electron microscopy observations, both high resolution and dark field images, reveal the effect of pinning on the domains and the boundaries of the charge density wave.

Domains, boundaries and pinning to defects are often evoked to explain various properties of CDW-materials, but these phenomena are less commonly observed directly. The layer compounds 1T-TaS_2 and 1T-$TaSe_2$ are good materials for such an observation because the strenght of the distortion permits characterization by electron microscopy. For observation, we used a conventional 120 kV electron microscope. Two types of images can be obtained : dark field micrographs, high resolution micrographs. These different aspects of microscopy are detailed by various authors [1,2,3]. We centered our interest on the commensurate phase of these two compounds ($T < 200$ K in TaS_2, $T < 473$ K in $TaSe_2$) and also on the nearly-commensurate phase 1T-TaS_2 which exists between 200 K and 355 K [4]. We have paid special attention to the effect of atomic lattice defects on CDW. Irradiation with electrons of very high energy (2 MeV) is a good means to create atomic defects in controlled proportions [5]. Introduction of atomic disorder decreases the transition temperatures between the different distorted phases [6]. Theoretically [7], it has been shown that atomic defects locally enhance the CDW but the screening conditions pin the CDW phase. Consequently there will appear domains where the CDW is coherent, separated by boundaries of phase mismash [8].

First, let us study the commensurate state, in the case of $TaSe_2$. The Ta atoms form star-of-David-shaped clusters of 13 atoms, and the crystal can be described by 3 different triclinic cells, related with a rotation symmetry of $\frac{2\Pi}{3}$ [9]. This situation is obtained because 3 different stackings are possible. High resolution microscopy allowed us to characterize two different types of boundaries between the domains observed on dark-field micrographs [2] : Fig. 1a shows a translation boundary, where a shift of the fringes is clearly visible between the two domains ; no perturbation of the fringes can be found out at boundaries between two parts of the crystal rotated of 120° from one another (fig. 1b). This strange result can be explained by considering the fact that different domains are present in the sample thickness ; the boundary generally does not cross the sample along the thickness. So it is difficult by high-resolution microscopy to obtain evidence of CDW perturbation at this type of boundaries even if its existence is not doubtful.

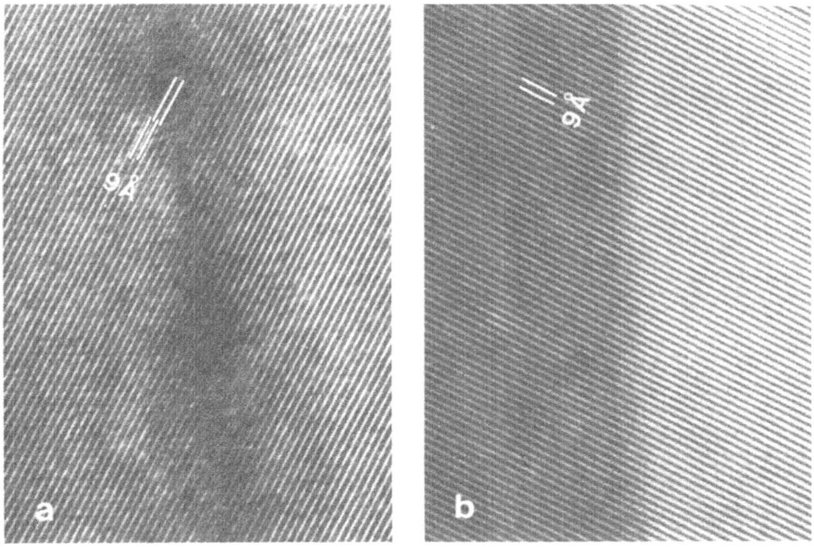

Fig. 1

Differents boundaries observed by high-resolution electron microscopy.
a) Translation boundary - b) "Stacking" boundary.

In order to study the evolution of domains under irradiation, we irradiated a thin sample in a high voltage electron microscope at room temperature and observed it, after different irradiation doses in a conventional electron microscope. The observation temperature is about 300 K, so the commensurate phase was maintained during all the experiments. In those conditions, fig. 2 shows that domains keep the same configuration and shape during irradiation, as long as the doses remain small ($< 6 \times 10^{-3}$ dpTa). The boundaries are affected by irradiation and become irregular, which proves that CDW defects are localized at domain boundaries, and that no perturbation is present inside domains. When the irradiation dose increases, dark-field micrographs of fig. 2 show a "granularity" which covers very rapidly the domains contrast observed before. Simultaneously, high-resolution micrographs show that the fringes, and, by consequence the CDW are strongly affected by irradiation : high irradiation doses induce a loss of coherence of CDW which increases with the density of defects.

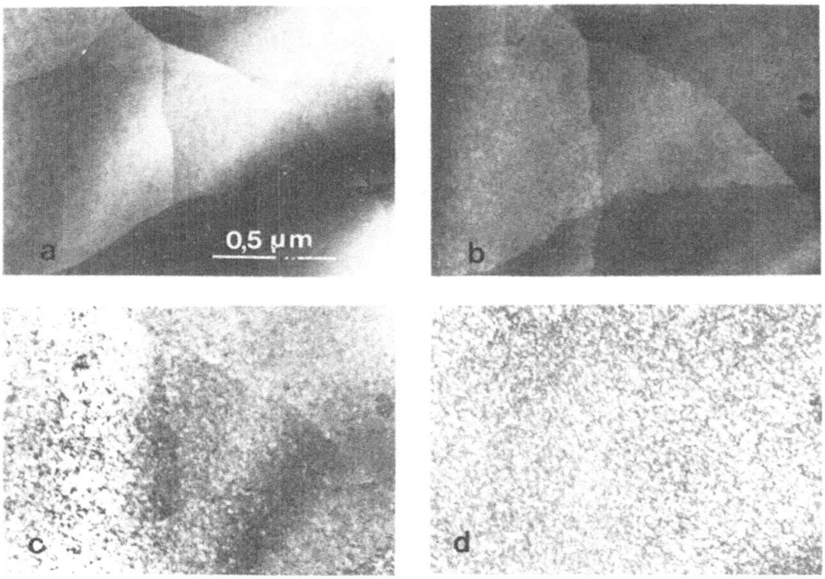

Fig. 2

Evolution with irradiation dose of dark-field micrographs, obtained at 300 K, in $TaSe_2$. a) Unirradiated. b) $2x10^{-3}$ dpTa. c) $8x10^{-3}$ dpTa. d) $1.1x10^{-2}$ dpTa.

The commensurate phase of 1T-TaS_2 is more disordered, already before irradiation, the domains are smaller than in 1T-$TaSe_2$ and the stacking has no clear periodicity [9]. The observations we have made on irradiated samples differ from those made on $TaSe_2$ because it has been possible to observe the domains only after having heated the sample up to room temperature. In these conditions, we cannot follow in detail the behaviour of individual domains but there is a general trend to decrease the domain size before the defects stabilize the non-commensurate phase down to the lowest obtainable temperature.

We also studied this non-commensurate phase, which exists only in 1T-TaS_2. No definitive structure has yet been found for this phase [9]. Recent papers [10] propose the existence of a new T-phase between commensurate and non-commensurate phases. Even if the real structure is not perfectly determined, we studied the irradiation defects, at room temperature on the non-commensurate phase $1T_2$-TaS_2. Dark-field micrographs exhibit dark contrasts called "defects" by Mutka et al. [6], the density of which increases with the irradiation dose, up to about $5x10^{-3}$ dpTa. For higher doses, the TaS_2 samples present the same behavior in the non-commensurate phase as those of $TaSe_2$ in the commensurate phase :apparition of granularity, perturbation of high-resolution fringes, rotation of satellite spots around the neighbouring Bragg spot in diffraction patterns. We determined the structure of the "defects" by high-resolution microscopy. The micrographs show that we can call them "CDW dislocation" because they correspond to a CDW phase shift of 2Π [2,11].

In situ observations show that the configuration of the CDW is controlled by the density of irradiation-induced defects. High-resolution microscopy permits to make evidence of CDW defects at the distortion scale : CDW dislocations, translation boundaries, loss of coherence of CDW when the irradiation dose increases... These observations corroborate the importance of the pinning of CDW to defects.

References

1. J. Van Landuyt, Physica 99B, 12-25, 1980.
2. G. Salvetti, Thesis Paul Sabatier University, Toulouse, France, 1984.
3. D.M. Bird et al., in this proceedings.
4. J.A. Wilson, F.J. Disalvo and S. Mahajan, Adv. Phys. 24, 117, 1975.
5. L. Zuppiroli, Rad. Effects 62, 53, 1982.
6. H. Mutka and N. Housseau, Phil. Mag. A47, 797, 1983.
7. T. Tsuzuki and S. Sasaki, Progress in Th. Phys. 65, 19, 1981.
8. H. Fukuyama and P.A. Lee, Phys. Rev. B17, 535, 1978.
9. R. Brouwer, Thesis University of Groningen, Netherlands, 1978.
10. S. Tanda, T. Sambongi, T. Tani and S. Tanaka, J. Phys. Soc. Jap. 53, 476, 1984.
11. G. Salvetti, C. Roucau, R. Ayroles, H. Mutka and P. Molinié, submitted to J. Phys. Lett.

STATIC PROPERTIES OF CDW SYSTEMS

ASPECTS OF STRONG ELECTRON-PHONON COUPLING RELATED TO THE CDW TRANSITION AT TEMPERATURES ABOVE IT

C.M. Varma
A.T. & T. Bell Laboratories, Murray Hill, NJ, USA

I. Introduction

The minimal aspects that a microscopic theory of the CDW state must address are

(i) A reasonable prediction of the Q-vector and of the transition temperature for a given compound;
(ii) the prediction of the correlation length of the CDW state. This is in turn related to the behavior of the specific heat at the transition.

Given a bandstructure for a transition metal compound (TMC), simple calculations of the electronic polarizability $X(q)$ do not have the quantitative and sometimes even the qualitative structure to yield a CDW transition. From the point of this talk they do not yield the observed phonon dispersion[1] of either the high temperature or the low temperature phase. A proper microscopic theory for electron-phonon interactions suitable for doing calculations in transition metals and compounds have been constructed with W. Weber[2]. This yields the right sort of phonon dispersion of either the high or the low temperature phase. However, if one calculates the transition temperature using weak-coupling theory, one gets answers which are an order of magnitude too high. This situation can be corrected by including fluctuations or more precisely anharmonic interactions[3].

In the most simplistic approximation, one can calculate the anharmonic interaction among phonons that are induced by electronic scattering as in Fig. (1) and calculate the fluctuations in a Gaussian approximation like the self-consistent harmonic approximation. These calculations yield the right sort of correlation length and the transition temperature and also the specific heat involved in the transition. In effect this represents a microscopic verification

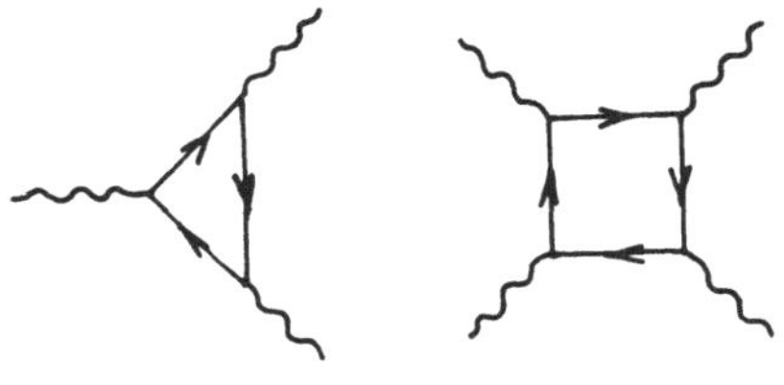

Fig. 1.

of the ideas of McMillan[4] that the major part of the entropy involved in the transition is due to the phonons.

In the course of this work, some questions arose about the validity of the Migdal approximation. In a wide variety of transition metals and compounds in which phonon anomalies (due to strong electron-phonon scattering) occur over wide regions of phase space, anomalous temperature dependences in the elastic constants, resistivity and magnetic susceptibility are well known. The suggestion here is that these high temperature properties anomalies are connected with the failure of the Migdal approximation. More physically, electrons and phonons in such situations can not be considered as independent entities whose residual interaction may be calculated in low order perturbation theory. Similar suggestion has also been made by Anderson and Yu[5].

II. Electron-Phonon Coupling - calculations of CDW parameters

Since the work on understanding the phonon dispersion and to a lesser extent that on anharmonic fluctuations has been well covered in the literature, only the principal ideas will be stated here.

The main difficulty in understanding the phonon anomalies and in particular their small correlation lengths in TMC are strong local-field effects in the electron-phonon interactions. The charge density in TMC varies strongly within a unit cell and one must calculate its variation due to ion-motion. Calculations starting from electron-phonon interactions in plane waves and reciprocal vector sets prove disastrous. Weber and I have adopted Friedel's[6] ideas and start from moving tight-binding basis sets in which the local field effects are naturally and simply calculated. We also need to make a novel grouping of various terms in the electron-phonon Hamiltonian so that short-range effects are explicitly separated from the non-local effects of electron-phonon coupling.

Traditional ideas of the shape of Fermi-surface needed for CDW transitions are revised with electron-phonon coupling calculated in this way. To illustrate, the vertices in Fig. (1) (with some drastic approximations) are given by

$$g_{\underset{\sim}{k}\underset{\sim}{k}'} \sim \frac{t}{a} (v_{\underset{\sim}{k}} - v_{\underset{\sim}{k}'}) \quad , \qquad (1)$$

where t is a relevant tight-binding transfer integral, a its characteristic fall-off distance including electron-electron scattering effects and $v_{\tilde{k}}$ the electronic velocity. The contribution of the electronic polarizability to the phonon frequency at $\tilde{q} = \tilde{k}-\tilde{k}'$ is

$$- \left(\frac{t}{a}\right)^2 \sum_{k'} \frac{(v_{\tilde{k}} - v_{\tilde{k}'})^2}{(\varepsilon_{\tilde{k}} - \varepsilon_{\tilde{k}'})} , \qquad (2)$$

If $\varepsilon_{\tilde{k}+\tilde{q}} \simeq \varepsilon_{\tilde{k}}$, anomaly is now to be expected for sharply rising energy dispersion in the direction $\tilde{q}$ and slowly varying dispersion transverse to $\tilde{q}$ unlike the result without considering eq. (1).

Using only short-range coupling parameters (only 4 for cubic crystals), phonon dispersion in nearly 20 metals and compounds has been calculated by W. Weber and by others using this method in rather detailed agreement with results from neutron inelastic scattering experiments.

Anharmonic vertices due to electron-phonon coupling (Fig.1.) can be calculated by extensions of this method without any new parameters. Using mode-mode coupling, the temperature dependent contribution of these fluctuations (Fig.2.) to the phonon energy has been calculated[3] for a simplified model for $TaSe_2$.

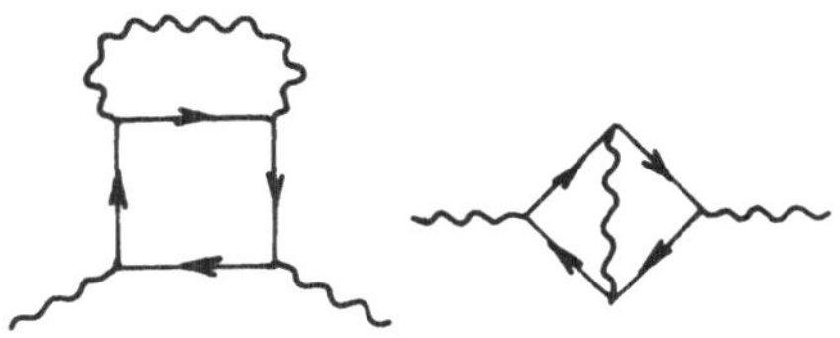

Fig. 2.

The contribution of these fluctuations is very large because they occur over a wide region in phase space and they stabilize the high temperature phase very strongly. In weak-coupling, where the entropy contributions is the only one considered, the transition temperature is predicted to be a few thousand degrees. With consideration of (anharmonic) fluctuations it is brought down to a few hundred degrees as observed.

III. Migdal Approximation

In the course of the above work, especially in connection with calculating processes like (2), questions arose about the validity of the Migdal approximation. As already mentioned TMC are beset with anomalies over their entire temperature range. Consider magnetic susceptibilities,

Fig. (3). A slowly varying temperature dependent contribution on the scale of the phonon energies is observed. Some of the most cherished theorems of condensed matter physics state that the magnetic susceptibility (or the static polarizability) is not renormalised by electron--phonon interactions. The renormalization rides with the chemical potential which itself moves with the perturbation so that no residual effect is left. Technically this is shown by a Ward identity relating the vertex to the self-energy. These theorems rely however, on the Migdal approximation which does not permit non-local (or interference) effects. It is natural to ask therefore if Migdal approximation is indeed valid.

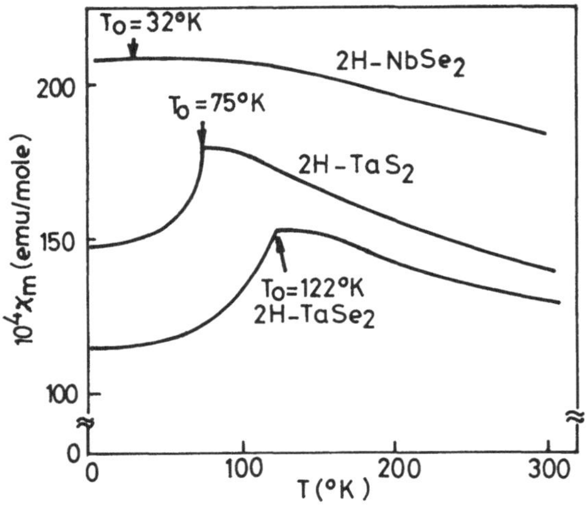

Fig. 3.

Migdal's theorem states that vertex corrections, the simplest of which is shown in Fig. (4), are at least of order $\lambda\sqrt{m_e/m_{ion}}$, where λ is the dimensionless coupling constant compared to the lowest order vertex. The argument is that the electron and phonon momentums have the same scale while the energy scales are very different. This makes $\partial\Sigma/\partial k$, where Σ is the electronic self energy a very small quantity anywhere on the Fermi-surface, (while $\partial\Sigma/\partial\omega$ is very large). In Fig. (4), $\underset{\sim}{p}$ may be choosen on the Fermi-surface, but for any $\underset{\sim}{q}$ such that $qv_F > \omega$, $\underset{\sim}{p}+\underset{\sim}{q}$ is far from the Fermi-surface; this introduces an energy denominator of order the electronic energy and hence a factor of $O(m_e/m_{ion})$. For $qv_F \ll \omega$, the Migdal approximation is not valid, as noted especially by Englesberg and Schrieffer[7]. But this region has little influence on most physical properties. It is easy to see that the argument must also break down for $q=Q$ such that

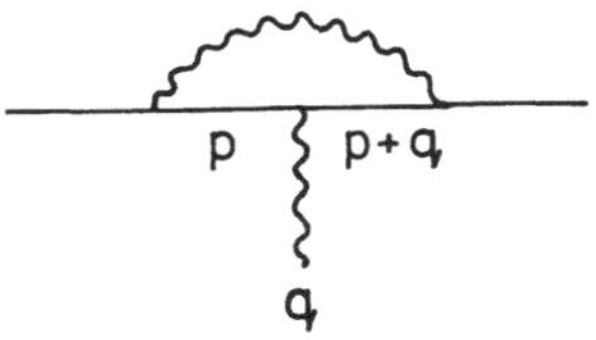

Fig. 4.

$\varepsilon_p = \varepsilon_{p+Q}$, as must happen near transition to a charge density wave state at Q . In this case $\Sigma(Q)$ must become very large. This alone would not be very interesting. It becomes interesting, however, if the correlation length of the transition is only a few lattice constants, so that the vertex corrections are important for a substantial region in phase space.

If we take as given that the phonons of a particular kind are soft over such a substantial region, no essential difference is introduced by taking them to be Einstein phonons of frequency $\Omega(T)$ and in the calculation of the self-energy, multiply by a phase factor at the end. The self energy in Fig. (5) can be calculated to be

Fig. 5.

$$\Sigma(\omega) = g^2\,\Omega\,\rho\,\{\Pi i((e^{-\beta\Omega}-1)^{-1} - (e^{\beta\Omega}-1)^{-1}) - \frac{1}{2}\int_{-\infty}^{\infty} d\omega'(e^{\beta\omega'}+1)^{-1}\frac{2\,\Omega}{(\omega+\omega')^2-\Omega^2}\} \tag{3}$$

which reduces to the logarithmic behavior of the real part calculated by Englesberg and Schrieffer at $T=0$. If $\varepsilon_{\underline{p}+\underline{Q}} = \varepsilon_{\underline{p}}$, the following pseudo Ward--identity is easily derived

$$\Gamma(p;\,\underset{\sim}{Q},\,q_o) = 1 - \frac{g^2\rho}{\Omega}\;\frac{\Sigma(p+q)-\Sigma(p)}{q_o} \tag{4}$$

Usually $g^2\rho/\Omega \approx \lambda \simeq 1$. Therefore eq. (4) shows the breakdown for such a vertex of the Migdal approximation.

For more general q around Q, I can show this only by an explicit calculation of the low order vertex (with propagator including the self-energy eq. (3)), but can motivate it physically. For a phonon soft over a substantial region ξ^{-1} around Q, if Σ at Q is large, so should it be in a region ξ^{-1} around Q; there should be only one correlation length scale in the problem. More physically, consider the region below T_c where a gap $\Delta >> T_c$ develops, implifying flat bands on a scale ξ^{-1} . Above the transition, in the fluctuation regime this gap is imaginary but exists over a substantially similar range. Enormous number of low energy coupled electron-phonon excitations coupling different parts of the Fermi-surface exist, as shown by the branch cuts in eq. (3); this leads to the breakdown of Migdal approximation. This means also that a quasi particle picture is not very good and that a Boltzmann-Landau type description of transport is probably not valid.

In an attempt to calculate the magnetic susceptibility including the self-energy, Fig. (6) and similar ladder diagrams appear to give a logarithmic contribution, which vanishes, for $\Omega/T \ll 1$ due to the Ward-identity. This may be related to the change in behavior in Fig. (3) which seems to appear at $\Omega \lesssim T$. The vanishing of such diagrams for $\Omega/T \ll 1$ is interesting as similar diagrams for the impurity scattering problem do not give any interesting contributions. Indeed for $\Omega/T \ll 1$, the phonon-electron problem is related to the impurity problem if Migdal approximation is invalid. This means that vertex corrections like in Fig. (7) may be very important[8]. These are yet to be evaluated.

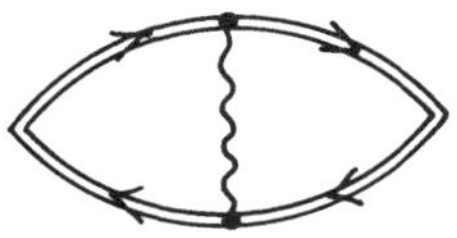

Fig. 6.

The connection to the impurity problem for $\Omega/T \ll 1$ may be relevant to the high temperature resistivity of transition metals and compounds with $\lambda \gtrsim 1$. The difference of course is that the density of both the elastic $(\Omega/T \ll 1)$ and inelastic $(\frac{\Omega}{T} \gtrsim 1)$ scatterers in the phonon problem increases with T. I suspect that the observed saturation resistivity is obtained by a scaling theory in the borderline regime in which $\ell_{elastic} \simeq \ell_{inelastic}$.

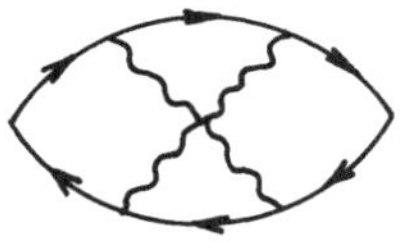

Fig. 7.

I wish to acknowledge discussions with J.R. Krishnamurthy and P. B. Littlewood. Discussion of the susceptibility data with F. diSalvo and the use of Fig. (3) is gratefully acknowledged.

References

1. D.E. Moncton, J.D. Axe and F.J. diSalvo, Phys. Rev. B14, 80 (1977)
2. C.M. Varma and W. Weber, Phys. Rev. B19, 6142 (1979)
3. C.M. Varma and A.L. Simons, Phys. Rev. Letters 51, 138 (1983)

4. W.L. McMillan, Phys. Rev. B16, 643 (1977)
5. P.W. Anderson and C. Yu, in the Proceedings of the International School of Physics, Varenna (1983); Phys. Rev.B29, 6165 (1984)
6. See for instance, S. Barisić, T. Labbe and J. Friedel, Phys. Rev. Letters 25, 919 (1970)
7. S. Englesberg and J.R. Schrieffer, Phys. Rev. 131, 443 (1963)
8. E. Abrahams, P.W. Anderson, D. Licciardello and T.V. Ramakrishnan, Phys. Rev. Letters 42, 673 (1979)

ELASTIC AND OTHER PROPERTIES AT THE COMMENSURATE-INCOMMENSURATE TRANSITION in 2H-$TaSe_2$

T. M. Rice and P. Prelovšek *)
Institut für Theoretische Physik, ETH-Hönggerberg,
CH-8093 Zürich, Switzerland.

The large anomaly in the Youngs modulus in the commensurate phase of 2H-$TaSe_2$ is attributed to a redistribution of the sizes of the orthorhombic domains. The restoring force against the redistribution is proportional to the domain-wall, or partial discommensuration energy and tends to zero as the commensurate-incommensurate transition is approached. The electrical response of 2H-$TaSe_2$ does not show the anomalous behavior seen in other systems. Very recently however the increase in the resistance due to discommensurations has been identified experimentally.

1. Introduction

The dynamical response and the non-linear response of charge density wave (CDW) systems are major topics of this conference reflecting the widespread interest in these questions in recent years. One of the central questions is the role of discommensurations (DC) in various CDW systems. DC may be mobile in certain cases and may carry a charge so that they may play an important role in electrical properties. Paradoxically the material in which the DC have been directly observed in some beautiful electron microscope studies, namely 2H-$TaSe_2$, is a CDW system in which neither non-linear conductivity nor sharp peaks in the noise spectrum have been observed. However it is clearly of interest to examine what effects can be observed in 2H-$TaSe_2$ and attributed to the DC.

McMillan [1] predicted that a commensurate (C) to incommensurate (I) transition occurs as a continuous transition through the generation of DC in the commensurate phase and this mechanism was established for the CI transition of 2H-$TaSe_2$ in a series of elegant electron microscope studies by, Fung et al [2], Chen et al [3] and, McKernan et al [4]. These and subsequent studies are reviewed in the talk by J. Steeds at this conference. Time-dependent studies [4] showed that at the CI transition the DC move after their generation due to the internal microscopic forces driving the incommensurability. This talk however, will concentrate on their response to externally applied forces.

Fung et al [2] established in their electron-microscope studies that the C-phase of 2H-$TaSe_2$ was not hexagonal, as originally believed, but orthorhombic and composed of a mosaic pattern of domains of different orientations of the orthorhombic axes. The domains are separated by domain walls. DC separate regions where the CDW are simply translated relative to the underlying crystal but the electron microscope studies [2-4] showed that DC are made from two domain walls relatively close to each other. Thus to the extent that the interaction between the domain walls can be neglected,DC can be regarded as the superposition of two of them and the energy per unit area of a domain wall (or partial DC; PDC) is just one-half that of the DC. Recently, Prelovšek and Rice [5] used this idea to explain the pronounced and puzzling anomaly in the Youngs modulus observed some years ago by Barmatz, Testardi and di Salvo [6] in the C-phase of 2H-$TaSe_2$ as the temperature is raised to the CI transition temperature. The anomaly was puzzling in that it appeared in a linear response function in the C-phase altho' the DC theory [1] shows only a non-linear object namely a DC, going soft (i.e. the generation energy going to zero). All the linear modes are unaffected as the CI transition is approached in the C-phase. This paradox was resolved by Prelovšek and Rice [5], who showed that there was a contribution to the elastic response due to a redistribution in size among the orthorhombic domains and that this contribution was inversely proportional to the domain wall energy per unit area. Since these domain walls are PDC their energy goes to zero as the CI transition is approached and hence the anomaly in Youngs modulus could be explained. Their theory will be reviewed and the implications for DC motion discussed in the following section.

A large electrical response is expected from DC in the case of CDW in one dimensional electronic systems since in this case the DC are charged [7], However, in 2H-$TaSe_2$ the electronic band structure is much more complicated and it is not obvious what, if any, charge will be associated with a DC [8]. Further any such charge will be screened by the large number of normal carriers remaining in the CDW state. Thus the question of what direct force can act on the DC in an electric field. To date repeated efforts [9] to observe a non-linear conductivity in 2H-$TaSe_2$ have failed. An indirect force from the scattering of normal carriers must be present. Very recent experiments by Jericho and Ott [10] have established the average reflection coefficient at a DC in 2H-$TaSe_2$. These experiments are described in detail elsewhere in the Proceedings and in Section 3; some theoretical comments on the implications of their results will be given.

2. Elastic Anomaly in the C-Phase of 2H-$TaSe_2$

The electron microscope studies of Fung et al [2] showed that the C-phase of the three coexisting CDW in 2H-$TaSe_2$ have orthorhombic symmetry. There are three possible orientations of the orthorhombic axes relative to the axes of the parent hexagonal structure. Fung et al [2] showed also that the C-phase took the form of textures made from an approximately double honeycomb network of each of the three orthorhombic domains. These domains are separated by domain walls which are partial DC. Two of these PDC in close proximity make up a regular DC i.e. the orthorhombic axis rotates inside each DC - a fact which makes it easy to observe in the electron microscope.

The general Landau free energy functional for the CDW state is complicated since there are no fewer than six complex order parameters describing the three CDW in each of the two layers [11] and the interested reader is referred to the original work for details. Prelovšek and Rice [5] considered the form of the Landau free energy when coupling to elastic strains $u_{\alpha\beta}(= \partial u_\alpha / \partial x_\beta)$ and stresses $\sigma_{\alpha\beta}$ are included. They constructed the lowest order CDW-strain coupling terms which are invariant under all symmetry operations of the hexagonal crystal. In terms of the symmetrized basal plane strains

$$\bar{u} = \frac{1}{2}(u_{xx} + u_{yy}); \quad r = \frac{1}{2}(u_{xy} - u_{yx}) \tag{1}$$

$$w = \frac{1}{2}(u_{xx} - u_{yy}); \quad d = \frac{1}{2}(u_{xy} + u_{yx})$$

and the corresponding stresses $\sigma_{\bar{u}}$, σ_r, σ_w and σ_d, the free energy can be expanded as

$$F = F_o - \frac{1}{2} w(\Phi_1 + \Phi_3 - 2\Phi_2) - \frac{1}{2} 3^{1/2} d(\Phi_3 - \Phi_1) + \frac{1}{2}\lambda(w^2 + d^2) \tag{2}$$

$$+ \frac{1}{2}\chi \bar{u}^2 - \sigma_w w - \sigma_d d - \sigma_{\bar{u}} \bar{u}$$

The coefficients λ and χ are elastic moduli. The y axis is chosen parallel to the Bragg vector $\vec{G}_2$. The coefficients Φ_j (the index j = 1, 2, 3 represents the three different orthorhombic domains) are derived from the CDW order parameters and through the interlayer coupling term will take the different values in the three orthorhombic domains [5]. The simplest choice is given in Table 1. Note the rotation r does not couple at all while the longitudinal strain $\bar{u}$ is zero in the absence of a hydrostatic stress. However the lowering of the symmetry from hexagonal to orthorhombic gives rise to spontaneous uniaxial strains w and d in the underlying hexagonal crystal. The magnitude of these spontaneous strains is easily found by minimizing Eq.(2) and is quoted also in Table 1.

Table 1: The values of the CDW order parameter terms Φ_j and the spontaneous uniaxial strains in the three different domains.

Domain	Φ_1,	Φ_2,	Φ_3	w	d
1	0	Φ_o	0	$-\Phi_o/\lambda$	0
2	0	0	Φ_o	$+\Phi_o/2\lambda$	$+3^{1/2}\Phi_o/2\lambda$
3	Φ_o	0	0	$+\Phi_o/2\lambda$	$-3^{1/2}\Phi_o/2\lambda$

In general the solution of the complete elastic problem is very complicated. One case is relatively simple, namely a regular texture composed from a regular double honeycomb arrangement of the three domains as shown in Fig. 1. This texture has no long range strains and has a topology similar to the textures observed in the electron microscope experiments. The experimental textures are not regular but in the absence of external stresses we must expect an equal area in each of the three domains, j = 1, 2, 3. The detailed distribution however will be determined by any macroscopic inhomogeneities in the internal stresses or the boundary conditions in the sample.

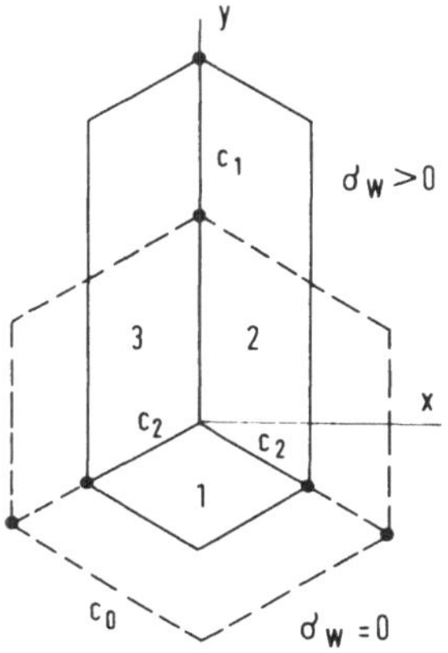

Fig. 1

Ideal double-honeycomb texture in the C-phase of 2H-$TaSe_2$. The domain structure is given in Table 1. The dashed and full lines represent the texture in zero and non-zero applied uniaxial stress σ_W, respectively.

Consider now what happens when an external uniaxial stress is applied to sample. Specifically let us consider the case, $\sigma_W > 0$. In this case there is a homogeneous term which favors domains 2 and 3 at the expense of a domain 1. There will be a free energy gain linear in the relative difference in the areas of the sample in domains 2 or 3 and in domain 1. This leads to a deformation of the regular texture as shown in Fig.1. The total free energy per hexagon unit can be expressed as

$$F_H = \frac{-3^{1/2}\Phi_o\sigma_W}{2\lambda}(c_1 c_2 - c_2^2) + 2 f_{PDC}(2c_2 + c_1) \tag{3}$$

where the lengths c_1, c_2 are shown in Fig. 1. The first term arises from the difference in area of domains 2, 3 and 1, while the second term is the energy of the domain walls or PDC. Two implicit assumptions are made in writing down Eq. (3), namely that the PDC can more (i.e. the applied stress is sufficient to overcome any pinning forces from impurities, etc. [12])and secondly that the number of hexagons as essentially topological objects is preserved. This second assumption means that the area of a hexagon is preserved and leads to the constraint $c_2^2 + 2c_1 c_2 = 3c_0^2$ where c_o was the length of the side of the regular hexagon. Minimizing Eq. (3) with respect to c_2 we arrive at a value for the deformation

$$c_1 - c_o = \frac{3^{1/2} \Phi_o c_o^2}{2 \lambda f_{PDC}} \sigma_W \qquad (4)$$

This deformation in turn leads to an average uniaxial strain $\bar{w}$ in response to the applied uniaxial stress σ_W

$$\bar{w} = \frac{1}{\lambda} (1 + \frac{3^{1/2} \Phi_o^2 c_o}{4 \lambda f_{PDC}}) \sigma_W \equiv \frac{\sigma_W}{\lambda_{eff}} \qquad (5)$$

Thus the effective elastic modulus λ_{eff} is changed by a term proportional to f_{PDC}^{-1}. But as remarked previously, a DC is essentially a superposition in close proximity of two PDC. This observation led Prelovšek and Rice [5] to write $f_{PDC} \approx \frac{1}{2} f_{DC}$ - the DC energy per unit area. Now it is just this DC energy which is going to zero as the temperature T approaches T_{CI} - the CI transition temperature. In the present case the vanishing of f_{DC} can be observed because there is a preexisting array of PDC and no nucleation processes are involved. Instead we have a distortion in response to the applied uniaxial stress which increases the length of the PDC array without changing its topology. Since in general the PDC array will minimize its energy subject to topological and other constraints, this increase in area response to an applied uniaxial stress will be quite general.

The result is an anomalous softening of the Youngs modulus proportional to f_{DC}^{-1} or $(T_{CI} - T)^{-1}$. Such an anomalous softening was observed by Barmatz, Testardi and di Salvo [6] in some of the earliest experiments on CDW system. In Fig. 2 we show their data and the fit achieved by Prelovšek and Rice [5] to their data. In the fit the form $f_{PDC} \sim (T_{CI} - T + \Delta T)$ was used. The offset temperature ΔT, which is found to be 6K, may represent an intrinsic discontinuity in the CI transition or the difference between f_{PDC} and $\frac{1}{2} f_{DC}$ or a mixture of both effects.

Barmatz, Testardi and di Salvo [6] observed an anomaly also in the internal friction coefficient, Q^{-1}. Since the most reasonable assumption is that the motion at very low frequencies (0,5 kHz) of a PDC is mainly relaxational, it is not surprising that a similar anomaly appears in the imaginary part of the response, Q^{-1}.

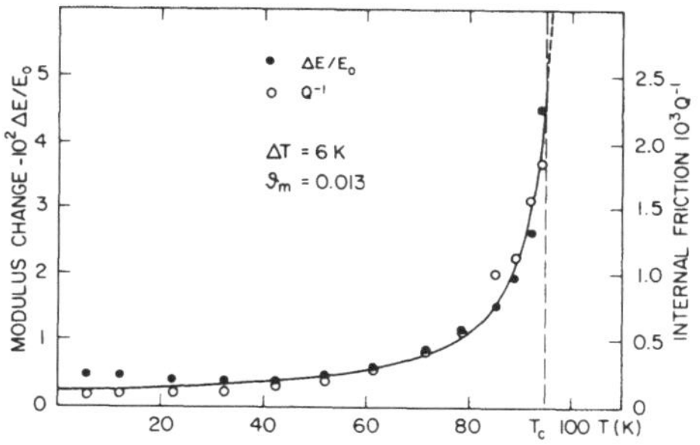

Fig. 2

The elastic modulus anomaly $\Delta E/E_0$ ($= \lambda_{eff}/\lambda - 1$) and the internal friction, Q^{-1}, as a function of temperature, T, for $T < T_{CI}$. Experimental points from Ref. 6 and the theoretical result (full curve) from Ref. 5.

Thus the experiments of Barmatz, Testardi and di Salvo [6] can be taken as confirmation that at low frequencies the PDC can move in a relaxational way and are not pinned even by the modest applied elastic forces in their vibrating reed experiments. It is interesting that their measurements are strongly frequency dependent. The anomaly at higher frequency (∿ 4kHz) was markedly smaller indicating either a strong increase of the fraction of effectively pinned PDC or at least a strong increase in the motional damping of PDC.

Another interesting observation [6] is the very sharp drop off the anomaly as the I-phase is entered. This is easy to explain, at least qualitatively, since in the I-phase the domain walls that should move are those between different orientations of the stripe array of parallel DC. From the electron microscope studies[2-4] one can see that such domain wall motion requires a rearrangement of the arrays of parallel DC and therefore has quite different characteristics to the motion of isolated PDC in the textures in the C-phase.

3. Electrical Properties of the DC in 2H-$TaSe_2$

In CDW systems with a one dimensional band structure such as CH_x [7,13] the DC have a charge which depends on the order of the commensurability. However if the electronic band structure is more complicated then the charge on a DC is much reduced and in the case of a CDW which couples closed Fermi surface pieces it vanishes altogether [8]. The DC charge in 2H-$TaSe_2$ is expected to be small [8] but a detailed estimate has not been made. Thus it is not clear how big the direct electrical force on a DC, or a PDC in 2H-$TaSe_2$ is. A separate question is the role of the large number of carriers remaining in the CDW state of 2H-$TaSe_2$ in screening any charge that may exist on a DC or PDC. Clearly there will be a complete screening of the DC or PDC charge. This however does not mean that there is no direct force on a charged DC or PDC. The problem is similar to that of electromigration of an ion in a metal. A particularly clear discussion has been given by Sham [14] who shows that the force can be written as a direct force and a force due to the electron wind.

As mentioned above it is not clear a priori how large the direct force on the DC and PDC in 2H-$TaSe_2$ should be. In experiments, its effect presumably would be to cause a depinning of the DC leading to a non-linear conductivity. However no such non-linear effects have been reported near the CI-transition, or at any other temperature in 2H-$TaSe_2$. Very recently Jericho and Ott [9] made a careful search for non-linear conductivity or an anomalous noise increase but to no avail. From these negative experiments we must conclude that the charge on DC or PDC is small,if it is non-zero at all.

The second effect that occurs in electromigration is an electron wind effect. This arises from the scattering of the electronic current off the perturbation - in this case a DC or PDC. Since a scattering process changes the electron momentum there is a net momentum transferred to the DC or PDC and hence a force exerted on the DC or PDC. The magnitude of this force depends on the reflection coefficient of the DC or PDC. Again little is known theoretically about such reflection coefficients, particularly in the complex band structure that prevails in 2H-$TaSe_2$. The microscopic theory of DC has been developed most fully for the case of CH_x. In this case the commensurability in 2:1 (it is 3:1 in 2H-$TaSe_2$) and both numerical [15] and analytic [16] treatments have been given. These calculations show that the DC (or soliton as it is referred to in the CH_x literature) causes a state to pulled down from the upper, or conduction band and pushed up from the lower, or valence, band. Thus the DC acts as an attractive potential for either electrons or holes with a depth of $\approx \Delta$, the energy gap parameter and a width $\approx \xi(= v_F/\Delta)$. The reflection coefficient of such a potential falls off rapidly for energies $\gtrsim \Delta$ away from the extremities of the conduction or valence bands. In 2H-$TaSe_2$ only a fraction of the Fermi surface is truncated by the CDW gaps. This means that in the parts of the Fermi surface that remain, the Fermi energy is either in the valence band below, or in the conduction band above the CDW energy gap. In both cases one expects the reflection coefficient to be small $\lesssim 10^{-1}$ for energies $\gtrsim \Delta$ away from the band gap. The total reflection coefficient R has to be averaged over the Fermi surface.

Recently Jericho and Ott [10] have examined carefully the resistance anomaly at the CI-transition using a shear strain to orient the domain structure. Details of these experiments will be given elsewhere [10] and here we will just quote some of their results. By cooling and heating the sample in an applied stress to orient the stripe phase [17] they found a larger anomaly in the resistance of the I-phase. By subtracting an extrapolated C-phase resistance they identified an additional resistance in the I-phase. Further the temperature dependence of this additional resistance agrees quite well with the temperature dependence of the number of DC per unit volume. Now in the stripe I-phase achieved on heating, the DC are parallel and separated by a distance $\gtrsim$ 300 Å. This separation is always much larger than the inelastic mean free path of the carriers estimated from the resistivity ($\sim$ 25 Å). Therefore there is no possibility for Bloch waves to be set up due to coherent scattering among the DC. The resistance is simply the classical series resistance for going through a set of parallel planar scatterers. From this result they obtain an average reflection coefficient which is small, $R \sim 0.4$. Such a value requires that for most of the Fermi surface in the C-phase the gap is not too far away in energy.

Following Sham's treatment [13] of the electromigration problem and with the simplest assumption of one carrier per Ta atom, we can estimate the size of the force from the electron wind as equivalent to the force on a charge e^* in a cross-sectional area of the wall containing one Ta atom with $e^* = e\,L\,\Delta\rho/\rho$. $\Delta\rho$ is the observed increase in resistivity over the resistivity ρ due to electron-phonon scattering. L is the separation between DC in units of the Ta-spacing. Using the measured values at $T \sim 110K$, $\Delta\rho/\rho \sim 1/30$ [9] $L \sim 60$ [18] we get $e^* \sim 2e$. However this estimate is too simple minded. Hall effect measurements [19] show that while 2H-$TaSe_2$ is p-type in the normal phase, the Hall constant drops rapidly at the onset of the CDW phase and is n-type below 90K. This shows that there is a mixture of electron and hole carriers in the CDW phase. The wind force has opposite signs for electron and hole carriers. Thus we can expect considerable cancellation and a reduction in the net force on a DC. The size of the reduction is difficult to estimate but even with a considerable cancellation we would expect from the measurements, an effective charge which could be of the order of a few tenths of the electron charge. The wind force therefore is substantial and not different in magnitude from a direct force. In the total force, the wind and direct forces may add or subtract compounding the difficulty of making an a priori estimate to be compared with the failure to see any depinning.

Acknowledgements

The authors are grateful to M. H. Jericho and H. R. Ott for many stimulating discussions of their work, prior to publication.

References

1. W.L. McMillan, Phys. Rev. B14, 1496 (1976)

2. K.K. Fung, S. McKernan, J.W. Steeds and J.A. Wilson, J. Phys. C14, 5417 (1981)

3. C.H. Chen, J.M. Gibson and R.M. Fleming, Phys. Rev. Lett. 47, 723 (1981) and Phys. Rev. B26, 184 (1981)

4. S. McKernan, J.W. Steeds and J.A. Wilson, Phys. Scripta 71, 74 (1982)

5. P. Prelovšek and T.M. Rice, Phys. Rev. Lett. 51, 903 (1983)

6. M. Barmatz, L.R. Testardi and F.J. di Salvo, Phys. Rev. B12, 4367 (1975)

7. M.J. Rice, A.R. Bishop, J.A. Krumhansl and S.E. Trullinger, Phys. Rev.Lett. 36, 432 (1976)

8. T.M. Rice, P.A. Lee and M.C. Cross, Phys. Rev. B20, 1345 (1979)

9. M.H. Jericho and H. R. Ott, private communication.

10. M.H. Jericho and H.R. Ott to be published.

11. M.B. Walker and A.E. Jacobs, Phys. Rev. B25, 4856 (1982)

12. T.M. Rice, S. Whitehouse and P. Littlewood, Phys. Rev. B24, 2751 (1981)

13. W.P. Su and J.R. Schrieffer, Phys. Rev. Lett. 46, 738 (1981)

14. L.J. Sham, Phys. Rev. B12, 3142 (1975)

15. W.P. Su, J.R. Schrieffer and A.J. Heeger, Phys. Rev. Lett. 42, 1698 (1979); Phys. Rev. B22, 2099 (1980)

16. H. Takayama, Y.R. Lin-Lin and K. Maki, Phys. Rev. B21, 2388 (1980)

17. D.B. McWhan and R.M. Fleming, 'Physics of Solids under High Pressure' (ed. J.S. Schilling and R.N. Shelton, North-Holland, Amsterdam, 1981) p. 219

18. R.M. Fleming, D.E. Moncton, D.B. McWhan and F.J. di Salvo, Phys. Rev. Lett. 45, 576 (1980)

19. H.N.S. Lee, M. Garcia, H. McKinzie and A. Wold, J. Solid St. Chem. 1, 190 (1970)

*) Permanent Address: Department of Physics
J. Stefan Institute
E. Kardelj University of Ljubljana
61000 Ljubljana, Yugoslavia.

CDW PHASE MODE INVESTIGATION IN THE FIR IN $K_{0.3}MoO_3$ AND BAND STRUCTURE CALCULATION

G. Travaglini and P. Wachter
Laboratorium für Festkörperphysik, ETH Zürich, 8093 Zürich, Switzerland

ABSTRACT

Optical reflectivity has been measured in the far infrared (FIR) region on the blue bronze $K_{0.3}MoO_3$ system using polarized light. At room temperature the reflectance spectrum polarized (P) parallel to the b-axis is metal-like while the perpendicular spectrum is semiconductor-like. For $T<T_c$ = 180 K the P∥b spectrum becomes also semiconductor-like with a very strong polarization dependent phonon spectrum. At hω < 8 meV a giant structure dominates the whole P∥b spectrum. The structure, reaching reflectivity values of 97% and having a width of 7 meV, is assigned to the phase oscillations of the pinned CDW. Optical constants are calculated by means of the Kramers Kronig relation. An LCAO calculation has been performed for one chain of $K_{0.3}MoO_3$: the results E(k), E_F, k_F, $D(E_F)$ m^*_e are compared with the experimental data.

REFELECTIVITY MEASUREMENTS

Reflectivity measurements of large single crystals of $K_{0.3}MoO_3$ [1] have been measured in an extended photon energy range from 12 eV down to 1 meV using linearly polarized light, in a temperature region between 5 and 300 K. As in ref. [2] the incident light was polarized parallel to the metallic b-axis and perpendicular to the [102] direction. In the far infrared (FIR) we have used a Bruker-Fourier spectrophotometer with TGS detectors down to 25 cm^{-1} and a liquid Helium cooled germanium bolometer from 100 to 8 cm^{-1}.

The whole reflectivity spectrum is represented in Fig. 1: one notes that for p∥b the 300 K spectrum is metal-like with a minimum due to plasma oscillation at about 1.85 eV and a reflectivity shoulder at about 0.15 eV.

The best samples show a very high reflectivity while other samples have in the free carrier absorption region a larger damping so that at room temperature broad phonon bands can also merge with the Drude part of the spectrum.

For $T<T_c$ = 180K the metallic reflectivity turns into a semiconductor-like spectrum, shown for T = 5K in the same figure; the shoulder at 0.15 eV has become a broad reflectivity maximum at 0.2 eV and typical phonon lines appear for photon energies below 0.12 eV. The insert of Fig. 1 presents the low temperature FIR spectrum in an energy expanded scale to give better evidence of the phonon spectrum. For $\hbar\omega<8$ meV a very high reflectivity peak reaching a value of 97% stands out from the

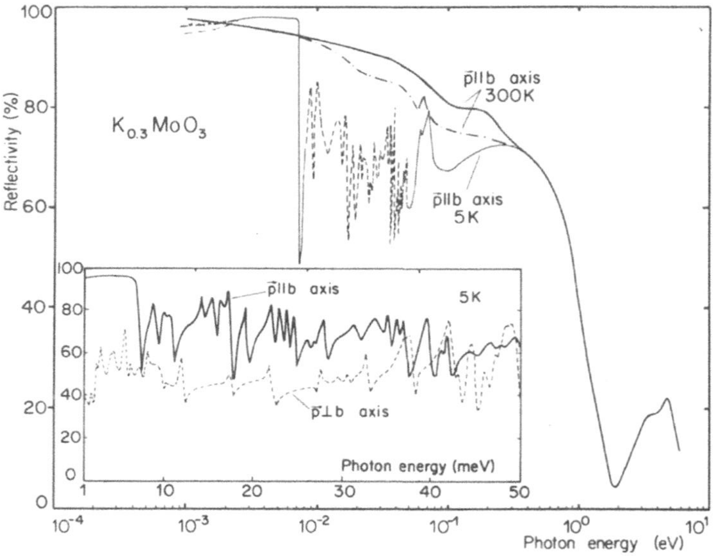

Fig. 1 Polarized reflectivity of $K_{0.3}MoO_3$ at 5 and 300K: to be noted the very strong structure in the FIR at 5K.

other maxima. It is to be noted that this peak does not exist in the perpendicular spectrum: the temperature dependence of this unusual maximum is shown in Fig. 2.

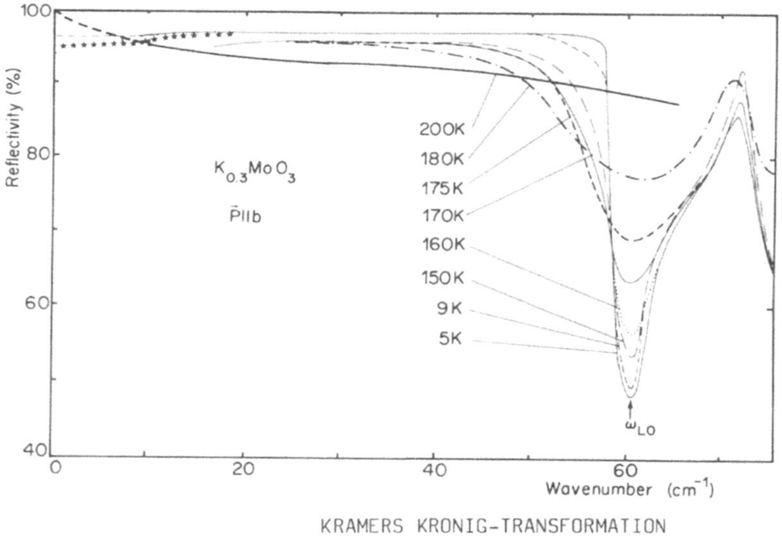

Fig. 2 Temperature dependence of the reflectivity observed in the FIR for light polarized parallel to the conducting axis.

KRAMERS KRONIG-TRANSFORMATION

The spectra have been analyzed by means of the Kramers-Kronig relation (R,θ) and discussed in terms of the dielectric functions $\varepsilon = \varepsilon_1 + i\varepsilon_2$ and of the real part of the optical conductivity σ_1. In correspondance with the giant reflectivity peak the dielectric functions show a very strong structure: ε_2 exhibits a large peak at

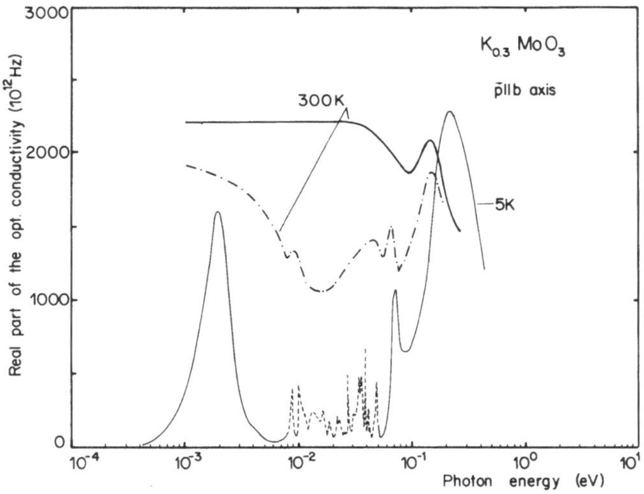

Fig. 3 Real part of the optical conductivity of the blue bronze at 5 and 300K: the FIR structure at 5K is the resonance of the oscillating CDW.

1.8 meV (ω_{TO}) which reaches a value of about 7000 and ε_1 has a very large dispersion in the same region. It intersects the abscissa with $d\varepsilon_1/d\omega>0$, yielding ω_{LO} at 7.4 meV. The ε_1, ε_2 values of the other phonon lines are around 100≈200. The dc values of ε_1 coming from this structure are very sensitive to the $\omega\to 0$ extrapolation. The derivation of ω_{TO} depends somewhat on the $\omega\to 0$ extrapolation. For all samples we found that $dR/d\omega$ is positive in the energy region between 10 and 15 cm^{-1}, yielding a transversal frequency of 1.8 meV and a "static" ($\hbar\omega$=1 meV) dielectric constant between 2000 and 3000. If we extrapolate the reflectivity curve for $\hbar\omega<15$ cm^{-1} with a zero slope towards $\omega=0$, the transversal frequency ω_{TO} will be reduced at most by a factor 6 (ω_{TO}=0.3 meV): ε_{stat} will be enhanced by a factor 10 and consequently, since ω_{LO} is independent from the extrapolation, the oscillator strength will also be enhanced. It is also possible that other structures are present in the reflectivity for $\omega\ll$THz, i.e. in the MHz or GHz region. Such structures will not influence at all the results in the FIR and we point out that the giant reflectivity structure is not a residue of this possible lower energy excitations. Nevertheless such structures will enhance the static dielectric constant by several orders of magnitude.

The real part of the optical conductivity σ_1 is plotted in Fig. 3. At room temperature the two kinds of samples present two different σ_1 spectra; the damping in the Drude part is in the dotted-dashed spectrum larger than in the continuous line spectrum so that phonon lines are visible at 300 K too. In both spectra the σ_1 maximum at about 0.15 eV is due to the Peierls Precursor. At 5K the σ_1 spectrum presents three strong structures: one located at 1.8 meV, the second centered at 0.07 eV and the third at 0.2 eV (Peierls Gap). The onset of the last peak in $\sigma_1(\omega)$ at the low energy side is about 100 meV. The strong line at 1.8 meV has a weight of about 10 to 20 times larger than the other phonon contributions in the energy region between 8 and 50 meV. The line does not exist in the perpendicular spectrum. The intensity of this mode is very strong temperature dependent; it becomes overdamped for $T \approx T_c$ and disappears for T>200 K. Such a large oscillator strength in the FIR region is quite unusual: a cluster-cluster oscillation ($K_3Mo_{10}O_{30}$ per cluster) is to be excluded since the red bronze $K_{0.33}MoO_3$, with a similar crystallographic structure [3], does not show a strong activity in the FIR region: the reflectivity of the red bronze reaches values of 30% for ω towards zero. It follows, that the interpretation of this structure with the model of a normal phonon excitation is quite inappropriate. The line with ω_{TO}=1.8 meV is therefore assigned to a pinned Fröhlich $2k_F$ phase mode.

If we assure that all the conduction electrons are condensed in the CDW, it turns out from $\omega_p^{CDW} = (4\pi e^2 N/m^*_{CDW}\varepsilon_{opt})^{1/2}$=7.4 meV, that $m^*_{CDW} \approx 900$ m_e taking 3 electrons per cluster and an estimated ε_{opt}=150 in the plasma frequency formula (if ε_{opt}=100 or 250, m^*_{CDW} will be ≈1200 or ≈600 m_e).

BAND STRUCTURE CALCULATION

Fig. 4 Energy dispersion in the Γ-X direction for the pseudo cell of $K_{0.3}MoO_3$ (see text)

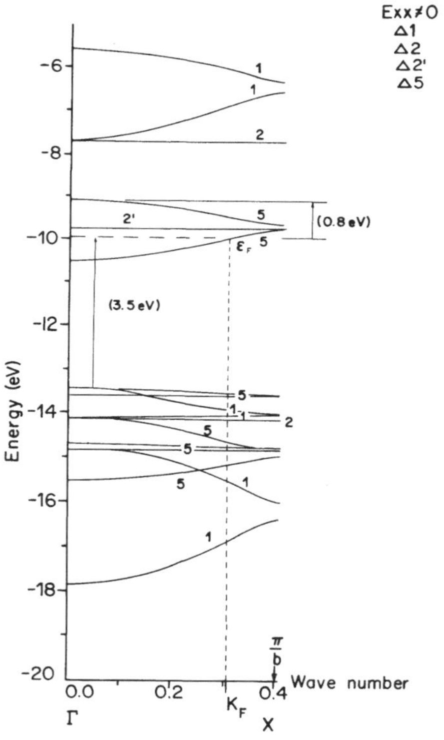

A LCAO calculation for one chain of $K_{0.3}MoO_3$ has been performed. To take into account the different octahedra distortions, the pseudo unit cell (p.u.c) for one chain (at 300 K) has been chosen to contain two Mo-Atoms ($Mo^{(2)}$ or $Mo^{(3)}$ site - see ref. [1]); the $Mo^{(1)}$ site has been neglected since $Mo^{(1)}$ has a 6+ valence. Since MoO_3 is an insulator the conduction electrons arise from K atoms so that the p.u.c. will contain 2 x 3/8 electrons. From crystallographic data [1] one can see that in the direction perpendicular to the b-axis 4 chains are connected together so that the calculated energy dispersion corresponds to the lower "bonding configuration" of the 4 chains. It is then to be expected that the conduction band will contain 4x(2x3/8)=3 electrons. The calculation has been started with the Mattheis [4] parameter for the d,p,s energies $\varepsilon d, \varepsilon p, \varepsilon_s$, nearest neighbor Vpdπ, Vpdσ, Vsdσ and second neighbor Wpp, Wsd interatomic matrix elements. The energy dispersion relation is depicted in Fig. 4. The occupied bonding (k=0) σ,π bands are separated from the conduction band by an energy gap of 3.5 eV. The conduction band arises from the Vpdπ* i.m.e., it has t_{2g} character and is two fold degenerate (Mo-dxz, dyz orbitals with b direction ≡ z) which implies 4 electrons for the whole band. The discrete level in the conduction band is due to the non coupled Mo-dxy orbital. The conduction band has a width of ≈1 eV and it is 3/4 full. The Fermi energy ε_F is about 0.58-0.65 eV and $k_F = 3/4(b^*/2)$. The conduction is mainly due to holes for a 3/4 full band in agreement with the positive thermopower at 300 K [5]. The integrated mass is about 0.9≈1.1 m_e and agrees with the experimental mass arising from the plasma frequency [2]. The interband transitions correspond with optically observed transitions [2]. The density of states $D(\varepsilon_F)$ is about 1.8≈2 states per eV per spin and the conduction band width ≈0.9 eV. With the knowledge of ε_F it is possible to calculate the mean field parameters; electron phonon coupling parameter $\lambda = 0.3$, mean field temperature $T_c^{MF} \approx$ 600-700 K (2Δ=0.2 eV) and $m^{*MF}_{CDW}(\lambda,\Delta)$=800 m_e: the effective mass agrees quite well with the experimental results.

With the knowledge of the CDW ω_{TO} it is possible after Grüner's et al. classical model [6] to estimate the treshold field E_c for the depinning of the CDW: for a sinussoidal pinning potential we found E_c≈170 KV/cm which is in contrast to the value obtained from the non Ohmic behaviour of the electrical conductivity: E_c= 0.1V/cm [7]. It is possible that the latter value is related to a motion of only a part of the CDW through the presence of dislocations or incommensurations in the CDW lattice. In this case the small electrical field E_c could be connected directly to the very strong mid-gap structure (0.07eV) present in σ_1.

In summary, we want to mention the similarity between the blue bronze CDW properties and those of KCP [8-11].

ACKNOWLEDGEMENT

The authors are very grateful to Professor T.M. Rice, Dr. D. Baeriswyl and Dr. R. Monnier for fruitful discussions.

REFERENCES

[1] J. Graham and A.D. Wadsley, Acta Cryst. 20, 93 (1966).
[2] G. Travaglini and P. Wachter, J. Phys. Soc. Jap. Suppl. A, 49, 869 (1980).
G. Travaglini, P. Wachter, J. Marcus and C. Schlenker, Solid State Commun. 37, 599 (1981).
[3] N.C. Stephenson, A.D. Wadsley, Acta Cryst. 18, 241 (1965).
[4] L.F. Mattheis, Phys. Rev. B6, 4718 (1972).
[5] R. Brusetti, B.K. Chakraverty, J. Devenyi, J. Dumas, J. Marcus and C. Schlenker in Proc. of the Annual Conference of Condensed Matter Div. Europhysical Society, Antwerp 1980.
[6] G. Grüner, A. Zawadowski and P.M. Chaikin: Phys. Rev. Lett. 46, 511 (1981).
[7] J. Dumas, C. Schlenker, J. Marcus and R. Buder: Phys. Rev. Lett., 50, 757 (1983).
[8] G. Travaglini and P. Wachter, Phys. Rev. B30, (1984).
[9] G. Travaglini, I. Mörke and P. Wachter, Solid State Commun. 43, 289 (1983).
[10] E.F. Steigmeier, R. London, G. Harbeke and H. Andersen, Solid State Commun. 17, 1447 (1975).
[11] P. Brüesch, S. Strässler and H.R. Zeller, Phys. Rev. B12, 219 (1975).

^{93}Nb NMR STUDY OF CDW IN $(NbSe_4)_{10/3}I$ SINGLE CRYSTAL

P. Butaud, P. Segransan, C. Berthier,
Laboratoire de Spectrometrie Physique, associe au C.N.R.S.
U.S.M.Grenoble, BP 87 ,38402 Saint-Martin d'Heres Cedex, France

A. Meerschaut,
Laboratoire de Physicochimie des Solides, associe au C.N.R.S.
Universite de Nantes, 44072 Nantes-Cedex, France

We have studied the ^{93}Nb NMR in single crystals of $(NbSe_4)_{10/3}I$ (1.8 mg) and $(NbSe_4)_3I$ (20 mg). In the first compound, the onset of the CDW has been followed through the temperature dependence of the (1/2,-1/2) transition. Using computer simulation of the lineshapes, we have determined the temperature dependence of the order parameter. Comparison of the spin-lattice relaxation rate in $(NbSe_4)_3I$ and $(NbSe_4)_{10/3}I$, indicate the fluctuations of the the pinned CDW are the dominant mechanism.

Within the last ten years, three classes of inorganic quasi-one dimensional conductors exhibiting charge density waves (CDW)on one hand and non-ohmic transport properties on the other associated with the depinning of the CDW from impurities have appeared:
the trichalcogenides of transition metal ($NbSe_3$, TaS_3)[1-4], the halogened transition metal tetrachalcogenides[5] ($(TaSe_4)_2I$, $(NbSe_4)_{10/3}I$ and the blue bronzes ($K_{0.30}MoO_3$, $Rb_{0.30}MoO_3$)[6,7].
Pulsed NMR has proved to be an powerful tool for studying at a microscopic level the static and dynamic properties of CDW's, [8-10], and more generally of incommensurate systems [11]. So there has been a strong temptation for a few years to perform NMR in the non-ohmic regim compounds. However clean experiments require a study of a truly single crystal, necessary for a clear interpretation of both NMR and transport measurements.
The halogened tetrachalcogenide family actually looks very appealing, since single crystals of a reasonable size can be grown for some of the members, and in the case of Nb compounds, the metallic chain can be studied directly. If we look more in detail, as we shall do further on, some dificulties arise due to the fact that in the normal metallic state a number of inequivalent sites may already be present, in particular in $(NbSe_4)_{10/3}I$. Despite this, we have undertaken a pulsed NMR study of ^{93}Nb in this compound; when necessary, we shall refer to a study of the parent compound $(NbSe_4)_3I$ [12]which presents an anomaly

in its transport properties around 275 K, but no non-ohmic behavior.[5]

Experiment

All spectra were recorded by analog Fourier transform using a boxcar integrator and sweeping the external magnetic field H_0[13], which was always kept perpendicular to the fourfold c-axis of the samples. The amplitude H_1 of the rotating rf-field was typically of the order of 50 G. In the case of $(NbSe_4)_{10/3}I$, we worked on a 1.8 mg single crystal, at 65 and 83 MHz, and only the (1/2,-1/2) transition was studied; the temperature of the sample was monitored to better than 0.1 K.

Results and discussion

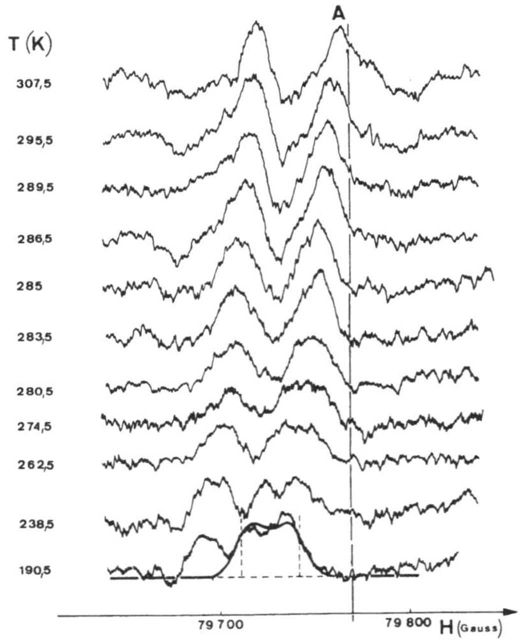

Fig.1 Temperature dependence of the lineshape of the (1/2,-1/2) transition. The solid lines are computer simulations of the frequency distribution due to the ICDW, covoluted with a Gaussian of width σ = 6.5 kHz.

In Fig.1 is shown the temperature dependence of the (1/2,-1/2) transition lineshape (at 83 Mhz) as a function of temperature. The lineshape above the onset of the incommensurate CDW (ICDW) at 285 K[14] already is a complex pattern, resulting from 10 inequivalent sites experiencing different magnetic and mainly quadrupolar hyperfine tensors (instead of 6 in $(NbSe_4)_3I$). This pattern is not fully understood yet, but we shall stress two points: first, the positions of the peaks are varying above and below 285 K, but these variations all exhibit a kink at this temperature. New measurements, (spin relaxation rate) are required to decide whether the origin of these variations is static or dynamic. Secondly, the peak labelled A (Fig.1) continuously broadens below 285 K until reaching the symmetric pattern expected from the frequency distribution resulting from an ICDW[8,11]. Assuming the amplitude of the distribution linear with the amplitude of the CDW, we have extracted the temperature dependence of the order parameter (Fig.2) from a computer simulation of the lineshape. If one compares these data with the measurements of Wang et al,[14] one note that the threshold field E_T varies as an exponential function of the order parameter, at least in the temperature range 190-240 K.

Finally, we shall comment on the difference in the spin-lattice relaxation rate (SLRR) in $(NbSe_4)_3I$ and $(NbSe_4)_{10/3}I$, which is an order of magnitude larger in this latter compound. (In both cases,the SLRR was only estimated from the amplitude of the signal as a function of the repetition of the rf-pulses). Since we expect the modulation of the quadrupolar tensors by the phonons to be very similar in both systems, we conclude that the SLRR in $(NbSe_4)_{10/3}I$ is dominated by the CDW fluctuations, according to the relation $1/T_1 = (A/\hbar)^2(\omega_p)^{-1}$, where A is the amplitude of the tansverse hyperfine field generated by the CDW, and $_p$ the pinning frequency. Although these quantities are not known yet, we note that reasonable values like $A/\hbar$ = 1 Mrad/s, ω_p = 10 Grad/s, give a good order of magnitude, i.e. a few milliseconds or less. This analysis neglect the effect of the electric field associated to the rf-pulses on the CDW; this point is not obvious and deserves further consideration.

In conclusion, we have demonstrated the feasibility of an NMR study of the ICDW in a tiny (0.3 mm^3) single crystal of $(NbSe_4)_{10/3}I$. These measurements open the way to further NMR sudies of the CDW dynamics in the non-ohmic regime.

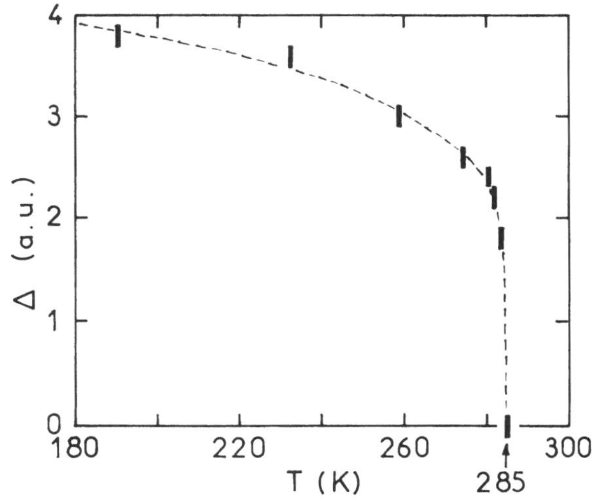

Fig.2 Temperature dependence of the order parameter Δ of the transition, assuming the amplitude of the frequency distribution associated with the ICDW linear with Δ.

<u>References</u>

[1] J. Chaussy, P. Haen, J.C. Lasjaunias, P. Monceau, G. Waysand, A. Waintal, A. Meerschaut, P. Molinie and J. Rouxel, Solid St. Commun., **20**, 759, (1976)

[2] N.P. Ong, Can. J. Phys. **80**, 757, (1983)

[3] G. Gruner, Physica **8D**, 1, (1983)

[4] H. Salva, Z.Z. Wang, P. Monceau, J. Richard and M. Renard, Phil.Mag.B. **49**, 385 (1984)

[5] P. Gressier, A. Meerschaut, L. Guemas, J. Rouxel and P. Monceau, Journal of Solid State Chemistry **51**, 2907, (1984)

[6] J.P. Pouget, J. Kagoshima, C. Schlenker, J. Marcus, J. Phys. (Paris)Lett. **44** L-113, (1983)

[7] J. Dumas, C. Schlenker, J. Marcus and R. Buder, Phys. Rev. Lett. **50**, 757, (1983)

[8] C. Berthier, D. Jerome, P.Molinie, and J. Rouxel, Solid State Commun.**19**, 131 (1976); C. Berthier, D. Jerome, and P Molinie, J. Phys.C:Solid State Phys. **11**, 797, (1978)

[9] B.H. Suits, S. Couturie, C.P. Slichter, Phys. Rev. B **23**, 5142, (1981)

[10] F. Devreux J. Phys.(Paris) **43** 1485, (1982)

[11] R. Blinc, Phys. Reports **79**, 331, (1981)

[12] P. Butaud, P. Segransan, C. Berthier, Proceedings of the XXIInd Congress Ampere (1984);to be published.

[13] W.G. Clark, Rev. Sci. Instrum., **35**, 316, (1964)

[14] Z.Z Wang, P. Monceau, M. Renard, P. Gressier, L. Guemas, A. Meerschaut, **47**, 439, (1983).

ELECTRONIC PROPERTIES AND Fe^{57} MÖSSBAUER MEASUREMENTS OF $T_{1+x}Nb_{3-x}Se_{10}$ WITH T = Fe, Cr

H. Gruber (a), E. Bauer (b), M. Reissner (c), W. Steiner (c)
Institut für Festkörperphysik, TU Graz, A-8010 Graz, Austria (a)
Institut für Experimentalphysik, TU Wien, A-1040 Wien, Austria (b)
Institut f. Angew. u. Techn. Physik, TU Wien, A-1040 Wien, Austria (c)

The metal-insulator transition of $Fe_{1+x}Nb_{3-x}Se_{10}$, $0.25 \leq x \leq 0.40$, and $Cr_{1+x}Nb_{3-x}Se_{10}$ was studied by measurements of resistivity, magnetic susceptibility and thermopower. The thermopower of the Fe containing compounds is positive in the measured temperature range from 5 to 300 K and the thermopower of the Cr containing compounds is negative at low temperatures. The Mössbauer spectra indicated a statistical distribution of Fe and Nb only in the octahedral chains for $Fe_{1+x}Nb_{3-x}Se_{10}$.

The resistivity of $Fe_{1+x}Nb_{3-x}Se_{10}$ increases by 9 orders of magnitude[1,2,3,4] and of $Cr_{1.6}Nb_{2.4}Se_{10}$ more than 7 orders[3]. This increase is supposed to originate from the atomic disorder due to Anderson localization[2]. X-ray diffraction showed the symmetry to be monoclinic and the structure to contain two trigonal prismatic as well as two octahedral chains[2,4,5]. X-ray scattering showed the CDW to be incommensurate with q (0.0, 0.27, 0.0) and the CDW vector is only slightly different from the high temperature CDW in $NbSe_3$. Recently, the first investigations of the physical properties[1,2,4,6,7] have been reported. We report measurements on single crystals and powder samples of the compounds $T_{1+x}Nb_{3-x}Se_{10}$ with T = Fe, (Fe, Ta), Cr, (Cr, Ti). Our crystals were prepared by heating the mixed powders of the alloys from the metallic components and Se[8]. For the compound $Fe_{1.33}Nb_{2.67}Se_{10}$ we found also a substructure at low temperatures with q (0.5, 0.33, 1) from our X-ray powder data in agreement with R.Moret[9]. Compared with the Fe-containing compounds the resistivity ratio ρ_T/ρ_{300} decreases drastically for $Cr_{1.45}Nb_{2.55}Se_{10}$ (Fig. 1). For a low Cr-content we found a resistivity minimum at 190 K [4]. Between 25 and 300 K $Cr_{0.38}Nb_{0.62}Se_3$ crystals exhibit a metallic conductivity with a nearly linear temperature dependence.

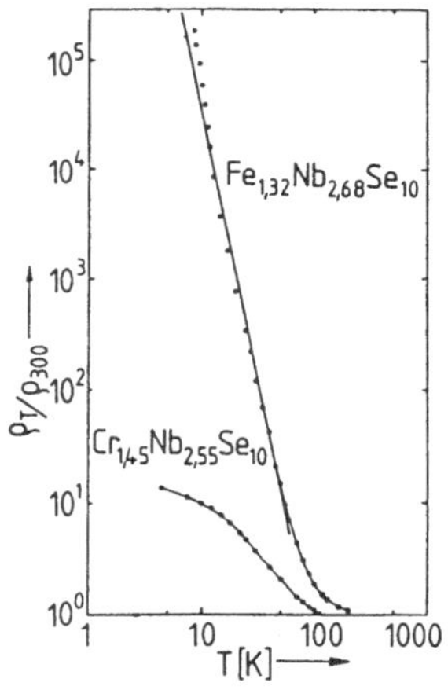

Fig. 1
Resistivity ratio versus temperature

The values of the magnetic susceptibility measured at 12.1 kÖe for $Cr_{1+x}Nb_{3-x}Se_{10}$ show a weak paramagnetic behaviour with increasing x in the measured temperature ranges. $TiCrNb_2Se_{10}$ can be fitted at $T < 100$ K with the Curie-Weiss expression (see Fig. 2) $\chi = C_g/(T+\Theta)+\chi_o$, C_g = Curie constant, Θ the Weiss temperature, χ_o = temperature independent term. For $Fe_{1+x}Nb_{3-x}Se_{10}$ at low temperature also a Curie contribution to χ appears in agreement with [4,6] (Fig.3).

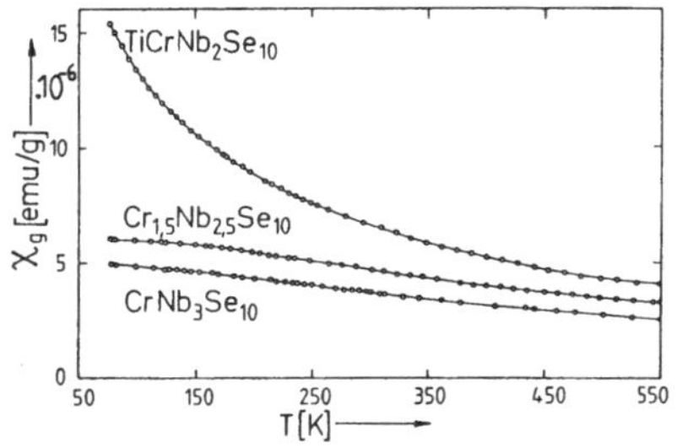

Fig. 2
Temperature dependence of the magnetic susceptibility between 77 and 550 K

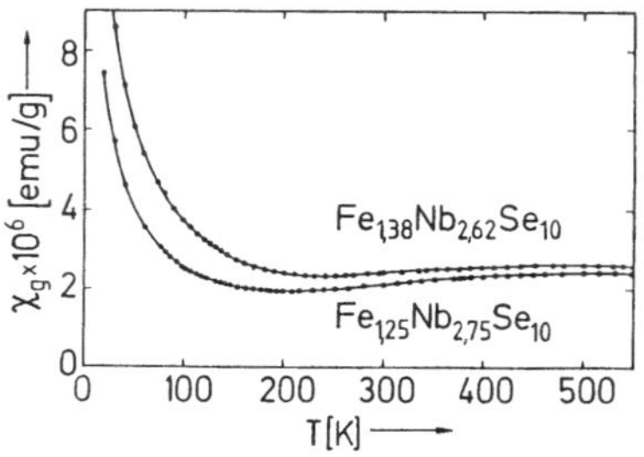

Fig. 3
Temperature dependence of the magnetic susceptibility between 77 and 550 K

For thermoelectric measurements needles of single crystals were used. The thermoelectric power of $Cr_{0.38}Nb_{0.62}Se_3$, $Cr_{1.33}Nb_{2.67}Se_{10}$ and of $Fe_{1.38}Nb_{2.62}Se_{10}$ changes from the negative values for $NbSe_3$ to positive values for $Fe_{1+x}Nb_{3-x}Se_{10}$. A partial substitution of Nb by Cr in $NbSe_3$ results a decrease of the negative thermopower of $NbSe_3$.

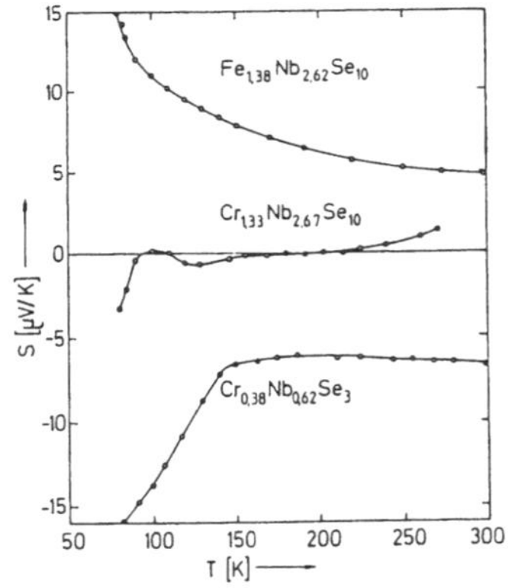

Fig. 4
Absolute thermopower of $Cr_{0.38}Nb_{0.62}Se_{10}$, $Cr_{1.33}Nb_{2.67}Se_{10}$ and $Fe_{1.38}Nb_{2.62}Se_{10}$ between 77 K and 300 K

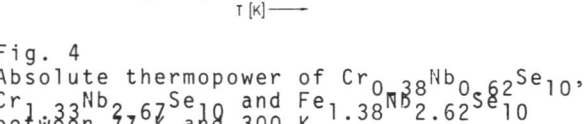

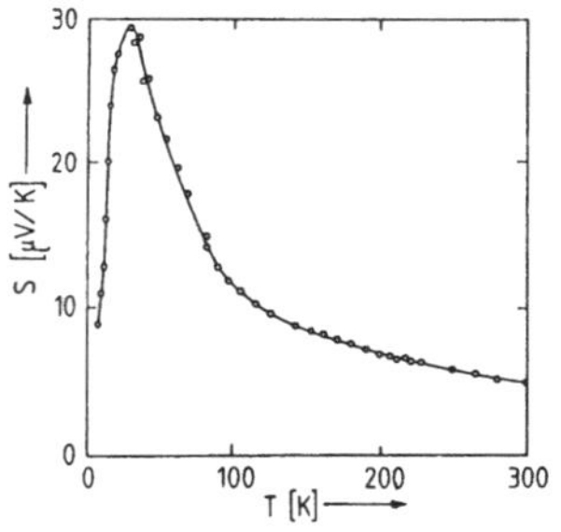

Fig. 5
Absolute thermopower of $Fe_{1.38}Nb_{2.62}Se_{10}$ between 4.2 K and 300 K

The thermopower of the Fe-containing compounds is positive from 5 to 300 K, for the Cr-containing compounds the thermopower is negative at low temperatures (Fig. 4). For $Fe_{1+x}Nb_{3-x}Se_{10}$ in the whole temperature range a positive thermopower has been found. The positive values increase with decreasing temperature down to 20 K, later on they fall down strongly till 5 K (see Fig. 5). This may be interpreted by a phonon drag term at low temperature. A possible explanation is that these compounds are semimetals which have a small number of electrons from a filled band overlapping with a higher band. This gives both electron and hole states. For $Fe_{1+x}Fe_{3-x}Se_{10}$ we have a hole dominated thermopower whereas $Cr_{1.33}Nb_{2.67}Se_{10}$ has a comparable number of electrons and holes and their carrier mobilities have similar values. The phase change affects the Fermi surface and because there are very few carriers this system should be extremly sensitive to small changes in the number of the delocalized electrons due to defects and impurities (see Fig. 4).

For the Mössbauer analysis we assumed: i) statistical substitution of Fe only in the octahedral chains, ii) center shift (Is) decreasing with increasing temperature due to the influence of second order Doppler shift, iii) quadrupole splitting (Q) mainly determined by the arrangement and number of surrounding Fe atoms. The intensity ratio of the subspectra for different environments were calculated by means of binomial distribution. Within these assumptions no fit was possible if only nearest neighbours were taken into account. An appropriate combination of the different environments which leads to the observed intensity ratio of .21, .22, .57 was only obtained if the contributions to the intensity resulting from some environments where Fe has one nearest Fe neighbour were added to the contributions where Fe has one or two nearest neighbours (Fig. 6). A consistent behaviour was only obtained if we assumed that Fe completely surrounded by Fe exhibits the smallest quadrupole splitting (Q). The assumption of a statistical substitution of Fe in the octahedral chains is valid for all temperatures and x = 0.25, 0.33, 0.40 for $Fe_{1+x}Nb_{3-x}Se_{10}$. However, an analogous assumption of a statistical distribution of Fe(Ta)only in the octahedral chains is not valid

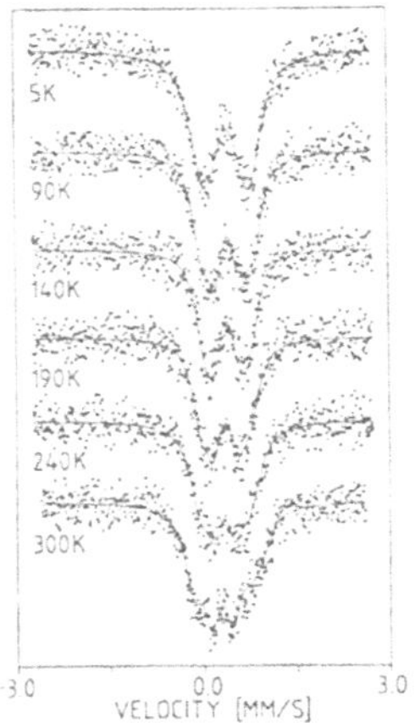

Fig. 6
Mössbauer spectra of $Fe_{1.25}Nb_{2.75}Se_{10}$ at various temperatures, full lines: fit according to a statistical distribution of Fe and Nb in the octahedral chains

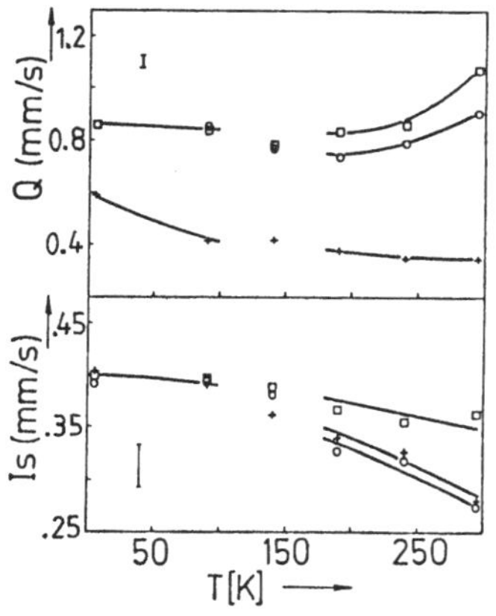

Fig. 7
Temperature dependence of (Q) and (Is) for $Fe_{1.25}Nb_{2.75}Se_{10}$. Intensity ratio for three subspectra according to a combination of nearest and next nearest environments: .21 (□), .22 (o), .57 (+)

for $Fe_{1.33}Ta_{0.67}Nb_2Se_{10}$. This is in agreement with the results of Meerschaut et al.[3]. Within this model the smallest value of Q is obtained for Fe completely surrounded by Fe. The values of both Q and Is at 140 K where the CDW appears in $Fe_{1+x}Nb_{3-x}Se_{10}$ are different regarding those for the other temperatures. Only minor different values of Is were measured below 140 K and the value of Q for the two subspectra representing Fe mainly surrounded by Nb coincide, whereas Q for Fe mainly surrounded by Fe is still different and shows a distinct temperature dependence (Fig. 7).

A strong distortion of the charge density of Fe caused by surrounding Nb atoms is suggested from the large differences in Q for different environments and the general decrease of Q with the number of Fe neighbours.

REFERENCES

1. S.J.Hillenius, F.V.Coleman, R.M.Fleming and R.J.Cava, Phys.Rev. B23 1567 (1981)
2. S.J.Hillenius and R.V.Coleman, Phys. Rev. B25, 2191 (1982)
3. A.Meerschaut, A.Ben Salem, L.Guemas, J.Rouxel, P.Monceau and H.Salva Proceedings Synthetic low dimensional Conductors and Superconductors J. Physique Colloq. 44, C3-1681 (1983)
4. H.Gruber, W.Sitte and H.Sassik, J. Physique 45, 1231 (1984)
5. A.Meerschaut, P.Gressier, L.Guemas and J.Rouxel, J. Mat. Res. Bull. 16, 1035 (1981)
6. R.J.Cava, F.J.Disalvo, M.Eibschütz, J.V.Waszezak, Phys. Rev. B27, 7412 (1983)
7. H.Gruber, M.Reissner and W.Steiner, Proceedings The Physics and Chemistry of low dimensional Synthetic Metals, J.Molecular Cryst., in print
8. H.Gruber, 2nd Austrian Scientific Exposition Vienna, Austria, Nov.1981, p.97, Ed.: Zentralsparkasse and Kommerzialbank Wien, Organizer: Bundeskonf. d. wissenschaftl.und künstler. Personals, A-1010 Wien
9. R.Moret, J.P.Pouget, A.Meerschaut and L.Guemas, J.Physique-LETTRES 44 (1983) L93

TRANSPORT AND MÖSSBAUER STUDIES OF THE PEIERLS TRANSITION IN Fe-DOPED $K_{0.30}MoO_3$

J.Y. Veuillen, R. Chevalier, D. Salomon, J. Dumas, J. Marcus and C. Schlenker*
Laboratoire d'Etudes des Propriétés Electroniques des Solides, C.N.R.S., B.P. 166, 38042 Grenoble Cedex, France
*Ruhr-Universität, Bochum, FRG.

Differential resistance dV/dI and ^{57}Fe Mössbauer studies in ^{57}Fe-doped $K_{0.30}MoO_3$ single crystals are reported. Large hysteresis in the dV/dI curves are found. The Mössbauer data are consistent with three different Fe sites. Anomalies in the isomer shift and in the quadrupole splitting appear at 180 K for one ^{57}Fe site.

The molybdenum blue bronze $K_{0.30}MoO_3$ has been shown to be a quasi-one dimensional metal at room temperature. It undergoes at 180 K a Peierls transition towards a semiconducting incommensurate charge density wave (CDW) state[1]. In the low temperature phase, non-linear conductivity due to the depinning of the incommensurate CDW has been observed[2]. The crystal structure is monoclinic[3]. It can be viewed as made of infinite chains // $\vec{b}$ of MoO_6 octahedra. Among the three inequivalent Mo sites, only the so-called Mo(2) and Mo(3) are involved in the infinite chains. X-ray refinements studies have established that most of the 4d electronic density is found on these sites[4]. In order to study the Peierls transition with a local probe, we have grown ^{57}Fe-doped $K_{0.30}MoO_3$ single crystals for Mössbauer studies. One should note that probing CDW transitions with the Mössbauer technique has received little attention up to now[5]. We report here transport properties and Mössbauer spectra of these crystals.

The single crystals used in this study, grown by electrolytic reduction of a K_2MoO_4 : MoO_3 melt are platelets // $\vec{b}$ and [102]. Doping with ^{57}Fe was obtained by incorporating $^{57}Fe_2O_3$ in the melt. Iron concentration in the crystals is $\sim$ 300 ppm ($\sim$ 0.1 at. %). From the resistivity data the Peierls transition is still found at 180 K in the doped samples with no significant broadening. The differential resistance dV/dI has been measured with a.c. lock-in technique using evaporated indium for the electrical contacts. The threshold electric field was found of the same order as in pure samples. Switching in the V-I curve has also been found in some doped samples. We have studied the thermal and electrical memory of the doped samples and found effects more pronounced than in the pure ones[6]. Fig. 1 shows the effect of the thermal and electrical history of a ^{57}Fe-doped sample on the low field resistance. When the sample is cooled from 300 to 77 K with an applied d.c. current, the Ohmic resistance is found larger than the resistance obtained with zero current cooling. We call the excess resistance the thermoremanent resistance (TRR). The TRR increases abruptly when the current applied during cooling is larger than the threshold current

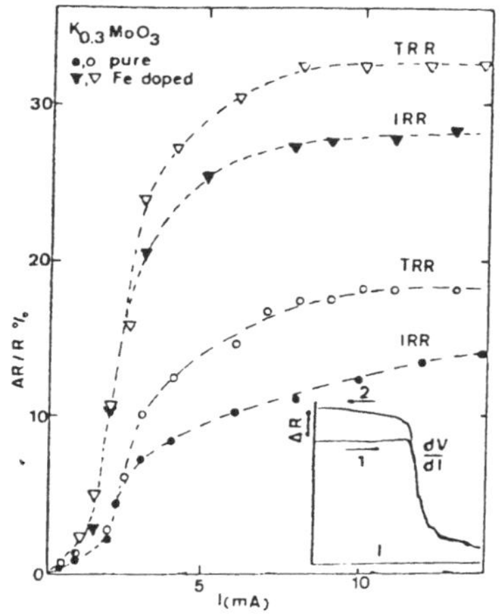

Fig. 1

Thermoremanent (TRR) and isothermal (IRR) remanent resistance of pure and ^{57}Fe-doped $K_{0.30}MoO_3$. The inset shows the hysteresis in the dV/dI vs I curve.

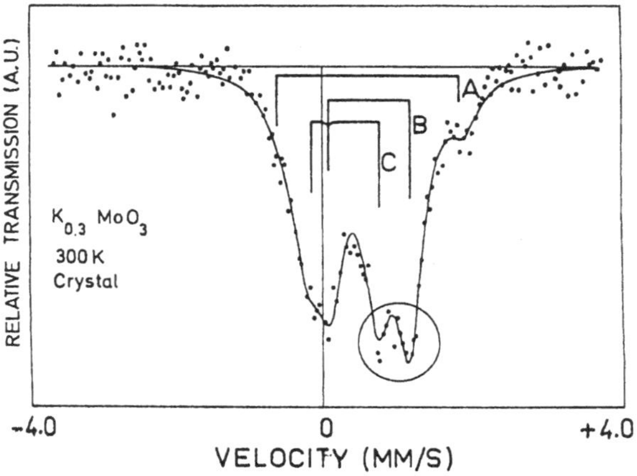

Fig. 2

Mössbauer spectrum of ^{57}Fe-doped $K_{0.30}MoO_3$ at 300 K. Circled region indicates the region of interest.

at 77 K. We have also defined an isothermal remanent resistance (IRR) obtained after a zero current cooling then sweeping the d.c. current up to a given value I and back to zero. The IRR also increases abruptly near the threshold current. We have proposed that the TRR and IRR were due to coupling of CDW domain boundaries to crystal defects[6]. These defects create localized levels in the Peierls gap and are responsible for the extrinsic nature of the low field resistivity. The larger TRR and IRR in Fe-doped samples would be due to additional pinning centers created by Fe impurities. For Mössbauer studies, the absorber was made of a mosaic of platelets with the γ beam perpendicular to the platelets. The source was $^{57}Co(Rh)$. At room temperature, the spectrum consists of three doublets characteristic of three distinct Fe sites A, B and C (Fig. 2). From these data, it may be concluded that the Fe atoms are substituted in the Mo octahedral sites. The intensity and the quadrupole splitting of the lines indicate that site A may correspond to the crystallographic site 1 and ^{57}Fe sites B and C to the sites 2 and 3.

We have studied the temperature dependence of the high velocity lines B_+ and C_+ of the doublets B and C. Spectra in the region of interest are shown in Fig. 3 at different temperatures. For temperatures above 180 K, B_+ has the largest amplitude while below 180 K, C_+ is the most intense. Experiments performed with a powder in the same velocity range did not show this effect. We therefore attribute the variation of the amplitude ratio of B_+ and C_+ vs T to an electric field gradient deorientation with opposite signs for doublets B and C. Powder spectra were recorded between 77 and 300 K. A fit was performed assuming the same linewidth for all the lines, and

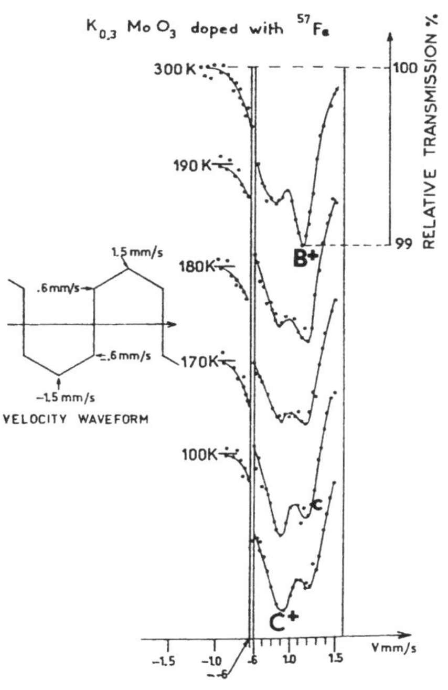

Fig. 3

Mössbauer spectra corresponding to the circled region of Fig. 1 at different temperatures.

the same amplitude for the two lines of each doublet. Isomer shift (I.S.) and quadrupole splitting (Q.S.) are shown in Fig. 4(a) and 4(b) respectively. From the I.S. data, doublet A corresponds to a Fe^{2+} site and doublet C to a Fe^{3+} site. Doublet A I.S. and Q.S. show an anomaly around 180 K, while doublets B and C are weakly temperature dependent. As the temperature dependence of the Q.S. reflects the local distorsion, these results indicate that the distorsion of the Mo(1)-O_6 octahedra may be larger in the low temperature phase. The temperature dependence of the I.S. indicates that a charge redistribution takes place and therefore that the onset of the CDW seems to be correlated with an electronic charge transfer between site 1 and sites 2 and 3.

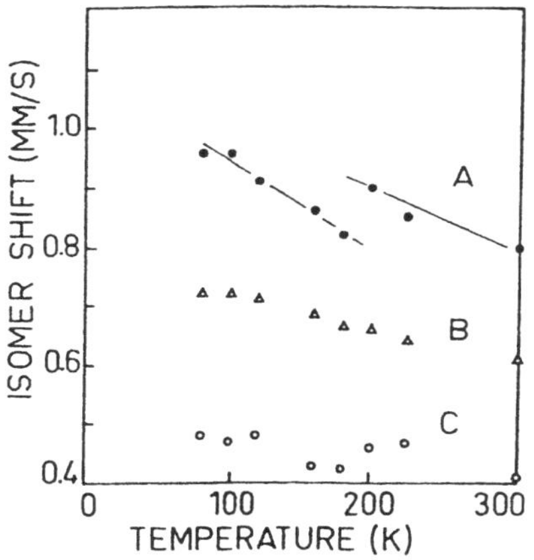

Fig. 4(a)

Isomer shifts for lines A, B, C as a function of temperature.

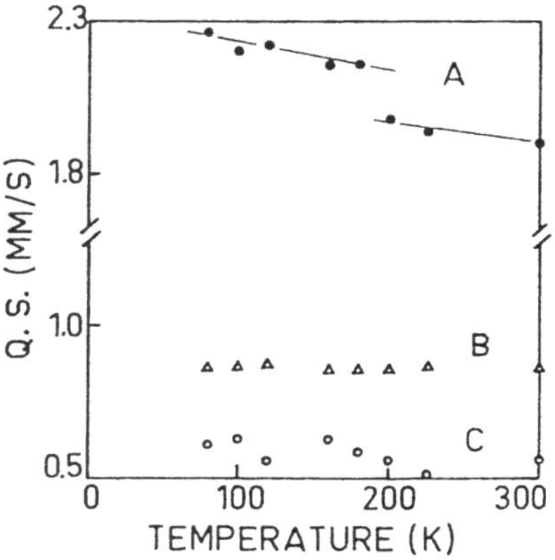

Fig. 4(b)

Quadrupole splittings for three lines A, B, C as a function of temperature .

REFERENCES

1. J.P. Pouget, S. Kagoshima, C. Schlenker and J. Marcus, J. Phys. Lett. 44, L-113 (1983).

2. J. Dumas, C. Schlenker, J. Marcus and R. Buder, Phys. Rev. Lett. 50, 757 (1983).

3. J. Graham and A.D. Wadsley, Acta Cryst. 20, 93 (1966).

4. M. Ghedira, J. Chenavas and M. Marezio, (to be published).

5. See for example, M. Eibschutz, D. Salomon and F.J. Di Salvo, Bull. Am. Phys. Soc. 29, 469 (1984) ; J.W. Brill, P. Boolchand and G.H. Lemon, Solid State Comm. 51, 9 (1984).

6. J. Dumas, A. Arbaoui, J. Marcus and C. Schlenker, in 'Proc. Int. Conf. on the Physics and Chemistry of low-dimensional synthetic metals', Abano Terme (Italy), (to be published).

CHARGE DENSITY WAVE INSTABILITIES IN QUASI TWO-DIMENSIONAL OXIDES η-Mo_4O_{11} AND γ-Mo_4O_{11}

H. Guyot, G. Fourcaudot, C. Schlenker
Laboratoire d'Etudes des Propriétés Electroniques des Solides, C.N.R.S., B.P. 166, 38042 Grenoble Cedex, France

Transport and magnetic properties have been studied comparatively in both metallic oxides η and γ-Mo_4O_{11}. Phase transitions occur at Tc_1 = 109 K and Tc_2 = 30 K in η-Mo_4O_{11} . γ-Mo_4O_{11} shows a semiconducting type behaviour below $T'c_1$ = 100 K. The transitions in both compounds are attributed to charge density wave instabilities.

The monoclinic η and orthorhombic γ-Mo_4O_{11} compounds are metallic oxides showing phase transitions. Detailed studies on η-Mo_4O_{11} single crystals have already been reported[1]. Powder data only are available for γ-Mo_4O_{11}[2]. We now report a comparative study performed on single crystals of both materials. The crystallographic structures of both compounds are built with infinite slabs of MoO_6 octahedra parallel to the (bc) plane, separated by MoO_4 tetrahedra[3,4]. In γ-Mo_4O_{11}, consecutive MoO_6 slabs are deduced by mirror symmetry.

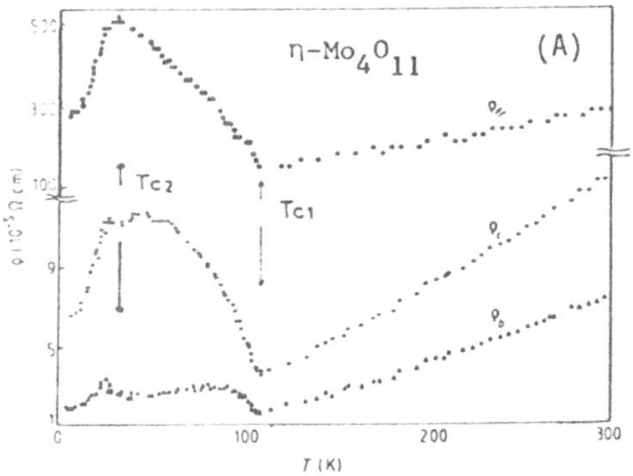

Fig. 1 A.
Electrical resistivity as a function of temperature for η-Mo_4O_{11} for different crystallographic directions. $\rho_{//}$ is measured along the a^*-axis.

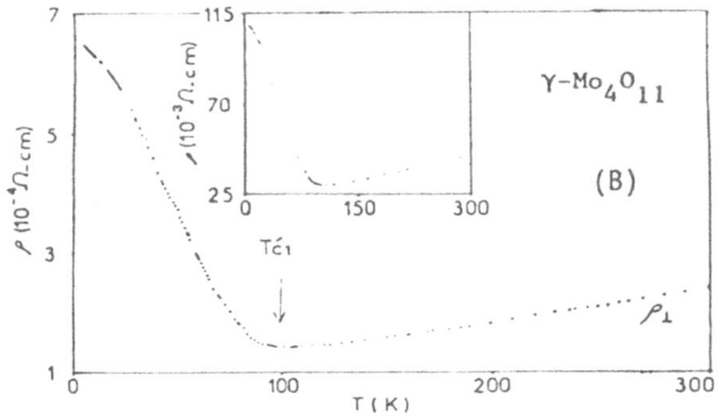

Fig. 1 B.
Electrical resistivity as a function of temperature for γ-Mo_4O_{11} . $\rho_\perp$ is measured in the bc plane and $\rho_{//}$ along the a^*-axis.

Fig. 1 shows the resistivity ρ vs temperature for both oxides in different directions. The strong anisotropy shows that they are quasi two-dimensional metals, the highest conductivity plane being (b, c). This is well-accounted by the crystal structure since the 4d conduction electrons are confined in the MoO_6 slabs[1]. Both compounds show a phase transition at $Tc_1 = 109$ K for η-Mo_4O_{11} and $T'c_1 = 100$ K for γ-Mo_4O_{11}. The anisotropy factors $\rho_{//}/\rho_{\perp}$ (see Fig. 1) are found to be in the range of 100 and weakly temperature dependent above Tc_1 ($T'c_1$). Below, the resistivity increases monotonously with decreasing temperature in γ-Mo_4O_{11} while it shows maxima in η-Mo_4O_{11}. In the η-compound, the data obtained along the monoclinic b-axis clearly show that a second transition takes place at $Tc_2 = 30$ K. One should note that $\rho_{//}/\rho_{\perp}$ is strongly temperature dependent below Tc_1 in η-Mo_4O_{11} only.

Thermopower measured as a function of temperature between 10 K and 300 K shows a similar behaviour for both compounds (Fig. 2). It is always n-type, linear with T above Tc_1 ($T'c_1$) and shows a bump below. In the case of η-Mo_4O_{11}, a slight anomaly is found below Tc_2.

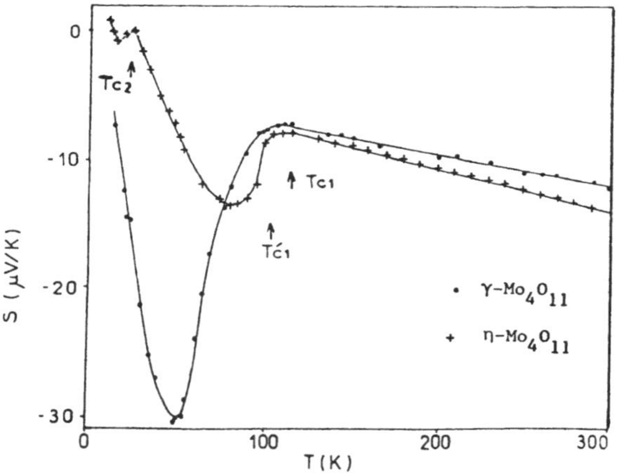

Fig. 2.
Thermopower measured along b, as a function of temperature for γ-Mo_4O_{11} *and* η-Mo_4O_{11}.

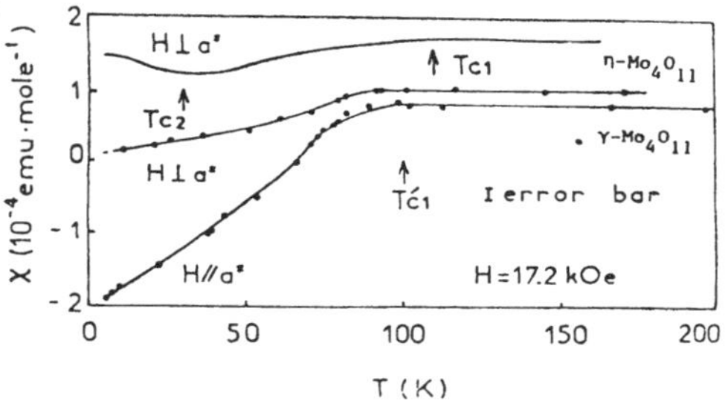

Fig. 3.
Magnetic susceptibility as a function of temperature. For γ-Mo_4O_{11}*, the anisotropy is shown. Absolute values are known with an accuracy of* $2x10^{-5}$ *emu/mole.*

The magnetic susceptibility χ, measured with a magnetic field of 17.2 kG, is plotted vs temperature on Fig. 3. A Curie type contribution, non negligible below 50 K, has been substracted from the data. Data obtained on γ-Mo_4O_{11} show that χ is strongly anisotropic below $T'c_1$. In both compounds, χ shows a plateau above Tc_1 ($T'c_1$). In γ-Mo_4O_{11} for $H // a^*$, a large decrease is found below and χ is negative at low temperatures. In η-Mo_4O_{11}, for $H \perp a^*$, χ shows a minimum at a temperature slightly larger than Tc_2. The order of magnitude of χ never exceeds 2×10^{-4} emu/mole.

All our data are consistent with CDW instabilities taking place in both compounds at Tc_1 ($T'c_1$). For η-Mo_4O_{11}, this has been corroborated by x--ray and electron diffraction studies showing below Tc_1, weak satellites at incommensurate positions along b^* ($q_b = 0.23\ b^*$ at 10 K)[5]. These CDW instabilities are clearly related to the quasi 2D electronic properties. In both compounds, the Fermi surface is expected to be strongly anisotropic and to show large portions parallel to the a^* axis leading to nesting properties. At Tc_1 ($T'c_1$), gap openings take place on parts of the Fermi surface, resulting in metal-metal transitions. In the case of γ-Mo_4O_{11}, the electrical resistivity data indicate that the gaps opening process goes on down to the lowest temperatures (4.2 K). For η-Mo_4O_{11}, both electrical resistivity and x-ray structural studies[5] show that the gap opens down to $\sim$ 50 K. The origin of the low temperature transition ($Tc_2 = 30$ K) is not clear at the moment. It could be due to a spin density wave instability.

The magnetic susceptibility includes in these compounds three contributions, the core diamagnetism, the Pauli conduction electrons paramagnetism and the Van Vleck orbital one. The decrease of χ below Tc_1($T'c_1$) could be due to a decrease of the density of states at the Fermi level. However, the strong anisotropy found for γ-Mo_4O_{11} suggests that the Van Vleck susceptibility may be more anisotropic in the low temperature phase.

The thermopower data show that the dominant carriers are electrons in all cases. However, detailed analysis of all the transport properties lead to the conclusion that several types of carriers, associated with both electrons and holes pockets, must be present. At least, a two-band model should be involved to account for the transport properties.

In the case of η-Mo_4O_{11}, the anisotropy of the Fermi surface is corroborated by magnetoresistance data. Also at low temperatures ($T < Tc_2$) both Shubnikov-De Haas and de Haas-Van Alfen oscillations are found with frequencies of 0.042, 0.31 and 2.5 MG[6]. One can evaluate from these frequencies, in a 2D model areas of the corresponding orbits: they are found to be of the order of 1 % of the first Brillouin zone, which shows that the gap openings at the transition leave very small electrons (and holes) pockets.

REFERENCES

1. Guyot H., Escribe-Filippini C., Fourcaudot G., Konaté K. and Schlenker C., 1983, J. Phys. C Letters 16, 1227-1232.
2. Gruber H., Haselmair H. and Fritzer H.P., 1983, J. Solid State Chem. 47, 84-91.
3. Kihlborg L., 1963, Ark. Kemi. 21, 365-377.
4. Ghedira M., Vincent H., Marezio M., Marcus J. and Fourcaudot G., 1984 (to be published in J. Solid State Chem.).
5. Ayrolles R., Roucau C., Pouget J.P., Guyot H., Schlenker C. (to be published).
6. Guyot H., Escribe-Filippini C., Englert Th., Schlenker C. (to be published).

THERMAL CONDUCTIVITY OF LAYERED DICHALCOGENIDES

M. Núñez Regueiro, J. Lopez Castillo and C. Ayache
Centre d'Etudes Nucleaires de Grenoble, SBT/Laboratoire de Cryophysique
85x, 38041 Grenoble Cedex, France

We have measured the thermal conductivity of 1T-TaS_2 and 2H-$TaSe_2$. We observe strong anomalies in both compounds at the charge density wave onset temperatures. The temperature dependence of the phonon thermal conductivity of 2H-$TaSe_2$ can be well fitted by a relaxation time constant above Tc and increasing under Tc up to its "normal" value. This implies that short range order is already present at room temperature, in accordance with Mc Millan's prediction.

Thermal conductivity studies of charge density wave compounds have been up to now neglected. It is however clear the need to look at heat transport (phonon transport in particular) in these materials where the $2k_F$ phonon freezes out in a permanent distorsion. In analogy with antiferroelectrics we can think that a lot of information about how the transition is felt by the phonons can be extracted from these type of measurements.

We have undertaken this work by using a recently developed technique for thermal transport measurements[1]. In these paper we report results obtained on 1T-TaS_2 and 2H-$TaSe_2$, two of the most interesting transition metal dichalcogenides. The first one shows an incommensurate charge density wave (ICDW) down to 350 K and a commensurate one (CCDW) under 200 K. In between these first order transitions, there exists a not yet well understood nearly comensurate(NCDW) phase. 2H-$TaSe_2$ has a transition to an ICDW at 122 K and the lock-in to the CCDW takes place at 80 K. While the first compound becomes semiconductor in the CCDW, 2H-$TaSe_2$ remains metallic down to low temperatures.

We show on figure 1 the thermal conductivity of 1T-TaS_2 between 80 and 400 K. The two first order transitions are clearly seen. However, while in the electrical conductivity the effect of both of them is to strongly decrease it, we observe here a decrease at 350 K opposite to a very sharp increase at 180 K. To our knowledge it is the first time that such type of dramatic effects are observed in a thermal conductivity measurement. We have also plotted on figure 1 the value of the electronic thermal conductivity obtained from the electrical conductivity through the Wiedemann-Franz relation. the decrease at 350 K is clearly due to the loss of electrical carriers at this transition. Evidently the increase at 180 K cannot be explained in the same way. It looks as if the causes of strong phonon scattering disappear at the transition, and that the sample recovers what would be its "normal" (umklapp, boundaries, etc.) thermal conductivity. However, the fact that the NCDW phase is not yet fully understood, the transiti-

on strongly first order, and the lack of measurements under 80 K does not allow further analysis. We shall see below that there is a similar effect in $2H\text{-}TaSe_2$ that can be explainedin terms of its better known properties.
The thermal conductivity of $2H\text{-}TaSe_2$ is shown on figure 2. As in $1T\text{-}TaS_2$ the conductivity decreases as we decrease the temperature. But when the ICDW transition temperature is attained it starts increasing in a smooth way as expected from a second order phase transition. The physical explanation is the same as with $1T\text{-}TaS_2$,i. e. the causes of phonon scattering disappear at the transition temperature.
It is interesting to compare the thermal conductivity of $2H\text{-}TaSe_2$ with its electrical resistance. In analogy with magnetic metals, the latter can be decomposed into two parts: the normal scattering due to phonons and impurities, and a "spin" dependent part[2]. This is constant (disordered spin resistivity) above the transition temperature and decreases roughly as $1-\Psi^2$ [3](where Ψ is the order parameter, the magnetization in the magnetic metal case) under the transition temperature.
If we suppose that the mean free path of the phonons has a similar dependence as the one of the electrons, we can calculate the phonon thermal conductivity using current expressions[4]. We obtain the fit shown on figure 2 for a Debye temperature of 500 K. This value is bigger than the one extracted from specific heat measurements, but we are here sampling only the phonons that actually contribute to the thermal conductivity, not the average over all modes that sees the specific heat. Above Tc as the mean free path is constant the shape of the curve is just proportional the the lattice specific heat . Under Tc the curve shows the gradual recovery of analso constant but longer (as is due to the boundaries) mean free path.
Thus the phonons and the electrons seem to be scattered in the same way for temperatures higher than Tc , but we must still precise the nature of the scatterers. We think that entities formed by the soft mode and condensing electrons ("pseudospins") exist even at high temperatures but in a disordered state, scattering the electrons and phonons. At the transition temperature they would order allowing the recovery of the normal mean free path. This type of order-disorder pseudospin transition, having the same effect on the thermal conductivity ,have been already seen in some antiferroelecrics[5]. This idea is actually very similar to what has been postulated by Mc Millan[6]. He advanced that the CDW transition is only the establishment of long range order as short range order exists in the material for temperatures higher than Tc. The short range order would form supercell clusters that would be able to scatter the electrons and the phonons.
In conclusion, the thermal conductivity measurements on $1T\text{-}TaS_2$ and $2H\text{-}TaSe_2$ indicate a pseudospin order-disorder nature for the CDW transition. The implications of this on the microscopic structure of the incommensurate phase and on the sliding modes of other chalcogenides must be carefully considered.
We gratefully acknwledge P. Molinié from the Laboratoire de Chimie des Solides de la Universite de Nantes for providing us with the samples.

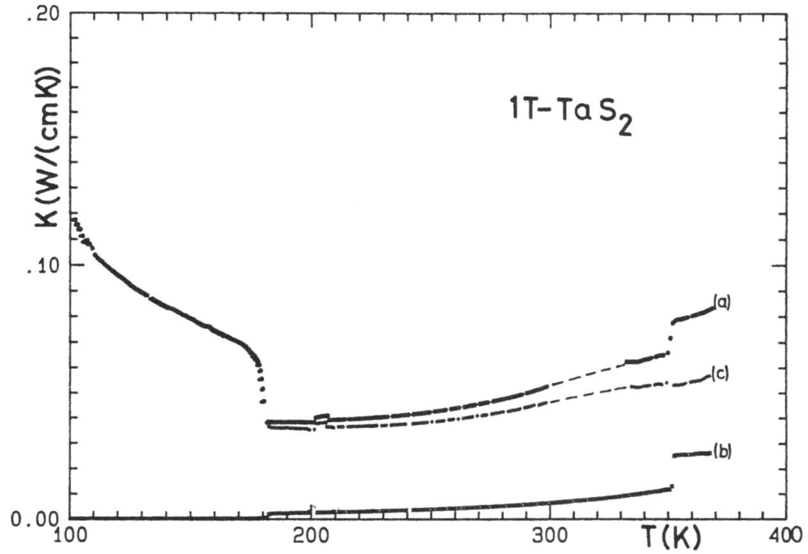

Figure 1: Thermal conductivity of $1T-TaS_2$
a) measured conductivity
b) electronic thermal conductivity calculated from the resistivity through the Wiedemann-Franz relation
c) phonon thermal conductivity a)-b)

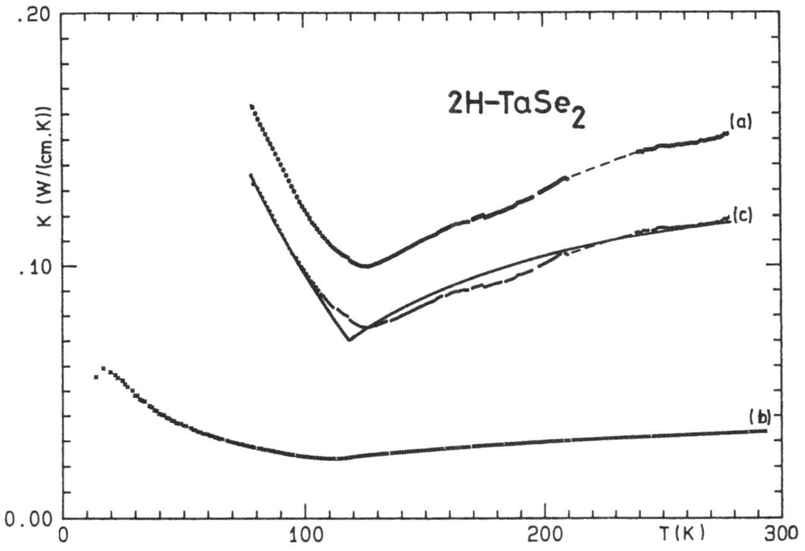

Figure 2: Thermal conductivity of $2H-TaSe_2$
a) measured conductivity
b) electronic thermal conductivity calculated from the resistivity through the Wiedemann-Franz relation
c) phonon thermal conductivity a)-b)
full line: phenomenological fit (see text)

References

1) J. Lopez Castillo and M. Núñez Regueiro, to be published.

2) M. Naito and S. Tanaka, J. Phys. Soc. Jpn. , 43, 1839 (1981)

3) P. G. De Gennes and J. Friedel, J. Phys. Chem. Solids, 4 , 71 (1958)

4) R. Berman ,Thermal Conductivity of Solids, Clarendon Press,Oxford, p. 37.

5) Y. Suemune, J. Phys. Soc. Jpn., 22 , 735(1967)

6) W. L. Mc Millan, Phys. Rev.,B16 ,643 (1977)

TUNNELING STUDY OF COMMENSURATE CHARGE DENSITY WAVE STATES IN 1T-TaS_2

H.Ozaki, T.Ohara, H.Fujimoto and H.Hotch
Department of Electrical Engineering, Waseda University
Ohkubo 3-4-1, Shinjuku-ku, Tokyo 160, Japan

The profile of the electronic density of states are observed in the commensurate CDW state of 1T-TaS_2 through the tunneling measurements. The temperature dependence of the electronic density of states accounts for the electrical properties of this material at low temperature.

Among many transition metal dichalcogenides, 1T-TaS_2 has attracted special attention since its commensurate (C-) charge density wave (CDW) phase is nonmetallic as well as it possesses the nearly commensurate (NC-) phase between the C- and the incommensurate phases. The transition to the C-CDW phase is explained by a Mott localization of conduction electrons[1], whereas the diverging increase of the electrical resistivity ρ at low temperature is ascribed to an Anderson localization of the residual carriers in the Mott-Hubbard band tail[2]. In the present study, the profile of the electronic density of states N(E) of the C-phase of 1T-TaS_2 is investigated through the tunneling measurements[3], in the temperature range of 4.2 – 63K.

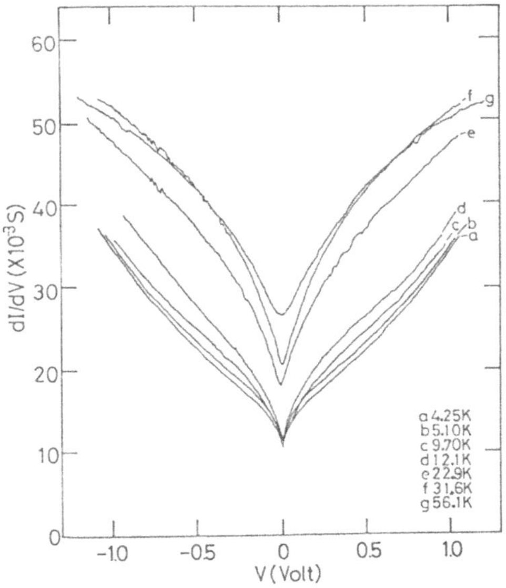

Fig. 1
The tunneling dI/dV vs. V characteristics of undpoed 1T-TaS_2 for several temperatures.

The tunneling barrier was prepared by oxidizing the aluminium film, which was made contact with the sample surface of several mm^2 to form a tunneling junction. Fig. 1 shows the tunneling dI/dV vs. V characteristics of undoped 1T-TaS_2 for several temperatures. The tunneling junction of two oxidized aluminium films, made contact together, shows a structureless dI/dV vs. V curve with a broad maximum at V=0 for all the temperatures between 4.2 and 300K. Hence, the profiles of the curves in Fig. 1, including the dip at V=0, reflect the N(E) of 1T-TaS_2. The negative V corresponds to the electron energy of 1T-TaS_2 below the Fermi level E_f. We will neglect here the energy dependence of the tunneling matrix, and thus in the following discussion regard the dI/dV at V as proportional to the electronic density of states N(eV) of the sample. Fig. 2 shows the magnitude of the decrease in dI/dV of undoped 1T-TaS_2 with respect to the value at 50.8K, which is denoted as $\Delta(dI/dV)_T$, for the two electron energies of $|E-E_f|=0$ and 0.6eV. Fig. 3 shows the tunneling dI/dV vs. V characteristics of 0.1%-W doped 1T-TaS_2 for the same temperature region as Fig. 1. W atoms act as an efficient dopant in 1T-TaS_2 to suppress the transition to the C-phase without introducing a strong substitutional disorder[4]. This sample does not make transition to the C-phase down to 4.2K. So, we can see the difference of the profile of N(E) between in C- and NC-phases at the same temperature. As shown in Fig. 3, the temperature dependence of dI/dV for the W doped sample is very small, which is reflected on the fairly constant ρ vs. T relation of this sample[4]. Typically, the zero bias tunneling resistance of an undoped sample increases by a factor of 1.5 across the transition from the NC- to the C-phase. So, it is shown by comparing Figs. 1 and 3 that the N(E) of the undoped sample is diminished not only near E_f but over the whole region of energy in Fig. 1, compared to that in the NC-phase, even at the highest temperature in Fig. 1. When the temperature decreases from higher side in Fig. 1, a marked change in N(E) occurs

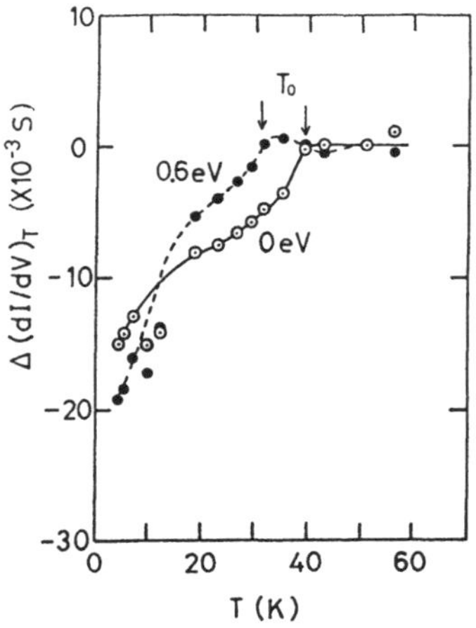

Fig. 2

The magnitude of the decrease in dI/dV with respect to the value at 50.8K for undoped 1T-TaS_2. The parameters of the curves are the electron energies with respect to E_f. T_0 indicates the onset of the decrease.

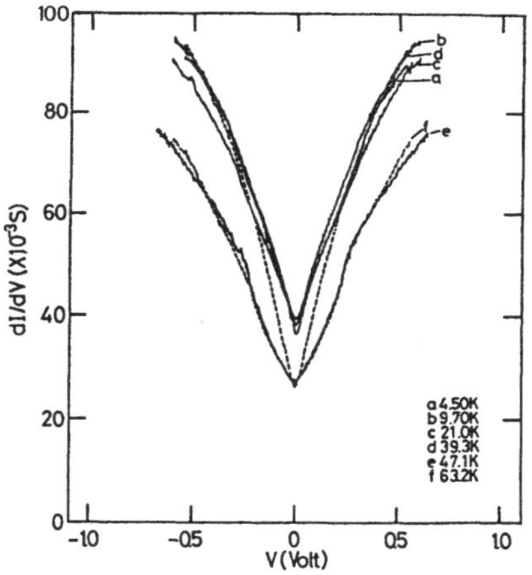

Fig. 3

The tunneling dI/dV vs. V characteristics of 0.1%-W doped TaS_2 for several temperatures.

in the temperature region of $31K \lesssim T_0 \lesssim 39K$, which is indicated in Fig. 2. Through this change, the dip of the N(E) at E_f (V=0) becomes sharp, and then, N(E) begins to decrease with further decrease of temperature as shown in Fig. 2. The rate of decrease in N(E) with decreasing temperature is high around 12 – 18K. The decrease of N(E) at 4.25K with respect to the value at 50.8K, $\Delta(dI/dV)_{4.25K}$, is shown in Fig. 4. It has a maximum at $|E-E_f| \simeq 0.6eV$ and extends to beyond $|E-E_f|=1eV$.

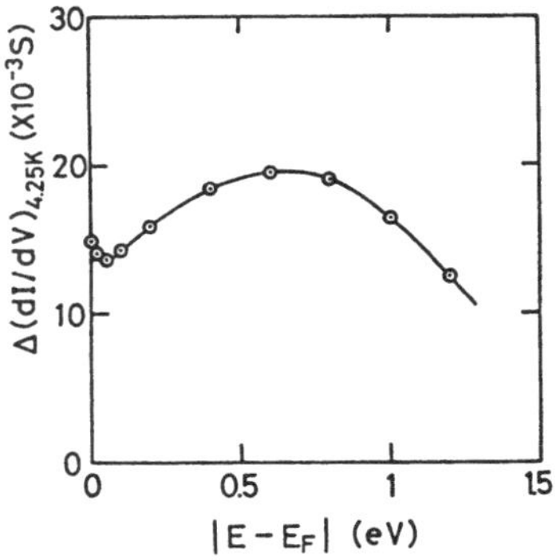

Fig. 4
The magnitude of the decrease in dI/dV at 4.25K with respect to the value at 50.8K, for undoped 1T-TaS_2, as a function of electron energy.

These behaviors of N(E) observed by tunneling measurements explain qualitatively the low temperature behaviors of electrical resistivity[2,5], Hall coefficient R_H[6,7] and Seebeck coefficient S[7], as follows. The mobility edge is supposed to lie somewhere below E_f around 50K, which gives rise to a positive R_H and S[6,7]. With decreasing temperature N(E) begins to decrease and it drives the mobility edge away from E_f, increasing R_H and S[6,7]. When the temperature goes down across ~18K, N(E) is diminished greatly over a wide range of energy as shown in Fig. 4. At this time, the mobility edge goes too far from E_f to contribute to the conduction and the crossover of conduction from the activation type at the mobility edge to the variable range hopping at E_f occurs. This crossover can be assigned to the peaks around 20K observed in R_H and S.[7]

The authors wish to thank K.Kawamoto, H.Wakamatsu, H.Sugawara and T.Hirota for their kind assistance in the experiments.

1. P.Fazekas and E.Tosatti: Phil.Mag. B39 (1979) 229; Physica 99B (1980) 183.
2. F.J.DiSalvo and J.E.Graebner: Solid State Commun. 23 (1977) 825. P.D.Hambourger and F.J.DiSalvo: Physica 99B (1980) 173.
3. H.Ozaki, T.Mutoh, H.Ohshima, A.Okubora and N.Yamagata: Physica 117B &118B (1983) 590. S.Noutomi, T.Futatsugi, M.Naito and S.Tanaka: Solid State Commun. 50 (1984) 181.
4. H.Fujimoto and H.Ozaki: Solid State Commun. 49 (1984) 1117.
5. F.J.DiSalvo, J.A.Wilson, B.G.Bagley and J.V.Waszczak: Phys.Rev.B 12 (1975) 2220.
6. R.Inada, Y.Ōnuki and S.Tanuma: Physica 99B (1980) 188.
7. T.Tani, K.Okajima, T.Itoh and S.Tanaka: Physica 105B (1981) 127.

GALVANOMAGNETIC PROPERTIES OF THE QUASI-TWO DIMENSIONAL PURPLE BRONZE $K_{0.9}Mo_6O_{17}$

E. Bervas, R.W. Cochrane*, J. Dumas, C. Escribe-Filippini, J. Marcus and C. Schlenker
LEPES - 25, Avenue des Martyrs, B.P. 166-38042 Grenoble-Cedex, France
*Département de Physique - Université de Montréal, Montréal, Canada.

Magnetoresistance temperature range 4.2 K - 200 K magnetic field up to 40 KG and Hall effect (50 K - 300 K, up to 10 KG) have been studied on the quasi two dimensional purple bronze $K_{0.9}Mo_6O_{17}$. The results are consistent with a quasi-cylindrical Fermi surface and with a CDW instability inducing both electrons and holes pockets below the CDW transition. At $T < T_c$ the data are well accounted by a two band model.

The purple molybdenum bronze $K_{0.9}Mo_6O_{17}$ is known to be a quasi two dimensional metal which shows an electronic transition at 120 K. The anomalies existing in the transport and other physical properties are characteristic of the opening of partial gaps at the Fermi surface and due to CDW instabilities [1].

The crystal structure space group $P\overline{3}$, can be described as being built with four layers of MoO_6 octahedra perpendicular to the trigonal c-axis, separated by KO_{12}, icosahedra linked to Mo O_4 tetrahedra [2]. The spatial separation between the $Mo^{5+}(4d^1)$ in the octahedral sites and the $Mo^6(4d^0)$ in the tetrahedral ones leads to a very anisotropic Fermi surface and to quasi two dimensional electronic properties. Both X-ray and electron diffraction studies have shown diffusions at room temperature which condense below the transition temperature of 120 K into very weak satellites, at commensurate position (1/2, 0, 0), (0, 1/2, 0) and (1/2, 1/2, 0). These data corroborate the model of a CDW instability [3].

The properties of $K_{0.9}Mo_6O_{17}$ are well described by a model of a quasi cylindrical Fermi surface with nesting properties inducing gap opening at $T < 120$ K and leading to a commensurate CDW state.

We now report Hall effect and magnetoresistance studies which are consistent with this picture [4].

The electrical resistivity has been measured in the temperature range 4.2 K-200 K in magnetic field B up to 40 kG (Fig. 1). B has no effect on

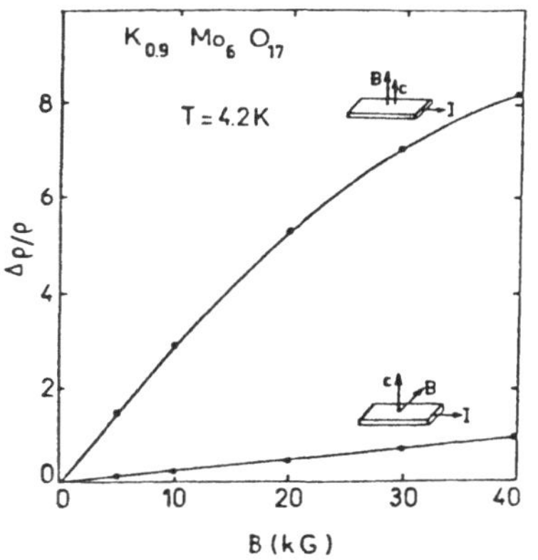

Fig. 1a.
Transverse magnetoresistance as a function of magnetic field for I // layers, B // c or B ⊥ c.

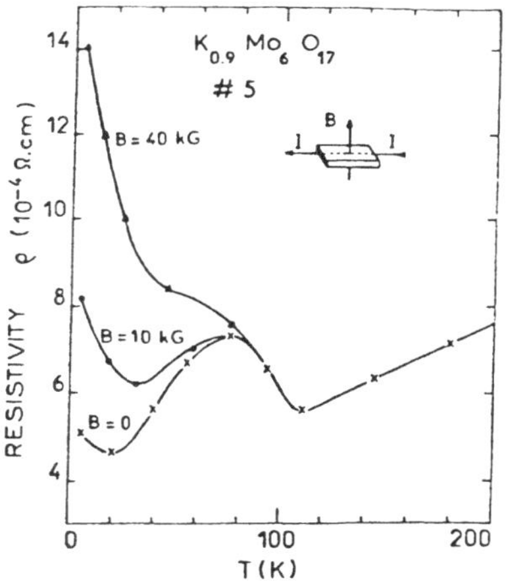

Fig. 1b.
Resistivity as a function of temperature I ⊥ c, B // c.

the transition temperature but increases dramatically the resistivity at low temperature (T < 90 K).

Fig. 1a shows that the magnetoresistance, always positive, is strongly anisotropic. The transverse value obtained with I parallel to the plane of the layers is found at 4.2 K to be roughly 10 times larger with B // c than with B ⊥ c. ρ does not follow a B^2 law and seems to saturate in the case B // c for B ~ 40 kG. For B ⊥ c ρ is linear with B . These results are consistent with a quasicylindrical Fermi surface and with open orbits along the c-axis [5]. The transverse magnetoresistance is plotted as a function of temperature for different values of B // c on Fig. 2b. One should note that Δρ/ρ becomes very large only below the CDW transition.

The Hall effect has been studied between 200 and 50 K in fields smaller than 10 kG.

Fig. 2a shows that Hall voltage V_H as a function of magnetic field for several temperatures $T < T_c$. For T > 75 K, V_H is negative and proportional

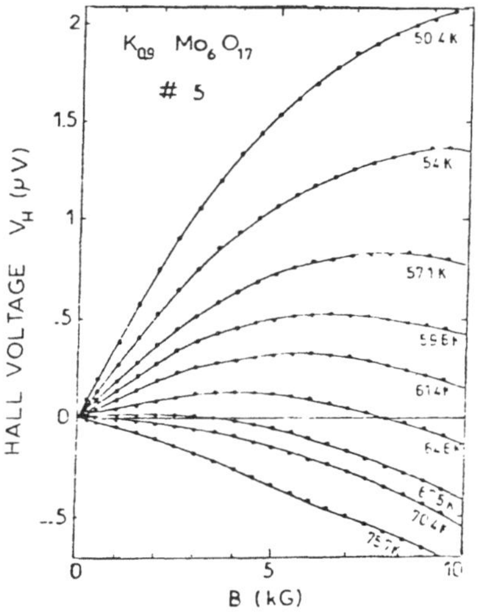

Fig. 2a.
Hall voltage as a function of magnetic field at different temperatures ($T < T_c$).

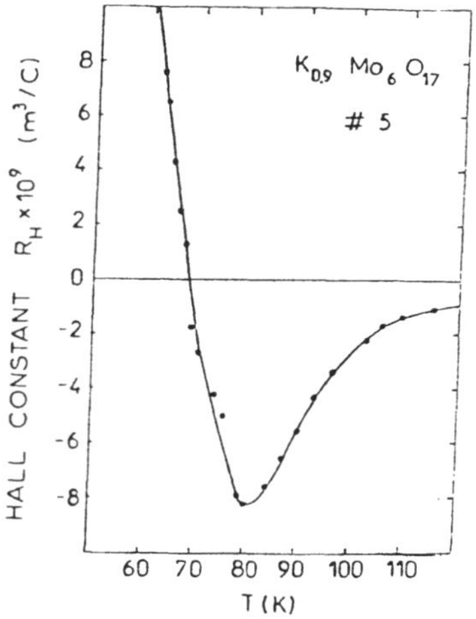

Fig. 2b.
Low field Hall constant versus temperature.

to B in the explored field range. Deviation from linearity increases with decreasing temperature. Also V_H changes sign in the vicinity of 70 K. The Hall constant R_H obtained in small fields is plotted as a function of temperature on Fig. 2b. $|R_H|$ increases steeply at $T < T_c$, goes through a maximum at ~ 80 K. At $T < 70$ K, R_H is clearly positive and increases dramatically with decreasing temperature. At T = 200 K the Hall constant is found to be 0.7×10^{-9} m^3/C. In a free electron picture, this corresponds to an electron concentration of 8.9×10^{21} cm^{-3}, which is in good agreement with the value of 8.3×10^{21} cm^{-3} obtained from the chemical formula by assuming a complete charge transfer from the K to the conduction band. The change of sign of R_H at 70 K clearly indicates that, while the dominant carriers are electrons in the high temperature phase, the gap openings at T_c leads to the formation of both electrons and holes pockets. A two band model [6] is therefore adequate to describe the Hall effect and low field magnetoresistance data. By combining both sets of data, one can evaluate the electrons (n) and holes (p) concentrations as well as their mobility. The results are shown on Fig. 3.

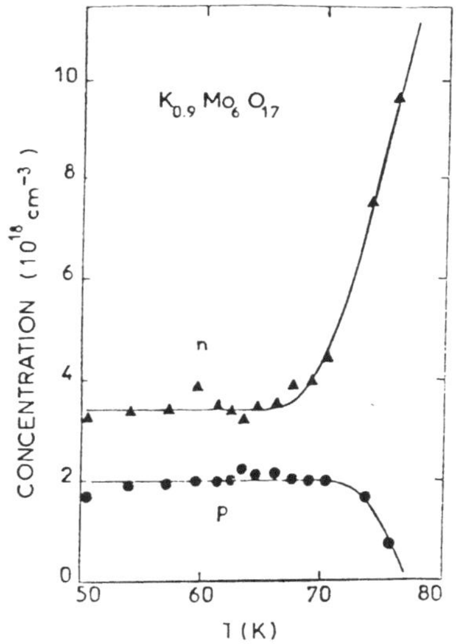

Fig. 3a.
Electrons (n) and holes (p) concentration vs temperature.

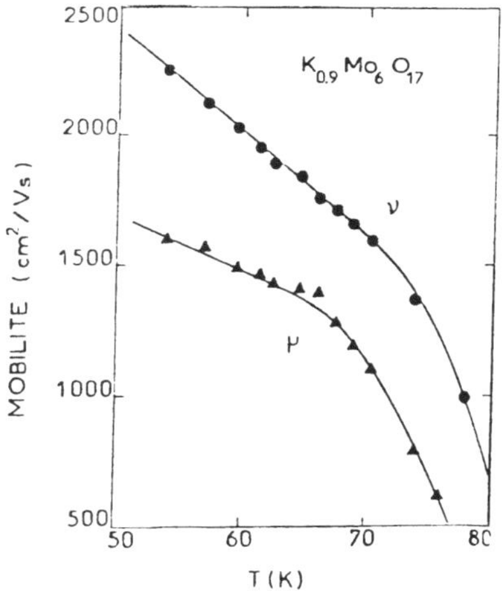

Fig. 3b.
Electrons (μ) and holes (ν) mobility vs temperature. Results obtained from a two band model.

The step decrease of n below T_c is consistent with gap openings on large parts of the Fermi surface at the transition. The increase of p shows that at the same time hole pockets are induced. These results also indicate that the gaps open down to ~ 70 K. This is consistent with the temperature dependence of the intensity of the satellites, as found by X-ray studies [3]. The increase of mobility with decreasing temperature shows clearly two-different regimes at $T < 70$ K and $T > 70$ K. This may be due to an increase of the electron phonon diffusion when the gap starts to close down at $T < T_c$

The behaviour of the magnetoresistance at higher fields and lower temperatures does not seem to be well accounted by a simple two-band model. Two possible mechanisms may be involved. In the first one, the existence of small gaps may allow for magnetic break-down [7] as it has been already suggested for similar results in the layered dichalcogenide 2H - $TaSe_2$ [8]. We rather propose that at $T < T_c$, large magnetic fields may induce further gap openings. Magnetization measurements even suggest that at $T < 30$ K, a spin density wave state may be stabilized under magnetic field [9].

REFERENCES

[1] R. Buder, J. Devenyi, J. Dumas, J. Marcus, J. Mercier and C. Schlenker, J. Physique Lettres 43, L59 (1982).

[2] H. Vincent, M. Ghedira, J. Marcus, J. Mercier and C. Schlenker, J. Solid State Chem. 47, 113 (1982).

[3] C. Escribe-Filippini, R. Almairac, R. Ayrolles, C. Roucau, K.Konaté, J. Marcus and C. Schlenker, à paraitre Phil. Mag. B (1984).

[4] E. Bervas, Thèse de Docteur Ingénieur, Université de Grenoble (1984).

[5] For the galvanomagnetic properties of metals, see for example J.P. Jan Solid State Physics 5, 1 (1957), E. Fawcett, Adv. in Physics 13, 139 (1964).

[6] See for example J.M. Ziman, Principles of the Theory of Solids, Cambridge University Press (1972).

[7] A.B. Pippard, The Dynamics of Conduction Electrons, Ed. Gordon and Breach (1965).

[8] N. Naito and S. Tanaka, J. of Phys. Soc. Japan 51, 228 (1982).

[9] J. Dumas, E. Bervas, D. Salomon, C. Schlenker and G. Fillion Proc. Int. Conf. Magnetism, Kyoto (1982).

NON-LOCAL ELASTIC FORCES IN CHARGE-DENSITY WAVE SYSTEMS

Dionys Baeriswyl* and Lars Kai Hansen
Physics Laboratory I, H.C. Ørsted Institute
University of Copenhagen, Denmark

Analytical expressions for the spectra of phase and amplitude modes of a one-dimensional charge-density wave are given. From these we derive an effective non-local Lagrangian for the phase fluctuations. The range of the non-local elastic forces is found to be of the order of the coherence length ξ . On this length scale the elastic potential turns out to be larger than the long-range Coulomb potential, when applied to KCP. Our results may have important implications for the theory of impurity pinning.

At low temperatures the relevant excitations of an incommensurate one-dimensional charge-density wave are fluctuations in the phase $\phi(x,t)$. They are frequently described by the Lagrangian

$$L = (v/4\pi)\int dx\,(u^{-2}\phi_t^2 - \phi_x^2) \tag{1}$$

which has been derived from microscopic theory by Brazovskii and Dzyaloshinskii.[1] In the context of the Peierls instability of a one-dimensional electron-phonon system v is the Fermi velocity and $u = (m/m^*)^{\frac{1}{2}}v$, where m is the electron mass and m^* an effective mass which includes the inertia of the phonon system. The form of Eq.(1) is directly related to the observation of Lee, Rice and Anderson[2] that an incommensurate Peierls-distorted system admits amplitude and phase fluctuations with spectra $\omega_+^2(q) = \lambda\omega_o^2 + \frac{1}{3}u^2q^2$ and $\omega_-^2(q) = u^2q^2$, respectively, λ being the dimensionless electron-phonon coupling and ω_o the bare phonon frequency at $2k_F$. In a more detailed analysis we have found[3] that this dispersion is only valid in a region $|q|\xi \ll 1$ where $\xi = v/\Delta$ is the coherence length and Δ the gap parameter. Moreover we have derived the following modified expression

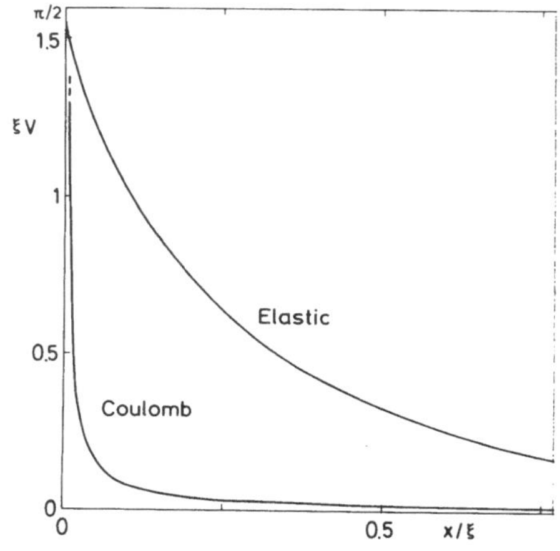

Fig.1
Comparison between the non-local elastic potential and the long-range Coulomb potential for the case of KCP.

$$\omega_{\pm}^2(q) = \lambda\omega_o^2\eta_q(\tanh\eta_q)^{\mp 1} \tag{2}$$

where $\sinh\eta_q = \frac{1}{2}\xi q$. It is valid for $|q|a << 1$, where a is the lattice constant. By using the relation $m^*/m = 4\Delta^2/(\lambda\omega_o^2)$ one recovers the expressions of Lee, Rice and Anderson in the limit $q \to 0$. In our derivation[3] we have treated the phonons as classical variables and used the adiabatic approximation. Furthermore we have linearized the bare electron spectra around $\pm k_F$ and neglected the dispersion of the bare phonon spectrum around $2k_F$.

The peculiar form of $\omega_{\pm}(q)$ has important consequences for the field-theoretical modeling of phase and amplitude fluctuations. Limiting ourselves to the phase variables we notice that the spectrum $\omega_-(q)$ can be derived from the Lagrangian

$$L = (v/4\pi)\{u^{-2}\int dx\,\phi_t^2 - \int dx\int dx'\phi_x V(x-x')\phi_{x'}\} \tag{3}$$

provided that we identify $V_q = \omega_-^2(q)/(u^2q^2)$. Due to the dispersion in $\omega_-(q)/q$ the potential $V(x)$ becomes non-local. We find the following simple expression

$$V(x) = \xi^{-1}\int_{2x/\xi}^{\infty} dy\; K_o(y) \tag{4}$$

where $K_o(y)$ is the modified Bessel function. The non-local nature of the elastic forces originates from virtual excitations of electron-hole pairs into extended band states. The form of the elastic term is the same as the long-range Coulomb interaction induced by phase fluctuations.[4] Furthermore $\xi V(x) \sim (\pi/2)^{\frac{1}{2}}(2x/\xi)^{-\frac{1}{2}}\exp(-2x/\xi)$ for $x >> \xi$ and therefore $\xi/2$ assumes the role of a screening length. The potential $V(x)$ is shown in Fig.1 and compared to the long-range Coulomb potential as given by Lee and Fukuyama[4]

$$V_C(x) = (4\pi)^{-1}(m^*/m)(d\omega_{pl}^*/v)^2|x|^{-1} \tag{5}$$

where d is the interchain distance and ω_{pl}^* is the plasma frequency associated with the phase mode. We have chosen the example of KCP with parameters $d = 9.87$ Å, $v = 7$eVÅ (assuming a free-electron band structure) and $m^*/m = 980$, $\omega_{pl}^* = 7.2$ meV (from the optical absorption experiments of Brüesch et al.[5]). Since the coherence length ξ in KCP is of the order of 100 Å we conclude that in this material, the long-range Coulomb interaction is largely dominated by elastic forces.

We anticipate that our results may have important consequences for the theory of charge-density wave transport. In particular the nature of the impurity pinning may be strongly modified. If the mean distance of impurities is smaller than ξ the elastic energy of Eq.(3) can be lowered by appropriately varying the signs of the gradients ϕ_x whereas the elastic energy of the local model depends only on ϕ_x^2.

A rough estimate of this effect can be obtained by considering a system of equally spaced impurities (at distance d_i). We choose the weak pinning limit where the phase is changed only little between consecutive impurities. In agreement with the theory of Fukuyama and Lee[6] we assume that the mean gradient $|\phi_x|$ is $1/(\alpha^{\frac{1}{2}}L_o)$ where $\alpha = 3/\pi^2$ and L_o is a characteristic length over which the phase looses memory. Furthermore, in order to gain elastic energy, we allow the phase gradient to change sign at every impurity site. This yields the following expression for the elastic energy per unit length.

$$K(L_o) = (v/4\pi)(\alpha L_o^2)^{-1} V_{\pi/d_i} . \qquad (6)$$

In the limiting case of dilute impurities ($d_i >> \xi$) $V_{\pi/d_i} \approx 1$ and we recover the result of reference 6. In the opposite limit ($d_i << \xi$), using the dispersion of Eq.(2) we,find

$$V_{\pi/d_i} = \lambda\omega_o^2(d_i/\pi u)^2 \ell n(\pi\xi/d_i) \qquad (7)$$

which shows that the elastic energy decreases with increasing impurity concentration. Adding the impurity potential energy[6] and minimizing with respect to L_o we find that the domain size L_o is strongly reduced as compared to its value in the local model. Therefore we conclude that the non-local nature of elastic forces enhances impurity pinning if the mean distance between impurities is smaller than the coherence length. This latter condition is satisfied in KCP where the disorder is usually attributed to the random distribution of Br atoms.

*Permanent Address: Seminar für theoretische Physik, ETH-Hönggerberg, CH-8093 Zürich.

References

1. S.A. Brazovskii and I.E. Dzyaloshinskii, Zh.Eksp.Teor.Fiz. 71, 2338 (1976) (Sov.Phys. JETP 44, 1233 (1976)).
2. P.A. Lee, T.M. Rice, and P.W. Anderson, Solid St. Commun. 14, 703 (1974).
3. L.K. Hansen and D. Baeriswyl, to be published.
4. P.A. Lee and H. Fukuyama, Phys.Rev. B17, 542 (1978).
5. B. Brüesch, S. Strässler, and H.R. Zeller, Phys.Rev. B12, 219 (1975).
6. H. Fukuyama and P.A. Lee, Phys.Rev. B17, 535 (1978).

DYNAMICS OF CHARGE DENSITY WAVES, THEORY

SOLITON MODEL OF CHARGE-DENSITY-WAVE DEPINNING

John Bardeen
Department of Physics
University of Illinois at Urbana-Champaign
1110 W. Green Street, Urbana, IL 61801 USA

and

J.R. Tucker
Department of Electrical Engineering
University of Illinois at Urbana-Champaign
1406 W. Green Street, Urbana, IL 61801 USA

The quantum tunneling model of depinning of charge-density waves in linear chain conductors can be simplified and made more concrete by reviving a soliton model similar to that studied in 1978 by Maki and by Larkin and Lee. They rejected a model of solitons on individual chains pinned by impurity fluctuations because the energy involved is far less than 1°K. However the transverse coherence distance includes 10^5 or 10^6 parallel chains. There is only one thermal degree of freedom for motion parallel to the chains in a domain of this area and a length containing a pinned soliton or phase kink. What is pinned is a parallel array of such phase kinks of average spacing L_d. The current acceleration, dJ/dt, from a field, E, by tunneling, is analogous to Josephson current flow across a tunnel junction from a phase difference.

During the past few months we have been attempting to develop a more detailed microscopic model of the theory of depinning of charge-density waves (CDW's) by quantum tunneling.[1] The ideas go back to early papers of Larkin and Lee[2] who suggested pinning of solitons by impurities and of Maki[3] who discussed depinning by soliton-anti-soliton creation in an electric field. The latter model was rejected when applied to quasi-1D conductors such as $NbSe_3$ because the soliton energy for a single chain is far less then 1°K.

In the present theory[4] we assume that what is pinned are phase-kinks extending over a phase coherent transverse area containing the order of 10^5 to 10^6 chains. There is only one thermal degree of freedom in a volume of this area and length L_d that contains a single phase kink. The amplitude that determines the tunneling probability is that for motion in the chain direction of individual electrons (or solitons) defined by the transverse wave vectors, k_y, k_z, of the 3D Fermi sea, with k_x in the chain direction. One may regard k_y, k_z as defining a single chain even though the electron density extends over the transverse area.

Acceleration requires displacement of the Fermi sea by a wave vector q, so that the 1D Fermi surface, $(-k_{Fx}, +k_{Fx}) \rightarrow (-k_{Fx} + q, +k_{Fx} + q)$. The Peierls gaps stay at the boundaries of the displaced Fermi sea and do not affect motion in the chain direction. This is also the case for a superconductor with a gap at the Fermi surface. The only thermal degree of freedom is the drift velocity $v_d = \hbar q/m$, where m is the band mass and v_d in turn determines the displacement of the Peierls gaps. In the absence of pinning, the equation of motion is

$$m \frac{dv_d}{dt} = \hbar \frac{dq}{dt} = e^* E \tag{1}$$

where $e^* = (m/M)e = 10^{-3}e$ and $M = m + M_F$ includes the Fröhlich mass, M_F, associated with ion motion.

In the original tunneling theory for depinning, the pinned ground state was modeled by a semiconductor with a small pinning gap, $E_{gap} = \hbar\omega_P$, at the Fermi surface, where $\omega_P/2\pi$ is the pinning frequency. Equation (1) is then replaced by

$$\hbar \frac{dq}{dt} = e^* EP(E) \quad , \tag{2}$$

where P(E) is the Zener tunneling probability across the pinning gap. One may apply (2) either to motion along x for a single (k_y, k_z) value (corresponding to a single chain) or to the motion of the 3D Fermi sea in a phase-coherent volume.

To show that (2) applies to coherent motion of a system of N-chains with wave functions specified by the single variable q, one may use the transfer-matrix formalism. Let T be the matrix element that adds δq to the wave vector of a single chain [(k_y, k_z) value]. The matrix element for transfer from an initial state

$$\Psi_i = \psi_{i1}\psi_{i2}\cdots\psi_{iN} \text{ (antisymmetrized)} \tag{3}$$

to a final state

$$\Psi_f = N^{-1/2} \sum_n \psi_{i1}\cdots\psi_{fn}\cdots\psi_{iN} \tag{4}$$

is

$$N \cdot N^{-1/2}T = N^{1/2}T \tag{5}$$

with a square NT^2. The total wave vector of the system of N chains is $Q = Nq$ so that

$$\hbar N \frac{dq}{dt} = e^{*}ENP(E) \quad , \tag{6}$$

which is the same as (2).

Actually, following Barnes and Zawadowski[5], it is the acceleration, dJ/dt, proportional to dq/dt, that is well-defined, not the individual tunneling events, in analogy with J being well defined, not the individual tunneling events that transfer pairs, in the Josephson effect.[5]

In the original model[1] P(E) was calculated from the Zener tunneling probability, $P(E) = \exp[-E_o/E]$, where

$$E_o = \frac{\pi(\hbar\omega_P)^2}{4v_F\hbar e^{*}} = \frac{\hbar\omega_P}{L_c e^{*}} \tag{7}$$

and where L_c is twice the Pippard coherence distance, ξ_o, or

$$L_c = 2\xi_o = 4v_F/(\pi\omega_P) \approx v_F/\omega_P \quad . \tag{8}$$

When the theory was first proposed (1979) little was known about the frequency response. In 1980, after the first measurements of $\sigma(\omega)$ were made by Grüner and associates at UCLA, it became evident that much smaller values of L_c were required than given by (8). In 1982 Wonneberger[6] showed that in addition to replacing e by e^{*}, it is necessary to take into account the effect of dissipation of wave vector to the macroscopically occupied phonon modes. By an extension of the Caldeira--Leggett theory of the effects of dissipation on tunneling rate, Wonneberger showed that the effect is to replace $2\xi_o$ by a length, L_c, of order c_o/ω_P. This revised theory led to a value of E_o given by an expression very similar to that derived by Maki[3] for the creation of a soliton, anti-soliton pair by tunneling in an electric field.

The length $L_c = c_o/\omega_P$ is the distance of phase-coherence in the chain direction in the Fukuyama-Lee-Rice[7,8] theory of weak pinning. In the pinned state the phase is adjusted in regions of volume proportional to

L_c^3 to maximize the pinning energy from impurity fluctuations. The phase may be adjusted by adding to the phase $2k_F x$ of a uniform CDW functions of the form

$$\phi_A = \frac{\pi}{2} \sin(\frac{\pi x}{L_d} + \phi_o) \tag{9a}$$

$$\phi_B = \pi - \frac{\pi}{2} \sin(\frac{\pi x}{L_d} + \phi_o) \tag{9b}$$

where ϕ_o is chosen to give maximum pinning energy. Here $L_d = \pi L_c$ corresponds to the length of the Lee-Rice domain and $2L_d = 2\pi c_o/\omega_p$.

In the example shown in Fig. 1, $\phi_o = 0$ and it is desired to have the phase $\phi = -\pi/2 \pmod{2\pi}$ at $x = -L_d/2$ and equal to $\pi/2 \pmod{2\pi}$ at $x = L_d/2$. This can be done with either ϕ_A or ϕ_B, which have opposite signs of charge (proportional to $\partial\phi/\partial x$).

When the CDW moves with a drift velocity, v_d, the phase ϕ goes to $2k_F(x-v_d t) = 2k_F x - \omega_d t$, where $\omega_d = 2k_F v_d$. As ϕ changes monotonically in time with current flow the phase alternates between A and B type solutions. This may be done, for example, with phason solutions of the form (valid for $\omega_d = \omega_p$):

$$\phi = -\omega_d t + \frac{\pi}{2} \sin\frac{\pi x}{L_d} \cos\omega_d t \; , \tag{10}$$

so that $\phi = \phi_A$ for $\omega_d t = 0 \pmod{2\pi}$ and $\phi = \phi_B$ for $\omega_d t = \pi \pmod{2\pi}$. The pinning energy is a minimum for integral values of π and is zero for half-odd integral values. The narrow band noise should have a period π, as in the theory of Barnes and Zawadowski.[5]

The phase variations in (9a) and (9b) as well as (10) are highly idealized, but suggest a model in which π-solitons (or π-phase kinks) of alternating sign are trapped by impurity fluctuations. The distance L_d between phase kinks would not have to have a fixed value but could have a distribution of values dependent on impurity fluctuations. A π-soliton of positive sign adds half a state (per chain) and one of negative sign subtracts half a state, as illustrated in Fig. 1.

Solitons may be regarded as corresponding to electron-hole quasiparticles in a semiconductor model of particles of mass $M_s = (mM)^{1/2}$ and Fermi velocity $c_o = (m/M)^{1/2} v_F$, with an energy gap at the Fermi surface

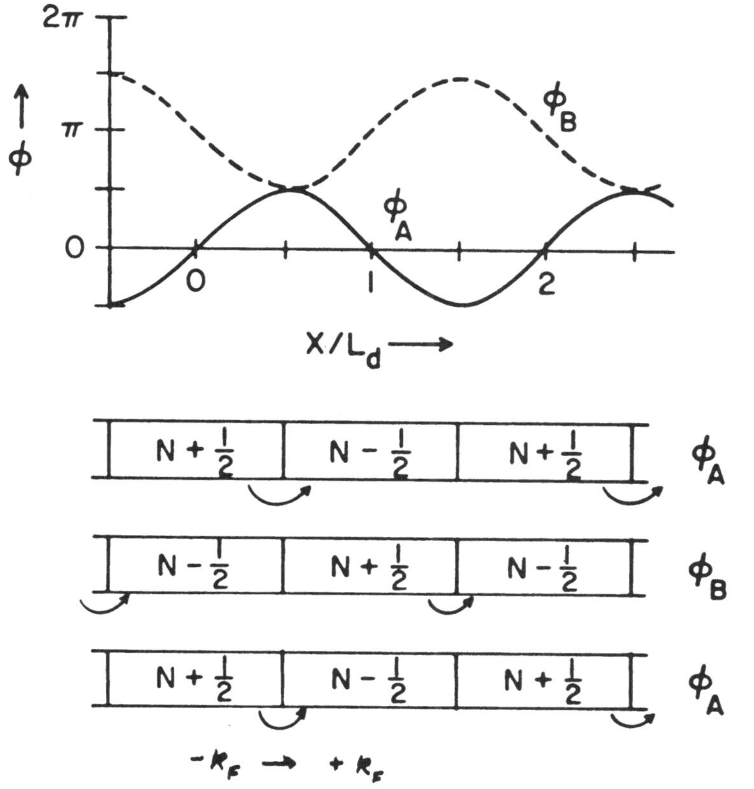

Fig. 1.
The proper phase to minimize the pinning energy can be fixed by π-solitons of alternating sign separated by an average spacing L_d. *In the example shown, the phase is fixed at* $\pi/2$ *for* $x/L_d = 1/2, 5/2$, *etc. and at* $-\pi/2$ *for* $x/L_d = -1/2, 3/2$, *etc. The phase can be adjusted to these values (mod* 2π*) by π--solitons of either sign, as shown by the functions* ϕ_A *and* ϕ_B. *When a current flows, the solutions alternate between* ϕ_A *and* ϕ_B *for every change of π in phase. A tunneling event removes an electron with wave vector* $-k_F$ *from one domain and places it with wave vector* $+k_F$ *in an adjacent domain, so that in effect* k_F *is added to the wave vectors in each of the two domains.*
(Reprinted from reference 4.)

$(-k_{Fx}, +k_{Fx})$, of $2E_\phi = (M/m)^{1/2}\hbar\omega_P$, where E_ϕ is the soliton energy. Acceleration of the system occurs by Zener tunneling across the semiconducting gap in an electric field.

An individual tunnel event that creates a π-soliton-antisoliton pair in the field in a region L_d is equivalent to adding a wave vector k_F to electrons of one spin or $2k_F$ to both spins in L_d for a given (k_y, k_z) . The wave vector $q_s = M_s v_d/\hbar$ is changed by:

$$\delta q_s = \pi/L_d \ . \tag{11}$$

The value of the drift frequency $\omega_d = 2k_F v_d = 2c_o q_s$ is changed by $2c_o\delta q_s = 2\pi c_o/L_d = 2\omega_F$. Thus if n_t is the number of tunnel events,

$$2v_F \frac{dq}{dt} = \frac{d\omega_d}{dt} = 2\omega_P \frac{dn_t}{dt} = \frac{2}{\hbar} e^* E \ v_F P(E) \ . \tag{12}$$

The total kinetic energy per spin in $2L_d$, the sum of $\frac{1}{2} M(\delta v_d)^2$ over the electrons, is the gap $(M/m)^{1/2}\hbar\omega_P$. With N coherent parallel chains, the change δq_s in a tunnel event is divided by N, or is $\delta q_s = \pi/(L_d N)$. Thus ω_d can be much less than ω_P .

If a π-soliton-antisoliton pair is created in L_d and the two are moving in opposite directions with the phason velocity, c_o, the current (both spins) is $2ec_o/L_d$. The density of electrons (both spins) is $2k_F/\pi$ per chain. The current should be the same whether it is calculated from soliton motion or from the drift velocity of the charge-density wave, δv_d . This requires that

$$\frac{2k_F \delta v_d}{\pi} = \frac{2c_o}{L_d} \tag{13}$$

or that $\hbar\delta q_s = m_s \delta v_d = \pi\hbar/L_d$, as in (11).

The tunneling probability P(E) for the event described in the preceding paragraph is the same as that for the creation of a pair of π-solitons in an electric field on a single chain as given by Maki,[3] with

$$E_o = \frac{\pi (M/m)(\hbar\omega_P)^2}{4e \ c_o \hbar \ e} \ . \tag{14}$$

This expression for E_o is the same as (7) with v_F replaced by c_o .

For both $NbSe_3$ and TaS_3, values of $L_c \sim c_o/\omega_P$ are consistent both with the maximum of the conductivity observed at high fields of frequencies as the tunneling probability $P \to 1$ and with the pinning frequency ω_P derived from the scaling relation between field and frequency in the photon-assisted tunneling theory. The drift velocity in the relaxation time approximation at high fields is $v_d = e^*\tau^*E/m$, where $e^*\tau^* \simeq e\tau$, the value expected in the absence of CDW formation. The conductance per chain,

$$\sigma_c = \frac{nev_d}{E} = \frac{2k_F}{\pi}\,\frac{e\,e^*\tau^*}{m} \quad , \tag{15}$$

may then be related to the drift frequency.

The drift frequency of the CDW, $\omega_d = 2k_F v_d$, is given by

$$\hbar\omega_d = e^* L_c E P(E) \quad , \tag{16}$$

so that as $P \to 1$,

$$e^* L_c = \frac{\hbar\omega_d}{E} = \frac{\pi\hbar\sigma_c}{e} \quad . \tag{17}$$

For $NbSe_3$ below T_1, the area per chain is about $25\text{x}10^{-16}$ cm^2 and room temperature bulk conductivity about $6\text{x}10^3$ $(\text{ohm cm})^{-1}$, giving $e^*L_cE \sim$ $\sim 3\text{x}10^{-19}$ ergs for $E \sim 1v/m$. This is consistent with $e^*/e = 10^{-3}$ and $L_c = 2\text{x}10^{-4}$ cm. In the relaxation time approximation, $L_c = 2v_F\tau^*$, and the value of τ^* is consistent with that derived from measurements of Reagor et al.[9]

If $L_c = c_o/\omega_P$, a simple relation may be derived between E_o and ω_P^2 :

$$\hbar\omega_P = e^* E_o L_c = e^* E_o (c_o/\omega_P) \quad , \tag{18}$$

or

$$\hbar\omega_P^2 = e^* E_o c_o \quad . \tag{19}$$

With $e^*/e = 10^{-3}$ and $c_o \cong 3\text{x}10^5$ cm/sec (corresponding to $v_F = 10^7$ cm/ /sec), the relation may be expressed in the form:

$$\nu_P = 100\,\sqrt{E_o}\ \text{MHz}, \quad E_o \text{ in V/cm} \quad . \tag{20}$$

This relation is found to hold approximately for ortho-TaS_3 and for $NbSe_3$ below both T_1 and T_2. From the scaling relation, $\sigma_{CDW}(\omega) \sim P(\omega) \sim$

~ $\exp(-\omega_p/\omega)$, so that ω_p can be estimated from the frequency for which $\sigma_{CDW}(\omega)$ drops to e^{-1} of its maximum value.

The main evidence for the tunneling model comes from application of photon-assisted tunneling theory to derive the frequency, bias and amplitude dependence of detection, mixing, harmonic mixing and harmonic generation from the observed nonlinear dc I-V characteristic and the scaling parameter derived from $\sigma_{CDW}(\omega)$. In the small signal limit, classical derivatives are replaced by quantum finite differences in which $\hbar\omega$ is taken as a quantum of energy. The scaling is $\hbar\omega = e^*EL_c$. In all cases, covering a wide range of parameter space, there is good semi-quantitative agreement between the theory and experiment and in some cases there is good quantitative agreement. In general, the results differ qualitatively from those expected from classical models[11] in which changes with frequency must come from distributions of relaxation times or of frequencies rather than quantum effects.

The main discrepancies occur when the bias is near threshold or below. It is found that in this region the response at low frequencies is considerably less than that expected. The expected ac signals may be regarded as current generators and it appears that these signals face an additional impedance that must be overcome to appear in the external circuit. In $NbSe_3$ and ortho-TaS_3, the impedance is bypassed at frequencies well below the pinning frequency.

REFERENCES

1. John Bardeen, Phys. Rev. Lett. 42 (1979) 1498; ibid 45 (1980); John Bardeen, Proceedings of the International School of Physics "Enrico Fermi," Varenna, Italy, (1983) to be published in Nuovo Cimento.
2. A.I. Larkin and P.A. Lee, Phys. Rev. B17 (1978) 1596.
3. K. Maki, Phys. Rev. Lett. 39 (1977) 46; Phys. Rev. B18 (1978) 1641.
4. For another discussion of the soliton model from a somewhat different point of view, see John Bardeen "Soliton Theory of Charge-Density Wave Depinning," Proceedings "International Conference on Low Temperature Physics - LT17," to be published in Physica B.

5. S.E. Barnes and A. Zawadowski, Phys. Rev. Lett. 51 (1983) 1003.

6. W. Wonneberger, Z. Phys. B50 (1983) 23.

7. H. Fukuyama and P.A. Lee, Phys. Rev. B17 (1978) 535.

8. P.A. Lee and M. Rice, Phys. Rev. B19 (1979) 3970.

9. David Reagor, S. Sridhar and G. Grüner "Internal Dynamics of CDW Transport in $NbSe_3$," these proceedings.

10. J. H. Miller, Jr., J. Richard, J.R. Tucker and John Bardeen, Phys. Rev. Lett. 51 (1983) 1592; J.H. Miller, Jr., J. Richard, R.E. Thorne, W.G. Lyons, J.R. Tucker and John Bardeen, Phys. Rev. B29 (1984) 2328 and to be published.

11. L. Sneddon, M.C. Cross and D.S. Fisher, Phys. Rev. Lett. 49 (1982) 292; L. Sneddon, Phys. Rev. B29 (1984) 719, 725; invited talks by D.S. Fisher and by L. Sneddon, these proceedings.

DYNAMICS OF INCOMMENSURATE STRUCTURES

Leigh Sneddon

Martin Fisher School of Physics, Brandeis University,
Waltham, MA 02254 USA

AC and DC dynamical properties of the incommensurate chain are determined by first reducing the DC dynamics to a purely static problem. The moving system is described by a static hull function which becomes singular, above the critical pinning strength, as the velocity approaches zero.

The AC/DC interference effects observed in CDW experiments are reproduced surprisingly well. The presence of sharp interference features in $NbSe_3$ is seen to depend on the screening effects of uncondensed electrons, while the qualitatively different behavior of TaS_3 is seen, for the first time, to be due to long-range Coulomb interactions in the CDW.

The observed scaling of field- and frequency-dependent conductivities is seen to occur in this classical model so that this effect can no longer be regarded as evidence of quantum tunnelling.

The case of infinite range interactions is solved exactly, using both analytic and graphical techniques. The ground states and all metastable states are identified. The AC response has a low-frequency singularity at threshold, but the dielectric constant is bounded, as seen in CDW experiments. The solution is also presented for the depinning transition; the sliding threshold; and the excitation spectra.

The discovery of electrical conduction due to sliding charge-density waves raised a wide range of questions concerning the dynamics of sliding incommensurate structures. This article reports progress in the analytic study of such dynamics and in the understanding of related experimental results.

The incommensurate systems studied here are extensions of the model of Frenkel and Kontorova,[1] and the dimensionless equations of motion can be written

$$\dot{u}_j = -\sum_p D_p u_{j-p} + f + P\sin(Hj + u_j), \tag{1}$$

where H is the lattice spacing, P is the strength of the pinning force, which has period 2π, and $H/2\pi$ is irrational.

It is known[2] that the deformations of a <u>stationary</u> incommensurate chain with nearest neighbor interactions can be written

$$u_j = \alpha + g(Hj + \alpha) \tag{2}$$

where g(x) is periodic: $g(x + 2\pi) = g(x)$, and α can be chosen to be the center of mass coordinate of the chain. Does this form also describe chains with more general interactions and, more importantly, does it describe moving chains? The answer, for dc motion, i.e. $\dot{\alpha} = v$, a constant, is "yes".

To see this we search for a solution to (1) of the form (2). For constant v such a solution is clearly supplied by the periodic solution to the nonlinear differential difference equation:

$$v\left(1 + \frac{dg}{dx}\right) = - \sum_p D_p g(x - Hp) + f + P\sin(x + g(x)) \qquad (3a)$$

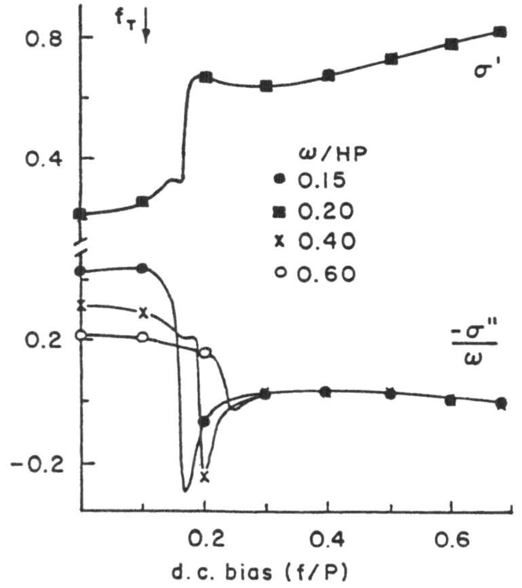

FIG. 1. ac response of sixfold-coordinated incommensurate chain, showing sharp interference features. The current, v, at which each feature occurs, depends linearly on frequency, ω, as seen in CDW experiments (Ref. 13).

This equation was solved in two ways. The first method[3] is to Fourier transform the x-variable to obtain a sequence of coupled equations for the Fourier components of the periodic function g(x). Truncating to a finite number, μ_{max}, of Fourier components and solving numerically then gives accurate results at all but the smallest values of the velocity, v. The procedure has the useful feature of representing the dc dynamic state by a static set of Fourier components. This means that the response to an ac perturbation on the sliding state can be determined using conventional linear response theory, without any numerical integration.

These solutions were tested in two ways. Firstly, the dc characteristic was calculated. With weak pinning, clear convergence, with increasing μ_{max}, to a linear response at f = 0 was found, with no threshold. For strong pinning the emergence of a threshold singularity with increasing μ_{max}, was clearly indicated. The results thus agree with present knowledge[2] at low velocities (and can readily be seen to be correct to all orders in perturbation theory at moderate and high velocities). Secondly, the solutions were tested for stability to small perturbations and were found to be dynamically stable.

To examine, in addition, the low velocity limit, equation (3a) was also solved exactly[4] for the case of infinite range interactions.

Exploiting the translational invariance of the infinite commensurate chain has thus allowed[3,4] the dc dynamics to be transformed to a purely static problem.

It is found that, in a moving system with strong pinning, a breaking of analyticity transition occurs in the new hull function, g, as the velocity approaches zero at threshold. The complicated time dependence of the $u_j(t)$ near threshold is expressed, by (2), completely in terms of the emergence of singularities in this new hull function.

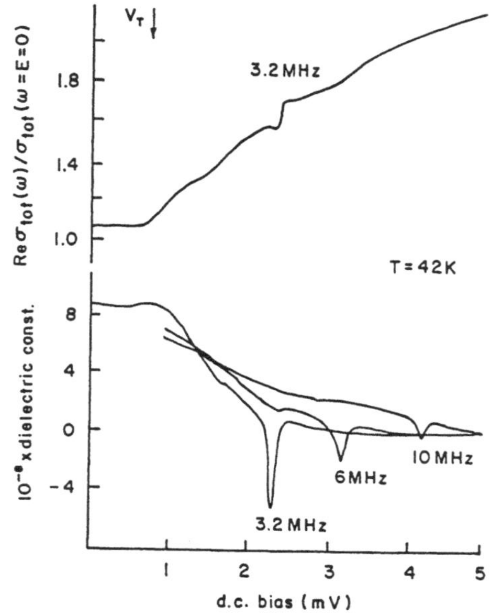

FIG. 2. ac response (Ref. 5) of $NbSe_3$.

Linear ac response, in the presence of a dc field, has been studied experimentally[5] in the CDW systems $NbSe_3$ and TaS_3. At fields a few times threshold low order perturbation theory[6] is not useful; but this region is the most commonly studied experimentally because the nonlinear effects are larger than in the high field region, and sample heating is not a problem. The Fourier truncation techniques were therefore used to determine the ac response of the sliding incommensurate chain near threshold.

The CDW's in $NbSe_3$ and TaS_3 are three-dimensionally coherent. One effect of higher dimensionality is to increase the coordination of the system. To mimic this increased coordination crudely, a six-fold coordinated chain was considered with $-D_{\pm 1} = -D_{\pm 2} = -D_{\pm 3} = 1/3$; $D_p = 0$, $|p| > 3$.

By considering a small perturbation about a static dc solution, the ac response, $\sigma(\omega) = \sigma' + i\sigma''$ was determined, for $H/2\pi = (\sqrt{5} + 1)/2$ plus any integer, and $P = 3.0$. The results (with $\mu_{max} = 15$) for σ' and the dielectric response $-\sigma''/\omega$ are shown in Fig. 1. The basic features in Figs. 1, 3 and 5 are preserved with increasing μ_{max}. The threshold force was estimated from the dc results.

Fig. 2 shows experimental results for $Re\sigma(\omega)$ and $\varepsilon(\omega)$ of the sliding charge density wave in $NbSe_3$. Fig. 1 is seen to account well for the voltage- and frequency-dependence of both components of the ac response. This may not have been expected since CDW dynamics are dominated by randomly positioned defects while the chain is in a periodic potential.

In experiments[5] performed on TaS_3 at 130K, the sharp interference features seen with $NbSe_3$ (Figs. 1 and 2) were not observed. TaS_3 becomes a semiconductor below the CDW transition, while $NbSe_3$ is metallic. At 130K the conductivity of TaS_3 has fallen 2 orders of magnitude from its value at the transition. As discussed earlier,[7] this reduces the screening capacity of the normal electrons and can allow long range Coulomb interactions of the CDW with itself.

The sliding dynamics of equation (1) with long range interactions: $-D_p = 2/N$ for all $p \neq 0$, was therefore determined. The results (with $\mu_{max} = 20$) are shown in Fig. 3, and can be compared with the experimental results in Fig. 4. Not only does including long range interactions account for the absence of interference features, but the properties of the incommensurate chain are seen to match those of TaS_3 extremely well. The <u>difference</u> between the a.c. properties of $NbSe_3$ and TaS_3 can now be understood for the first time, as being due to the presence in TaS_3, as suggested earlier,[7] of long range Coulomb interactions of the CDW with itself.

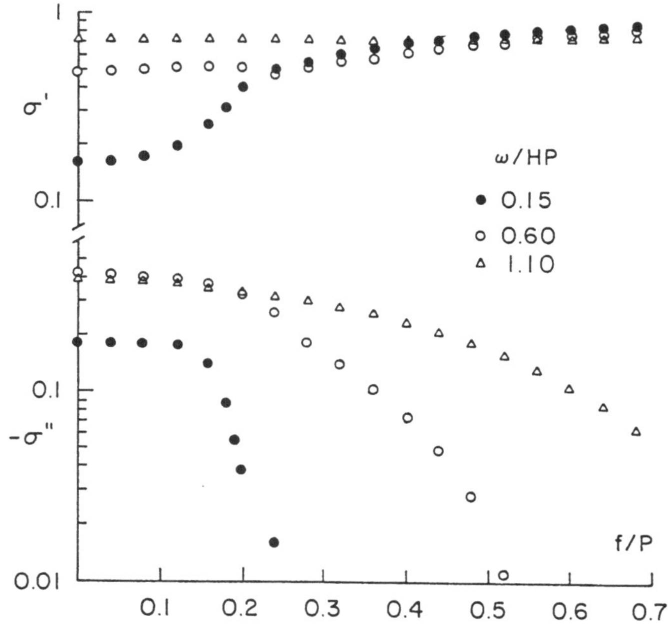

FIG. 3. ac response of incommensurate chain with infinite-range interactions; c.f. Fig. 4 and text.

The ac response was also determined with f = 0, and compared to the dc conductivity v/f. The results (with μ_{max} = 20) are shown in Fig. 5 for long range interactions. Similar results were obtained for the six-fold coordinated chain. The experimentally observed[8,9] scaling, of field and frequency-dependent conductivities, is thus exhibited by this classical model, and can no longer be regarded[9] as evidence for a quantum mechanical theory of CDW conductivity.

It is interesting to speculate that the detailed form of the potential becomes less important as one approaches threshold. In any case, the comparison of theory with experiment seen in Figs. 1 - 5 shows that, in fields comparable to threshold, the incommensurate chain gives a much better picture of CDW dynamics than might have been suspected.

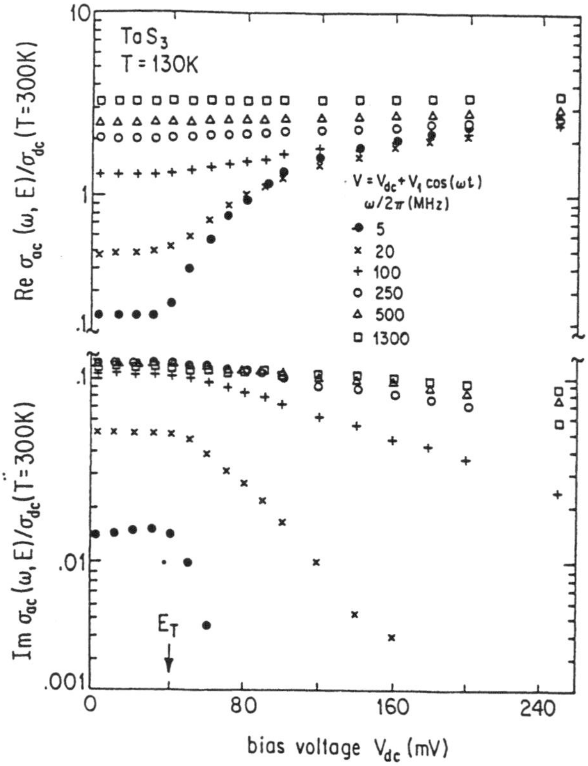

FIG. 4. ac response (Ref. 5) of TaS_3.

For the case of infinite range interactions an exact solution was also obtained. In this case (3a) reduces to a nonlinear differential equation

$$v(1 + \frac{dg}{dx}) = P\sin(x + g(x)) - g \tag{3b}$$

where the applied force is given by

$$f = - \frac{1}{2\pi} \int_{-\pi}^{\pi} g(x)dx \tag{4}$$

Fisher[10] has also studied a more general problem where P is replaced by a randomly distributed variable. For fixed P, however, a simple solution is possible and some new results can be obtained.

When $v = 0$, (6) is a transcendental equation which can be solved graphically. For $P < 1$ there is a unique solution (Fig. 6a). It is continuous and odd so that, using (4), when $v = 0$, $f = 0$, and there is no sliding threshold. For $P > 1$ there are multiple, discontinuous solutions, g, many with non-zero means (Fig. 6b, c). The threshold force is clearly

$$f_T = \max_{\{g\}} - (2\pi)^{-1} \int_{-\pi}^{\pi} g(x)dx \; . \qquad (5)$$

Thus the critical value of P defining the depinning transition below which the sliding threshold and multiplicity of solutions disappear, is immediately seen to be $P_c = 1$.

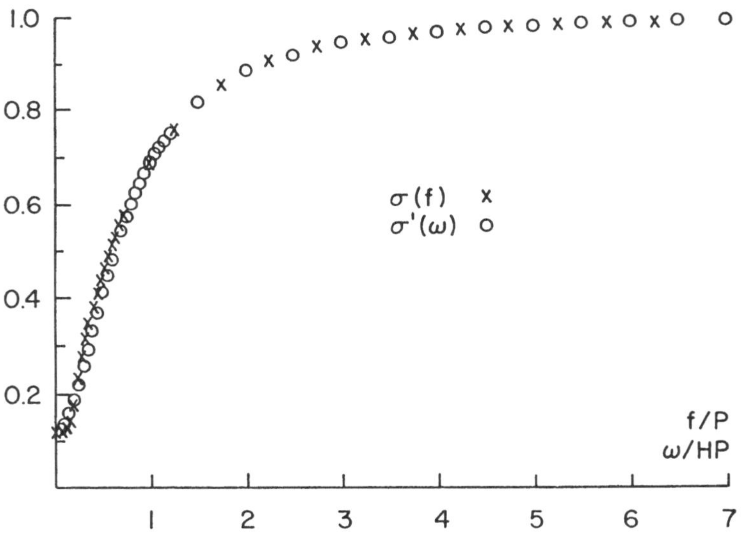

FIG. 5. Scaling of field-dependent (crosses) and frequency-dependent (circles) conductivities in the classical chain.

The threshold force can be determined graphically, using (5), for all P. Further, in the limit $P \to P_c^+$, $f_T \alpha (P - P_c)^{\psi_T}$, where $\psi_T = 2$. It is also immediately clear from Fig. 6b that at $f = f_T$ there is only one stationary state,[10] g_T.

Turning to dc dynamics, $\alpha = vt$, $v \neq 0$ and g is continuous. As $v \to 0$, g will approach g_T as $f \to f_T$. For x away from the critical value x_T, $g - g_T = O(v)$. In the vicinity of x_T, however, putting $x = x_T + y$; $x + g(x) = x_T + g_T(x_T^-) + h(y)$ and considering (3b) in the limit of small y and h(y) gives $v\, dh/dy = y + ah^2$ where $a = -g_T(x_T^-)/2$. Transforming by

$$dw/dz = -\alpha h(\beta z)\, w(z),$$

where $\alpha = (a^2/v)^{1/3}$ and $\beta = (v^2/a)^{1/3}$, gives $w'' = -zw(z)$, the solutions of which are the Airy functions $Ai(-z)$ and $Bi(-z)$. One finds then that the limiting value of $y = x - x_T$ for finite h, as $v \to 0$, is $(v^2/a)^{1/3}z_0$ where z_0 is the first zero of $Ai(-z)$. This result is seen graphically to give a dominant contribution $\sim v^{2/3}$ to $f - f_T$ in (4). Thus $v = B(f - f_T)^{3/2}$ where, as $P \to P_c^+$, $B\alpha(P - P_c)^{-1/2}$. Thus the depinning transition, and the 3/2 threshold exponent with a coefficient which diverges as $P \to P_c^+$, in agreement with Ref. 10, can be obtained quite straightforwardly for the incommensurate chain.

Further, it is possible to determine the energy and stability of each stationary state and thus specify, for $P > P_c$, which is the ground state, which are the metastable states, and which are the unstable states. The energy corresponding to equation (1) is $H = \sum_j P\cos(Hj + u_j) + (4N)^{-1} \times \sum_{ij}(u_i - u_j)^2$. Using (2) and choosing g to minimize H shows that the ground state has a single discontinuity, which moves from 0 to x_T as f increases from 0 to f_T (Fig. 1b).

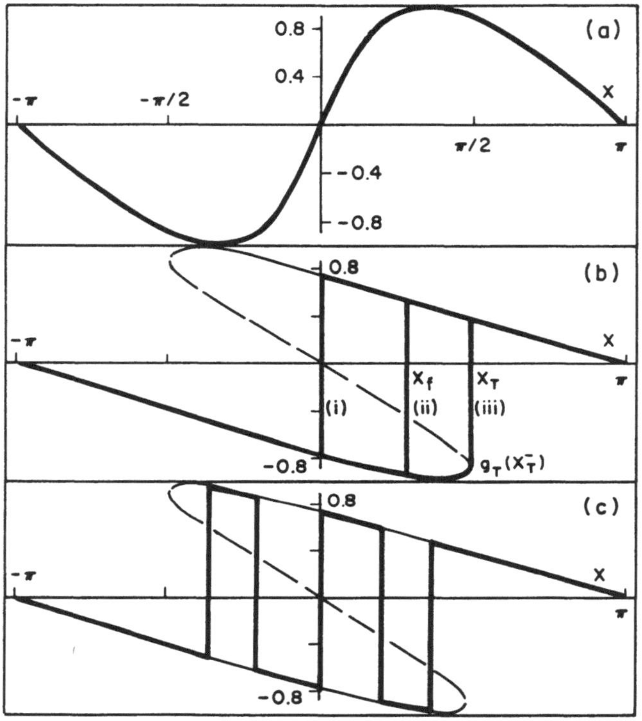

FIG. 6. Stationary states g(x). (a) P=0.5. (b) P=3, ground states for f=0, (i); $0<f<f_T$, (ii); $f=f_T$, $g=g_T$, (iii); (c) P=3, a metastable state.

An exact linear stability analysis was performed and it was found that any g which occupies a finite part of the middle branch (dashed line in Fig. 6) will be unstable. Further, it was shown that any state g which avoids the "unstable" middle branch is locally stable. An example of a metastable state is shown in Fig. 6c.

Since the Green's function is known, the exact density of relaxational [(1) is massless] excitations, $\rho(\lambda)$ can be determined in the usual way. The results are shown in Fig. 7. The isolated excitation at $\lambda = 0$ for $P < P_c$ is the sliding mode of the unpinned chain.

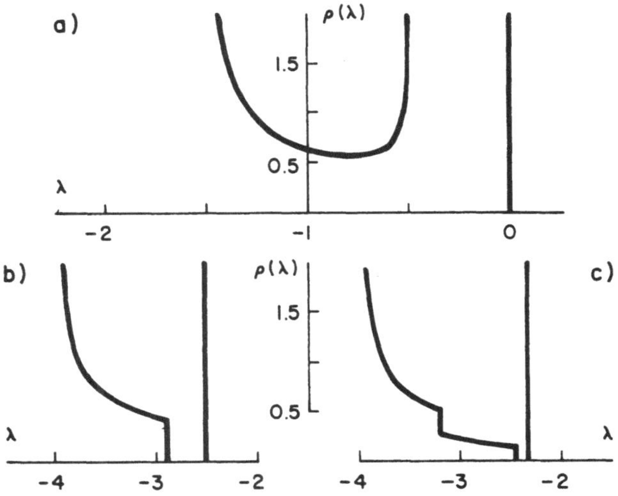

FIG. 7. Excitation spectra $\rho(\lambda)$; (a) P=0.5, f=0; (b) P=3, f=0; (c) P=3, $0<f<f_T$.

Finally, knowing the Green's function gives the exact linear ac response of the pinned lattice. The result may be compared with CDW experiments for both the real and imaginary parts of $\sigma(\omega)$ for all $|f| < f_T$ and for all ω. The single particle model[11] is easily seen to have the property $\lim_{\omega \to 0} - \mathrm{Im}\sigma/\omega \to \infty$ as $f \to f_T$. No such divergence has been observed experimentally. As $\omega \to 0$ and $f \to f_T$, the exact solution gives a dielectric constant, $\varepsilon = i\sigma/\omega$, which remains finite near f_T even as $\omega \to 0$. The entire spectral weight of the single particle model is at one frequency which approaches zero at threshold. The pinned many-body system considered here has a broad spectrum of excitations, only one edge of which approaches zero at threshold. This is the first explicit demonstration that in such a case the singularity in σ is weaker than in the single particle case and ε can remain finite, as is observed experimentally.[5]

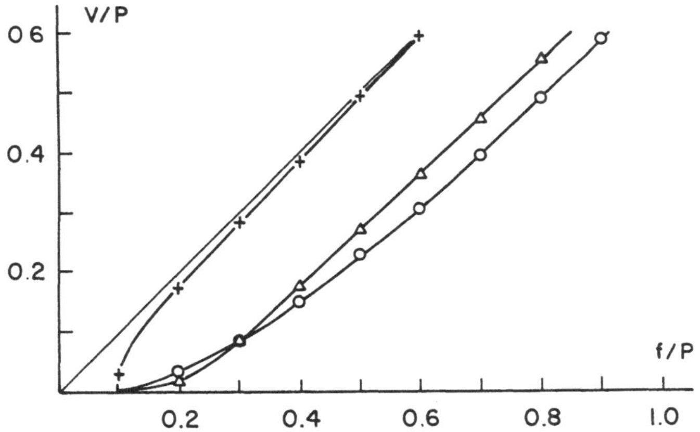

FIG. 8. dc characteristic for P=1.5(O); also (with matched thresholds and large v slopes) for single-particle model (Ref. 11) (+) and $NbSe_3$ CDW (Ref.12) (Δ).

For constant $f > f_T$ outside the threshold region, the

truncation procedure[3] mentioned earlier gives an accurate solution of equation (1), including the response to small ac perturbations. As seen for example in Fig. 8, this combines with the present results for $f \lesssim f_T$ to give a complete steady state solution of this system.

The author thanks E. Gross for stimulating discussions and the Air Force Office of Scientific Research for support under Grant No. 84-0014.

REFERENCES

1. Y. I. Frenkel and T. Kontorova, Zh.E.T.F. 8, 1340 (1938).
2. M. Peyrard, S. Aubry, J. Phys. C16, 1593 (1983).
3. L. Sneddon, Phys. Rev. Lett. 52, 65 (1984).
4. L. Sneddon, Phys. Rev. B30 (Rapid Communications) Sept. 1 (1984).
5. A. Zettl, G. Grüner, Phys. Rev. B29, 755 (1984).
6. L. Sneddon, M. Cross, D. Fisher, Phys. Rev. Lett. 49, 292 (1982).
7. L. Sneddon, Phys. Rev. B29, 719 (1984).
8. G. Grüner, Molec. Cryst. Liq. Cryst. 81, 17 (1982), A. Zettl, et. al. Phys. Rev. B26, 5773 (1982), G. Grüner et. al. Phys. Rev. B24, 7247 (1981).
9. J. Bardeen, Phys. Rev. Lett. 45, 1978 (1980), A. Zettl and G. Grüner, Phys. Rev. B25, 2081 (1982), J. Bardeen, Mol. Cryst. Liq. Cryst. 81, 1 (1982), J. H. Miller, J. Richard, J. R. Tucker, J. Bardeen, Phys. Rev. Lett. 51, 1592 (1983).
10. D. S. Fisher, Phys. Rev. Lett. 50, 1486 (1983); and unpublished.
11. G. Grüner, A. Zawadowski, P. Chaikin, Phys. Rev. Lett. 46, 511 (1981).
12. R. Fleming, Phys. Rev. B22, 5606 (1980).
13. A. Zettl, private communication.

SOME PROBLEMS ARISING FROM ELECTROSTATIC POTENTIAL IN CDW BEHAVIOR

M. Renard

Centre de Recherches sur les Très Basses Températures, C.N.R.S., B.P. 166 X, 38042 Grenoble cedex, France.

Charge conservation property implies that if the current carried by the motion of a CDW is just

$$j = - n_o e v$$

charges have to be considered as completely liked to the phase. A deformation of the phase has to be associated with an uncompensated charge. The consequences would be the replacement of the soft phason mode by a plasmon like mode, and of the elastic response by an electrostically screened one. Preliminary results on a microscopic calculation, show that the phason exists and does not carry a K component for the density of charges.

1°) PHENOMENOLOGICAL APPROACH

It is generally admitted that in the case of motion of a charge density wave the associated current j, n_o being the number of electrons in the band, and v the velocity of the wave, is

$$j_{CDW} = - n_o e v \qquad (1)$$

At finite temperature some excitation counter-current may change the n_o value. Let us call ϕ the departure of the local phase ψ , from the ideal value $(q_o r)$:

$$\psi = \phi(r,t) + q_o r \qquad (2)$$

(1) is equivalent to :

$$j_{CDW} = + \frac{n_o e}{q_o} \frac{\partial \phi}{\partial t} \qquad (3)$$

If ϕ behaves as a well defined function, i.e. if its motion is conservative, in the case of a distorted ϕ , one must have

$$\rho_{CDW} = - \frac{n_o e}{q_o} \frac{\partial \phi}{\partial x} \qquad (4)$$

In order to save electric charge conservation :

$$\frac{\partial j}{\partial x} + \frac{\partial \phi}{\partial t} = 0$$

Equation (4) is consistent with the fact that $q_o = 2 k_F$ is given by the number of electrons. For a slowly varying ϕ , we can define a local q value by :

$$q = q_o + \frac{\partial \phi}{\partial x}$$

and since the gap appears at $\frac{q}{2}$, we have a local number of electrons n such that :

(5) $$\left(\frac{n}{n_o}\right) = \frac{q}{q_o} = 1 + \frac{1}{q_o}\frac{\partial\varphi}{\partial x}$$

which is equivalent to (4) : in other words, everything looks consistent with the assumption that the band charge is strictly liked with the phase of CDW.

This assumption may be relaxed if for example ϕ does not behave as a true function. This would be the case if for example one wavelength of CDW was suddenly evaporated, to give one more electronic excitation, without the corresponding hole. This process conserves the charge, but is highly unprobable unless $\Delta \to 0$, since it corresponds to a change in the q value of $\frac{4\Pi}{L}$, and the corresponding wave functions are :

Before evaporation :

$$\frac{1}{\sqrt{2}}\left[|k_F> + |-k_F>\right] \prod_{k<k_F-\frac{2\pi}{L}} \left[v_k|k> + u_k|k-q>\right] = \psi_1$$

After evaporation :

$$[\varphi_{\text{Excitation}}] \prod_{k<k_F-\frac{2\pi}{L}} \left[v_k|k> + u_k|k-q+\frac{4\pi}{L}>\right] = \psi_2$$

A phonon field may well change the momentum and energy of the condensed $\pm k_F$ state to give the $\varphi_{\text{excitation}}$. But the two infinite products π_1, and π_2 are such that:

$$<\pi_1|\pi_2> = \left[\left(\prod_k(v_k^2) + \prod_k(u_k^2)\right)\right] = 0$$

This is 0 unless all $v_k^2 = 1$ for $k < k_F$, which is the case if $\Delta \to 0$.

2°) REMARKS ON PHASONS AND ELASTIC BEHAVIOR

Following Mac Millan we can write a phason mode as an expression for the electronic charge:

(6) $$\rho = \rho_o + \rho_q \cos\left[qr + \alpha\cos Kr\right]$$

$$K << q, \quad \alpha K << q$$

which corresponds to Fourier components of ρ_q over $(q\pm K)$ wave vectors. This expression is inconsistent with the preceeding remarks, which ask for a ρ_o modulation, with $\pm$ K components. Moreover Mac Millan claims that if $K \to 0$ the corresponding energy for the phase deformation tends to zero giving rise to a soft phason mode, and to an elastic behavior for this kind of deformation. Such an elastic behavior is essential to understand the pinning properties (2) (3).

If we believe that the phase is linked with the band charge, to the contrary, electrostatic interaction will give a plasma mode of finite frequency for K = O, and mechanical properties governed by a screening length due to competition between electric and elastic forces. This screening length is expected to be of the order of magnitude of the free electrons Thomas-Fermi length since kinetic energy is much higher than condensation energy per electron.

3°) MICROSCOPIC CONSIDERATIONS

The wave functions for the fundamental ψ or excited ϕ states in the CDW systems are linear combinations of $|k>$ and $|k + nq>$ states. If $\frac{\Delta}{E_F} << 1$ one may choose $n = \pm 1$.

A K component of a perturbational potential W, will mix ψ_k and ϕ_{k+K}, giving first order oscillations in ρ , with wave vectors $\pm$ (K $\pm$ nq).

So that we have to treat all together $\pm$ K, $\pm$ (K + nq) components of W, since the resulting electronic density, have the same components.

We have achieved a self consistent perturbational calculation with Z.Z. Wang, and are able to give some preliminary results.

The internal energy of the deformed system measured from the fundamental state, is expressed in the form :

$$E_{(K)} = \frac{1}{2} [A_1\rho_1^2 + A_2\rho_2^2 + A_3\rho_3^2]+B_{12}\,\rho_1\rho_2 \cos \phi_{12} + B_{13}\rho_1\rho_3 \cos \phi_{13} - B_{23}\rho_2\rho_3 \cos \phi_{23}$$

where :

$$\begin{cases} \rho_k = \rho_1\, e^{i\psi_1} \\ \rho_{k+q} = \rho_2\, e^{i\psi_2} \\ \rho_{k-q} = \rho_3\, e^{i\psi_3} \end{cases} \qquad \begin{cases} \phi_{12} = \psi_1 - \psi_2 \\ \phi_{13} = \psi_1 - \psi_3 \\ \phi_{13} = \psi_2 - \psi_3 \end{cases}$$

So that : $\phi_{13} = \phi_{12} + \phi_{23}$

The A_i and B_{ij} being positive functions of K, $A_1 \to \infty$ as K $\to$ O, the other coefficients keeping finite values.

So we get two kind of modes : (K $\to$ O)

- A nearly K pure divergent eigenmode with an ε_1 eigenvalue following a Thomas-Fermi screening law : $|k_s \sim 6\ \mathring{A}$ for $NbSe_3|$

$$\rho_K = \frac{-N(o)\ W_K}{1 + \dfrac{4\pi e^2\ N(o)}{\varepsilon_o k^2}}$$

- Two Mc Millan like modes ρ_{q+k} and ρ_{q-k} rich, with small admiture of $\rho_k \to 0$ as $K \to 0$ depending essentially on ρ_2, ρ_3, and due to the $(-B_{23}\rho_2\rho_2 \cos\phi_{23})$, having a low energy for $\phi_{23} = 0$: $\rho_{q-k} = \rho_{q+k}$: Phason mode, and a higher energy for $\phi_{23} = \pi$! Amplitude mode. An essential feature of the phason mode is that ε does not tend to zero as $K \to 0$. This is probably due to the vanishing admixture of divergent K state.

The amplitude of $\rho_2 \neq \rho_3$, and a good approximation for small K is

$$\varepsilon = \rho_2^2\ a_o\ (1 + L^2K^2)$$

L is a length of the order of three time the CDW wavelength (40 Å).

Coming back to equation (6)

$$\rho_2 = \rho_3 = \alpha\rho_q$$

For an arbitrary phase $\phi(r)$ with Fourier components $\phi_K = \alpha_k = \dfrac{\rho_2(k)}{\rho_q}$ we get a deformation energy

$$\varepsilon = \rho_q^2\ a_o\ \phi_k^2\ (1 + L^2K^2)$$

The associated generalized force is :

$$F_k = 2\ \rho_q^2\ a_o\ \phi_K\ (1 + L^2k^2)$$

In the absence of an external force, $\phi(r)$ obeys to the equation :

$$\phi - L^2\ \frac{\partial^2\phi}{\partial x^2} = 0$$

L appears as a screening length. The ϕ term may appear as curious, but its meaning is the difference between the local phase value, and the ideal value it would have, when extrapolated from unperturbed zones (equation 2).

Since for small K, the ρ_k component is vanishingly small, contrarily to the assumptions of the first part of this article, long wavelength deformations of the wave does not carry a charge. The only mode able to link with an external field is the translational Fröhlich mode. The charges appearing only on distances of the order of the Thomas-Fermi screening length for the corresponding metal, near the ends of the samples : in weak fields this corresponds to very high ε values as measured by Grüner (4).

REFERENCES

(1) W.L. Mc Millan, Phys. Rev. B 14 (76) 1496.
(2) P.A. Lee and T.M. Rice, Phys. Rev. B 19 (79) 3970.
(3) L. Sneddon, M.C. Cross and D.S. Fisher, Phys. Rev. Lett. (49) 1982, 292.
(4) G. Grüner, Physica, 8-D, 1983, 1.

THE SINGLE DOMAIN MODEL OF CHARGE-DENSITY WAVE TRANSPORT

R. A. Klemm, Mark O. Robbins, and J. R. Schrieffer†
Corporate Research Science Laboratories
Exxon Research and Engineering Company
Clinton Township, Route 22 East, Annandale NJ 08801 USA

The classical model of Efetov and Larkin and of Fukuyama and Lee for the pinning of the charge-density wave phase ϕ by impurities is employed to study the collective motion of the charge-density wave within a coherent region or domain. The static coherence length ξ_s of the order parameter in the presence of impurities is closely related to the Lee-Rice domain size, and is taken to be the natural cutoff in the problem. An equation of motion for the collective coordinate Δ is found by assuming dissipative dynamics, and by calculating the impurity pinning force $F(\Delta)$ within the adiabatic approximation. We find that $F(\Delta)$ can be written as a Fourier series in Δ, where the amplitudes A_n and phases θ_n of the nth term in the series depend strongly upon the particular impurity configuration. The probability functions for the A_n and θ_n values are found from correlation functions of $F(\Delta)$ using perturbation theory and by an iteration procedure. We find that the threshold fields for different directions are different, and that metastable states exist within a domain. The spectrum of narrow band noise is determined, and it is suggested how this work might be extended to include the coupling of domains. In addition, the model has been extended to include non-adiabatic corrections to F, which are found to be significant. This extended model gives I/V curves in agreement with low-temperature experiments on some crystals, and with the high-field and high frequency behavior of the time-averaged current.

A number of authors(1-9) have taken the classical approach to the dynamics of a charge-density wave (CDW), using an effective Hamiltonian in which the single electron coordinates are integrated out in the adiabatic approximation in favor of the classical phase field $\phi(\vec{r},t)$ of the CDW. The electron density is expressed in the form $\rho(\vec{r},t)=\rho_0+\rho_1\cos[\vec{Q}\cdot\vec{r}+\phi(\vec{r},t)]$, where $\lambda=2\pi/|\vec{Q}|$ is the CDW wavelength and fluctuations of the amplitude ρ_1 are presumed to be small. The validity of this scheme was discussed by Efetov and Larkin (EL)(1) and by Fukuyama and Lee(2).

A concept important in understanding CDW motion is the static correlation length ξ_s of the order parameter $\psi(\vec{r}) = \rho_1\exp[i\phi(\vec{r})]$. EL(1) found that the impurity average of $\psi(\vec{r})\psi^*(\vec{r}\,')$ is given by $\rho_1^2\exp[-|\vec{r}-\vec{r}\,'|/\xi_s]$, where ξ_s^{-1} is proportional to the concentration c of impurities. An equivalent definition of ξ_s was presented heuristically by Lee and Rice (LR)(3), who required that ξ_s be the size of a domain over which the root mean square fluctuation of the phase be of order 2π. Since phases are defined up to an integer multiple of 2π, they suggested that domains of size ξ_s should in some sense behave as dynamical objects. Presumably interactions with neighboring domains lead to the overall collective behavior of the system. This is not to imply that the domains have sharp boundaries. They are fuzzy regions over which the phase fluctuations are of order 2π.

In the classical model of EL and FL, the effective Hamiltonian is found to be of the form(1,2),

$$H=C\int\{(\dot{\phi})^2/2+(\nabla\phi)^2/2+V_0\delta n(\vec{r})\cos[\vec{Q}\cdot\vec{r}+\phi(\vec{r},t)]+e^*E\phi(\vec{r},t)\}d^3r, \qquad (1)$$

where we have chosen time units such that the speed of harmonic phase fluctuations (phasons) is unity. The constant $C= E_F k_F/(2\pi^4)$ where $E_F = h^2k_F^2/2m$ and k_F is the Fermi wave number. The impurity pinning potential V_0 has the units of length so that CV_0 is the actual pinning energy, typically 0.01-0.1 eV. The impurity density is given by $n(\vec{r}) = \Sigma_i\delta(\vec{r}-\vec{r}_i)$ with $\delta n(\vec{r}) = n(\vec{r})-c$. Finally, the last term in eq. (1) accounts for the coupling to an external electric field applied along the direction of the CDW wave vector $\vec{Q}$, $E=\vec{E}\cdot\vec{Q}$, with $e^*= -\rho_0 e/QC$, where $-e$ is the electronic charge.

The equation of motion for ϕ is

$$\ddot{\phi}+\dot{\phi}/\tau-\nabla^2\phi-V_0\delta n(\vec{r})\sin[\vec{Q}\cdot\vec{r}+\phi(\vec{r},t)] = e^*E(\vec{r},t) , \qquad (2)$$

where we have included the damping force $-\dot{\phi}/\tau$ to account for dissipative coupling to the electrons and phonons. For CDW frequencies $\omega \ll 10^8 \sec^{-1}$, the inertial term $\ddot{\phi}$ can be neglected compared to $\dot{\phi}/\tau$. In deriving Eq. (2), we have assumed there are no forces resulting from the sample boundary, i.e., $\int d\vec{s}\cdot\vec{\nabla}\phi=0$.

We define the collective coordinate Δ to be the spatial average of ϕ over a region of size ξ,

$$\Delta(t)\equiv\langle\phi\rangle_\xi=\int^\xi d^3r\phi/\int^\xi d^3r. \qquad (3)$$

By averaging eq. (2) over this region, we obtain

$$\dot{\Delta}/\tau = e^*E + F(t) \quad . \qquad (4)$$

We also have an equation for $\delta\phi= \phi-\Delta$,

$$\delta\dot{\phi}/\tau-\nabla^2\delta\phi = f(\vec{r},t)-F(t), \qquad (5)$$

where

$$f(\vec{r},t)=V_0\delta n(\vec{r})\sin[\vec{Q}\cdot\vec{r}+\Delta+\delta\phi] \qquad (6)$$

and

$$F(t) =\langle f(\vec{r},t)\rangle_\xi \quad . \qquad (7)$$

The assumed vanishing of $\int^{\xi} d\vec{s}\cdot\vec{\nabla}\phi$ implies that the CDW in this region does not experience forces from regions outside ξ. This is consistent with our approach of treating each region (or domain) as being independent to zeroth order, and then including forces between domains. It is these boundary terms which exert elastic forces between domains.

Before we proceed with the perturbation theory, we must examine its validity. Clearly, if $\delta\phi$ is larger than 2π, it is inappropriate to expand $\sin[\vec{q}\cdot\vec{r}+\Delta+\delta\phi]$ in powers of $\delta\phi$. We therefore limit the size ξ of the domain such that the mean square fluctuation of the phase $\delta\phi$ is of order $(2\pi)^2$,

$$\overline{\langle(\delta\phi)^2\rangle}_{\xi} = (2\pi)^2\gamma \quad , \tag{8}$$

where γ is a parameter of order unity. The above equation serves as a definition of the length ξ. We have performed the perturbation theory in the static limit, and find that only the diagram pictured in Fig. 1 contributes to $\overline{\langle(\delta\phi)^2\rangle}_{\xi}$. All higher order diagrams vanish for reasons that are similar to those given by EL. The result is

$$\xi = (4\pi)^3\gamma/(2cV_0^2). \tag{9}$$

Since the EL result for ξ_s is[1]

$$\xi_s = 16\pi/(cV_0^2) \quad , \tag{10}$$

we see that $\xi/\xi_s = 2\pi^2\gamma$, so that $\xi=\xi_s$ when $\gamma = (2\pi^2)^{-1}$.

Fig. 1. Bubble diagram of value B. The lines are Green's functions.

We now proceed with the calculation of the pinning force F using perturbation theory in the pinning potential V_0. We write

$$F = \frac{1}{V}\sum_{n=1}^{\infty} f_n(t)V_0^n, \tag{11}$$

where $V=4\pi\xi^3/3$ is the volume of a domain. To order V_0 we have

$$f_1 = V\langle\delta n(\vec{r})\sin(\vec{Q}\cdot\vec{r}+\Delta)\rangle_{\xi} = \sum_i \sin(\vec{Q}\cdot\vec{r}_i^{\,a}+\Delta) = A_{1a}\sin(\Delta+\theta_{1a}), \tag{12}$$

where

$$A_{1a} = \{\sum_{ij}\cos[\vec{Q}\cdot(\vec{r}_i^{\,a}-\vec{r}_j^{\,a})]\}^{1/2} \quad , \qquad \theta_{1a} = \tan^{-1}\left[\frac{\sum_i \sin\vec{Q}\cdot\vec{r}_i^{\,a}}{\sum_i \cos\vec{Q}\cdot\vec{r}_i^{\,a}}\right] \quad , \tag{13}$$

and the sums are over the impurities in the domain. We note that A_{1a} and θ_{1a} depend strongly upon the particular impurity configuration a of impurities. However, we may find a typical value for A_{1a} by averaging A_{1a}^2. We find $\overline{A_{1a}^2}=N_i$, where N_i is the number of impurities in the domain.

This particular number would be reliable if $\overline{A_{1a}^4} - (\overline{A_{1a}^2})^2 \ll (\overline{A_{1a}^2})^2$. However, the fluctuations in A_{1a}^2 are of the same order as the mean. Nevertheless, to this order in perturbation theory, one can readily obtain the probability distribution P for A_{1a} by calculating $\overline{A_{1a}^{2n}}$,

$$\overline{A_{1a}^{2n}} = n!N_i^n[1 + O(N_i^{-1})]. \tag{14}$$

This implies

$$P(A_{1a}) = \frac{2A_{1a}}{N_i} e^{-A_{1a}^2/N_i}, \tag{15}$$

which is the probability that a random walker in two dimensions is at the distance A_{1a} from the origin after N_i steps.

Similarly, it can be shown that to second order in V_0, the contribution f_2 to the adiabatic pinning force is given by

$$f_2 = A_{2a}\sin(2\Delta + \theta_{2a}). \tag{16}$$

To this order in perturbation theory, A_{2a}^2 equals one-half the diagram pictured in Fig. 1. The fluctuations in A_{2a}^2 are more complicated than the fluctuations in A_{1a}^2, since

$$\overline{A_{2a}^4} = B^2/2 + S, \tag{17}$$

Fig. 2. Diagrams contributing to $\overline{A_2^4}$ to leading order in perturbation theory.

where B^2 and S are the diagrams pictured in Fig. 2.

In general, it can be shown that the adiabatic pinning force F can be written as

$$F = \sum_{n=1}^{\infty} A_{na} \sin(n\Delta + \theta_{na}), \tag{18}$$

where $A_{na} > 0$ and the θ_{na} depend strongly upon the impurity configuration. We thus have an equation of motion for Δ of the form,

$$\dot{\Delta}/\tau = e^*E + \sum_{n=1}^{\infty} A_{na}\sin(n\Delta + \theta_{na}). \tag{19}$$

If for some time t^*, $\dot{\Delta}(t^*)=0$, $\Delta(t)$ will remain a constant thereafter. Thus, for electric fields such that the right hand side of Eq. (19) vanishes for some $\Delta(t^*)$, there will be no time-average current in the domain. The right hand side of Eq. (19) vanishes below the threshold field for a field in the ($\pm$) directions, given by

$$e^*E_{T\pm} = \begin{bmatrix} -\min_\Delta \\ \max_\Delta \end{bmatrix} \{\sum_{n=1}^{\infty} A_{n\alpha}\sin(n\Delta + \theta_{n\alpha})\}. \tag{20}$$

In the above equations, each of the $A_{n\alpha}$ and hence $e^*E_{T\pm}$ are proportional to $(cV_\circ^2)^2$,

$$e^*E_{T\pm} = K_\pm(cV_\circ^2)^2, \tag{21}$$

where $K_\pm$ is a constant of order unity that depends upon the impurity configuration α.

As we stated previously, each of the $A_{n\alpha}$ and $\theta_{n\alpha}$ in the Fourier series for the pinning force F depend strongly upon the particular impurity configuration in the domain. However, we showed that to leading order in perturbation theory in $V_\circ$ (and hence γ), we could estimate $A_{1\alpha}$ from $\overline{A_{1\alpha}^2}$. This suggests that to arbitrary order in perturbation theory, we can estimate the amplitudes $A_{n\alpha}$ from the average force-force correlation function $\overline{F(\Delta_1)F(\Delta_1+\Delta)}$. It is easy to show that

$$\overline{A_{n\alpha}^2} = 4\int_0^{2\pi} \frac{d\Delta}{2\pi} \cos(n\Delta)\ \overline{F(\Delta_1)F(\Delta_1 + \Delta)}\ , \tag{22}$$

since $\overline{F(\Delta_1)F(\Delta_1+\Delta)}$ is independent of Δ_1, due to translational invariance. The width of the distribution is obtained from $\overline{A_{n\alpha}^4}$, which is related to the four-force correlation function $\overline{F(\Delta_1)F(\Delta_1+\Delta)F(\Delta_2+\Delta)F(\Delta_2)}$, for example. We can similarly obtain an estimate of the typical phase values from the three-force correlation function,

$$\begin{aligned}&\overline{A_{n\alpha}A_{m\alpha}A_{n+m,\alpha}e^{i(\theta_{n\alpha}+\theta_{m\alpha}-\theta_{n+m,\alpha})}} \\ &\quad = 8i\int_0^{2\pi}\frac{d\Delta}{2\pi}\, e^{-in\Delta}\int_0^{2\pi}\frac{d\Delta'}{2\pi}\, e^{-im\Delta'}\overline{F(\Delta_1)F(\Delta_1+\Delta)F(\Delta_1+\Delta')}.\end{aligned} \tag{23}$$

Hence, we may determine the probability distributions for the A_n's and θ_n's by calculating the appropriate correlation function of F.

To order $V_\circ^2$, the force-force correlation function is found to be

$$\overline{F(\Delta_1)F(\Delta_1+\Delta)} = V^{-1}[(cV_\circ^2/2)\cos\Delta+(cV_\circ^2/2)^2B(\cos2\Delta-\cos\Delta)+O(cV_\circ^2)^3], \tag{24}$$

where B is the bubble diagram shown in Fig. 1. Since $B=\xi/4\pi$, both terms in Eq. (24) are of the same order in $cV_\circ^2$, i.e. $(cV_\circ^2)^4$.

We are now in a position to evaluate A_{na}^2 in lowest order. We have

$$\overline{A_{na}^2} = g\frac{(cV_\circ^2)^4}{\tilde{\gamma}^{4-n}} \sum_{m=0}^{\infty} C_{nm}\, \tilde{\gamma}^m, \tag{25}$$

where $g=3/(\pi^4 2^{17})$, and $\tilde{\gamma}=\pi^2\gamma$. For example, $C_{10}=1$, $C_{11}=-4$, and $C_{20}=4$ can be obtained from Eqs. (22) and (24). With these few coefficients, we can already see that this perturbative approach runs into difficulty on the scale of a LR domain: For $\tilde{\gamma}>1/4$, the first two terms in the series [eq.(25)] for $\overline{A_{1a}^2}$ give $\overline{A_{1a}^2}<0$, which is impossible. Since a LR domain corresponds to $\tilde{\gamma}=1/2$, this breakdown occurs within a LR domain.

Let us now turn our attention to the phases θ_{na}. In eq. (23) we showed that the three-force correlation function could be employed to calculate impurity averages involving the phases. The lowest order diagram is the three-sided loop of value L pictured in Fig. 3. To lowest order in perturbation theory,

$$\overline{A_1^2 A_2 e^{i(2\theta_1-\theta_2)}} = \frac{-2L(cV_\circ^2/2)^3}{V^2} \tag{26}$$

$$\overline{A_1 A_2 A_3 e^{i(\theta_1+\theta_2-\theta_3)}} = \frac{3L(cV_\circ^2/2)^3}{V^2}\ . \tag{27}$$

Fig. 3. Loop diagram of value L contributing to the three-force correlation function.

We find $L=3V/(64\pi^2)$, and hence these terms are of order $(cV_\circ^2)^6$, as expected since each A_{na} is of order $(cV_\circ^2)^2$. In addition, $\overline{F(\Delta_1)F(\Delta_1+\Delta)F(\Delta_1+\Delta')F(\Delta_1+\Delta'')}$ to lowest order yields

$$\overline{A_1^3 A_3 \cos(3\theta_1-\theta_3)} = 6(cV_\circ^2/2)^4 S/V\ , \tag{28}$$

where S is the square pictured in Fig. 2. These lowest order terms do give us some valuable information regarding the allowed θ_{na} values. Since the imaginary parts of eqs. (26) and (27) vanish, we may treat the $x_{na} = \theta_{na}-n\theta_{na}$ values as being given by a probability distribution $\mathcal{P}_n(x)$ that is a sum of symmetric distributions $\mathcal{P}_{in}(x)$ centered about 0 and π,

$$\mathcal{P}_n(x_{na}) = a_n\mathcal{P}_{1n}(x_{na}) + (1-a_n)\mathcal{P}_{2n}(x_{na}-\pi), \tag{29}$$

where $1>a_n>0$. The signs of the right hand sides of eqs. (26) - (28) imply $a_2<1/2$ and $a_3>1/2$, to this order in perturbation theory. We could in principle calculate a succesive hierarchy of such force correlation functions, obtaining the a_n. One could also obtain information about the widths of the $\tilde{P}_{in}$ distributions from higher order force correlation functions (i.e. four or more forces averaged together and integrated).

As we stated following eq. (25), the perturbation approach breaks down for spatial averages on the size of a LR domain. That is, the perturbation approach converges for $\xi/\xi_s \ll 1$, but is not well behaved in the interesting regime $\xi/\xi_s \sim 1$. In order to investigate the region $\xi/\xi_s \sim 1$, we have developed an <u>iteration</u> solution to the problem.

To illustrate the iteration procedure, let us reexamine the force F,

$$F = V_\circ \langle \delta n(\vec{r})\sin(\vec{Q}\cdot\vec{r} + \Delta + \delta\phi)\rangle_\xi, \tag{30}$$

where

$$\delta\phi(\vec{r},\Delta)=V_\circ \int d^3r' \tilde{G}(\vec{r},\vec{r}\,')\delta n(\vec{r}\,')\sin[\vec{Q}\cdot\vec{r}\,'+\Delta+\delta\phi(\vec{r}\,',\Delta)]. \tag{31}$$

In eq. (31), $\tilde{G}(\vec{r},\vec{r}\,')=G(\vec{r},\vec{r}\,')-\langle G(\vec{r},\vec{r}\,')\rangle_\xi$ and $G(\vec{r},\vec{r}\,')=[4\pi|\vec{r}-\vec{r}\,'|]^{-1}$ (10) The iteration approach we employ is as follows. The zeroth order approximation is

$$F_0 = V_\circ \langle \delta n(\vec{r})\sin(\vec{Q}\cdot\vec{r}+\Delta)\rangle_\xi, \quad \delta\phi_0=0, \tag{32}$$

which results in the Grüner, Zawadowski, and Chaikin (GZC) [4] model. The first iteration gives

$$F_1 = V_\circ \langle \delta n(\vec{r})\sin[\vec{Q}\cdot\vec{r}+\Delta+\delta\phi_1(\vec{r},\Delta)]\rangle_\xi, \tag{33}$$

where $\delta\phi_1(\vec{r},\Delta)$ is the first order iteration for $\delta\phi$ obtained by setting $\delta\phi=\delta\phi_0=0$ on the right hand side of eq. (31). The nth iteration is given by

$$F_n = V_\circ \langle \delta n(\vec{r})\sin(\vec{Q}\cdot\vec{r}+\Delta+\delta\phi_n)\rangle_\xi \tag{34}$$

and

$$\delta\phi_n(\vec{r},\Delta) = V_\circ \int d^3r' \tilde{G}(\vec{r},\vec{r}\,')\delta n(\vec{r}\,')\sin[\vec{Q}\cdot\vec{r}\,'+\Delta+\delta\phi_{n-1}(\vec{r}\,',\Delta)]. \tag{35}$$

We have employed this iterative procedure to find the appropriate correlation functions to first order in the iteration. We find

$$\overline{A_n^2} = \frac{cV_\circ^2 e^{-\tilde{\gamma}}}{12V}\{7[I_{n-1}(\tilde{\gamma})-I_{n+1}(\tilde{\gamma})] -4[I_{n-1}(\tilde{\gamma}/4)-I_{n+1}(\tilde{\gamma}/4)] +8[(-1)^n -1]\sum_{m=0}^{\infty}(-1)^m[I_{n+2m+1}(\tilde{\gamma})-I_{n+2m+1}(\tilde{\gamma}/4)]\}, \tag{36}$$

where $\tilde{\gamma}=\pi^2\gamma$, and $I_n(z)$ is a Bessel function. A similar but more complicated expression is obtained for the averages involving the phases. We note that the above expression appears to be convergent for all γ. We expect, however, that the procedure will not be convergent as $\gamma\to\infty$, as that would correspond to an infinite domain. The procedure does appear to be convergent for $\tilde{\gamma}\sim 1$, however.

For $\xi/\xi_s \sim 1$, the force F contains many harmonics, as all of the A_{na} are comparable in magnitude. We have investigated the narrow band noise spectrum that arises from a few terms in the Fourier series for F, and find that variations in the relative phases $\theta_{na}-n\theta_{1a}$ result in strong variations in the resulting amplitudes of the narrow band noise. In addition, the large number of terms in the Fourier series for the force F results in a force with many minima and maxima in $F(\Delta)$, resulting in metastable states.

While the results discussed so far have considerable experimental support, we also view this work as an important ingredient of a more complete theory which includes domain-domain interactions. We propose that the motion of a collection of domains indexed by i be described by the effective Hamiltonian dynamics,

$$H = \sum_i V_i(\Delta_i) + \sum_{i<j} U(\Delta_i,\Delta_j) + C \sum_i e^*E\Delta_i, \qquad \dot{\Delta}_i/\tau = -\frac{\partial H}{\partial \Delta_i} . \tag{37}$$

The one-body potential V_i is the single domain potential obtained from $F_i(\Delta_i)$ by $F_i(\Delta_i)=-dV_i/d\Delta_i$, and U is a coupling energy primarily between neighboring domains.

Fisher[9] has studied a model that is similar in appearance to the above, but which differs in several important respects. Instead of the collective domain coordinates, he treats the phases of the CDW at the impurity sites in the above fashion, so the number of terms in the sums differ by $\sim 10^{10}$. Second, the form of his potential is the simple cosine, which does not include metastable states. In addition, the amplitudes of our potentials are all explicitly proportional to $(cV_0^2)^2$, in agreement with experiment.

Finally, we consider the dynamics of a sample of size on the order of a single LR domain. We have generalized the above treatment somewhat to include the leading order non-adiabatic corrections to the pinning force F. We consider a cube of volume L^3, measure length in units of L, and time in units of $t_L=L^2/\tau$, the time for information to traverse a domain. We choose periodic boundary conditions. The Green's function is of the form

$$G(\vec{r},t)=\Theta(t)(4\pi t)^{-3/2}\exp(-r^2/4t). \tag{38}$$

To second order in the pinning potential V_0, the equation of motion for Δ becomes

$$\dot{\Delta}(t) = \tilde{E} + \mathrm{Im}\{Ae^{i\Delta(t)} + \int dt' \sum_{s=\pm 1} B^s(t-t')e^{i[\Delta(t')+s\Delta(t)]}\}, \qquad (39)$$

where $\tilde{E} = e^*EL^2$ and

$$A = V_0 L^{-1} \int d^3r\, e^{i\vec{Q}\cdot\vec{r}'} \delta n(\vec{r}), \qquad (40)$$

$$B^s(t) = V_0^2 (2L)^{-1} \int d^3r d^3r' \delta n(\vec{r}) \delta n(\vec{r}') G(\vec{r}-\vec{r}', t) e^{i\vec{Q}\cdot(\vec{r}'+s\vec{r})}. \qquad (41)$$

We note that A and B^+ play essentially the same role as in the adiabatic theory, contributing the leading terms to $F(\Delta)$. The quantity $B^-(t)$, however, did not enter the calculation in the adiabatic approximation, due to the vanishing of $\sin[\Delta(t)-\Delta(t')]$ as $t \to t'$. This term greatly effects the d. c. conductivity, as its impurity average is non-zero,

$$\overline{B^-(t)} = \frac{L}{2\eta\xi_s} G(0,t), \qquad (44)$$

where $\eta = 8.266/(2\pi)^6$. Since $B^-(t) \propto (t/t_L)^{-3/2}$ for $t/t_L \ll 1$, the B^- term becomes important, renormalizing the effective damping time τ to $\tau(\omega)$, where $\tau(\omega) \propto \omega^{1/2}$ in the large ω limit. This behavior results in the large field behavior of the d.c. conductivity $\sigma(E) \to \sigma_\infty - CE^{-1/2}$ predicted by Sneddon, Cross, and Fisher(6). We also find $\sigma(\omega) \to \sigma_\infty - C'\omega^{-1/2}$ for large frequencies.

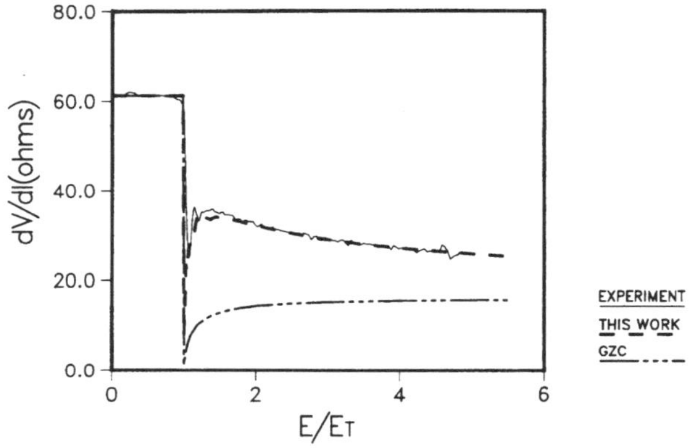

Fig. 4. dV/dI generated from eq. (39) versus field for a single domain. For comparison, the prediction of GZC (4) and experimental data on a small sample of $NbSe_3$ at 48K (11) are plotted.

We have performed numerical studies of eq. (39) by assigning values to A and $B^{\pm}$ both from their impurity-averaged moments and by assigning impurities at random in the cube. The resulting differential resistance dV/dI is plotted versus field in Fig. 4 for a sample with $L=\xi_s/2\pi$. For comparison with experiment a parallel resistance due to normal electrons has been included. Results from the GZC model and experiments on a 2mmx10μx10μ sample of $NbSe_3$ at 48K are also shown (11). The results are in good agreement with the data. It should be noted, however, that not all samples show this behavior at low temperature. Within our model, the kink in dV/dI near threshold is a finite size effect, indicating the sample contains a small number of domains. If the calculation is extrapolated to infinite volume, dV/dI decreases monotonically from threshold. This behavior is seen in many samples. We expect that extending the calculation to higher order in V_0 will change the quantitative details but not the qualitative features of these results.

REFERENCES

† main address: Institute for Theoretical Physics, University of California, Santa Barbara, California 93106.

1. K. B. Efetov and A. I. Larkin, Zh. Eksp. Teor. Fiz. 72, 2350 (1977) [Sov. Phys. JETP 45, 1236 (1977)].
2. H. Fukuyama and P. A. Lee, Phys. Rev. B27, 535 (1978).
3. P. A. Lee and T. M. Rice, Phys. Rev. B19, 3970 (1979).
4. G. Grüner, A. Zawadowski, and P. M. Chaikin, Phys. Rev. Lett. 46, 511 (1981).
5. R. A. Klemm and J. R. Schrieffer, Phys. Rev. Lett. 51, 47 (1983) and to be published.
6. L. Sneddon, M. Cross, and D. Fisher, Phys. Rev. Lett. 49, 292 (1982).
7. L. Pietronero and S. Strässler, Phys. Rev. B28, 5863 (1983).
8. J. B. Sokoloff, Phys. Rev. B23, 1992 (1981).
9. D. S. Fisher, Phys. Rev. Lett. 50, 1486 (1983) and to be published.
10. We have taken $\tilde{G}(r)$ to be of the form $\Theta(\xi-r)/4\pi r$ for simplicity.
11. J. Stokes, private communication.

ON THE MICROSCOPIC THEORY OF KINETIC PHENOMENA IN PEIERLS CONDUCTORS.

S.N.Artemenko, A.F.Volkov
Institute of Radioengineering & Electronics of the Academy of Sciences of the USSR, 103907 Moscow, Marx av. 18, USSR

Kinetic phenomena in conductors with a charge density wave (CDW) are investigated by means of equations for the Green functions. We find the response of a Peierls conductor (PC) to applied electric field and analyse different nonlocal effects, such as the appearance of voltage in a region, where total current is absent, the influence of the CDW deformation on the pinning threshold field, the generation of domain walls (phase solitons) by the current etc. The CDW motion and its contribution to the conductivity of the PC is studied in the case, when a lattice of amplitude solitons is present in the PC. The system of two weakly coupled PC is considered.

1. Introduction.

It has been well established that kinetic properties of quasi-one-dimensional conductors below the Peierls transition temperature are to a great extent determined by the CDW /1-3/. The usual kinetic equation is not valid for the description of transport phenomena in a PC, because the conductivity of a PC is determined not only by quasiparticles but also by the CDW motion. The situation here is very similar to the case of superconductors, where microscopic equations for the Green functions were used successfully for the description of kinetic phenomena /4-6/. The derivation of these equations is based either on the method of analytic continuation of the response calculated with the help of the Matsubara technique /4,5/, or on the Keldysh method /6/.

The approach developed for superconductors was applied to PC. The analytic continuation method was used by Gor'kov and Dolgov /7/. We have derived the kinetic equations for the Green functions using the Keldysh method /8/. (In the case of small frequencies and gradients the CDW transport may be described by quasiclassic kinetic equation obtained in Ref.9 and used for the calculation of the Hall current in a PC). By means of these equations, we consider here a number of effects in the PC: some of them were observed experimentally, the others can be studied in future experiments. We have derived equations for the Green matrix functions 4x4, whose elements are $\check{g}_{11} = \hat{g}^{R}$, $\check{g}_{22} = \hat{g}^{A}$, $\check{g}_{12} = \hat{g}$, where $\hat{g}^{R(A)}$ is the retarded (advanced) Green function integrated over the longitudinal momentum, $\hat{g}$ is the Green function describing the kinetics of the system and related to the quasiparticles distribution function, $\hat{g}^{R(A)}$ and $\hat{g}$ are matrices, as well. The equation for $\check{g}$ has the form /8/

$$\tilde{\varepsilon}\,\check{\sigma}_z\check{g} - \check{g}\sigma_z\tilde{\varepsilon}' + [\check{\Delta} - \varphi\check{\sigma}_z, \check{g}] + iv\,d\check{g}/dx + (i/2)v_\perp\nabla_\perp[\check{g},\check{\sigma}_z]_+ = [\check{\Sigma},\check{g}]_- \quad (1)$$

where $\check{\Sigma}$ is the mass operator describing the elastic scattering (by impurities or by phonons). This equation is derived in the mean-field approximation and is valid, therefore, when the 3-D effects in the electron and phonon spectra are sufficiently large. The CDW phase on different chains is considered to be correlated.

The current and charge density are expressed in terms of g by

$$j = \int d\varepsilon\ Sp\langle v\hat{g}\rangle\ , \quad \rho = (k_0^2/4\pi)[Sp\int d\varepsilon\, \hat{\sigma}_z\langle\hat{g}\rangle - \varphi] \qquad (2)$$

where $\langle\ldots\rangle$ means averaging over transverse momentum $p_\perp$.

The relation connecting the amplitude Δ and the phase χ of the order parameter with the nondiagonal components of $\hat{g}$ follows from equations of motion for the phonon operators: it has the form:

$$\Delta \exp(i\chi) = \lambda/4(1 - \omega^2/\omega_Q^2)^{-1}\int d\varepsilon\ Sp(\hat{\sigma}_x - i\,\hat{\sigma}_y)\langle\hat{g}\rangle. \qquad (3)$$

The effect of CDW pinning is taken into account in a simple model, that can be rigorously justified for the commensurate case. In this case an addittonal term $\sim (\Delta/\varepsilon_0)^{n-2}\sin(n\chi)$ should be added to the right-hand side of equation for χ (n is the commensurability order, ε_0 is a large energy of the order of the electron bandwidth).

2. Motion and deformation of the CDW in nonuniform electric field.

It has been shown recently that nonuniformity caused by contacts influences greatly the motion of the CDW /11-16/. Recently, a finite voltage across a region with zero total current was observed /15/. The current was flowing between the contacts 2,3, and the voltage V_{jk} was measured between the contacts i and k (i,k = 1,2,3,4: see Fig.). If the

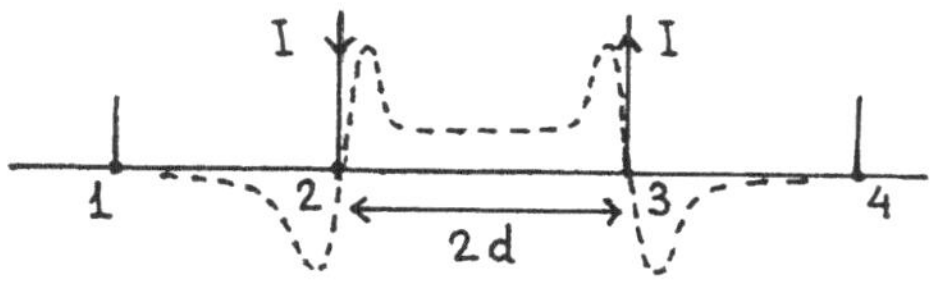

current exceeds a threshold value, then voltages V_{12} and V_{34} appear in the regions with zero total current. Their values are of the order of V_{23} and their signs are opposite to that of V_{23}. In addition, the increase of the conductivity in the regions (1,2) and (3,4) was observed. An explanation for this effect was suggested in Ref.16. It was shown that the CDW proceeded sliding in the regions (1,2) and (3,4). In these regions the CDW current and opposite to it quasiparticle current are flowing; these currents compensate each other. According to estimate for TaS_3, these currents and the electric field decay over a macroscopic length of the order of 100 mcm, in agreement with the experimental data /15/.

To find the spatial distribution of the electric field E in the sample we shall use an equation for the phase of the CDW and the expression for the longitudinal current. They can be obtained from the expression for the Green function $\hat{g}$ /10/ and have the form

$$-(D/2)\chi'' - \gamma\dot{\chi} = El - E_0 l\cdot\sin(n\chi), \qquad (4)$$

$$I = I_q + \sigma_2\dot{\chi}/l, \quad (d/dt - D_1 d^2/dx^2)I_q = \sigma_1 dE/dt, \qquad (5)$$

where D is the diffusion constant, E_0 is the pinning threshold field in the uniform case. If the bending of the Fermi surfaces is small ($\eta \ll T$) and $T \ll \Delta$ (just in this case the contribution of the CDW to the current is large), we get $\beta \simeq (T/\Delta)\exp(-\Delta/T)$, $\gamma \simeq (\Delta/T)\exp(-\Delta/T)$, $\sigma_1 = \beta\sigma_N$, $\sigma_2 = \sigma_N$, σ_N is the conductivity in the normal state (i. e. in the state with $\Delta = 0$). Eq.(5) is obtained approximately (with the accuracy to a constant factor) from the quasiparticles diffusion equation, replacing the energy-dependent diffusion coefficient by a constant D. In the limit of stationary or uniform field E, this equation coincides with that obtained in Ref.10. From Eq.(5) we obtain an expression for the current with allowance for the CDW deformation

$$I = \sigma_N\beta E - \sigma_N D_1\chi''/l + \sigma_N\dot{\chi}/l \qquad (6)$$

With the aid of Eq.(6) we eliminate E from Eq.(4) and obtain an equation for the phase

$$-D_1\chi'' + \dot{\chi} + \beta E_0 l\cdot\sin(n\chi) = Il/\sigma_N \qquad (7)$$

Similar equation was derived in Ref.10 on the ground of a microscopic theory, and in uniform case in Ref.17 on the ground of phenomenological considerations. This equation is known as classical model equation for the CDW transport.

Let us consider the case of large enough currents when pinning plays no role and the last term in the left-hand side of the Eq.(7) can be neglected. Then, the equation for χ has the form of the diffusion equation with a source acting in the region $|x|<2d$ where $I = 0$.

When a long current pulse is applied, the phase starts to increase in time in the region (2,3). This results in the appearance of the phase gradient near points 2 and 3, which propagates outwards the region (2,3) in a diffusive way. The spatial dependence of E in this case is shown by dotted line in Fig. The length, over which E penetrates into the region (2,3) increases with time, over the same length the velocity of the CDW being different from zero. The voltage $V_{out} = V_{12} + V_{34}$ increases in time too. The further evolution of E and V_{ik} depends on the value of I.

At sufficiently large currents I or distances d, when $I > I_c \simeq \sigma_N \Delta/d$ (just this case corresponds to the experimental conditions of Ref.15), an increase in $d\chi/dx$ leads to a suppression of the gap at points 2 and 3. This happens when $V_{out} \simeq \Delta$. (Note that the voltage V_{23} is larger than the voltage, which would be in the case of uniform CDW sliding, the last coincides with V_{14}). As a result, phase-slip centres (PSC) appear in the system /11-12/ and further increase of χ will be stopped. To analyze this situation it would be necessary to solve Eq.(1) taking into account variations of Δ . Up to now such a problem at low temperatures was not solved.

If the current is sufficiently small (but still larger than its threshold value), so that $I < I_c$, then the perturbations of χ propagating from contacts 2 and 3 meet each other, and their mutual supression begins. The field E starts to diminish as $1/\sqrt{t}$, spreading into the regions (1,2) and (3,4). This process will stop when CDW is fixed by pinning. The voltages are approaching to their asymptotic values $-V_{out} \simeq V_{23} = V_0 (\Delta/T)^{3/2}$, $V_{14} = \gamma V_0$, where $V_0 = 2dI/\sigma_N$. One can see that the voltage $V_{34} = V_{12} \simeq (1/2)V_{23}$ turns out to be much larger than V_{14} . If the CDW is not deformed, the voltage V_{14} would be equal to γV_0.

Thus, the spatial distribution of E depends on the value of the current and on the distance between contacts. It can be easily shown that if the current I is the alternating one, the field distribution and the voltages V_{23} and V_{out} must be frequency dependent with the characteristic frequency $\omega_0 = D/d^2$/16/. The penetration depth of E into external regions turns out to be $L_E = (D_1/\omega)^{1/2}$. Again, the voltage V_{23} is larger than it was in the case of uniform CDW sliding ($V_{23} = V_0 D/D_1$ at $\omega < \omega_0$), and V_{14} coincides with the voltage which would be in the absence of CDW deformation.

Consider now the effect of the pinning on the phenomena under consideration. We use a model rigorously justified in the case of the commensurability pinning. It follows from Eq.(7) that at small currents the phase χ remains stationary but it deforms in the region (2,3) and decays to zero over the length $L = (D_1/\beta E_0 l)^{1/2}$ equal to the dimension of the phase soliton /10/. The relation between the current and the voltage is changed due to deformation of the CDW

$$I = \sigma_1 V_{23}/(2d - L(1 - \exp(-2d/L))) \qquad (8)$$

From Eq.(8) it follows that the deformation changes the temperature dependence of the conductivity $\sigma_{23} = 2dI/V_{23}$ measured between contacts 2 and 3. As noted in Ref.10, the soliton dimension increases exponentially due to the screening effect when the temperature decreases ($\beta \to 0$ at $T \to 0$). If the spacing between contacts $2d \gg L$, the deformation of the CDW is not important, and $\sigma_{23} = \sigma_1$ as in the uniform case. As the temperature decreases, L becomes larger than 2d. From Eq.(8) it follows that $\sigma_{23} = \sigma_1 L/d$. The temperature dependence of σ_{23} is changed, in particular, the activation energy becomes two times smaller than Δ.

The increase of the current leads to an increase in CDW deformation, and at a certain magnitude of $I = I_T$, there is no static solution for the phase decaying to zero at $x \to \infty$. The calculation shows that the magnitude of I_T depends on the spacing between contacts 2 and 3, and increases when the spacing diminishes. At $L \ll d$, I_T is close to the threshold value in uniform case, $I_T = \sigma_1 E$, and at $L \gg d$, $I_T = \sigma_1 E \cdot (2L/d)$.

In real crystals, such effects may happen due to the deformation of the CDW caused not only by contacts, but by impurities and defects. In principle, this deformation might be a reason for a change of the temperature dependence of the conductivity in TaS_3 at fields E below a threshold value E_T at $T < 100$ k, and for an increase of E_T at low temperatures /3/.

If $I > I_T$, the static solution of Eq.(7) has the form of domain walls - a chain of phase solitons. The period of the chain depends on $I - I_T$. The formation of this chain at the currents slightly exceeding I_T is presumed to occur through the consequent creation of the phase solitons at points 2 and 3, and their propagation to the infinity. At large currents, spatial oscillations of the phase gradient become small, and the dependence $\chi(x)$ approaches to the static solution of the diffusion equation.

3. Amplitude solitons in incommensurate Peierls conductor.

In the foregoing section it was assumed that the amplitude of the order parameter did not depend on coordinates. But it is known that under certain conditions amplitude solitons may appear in the PC /18,19/. The presence of such solitons in polyacetylene, where period is doubled, has been confirmed by many experiments. In principle, the amplitude solitons may appear also in the case when the period of the CDW differs from the doubled period of the lattice /18/. Some experimental evidences were reported in favour of the presence of the amplitude solitons in TaS_3 /20/.

It was shown in Ref.21-22 that in conductors with the CDW and with strong enough interaction between chains the new phase may exist, which contains domain walls formed by amplitude solitons. In these papers the case of period doubling is investigated, when the allowed zone appears at the center of the energy gap, and the order parameter $\Delta = 0$ at the center of the soliton. Here we consider the case of the incommensurate CDW and show that in this case, generally speaking, the allowed zone is not located at the midgap, and Δ differs from zero everywhere. We consider also a problem concerning the CDW sliding in the presence of the amplitude solitons.

To describe the equilibrium state with amplitude solitons one has to solve Eq.(1) for $g^{R(A)}$ and then with the aid of the self-consistency equation (3) to find $\Delta(x)$. Solutions describing soliton states have the form

$$g^{R(A)} = (\tilde{\varepsilon}(\tilde{\varepsilon} - c) - (\Delta_+^2 - \Delta_-^2 - |\Delta(x)|^2)/2)/B \quad , \quad g^{R(A)} = \hat{g}_{11}^{R(A)},$$

$$f^{R(A)} = ((\tilde{\varepsilon} - c)\Delta(x) + V\Delta'(x)/2)/B \quad , \quad f^{R(A)} = \hat{g}_{12}^{R(A)}, \tag{9}$$

$$B = ((\tilde{\varepsilon} + i0)^2 - \Delta_+^2)^{1/2} \cdot ((\tilde{\varepsilon} + i0 - c)^2 - \Delta_-^2)^{1/2}, \quad \tilde{\varepsilon} = \varepsilon - \eta(p_\perp),$$

$$\Delta = \exp(i\chi)(c^2 + \Delta_k^2 \mathrm{sn}(\Delta_k x/(kv), k))^{1/2}, \chi' = c(b - \frac{\Delta_+ - \Delta_-}{|\Delta|^2})$$

$$\Delta_k^2 = (\Delta_+ - \Delta_-)^2 - c^2, \quad k^2 = \Delta_k^2/((\Delta_+ + \Delta_-)^2 - c^2), \quad c = \cos\Theta. \tag{10}$$

Parameter k determines the period of the structure. Energies Δ_+ and Δ_- determine the boundaries of allowed zones $\varepsilon > \Delta_+$ and $|\varepsilon - c| < \Delta_-$. The parameters entering in Eqs.9-10 and the position of the Fermi level are determined by the conditions of the energy minimum and by the self-consistency condition.

The chains of the amplitude solitons with c = 0, in which the phase is changed by π, and $\Delta = 0$ at the center of solitons, were found in Ref.24. Solitons with $c \neq 0$ describing isolated solitons (this case corresponds to $k \to 1$, $\mathrm{sn}(\Delta_k x/kv, k) \to \tanh(\Delta x/v \cdot \sin\Theta)$ were analized in Ref.18. It was shown there that c = 0 both in the case of period doubling and in the incommensurate case. We generalize this result and show that with allowance for transverse dispersion of electronic spectrum, $\eta(p_\perp)$, generally speaking, $c \neq 0$.

In an isolated soliton there is a localized state with the energy $\varepsilon = \cos\Theta$. The total change of the phase over the soliton is 2Θ. To determine Θ we substitute g from Eq.(9) at $k \to 1$ into the Eq.(3). We get

$$\int d\varepsilon \langle \tanh(\varepsilon/2T)(\frac{1}{\xi^R(\tilde{\varepsilon}-c)} - \frac{1}{\xi^A(\tilde{\varepsilon}-c)})\rangle = 0, \tag{11}$$

$$\xi^{R(A)} = \pm((\tilde{\varepsilon} \pm i0)^2 - \Delta^2)^{1/2}$$

Let us consider the case $T \ll \eta, \Delta$. If one neglects the terms of the order of $\exp(-\Delta/T)$, the self-consistency condition for the phase takes the form $n(c + \eta) = \Theta/\pi$, $n(\varepsilon) = (1 - \tanh(\varepsilon/2T))/2$ is the Fermi distribution function. When $c = 0$, Eq.(13) has, except spatially uniform solutions with $\Theta = 0$ and $\Theta = 2\pi$, only one more solution corresponding to the soliton with $\Theta = \pi/2$. If $\eta = 0$ this equation has a solution with the values of θ determined by the form of $\eta(p_\perp)$. At a given function $\eta(p_\perp)$ one can get several solutions for θ corresponding to different positions of soliton energy levels.

The energy of the domain wall is lowered due to the dependence of the soliton energy on η. For the appearance of the domain wall its energy must be negative /23/.

Energy density of a soliton wall per single chain can be found from the expression

$$w = \omega_Q^2 \Delta^2/2g^2 + 2/(\pi v)\int \varepsilon \langle (g^R - g^A) \rangle n(\varepsilon) d\varepsilon/2 \qquad (12)$$

At low temperatures $T \ll \eta$, Δ this equation is reduced to

$$w = (2\Delta/\pi) \sin \theta + 2\langle \eta \cdot n(c + \eta) \rangle \qquad (13)$$

To illustrate the possibility of the appearance of domain walls, in which $c \neq 0$, we consider the simplest model for the electronic spectrum in transverse direction. Let us assume that the electron bandwidth in transverse direction is equal to $2\varepsilon_1$ and the function $\eta(p_\perp)$ has steplike form: $\eta = 2\varepsilon_1(1-s)$ on a part of the Fermi surface, the square of which is sS (S is the total square of the Fermi surface), and $\eta = -2\varepsilon_1 s$ on another part of the Fermi surface. Such a form of $\eta(p_\perp)$ satisfies a necessary condition following from the minimum of the free energy. From the self-consistency equation we find in this case three solutions for θ

$$\theta_1 = \cos^{-1}(2\varepsilon_1(1-s)/\Delta), \quad \theta_2 = \pi s, \quad \theta_3 = \cos^{-1}(-2\varepsilon_1 s/\Delta) \qquad (14)$$

If the parameters s and ε_1/Δ satisfy the condition $\theta_1 < \theta_2 < \theta_3$, then all three solutions exist. If it turns out that $\theta_1 > \theta_2$ ($\theta_2 > \theta_3$), then only solution θ_3 (or θ_1) exists. Depending on the value of s and ε_1/Δ, the energy for each of these solutions might be negative at $\varepsilon_1 < \Delta$. It means that with increasing of the interchain interaction the domain phase may appear earlier then the PC becomes a semimetal at $\varepsilon_1 \geqslant \Delta$. It is worth to note that the case $\theta = \pi/2$ is realized only for symmetrical form of η corresponding to $s = 1/2$. In the case of arbitrary $\eta(p_\perp)$ it may turn out that $w > 0$ up to the value $\varepsilon_1 = \Delta$: in this case the PC becomes a semimetal earlier than the amplitude solitons are formed.

In Ref.23 the contribution to the conductivity caused by electrons in soliton zone and by motion of domain structure was calculated. Here we present the results of our investigations of the CDW motion and its contribution to the conductivity of a PC in the presence of a fixed domain chain formed by the amplitude solitons. Performing the calculations, we have used Eq.(1) linearized with respect to the field E and to the phase χ, the solution for which we looked for in the form $\chi(x,t) = \dot{\chi} t + \delta\chi$, where $\dot{\chi}$ = const, and $\delta\dot{\chi} = 0$. As a result of cumbersome calculations we get for the current

$$j = \sigma_1 \bar{E} + \sigma_2 \dot{x}/l \qquad (15)$$

where

$$\sigma_1 = \sigma_N \int \alpha \, d\varepsilon \langle \nu_2/\nu_* \rangle, \quad \sigma_2 = \sigma_N (1 - \int d\varepsilon \, \alpha \langle \nu_2 G/2\nu_* \rangle),$$

$$\nu_* = (\nu_2/2)(\overline{\langle G \rangle/G} + 1) + (\nu_1/2)(\overline{\langle G \rangle/G} - 1) + 2i\overline{\Delta F/G},$$

$$G = g^R - g^A, \; F = f^R + f^A, \; \alpha = 1/(2T\cosh^2(\varepsilon/2T)),$$

$1/\nu_1$ and $1/\nu_2$ are the forward and backward scattering times. The bar means spatial averaging. Taking into account that $F \sim \nu$, one can see that $1/\nu_*$ has the meaning of the effective relaxation time. The value of σ_2 at low temperatures and low density of solitons equals approximately to σ_N.

From the selfconsistency condition, one gets

$$\gamma \dot{x} = \bar{E} l, \; \gamma = (\sigma_N/2\sigma_2) \int d\varepsilon \, \alpha (\overline{\langle G^2 \rangle} - (\nu_2/\nu_*) \langle \bar{G} \rangle^2) \qquad (16)$$

We avaluate the magnitudes of σ_1 and γ neglecting the interaction between chains (i.e. considering the case $\eta \to 0$). The contribution to σ_1 is caused by quasiparticles with the energies $\varepsilon > \Delta$ and $\varepsilon < \Delta_-$. Quasiparticles with the energies $\varepsilon < \Delta_-$ (i.e. those moving inside the soliton zone) at low temperatures give the contribution to σ_1 of the order of

$$\sigma_{1s} = \sigma_N (\Delta_-^2/\Delta^2) \cdot \begin{cases} 1, & \text{at } T \ll \Delta_- \\ (\Delta_-/T), & \text{at } T \gg \Delta_- \end{cases} \qquad (17)$$

The expression for the contribution to γ due to quasiparticles having energies $\varepsilon < \Delta_-$ is equal at $\nu \ll \Delta_-$ approximately to

$$\gamma_s = n\xi (\Delta^2/\Delta_- T) \cdot \begin{cases} (T/\Delta_-), & \text{at } T \ll \Delta_- \\ 1, & \text{at } T \gg \Delta_- \end{cases} \qquad (18)$$

where n is the soliton concentration on a single chain, $\xi = v/\Delta$ is the soliton dimension. At $\nu \gg \Delta_-$ the soliton bandwidth Δ_- in Eq.(17) should be replaced by ν.

Analyzing Eqs.(15-18) and taking into account that $\Delta_- \sim \exp(-1/n\xi)$ at $n\xi \ll 1$, one can easily find that amplitude solitons contribute relatively more to the friction constant than to the conductivity σ_1. This means that if $T, \eta \ll \Delta$, a certain temperature range exists, in which the conductivity σ_1 is determined by excitations with $\varepsilon > \Delta$ and depends on temperature exponentially. On the other hand, γ is determined by localized states in the energy gap; therefore, the contribution to the conductivity due to the CDW motion does not depend on T exponentially at low temperatures. Similar temperature dependence of the conductivity, at the fields above and below the threshold value, was observed in TaS_3 /3/. Only at rather low temperatures, the contribution of soliton states to the conductivity σ_1 becomes noticeable. Note that in real crystals the conductivity may be determined not by zone mechanism, as assumed above, but by hopping mechanism of charge transfer /26/.

4. Weakly coupled Peierls conductors.

It is well known that in some sense the properties of a PC and a superconductor are similar. As in a superconductor, there is the energy gap in a PC determined by the amplitude of the order parameter. The CDW in a PC plays the role of the condensate. As is well known, interesting phenomena, discovered by Josephson, occur in weakly coupled superconductors. It is of interest to investigate the effects arising in weakly coupled PC. Such a weak link may be created artificially (tunnel junction or microbridge), or may appear due to defects or microcracks in a crystal. In this section we present the results of theoretical investigations /25/ of the effects arising in weakly coupled PC.

In the case of a tunnel junction we get for the tunneling current

$$I = \frac{1}{2R}\int d\varepsilon(\tanh(\varepsilon_+/2T) - \tanh(\varepsilon_-/2T))N_1(\varepsilon_+)N_2(\varepsilon_-)(1 + \frac{T_o^2}{T_o^2+T_Q^2}\frac{\Delta_1\Delta_2}{\varepsilon_+\varepsilon_-}\cos(\chi_1-\chi_2)) \tag{19}$$

where $\varepsilon_\pm = \varepsilon \pm V/2$, R is the junction resistance at $\Delta = 0$, V is the voltage across the barrier, T_o and T_Q are the matrix elements for tunneling with and without change of the momentum by the vector Q, $N_{12}(\varepsilon)$ are the density-of-states in electrodes 1 and 2. The first term in Eq.(19) corresponds to the quasiparticle current and the second one corresponds to $Im(I_c(V))$ term in Josephson current. It differs from zero at $V \neq 0$ and $\Delta \neq 0$. This component of the current will oscillate in time if an additional current is passed through one of the electrodes, and the CDW is sliding in this electrode. Note that this effect can be observed only if the conducting chains in both electrodes are parallel to each other.

The density-of-states $N(\varepsilon) = (g^R - g^A)/2$ can be found with the aid of Eq.(9). When T = 0 and there are no amplitude solitons in crystals, the current (19) is not equal to zero only if $V > 2\Delta$. When amplitude solitons exist in a crystal, $N(\varepsilon) \neq 0$ at $\varepsilon < 2\Delta$ and, therefore, $I \neq 0$ at $V < 2\Delta$.

Similar results are obtained in the case of a weak link with direct conductivity (for example, of a microbridge prepared on the base of a PC). The main distinction consists in the deviation of the I(V) curve from Ohm's low at high voltages $V \gg 2\Delta$: $I = V/R - I_o \mathrm{sign}(V)$, where I_o is a constant /25/.

5. Conclusion.

The theory presented above allows to study a number of different phenomena in a PC, such as different soliton states, influence of solitons on the conductivity of a PC and on the friction of the CDW, nonlocal effects arising in a sample with nonuniform current distribution, generation of the phase solitons at the contacts, electron tunneling in junctions formed by conductors with the CDW etc. From the analysis, carried out above, follows that nonuniform deformation of the CDW changes greatly the total conductivity of the PC in comparison with the case of uniform CDW. This fact should be taken into account in interpreting experimental results, because contacts and defects may lead to the CDW nonuniformity spreading over macroscopic distancies.The lack of necessary information on properties of the PC (for example, the electronic spectrum of many compounds is not yet known in detail) hinders the quantitative comparison of the theoretical results with the available experimental data. In addition, we did not take into account, in the explicit form, the impurity pinning which plays an important role in transport phenomena in the PC. Therefore, the model adopted here needs to be improved. Nevertheless, this model even in the present state can be considered as a basis for the description, at least qualitative, of different properties of a PC in a wide range of parameters.

1. P.A.Lee, T.M.Rice, P.W.Anderson, Sol.St.Commun. 14, 703 (1974).
2. N.P.Ong, P.Monceau, Phys. Rev. B16, 3443 (1977).
3. G.Gruner, Physica 8D, 1 (1983).
4. L.P.Gor'kov, G.M.Eliashberg, Sov. Phys. JETP 27, 328 (1968).
5. G.M.Eliashberg, ibid. 34, 668 (1972).
6. A.I.Larkin, Yu.N.Ovchinnikov, ibid. 46, 155 (1977).
7. L.P.Gor'kov, E.P.Dolgov, ibid. 50, 203 (1979).
8. S.N.Artemenko, A.F.Volkov, ibid, 53, 1050 (1981).
9. S.N.Artemenko, A.N.Kruglov, Fiz. Tverd. Tela 26, 2391 (1984).
10. S.N.Artemenko, A.F.Volkov, JETP Lett. 33, 147 (1981): Sov. Phys. JETP 54, 992 (1981).
11. L.P.Gor'kov, Pis'ma Zh. Eksp. Teor. Fiz. 38, 76 (1983): Zh. Eksp. Teor. Fiz. 86, 1818 (1984).
12. N.P.Ong, G.Varma, Phys. Rev. B27, 4495 (1983).
13. G.Mihaly, Gy.Hutiray, L.Mihaly, Phys. Rev. B28, 4896 (1983).
14. J.C.Gill, Sol. St. Commun. 44, 1041 (1982).
15. Yu.I.Latyshev, Ya.S.Savitskaya, V.V.Frolov, Pis'ma Zh.Eksp. Teor. Fiz. 40, 72 (1984).
16. S.N.Artemenko, A.F.Volkov, ibid. 40, 74 (1984).
17. G.Gruner,A.Zawadowsky,P.Chaikin, Phys. Rev. Lett. 46, 511 (1981).
18. S.A.Brazovskii, Pis'ma Zh. Eksp. Teor. Fiz. 28, 656 (1978): Zh. Eksp. Teor. Fiz. 78, 677 (1980).
19. W.P.Su,J.R.Schrieffer,A.J.Heeger, Phys. Rev. Lett. 42, 1698 (1979).
20. M.E.Itkis, F.Ya.Nad', Pis'ma Zh. Eksp. Teor. Fiz. 39, 373 (1984).
21. S.A.Brazovskii,L.P.Gor'kov,J.R.Schrieffer,Phys.Scripta 25,423 (1982)
22. S.A.Brazovskii, L.P.Gor'kov, A.G.Lebed', Zh. Eksp. Teor. Fiz. 83, 1198 (1982).
23. A.G.Lebed', Zh. Eksp. Teor. Fiz. 86, 1553 (1984).
24. S.A.Brazovskii, S.A.Gordyunin, N.N.Kirova, Pis'ma Zh. Eksp. Teor. Fiz. 31, 486 (1980).
25. S.N.Artemenko, A.F.Volkov, Pis'ma Zh. Eksp. Teor. Fiz. 37, 310 (1983): Zh. Eksp. Teor. Fiz. 87, 695 (1984).
26. S.K.Zhilinskii,M.E.Itkis,F.Ya.Nad', Phys.Stat.Sol.(a) 81,367 (1984)

NEAR COMMENSURABILITY EFFECTS ON CHARGE DENSITY WAVE DYNAMICS

Baruch Horovitz[+]
Department of Physics, Ben-Gurion University
Beer-Sheva 84105 Israel

Nearly commensurate charge density waves are described by a dilute phase-kink lattice. Linear response analysis shows that the shape of the AC response depends on the kink density and corresponds to a distribution of crossover frequencies. For nonlinear response a DC field E_A is defined where the sliding kink lattice reaches the phason velocity. For $E > E_A$ and for some range of fields and damping constants the sliding system is unstable and the phenomena of narrow band noise is observed. The results also correspond to unusual flux lattice dynamics in superconducting films.

I. INTRODUCTION

The family of compounds exhibiting a sliding charge density wave (CDW) has considerably grown in recent years. In addition to the veteran $NbSe_3$ and TaS_3 new compounds such as $K_{0.30}MoO_3$, $Rb_{0.30}MoO_3$, NbS_3, $(TaSe_4)_2I$ and $(NbSe_4)_{10}I_3$ have appeared as presented in this volume. Some of the common features of these compounds are: a) Nonlinear conductivity when the electric field E exceeds a threshold value E_c. b) Narrow band noise, i.e. an AC response to a DC field, for $E > E_c$. c) Most of these compounds are very close to being 4-fold commensurate. E.g. $NbSe_3$ exhibits two CDW transitions at $T_1 = 142°k$ and $T_2 = 59°k$ with wavevectors[1] $\vec{q}_1 = (0,0.2412,0)$ and $\vec{q}_2 = (0.5,0.2604,0.5)$ respectively. Some compounds even become commensurate at lower temperatures (Orthorhombic TaS_3[2], $K_{0.30}MoO_3$[3]).

The aim of the present work is to study the near commensurate situation. Impurity scattering is not included explicitly; it provides a homogenous damping mechanism.

Consider a CDW of the form $\sim \cos[(2\pi x/a+\psi)/4]$ where a is the lattice constant and x is the chain direction. Assume first that the CDW is commensurate, i.e. the phase ψ is constant. (An M-th order commensurability is treated similarly with $\psi/4 \rightarrow \psi/M$.) A change of ψ by 2π is equivalent to a translation $x \rightarrow x + a$ and leads therefore to a degenerate ground state; this implies an energy term $(4\pi)^{-1}E_1\cos\psi$ per unit length[4]. The force E_1 is determined by the scattering sequence $k_F \rightarrow 3k_F \rightarrow 5k_F \rightarrow 7k_F(= -k_F) \rightarrow k_F$ where $k_F = \pi/4a$ is the Fermi wavevector. Since $3k_F$ and $5k_F$ are high energy states $E_1 \sim \Delta^4/aW^3$ where 2Δ is the gap in the electron spectrum and W is of order of bandwidth[4,5]. This should be compared with the condensation energy for an incommensurate CDW which is $\sim \Delta^2/aW$. Since $\Delta \ll W$ E_1 has a negligible effect on the amplitude Δ. A possibility that E_1 is even smaller than the above estimate[5,6] is discussed in Section 4.

The low-lying excitations are obtained by allowing a space and time dependent $\psi(x,t)$. In particular the charge and current expectation values to lowest order in space and time derivatives are[4,5]

$$\rho(x,t) = \psi'(x,t)/4\pi \qquad j(x,t) = -\dot{\psi}(x,t)/4\pi. \tag{1}$$

These results can also be obtained by simple charge conservation and energy arguments[7,8]. (Note that the conventional CDW phase is $\psi/4$.)

A well-defined expansion in derivatives leads to an effective Lagrangian density[5] in presence of an external electric field $E(t)$

$$\mathcal{L}\{\psi\} = \frac{1}{4\pi}\{[\frac{M^*}{m}\dot{\psi}^2(x,t) - v_F^2\psi'^2(x,t)]/16v_F + E_1\cos\psi(x,t) - E(t)\psi(x,t)\} \tag{2}$$

where v_F is the Fermi velocity ($\sim Wa$), $M^*/m = 1 + 4\Delta^2/\lambda\omega_o^2$ is the Frohlich mass ratio, ω_o the phonon frequency and λ the dimensionless electron-phonon coupling. The small oscillation spectrum with $\psi \sim \exp(i\omega t - iqx)$ is

$$\omega^2 = \omega_o^2(1+q^2\xi^2) \tag{3}$$

where $\omega_o^2 = 8v_F E_1 m/M^*$ is the pinning frequency and ξ is the coherence length with $\omega_o^2\xi^2 = v_F^2 m/M^*$.

The incommensurate situation is described by the same Langrangian (2) except that $\psi(x,t)$ has boundary conditions such that its average slope gives the deviation from commensurability. The nonlinear $\cos\psi$ term modulates this slope such that locally ψ is a multiple of 2π and the 2π change (a phase kink) is localized. The resulting structure is a phase kink lattice with a density n_k of kinks[9]. The boundary condition is then

$$\int \psi'(x,t)dx/L = 2\pi n_k. \tag{4}$$

For $NbSe_3$ $n_k = -0.035/a$ and $n_k = 0.042/a$ for the two CDW's respectively. TaS_3 is closer to being commensurate with $n_k = 0.02/a$ below 210°K and becomes commensurate ($n_k=0$) below 130°K[2]. The phase kinks are considered to be two-dimensional walls which are deformable in the x direction; a zero temperature theory is then appropriate.

The basic equation of motion studied here is

$$\frac{1}{\omega_o^2}\ddot{\psi} + \frac{1}{\omega_c}\dot{\psi} - \xi^2\psi'' + \sin\psi = E(t)/E_1 \tag{5}$$

This equation follows from the Lagrangian (2) except for the $\dot{\psi}$ term which represents resistive terms due to impurities and phonon scatterings[10]. In the next sections the linear and nonlinear response of this equation with the special boundary condition (4) is studied.

II. LINEAR RESPONSE

Consider the AC conductivity with $E(t) = E\exp(i\omega t)$. For a low kink density $\bar{n}_k = n_k\xi << 1$ we can consider each kink separately. The results of the calculation[11] are summarized here: The conductivity is written in the form $\sigma(\omega) = A[\bar{\sigma}_T(\omega) + \bar{\sigma}_c(\omega)]$ with $A = e\rho_{eff}\omega_c/(4\pi SE_1)$ where ρ_{eff} is the fraction of total electron density which participates in the CDW transport and S is the area per conducting chain. The response is separated into conductivity $\bar{\sigma}_T(\omega)$ of the kink's localized translation mode and $\bar{\sigma}_c(\omega)$ of the extended continuum modes. We find for strong damping $(\omega_o >> \omega_c)$ $\bar{\sigma}_T(\omega) = \pi^2\bar{n}_k/2$ and

$$\mathrm{Re}\bar{\sigma}_c(\Omega) = \frac{\Omega^2}{1+\Omega^2}(1-2\bar{n}_k) - 4\bar{n}_k\Omega^2\int_0^\infty \frac{3\varepsilon_k^4+\Omega^2}{\varepsilon_k^4(\varepsilon_k^4+\Omega^2)^2}\,\frac{kdk}{\tanh\frac{1}{2}\pi k}$$

$$\mathrm{Im}\bar{\sigma}_c(\Omega) = \frac{\Omega}{1+\Omega^2}(1-2\bar{n}_k) - 8\bar{n}_k\Omega\int_0^\infty \frac{\varepsilon_k^2}{(\varepsilon_k^4+\Omega^2)^2}\,\frac{kdk}{\tanh\frac{1}{2}\pi k}$$

where $\Omega = \omega/\omega_c$ and $\varepsilon_k^2 = 1 + k^2$.

For $\bar{n}_k = o$ there is a single crossover frequency ω_c . When $\bar{n}_k \neq o$ all crossover frequencies $\omega_c(1+k^2)$ are integrated (Eq.6) with the effect of broadening the response. The reason is that the kinks break the translation invariance of the system and then all wavevectors k couple to a spatially uniform electric field. This effect is also shown in Fig.1.

An effective distribution in ω_c when $n \neq o$ is predicted by Eq. (6). The AC conductivity of TaS_3 was in fact interpreted by a distribution in ω_c[12]; this distribution, however, has a significant component at $\omega < \omega_o$ indicating the relevance of random impurities for the very low frequency response. If purer samples become available, it would be important to study their AC response. In particular when TaS_3 or $K_{0.30}M_oO_3$ become more commensurate (as function of temperature) the effects of Eq. (6) should be more apparent for sufficiently pure samples.

As TaS_3 or $K_{0.30}MoO_3$ become commensurate we predict an increase and slight sharpening of the AC response (Fig.1). An overall increase may also be due to an increase in the number of conducting chains[2] or a change in the gap which affects E_1 . However, the sharpening feature is uniquely related to the absence of phase kinks in the commensurate phase.

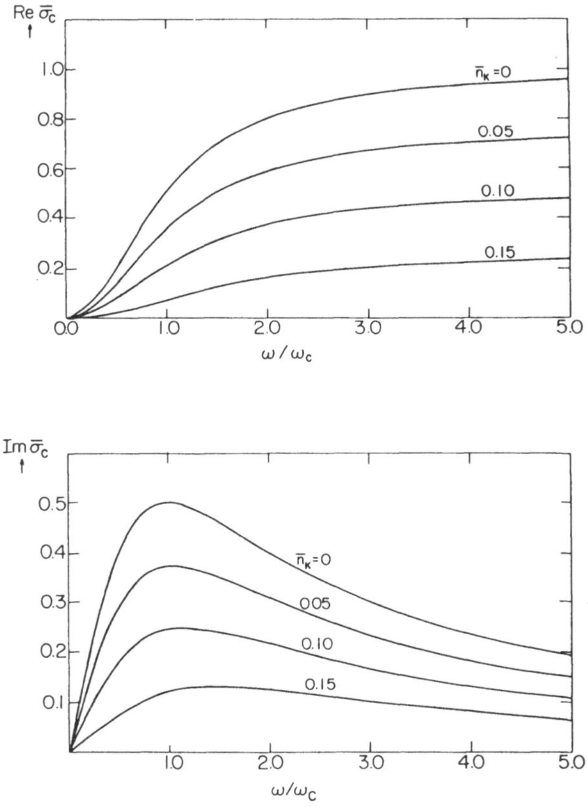

Fig. 1

Real part (top) and imaginary part (bottom) of the frequency dependent conductivity. $\bar{\sigma}_c = \bar{\sigma} - \bar{\sigma}_T$ excludes the frequency independent contribution of the kink's translation mode. The curves are for various values of kink density $\bar{n}_k = n_k\xi$.

III. NONLINEAR RESPONSE

Consider the current $j = \langle\dot{\psi}\rangle/4\pi$ where $\langle\ \rangle$ is a space average. This describes the measured current since the contact is a macroscopic object and therefore measures a space average of Eq.(1).

The commensurate overdamped system has a well known[13] critical field $E = E_1$, below which $j = o$ while for $E > E_1$ $j \neq o$ and is time dependent. The DC part is described by

$$\langle\dot{\psi}\rangle_{DC} = \omega_c[(E/E_1)^2 - 1]^{1/2} \tag{7}$$

while the AC part has a fundamental frequency $\omega_f = \langle\dot{\psi}\rangle_{DC}$.

The basic question now is whether similar phenomena happen when n_k is small but finite. Since the kink lattice is invariant under translation even weak fields result in $j \neq o$. In fact the linear response of the previous section yields $\langle\dot{\psi}\rangle = \frac{1}{2}\pi^2\bar{n}_k\omega_c E/E_1$. This current is considered as part of the experimentally measured Ohmic current and we look for nonlinearity at higher fields.

The moving kink lattice solution was previously studied in the context of Josephson junctions[14]. Moving solutions of the form $\psi(x-vt)$ were found in the velocity range $o < v < \omega_o\xi$. For the CDW problem we need to study also $v > \omega_o\xi$, i.e. velocities higher than the phason velocity. It has been expected that such solutions are unstable and become a non-moving solution $\psi(x,t)$[15].

A moving solution has only a d.c. current. This follows from Eq.(4) and

$$\langle\dot{\psi}\rangle = \int \dot{\psi}(x-vt)\,dx/L = -2\pi v n_k. \tag{8}$$

A non-moving solution $\psi(x,t)$ has in general a time dependent average $\langle\dot{\psi}\rangle$. Thus a transition from moving to non-moving solution results in the appearance of narrow band noise.

When $v = \omega_o\xi$ the second derivatives in Eq.(5) cancel and the leftover terms are just those of the overdamped pendulum. Therefore the currents of Eqs.(7,8) coincide when $v = \omega_o\xi$ and determine a field E_A,

$$E_A = E_1[1 + (2\pi n_k\xi\omega_o/\omega_c)^2]^{1/2} \tag{9}$$

Perturbation expansion around this solution shows that moving solutions yield a $j(E)$ relation which is tangent to the parabola of Eq.(7) (see Fig.2). Also a high field expansion around $\psi = 2\pi n_k x + \omega_c Et/E_1$ can be done. The result is a moving solution with a $j(E)$ relation approaching the parabola Eq.(7) from above.

To search for non-moving solutions we have performed a numerical study of Eq.(5)[16]. It was found that for a given $\bar{n}_k$ a finite regime of E/E_1 and ω_c/ω_o values exist in which moving solutions are unstable and narrow band noise exists. An example is shown in Fig.2.

A first order transition occurs slightly above E_A into a non-moving solution, the double line in Fig.2. As the field increases the time dependent part of $\langle\dot{\psi}\rangle$ decreases and the transition into a moving solution at high fields seems to be continuous.

The instability is related to formation of kink-antikink pairs in the almost commensurate regions of the kink lattice. The original kink annihilates the newly generated antikink; the new kink thus replaced the old kink so that the net kink propagation is highly nonlinear[16].

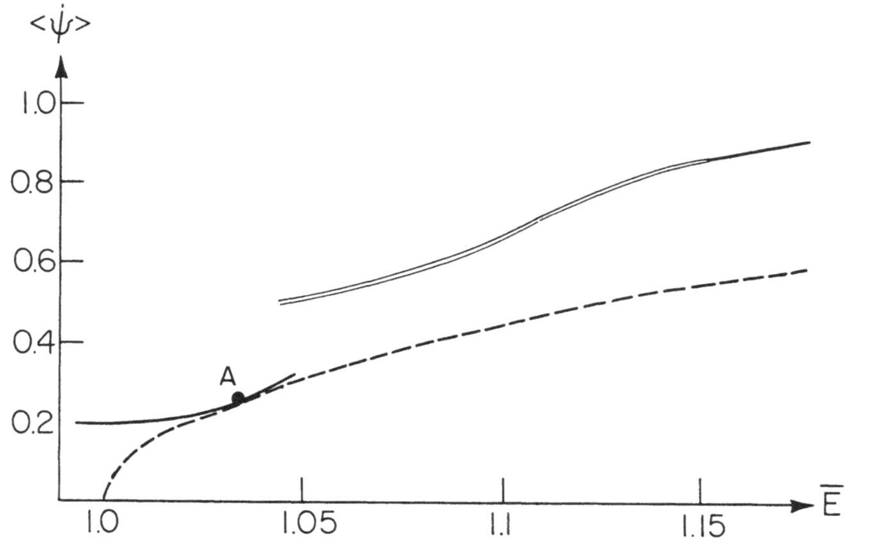

Fig. 2

DC part of $\langle\dot{\psi}\rangle$ (~ current) versus field $\bar{E} = E/E_1$. The dashed line is the parabola of the overdamped pendulum Eq.(7). The full lines are numerical solutions[16] with $\bar{n}_k = 1/24$, $\omega_c/\omega_o = 1$. The lines are tangent at point A where the field is E_A (Eq.9). The double line corresponds to a non-moving solution with narrow band noise.

IV. DISCUSSION

The linear response analysis of Section II shows that near-commensurability is an important factor. The non-linear response of Section III does not account quantitatively for the data. Eq.(5) is a simplified description of the nonlinear regime; its purpose here is to show that near commensurability can provide a nonlinear force for generating narrow band noise.

The threshold field of TaS_3 becomes smaller in the commensurate phase below 130°K[2]. This surprising result is in fact consistent with our model. The critical field $E_c \gtrsim E_A$ is higher than that of the commensurate situation (Eq.9).

The commensurate compounds TaS_3 and $K_{0.33}MoO_3$ also show that commensurability is an inevitable ingredient in the nonlinear response. The value of $E_1 \sim \Delta^4/aW^3$ seems, however, too large to account for the experimental values[2,5]. A situation where E_1 is smaller is the tight binding scheme for a 4-fold commensurate system[5,6]. The electron-phonon coupling for scattering electrons $k - q \to k$ is $g_{kq} \sim \sin(\frac{1}{2}qa)\cos[(k-\frac{1}{2}q)a]$. For the scattering $k_F \to 3k_F$ and $5k_F \to -k_F$ when $k_F = \pi/4a$ this coupling vanishes. E_1 is then obtained by integrating nearby scatterings which involve an additional $\int^{\Delta} \varepsilon^2 d\varepsilon/\Delta \sim \Delta^2$ factor, i.e. $E_1 \sim \Delta^5/aW^6$. There is, however, a correction from states deeper in the band with energies $\gg \Delta$. The coupling of these states is nonzero and then $E_1 \sim \Delta^3/aW^4$, although with a small numerical coefficient[6]. Such corrections are sensitive to the band structure and even depend on the occupation of other bands.

Recent experiments suggest that the narrow band noise is a surface[17] or a finite size[18] effect. If these results are valid they contradict the above theory which yields a bulk effect. In the latter case the theory should be taken as a prediction, that for some range of parameters a bulk narrow band noise is possible.

The phenomena of narrow-band noise has also been observed in flux lattice dynamics in a superconducting film[19]. In this related system[20] one can controll both random and commensurability effects, the latter by modulating the thickness of the film. Data on films with only random potentials shows the inverse effect, i.e. DC steps in an external AC field, but the direct effect of narrow band noise has not been seen[21]. With thickness modulation narrow band noise is present in both commensurate and near commensurate situation[19]. This unusual dynamics in the near-commensurate system is consistent with the model presented here. Its further experimental and theoretical study, interesting in their own right, will also shed more light on nonlinear CDW dynamics.

Acknowledgements: Parts of this work are collaborations with M. Weger, S.E.Trullinger, A.R. Bishop and P.S. Lomdahl.

+On leave from the Weizmann Institute, Rehovot, Israel.

REFERENCES

1) R.M. Fleming, C.H. Chen and D.E. Moncton, J. de Physique Colloque 44, C3-1651 (1983)
2) Z.Z. Wang, H. Salva, P. Monceau, M. Renard, C. Rouceau, R. Ayroles, F. Levy, L. Guemas and A. Meerschaut, J. Physique Lett. 44, L-311 (1983)
3) C. Schlenker, J. Dumas and J.P. Pouget, in proc. of Int. Conf. Synthetic Metals, Mol. Cryst. Liq. Cryst. (to be published)
4) P.A. Lee, T.M. Rice and P.W. Anderson, Solid State Commun. 14, 703 (1974)
5) B. Horovitz and J.A. Krumhansl, Phys. Rev. B29, 2109 (1984)
6) Y. Ohfuti and Y. Ono, Solid State Commun. 48, 985 (1983)

7) T.M. Rice, Solid State Commun. 17, 1055 (1975)

8) B. Horovitz, in Solitons, Ed. by S.E. Trullinger and V. Zakharov (North Holland, to be published)

9) W.L. McMillan, Phys. Rev. B14, 1496 (1976)

10) S.N. Artemenko and A.F. Volkov, JETP Lett. 33, 147 (1981)

11) B. Horovitz and S.E. Trullinger, Solid State Commun. 49, 195 (1984)

12) W.Wu, L. Mihály, G. Mozurkewich and G. Grüner, in this volume

13) G. Grüner, A. Zawadowski and P.M. Chaikin, Phys. Rev. Lett. 46, 511 (1981)

14) P.M. Marcus and Y. Imry, Solid State Commun. 33, 345 (1980)

15) M. Weger and B. Horovitz, Solid State Commun. 43, 583 (1982)

16) A.R. Bishop, B. Horovitz and P.S. Lomdahl (to be published)

17) N.P. Ong, G. Varma and K. Maki, Phys. Rev. Lett. 52, 663 (1984)

18) G. Mozurkewich and G. Grüner (to be published)

19) P. Martinoli, H. Beck, G.A. Racine, F. Patthey and Ch. Leemann, in this volume; P. Martinoli, O. Daldini, C. Leeman and E. Stocker, Solid State Comm. 17, 205 (1975)

20) A. Schmid and W. Hauger, J. Low Temp. Phys. 11, 667 (1973)

21) A.T. Fiory, Phys. Rev. Lett. 27, 501 (1971)

SHIFT IN THE LONGITUDINAL SOUND VELOCITY DUE TO SLIDING CHARGE DENSITY WAVES.

S.N. Coppersmith, Brookhaven National Laboratory, Upton, New York 11973 and C.M. Varma, AT&T Bell Laboratories, Murray Hill, New Jersey 07974.

The nonlinear conductivity observed for moderate electric fields in $NbSe_3$, TaS_3, $(TaS_4)_2I$, and $K_{0.3}MoO_3$ below the charge density wave (CDW) transition is believed to be due to the sliding of the CDWs. The sliding motion leads to a Doppler shift of the x-ray diffraction peaks, but this effect has not yet been resolved. We show here that besides the Doppler shift, a sliding incommensurate CDW causes a change in the longitudinal sound velocity of the crystal that is linear in the CDW velocity. The resulting anisotropic shift is estimated in a mean field approximation and found to be experimentally observable.

Some compounds exhibit charge density waves (CDWs) below a transition temperature T_c and also display nonlinear conductivity for small electric fields (on the order of 1V/cm).[1] This nonlinear conductivity has been interpreted as arising from "excess" current caused by sliding of the CDW. The small magnitude of the threshold field E_t supports this view because the electric field energy is small compared to the Fermi energy E_f, making a change in the number of free carriers unlikely. Experimentally, it has been shown that the x-ray diffraction peaks from the CDW do not lose intensity in the field, so that the nonlinearity is not caused by conversion of condensate electrons to normal electrons.[2]

However, to date there has not been an independent experimental verification that the excess conductivity for fields above E_t is due to sliding motion of the CDW. The most direct measurement would resolve the Doppler shift of the x-ray diffraction superlattice peaks when a CDW with wavevector $\vec{Q}$ moves with finite velocity v, causing the elastic peak at $(\vec{Q}, \omega=0)$ to change to an inelastic peak at $(\vec{Q}, \omega=\vec{Q}\cdot\vec{v})$, but so far the shift is below the experimental resolution.[3]

In this paper we show that for an incommensurate CDW the motion induces changes in the longitudinal sound velocity of the crystal that could be measured in ultrasonic experiments. The change in the sound velocity is proportional to the sliding velocity, and the resulting anisotropy should make the shift easier to resolve. The size of the effect is estimated and shown to be accessible to present ultrasonic techniques.

The effect is estimated using a very simple mean field approximation in which the CDW amplitude is fixed and the phase ϕ varies sinusoidally, $\phi = \vec{Q}\cdot\vec{x}$. Impurities and thermal effects that induce fluctuations in the phase are ignored. It is straightforward to generalize the discussion to allow for harmonics of the wavevector $\vec{Q}$. One can imagine starting with a microscopic model involving electron-phonon coupling and solving for the equilibrium CDW distortion. For instance, in one dimension one could use the Fröhlich Hamiltonian[4]

$$H = \sum_k \left[\frac{\hbar^2 k^2}{2m} b_k^+ b_k + \hbar sk\ a_k^+ a_k \right] + \sum_{k,q} \lambda_k \left[a_k^+ b_q^+ b_{q+k} + h.c \right] \qquad (1)$$

and solve the gap equation in the mean field approximation to find the CDW amplitude at wavevector $Q = 2k_F$. Here, the a^+'s and a's are phonon creation and annihilation operators, the b^+'s and b's create and annihilate electrons, m is the electron mass, and s is the speed of sound. Regarding this amplitude as fixed, one then evaluates the resulting effective Hamiltonian for the low-frequency phonons. A static CDW thus induces a potential on the ions of the form $V_0 \cos\vec{Q}\cdot\vec{x}$. One can evaluate V_0 for the one-dimensional jellium model, but here it will be estimated by using experimentally obtained values of the mean ionic displacements in the CDW state. The change in the phonon frequencies caused by a periodic potential sliding with velocity v is calculated to second order in perturbation theory for small V_0, and it is shown that there is a contribution linear in v.

The simplest case involves phonons parallel to the CDW wavevector, for which one can model the phonons as arising from a one-dimensional chain of ions. For a CDW sliding with velocity v, the classical equation of motion[5] for the displacement x_j of the jth ion can be written

$$m\ddot{x}_j = -\sum_k D_{jk}x_k + QV_0\sin Q(x_j + vt) . \qquad (2)$$

The dynamical matrix D_{jk} describes the ion-ion interactions in the harmonic approximation, and m is the ion mass. We assume D_{jk} describes phonons so that it is a function of j-k, it falls off sufficiently quickly with distance and that it is symmetric. This equation is nonlinear, so we proceed by assuming that V_0, and hence the distortions, are small. However, one must allow for the fact that the lowest energy state of the chain is distorted; so one writes the position x_j of the jth ion as $x_j = aj + \delta_j(t) + u_j(t)$, where a is the lattice constant, $\delta_j(t)$ is the forced distortion, and $u_j(t)$ describes the phonons. To first order in V_0 the distortion $\delta_j^{(1)}(t)$ is

$$\delta_j^{(1)}(t) = \frac{QV_0}{m(Qv)^2 - D(q=Q)} \sin Q(aj+vt). \qquad (3)$$

Here, $D(q) = \sum_j D_{oj}e^{iqaj}$. To second order in V_0, one finds that the

$$u(q,\omega) = \frac{1}{2\pi N}\sum_j \int dt\, e^{i(qaj-\omega t)} u_j(t) \text{ obey}$$

$$m\omega^2(q)\, u(q,\omega) = D(q)\, u(q,\omega) - \frac{1}{2} Q^4 V_0^2 \frac{u(q,\omega)}{m(Qv)^2 - D(q=Q)} + \frac{1}{2} V_0 Q^2 \left[u(q+Q, \omega-Qv) + u(q-Q, \omega+Qv)\right] . \qquad (4)$$

Equations (4) are coupled linear equations that can be solved by iteration for small V_0 to yield

$$m\omega^2(q) = D(q) - \frac{1}{2} V_0^2 Q^4 \frac{1}{m(Qv)^2 - D(q=Q)}$$
$$+ \frac{1}{4} V_0^2 Q^4 \left[\frac{1}{m(\omega+Qv)^2 - D(q-Q)} + \frac{1}{m(\omega-Qv)^2 - D(q+Q)} \right] \quad (5)$$

We assume the unperturbed system has reflection invariance, so that $D(k) = D(-k)$. This expression can be evaluated in the limit $q \to 0$, $\omega \to sq$ (s is the speed of sound), $v \to 0$ to yield

$$s = s_0 - \frac{Q^4 V_0^2}{D^3(Q)} QD'(Q)v + O(v^2) \quad (6)$$

Changes in s that are independent of v are accounted for in s_0, which is the speed of sound when $v = 0$, and $D'(q) = dD(q)/dq$.

Alternatively, one can evaluate the phonon frequencies using Green's function techniques. It is again necessary to allow for the distortions that occur in the lowest energy state, Fig. 1a, in order to ensure stability. The diagrams that lead to equation (5) are all represented in Fig. (1). The first order scattering process, not shown, from (q, ω) to $(q \pm Q, \omega \mp Qv)$ is responsible for the Doppler shift in the Bragg peak mentioned earlier but does not affect the sound velocity. A pictorial description of the mechanism is shown in Fig. 2; the phonon at (q, ω) scatters off the distortion at $(q+Q, \omega(q+Q) - Qv)$ and $(q-Q, \omega(q-Q) + Qv)$. The usual denominators of second order perturbation theory are then slightly shifted in different directions.

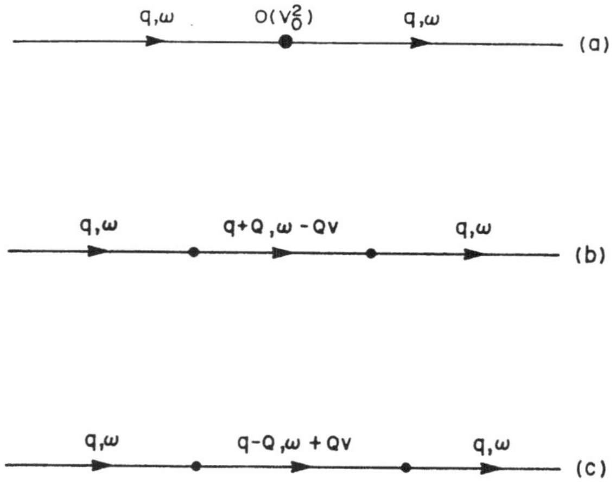

Figure 1. Second order diagrams that contribute to the sound velocity shift. Diagrams 1b and 1c contribute the term linear in the CDW velocity.

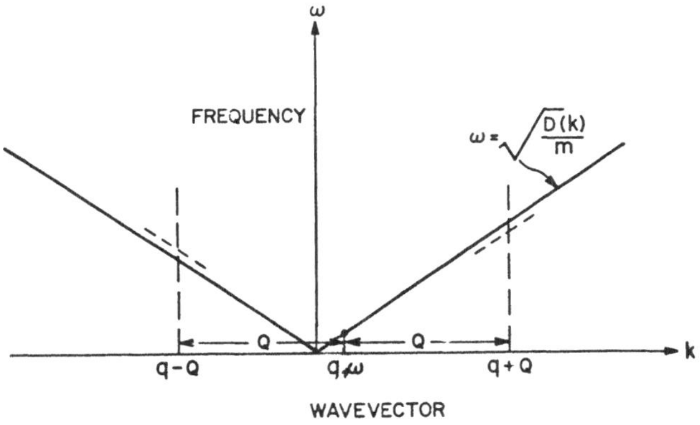

Figure 2. Schematic representation of mechanism causing linear velocity shift. There are momentum transfers from q to q + Q and q − Q, and the finite velocity causes different energy shifts for the two terms.

In the calculations discussed above the charge density wave is assumed to slide as a rigid body for fields above threshold. This assumption is only valid well above threshold. Near threshold the internal degrees of freedom of the charge density wave can not be ignored. The qualitative aspect of our result, the linear shift, which arises from symmetry breaking due to a moving density wave will persist however, although there may well be enhanced damping of the sound mode.

The expression (6) for $\delta s = s - s_0$ can be written in terms of the magnitude of the distortion $|\delta_j| \simeq \frac{QV_0}{D(Q)}$ as

$$\delta s = |\delta_j|^2 \, Q^3 \, \frac{D'(Q)}{D(Q)} \, v \, . \tag{7}$$

This expression for the shift displays several interesting features. First, it is proportional to $D'(Q)$, which is finite for an incommensurate CDW but is zero for a commensurate CDW. In the commensurate case, umklapp scattering within the unit cell must be considered, rendering the treatment described here inadequate, but we do not expect a shift linear in v to appear. Assuming that $D'(Q)$ is not anomalously large or small (so $QD'(Q) \sim D(Q)$), one finds δs is on the order of $Q^2 |\delta_j|^2 v$. Note that $|\delta_j|^2 Q^2$ is the dimensionless measure of the lattice distortion due to the charge density wave. The velocity shift obtained here can then be looked on as arising from the motion of this distortion at velocity v. Experimentally, $|\delta_j|$ is found to be approximately 5% of the lattice constant [6], so for $Q \sim \pi$, drift velocities on the order of 10 cm/sec and sound velocities of about 10^5 cm/sec, one finds

$$\delta s/s \approx 10^{-6} , \qquad (8)$$

which is large enough to be resolved experimentally.[7]

Since the shift is linear in v, changing the direction of the electric field driving the CDW should alter the shift. It is emphasized that this shift is added to the change induced by a static CDW.

In summary, we have shown that a moving CDW causes an anisotropic shift in the longitudinal sound velocity of the crystal that is linear in the CDW velocity. The effect provides a means to obtain independent experimental evidence for sliding CDW conductivity in crystals such as $NbSe_3$, $K_{0.3}MoO_3$, TaS_3 and $(TaS_4)_2I$.

At a conference where preliminary results of this work were reported,[8] we learnt of experiments[9,10] in which a change of the elastic modulus due to sliding charge density wave were observed. However, the magnitude of the observed effects are much larger than estimated here because of geometry and phason distortion effects. Detailed comparison with experiment must await more complete understanding of these factors.

Acknowledgement

Work at Brookhaven supported by the Division of Materials Sciences U.S. Department of Energy under Contract No. DE-AC02-76CH00016.

References

1. See, i.e., N. P. Ong, Can. J. Phys. 60, 757 (1982).
2. R. M. Fleming, D. E. Moncton, and D. B. McWhan, Phys. Rev. B18, 5560 (1978).
3. D. Moncton, private communication.
4. H. Fröhlich, Proc. Roy. Soc. (London) A23, 296 (1954).
5. The phonon eigenvalues are insensitive to whether a classical or quantum Hamiltonian is used.
6. N. V. Smith and S. D. Kevan, to be published.
7. B. Golding, private communication.
8. S. N. Coppersmith and C. M. Varma, Bull. Am. Phys. Soc. 29, 357 (1984).
9. G. Mozurkewich, P. M. Chaikin, W. G. Clark, and G. Grüner, Bull. Am. Phys. Soc. 29, 469 (1984).
10. J. W. Brill and W. Roark, Bull. Am. Phys. Soc. 29, 470 (1984).

MICROSCOPIC LOCAL MECHANISMS FOR "NOISES" GENERATED BY MOVING CDW

L.Gor'kov
L.D.Landau Institute for Theoretical Physics,
142432 Chernogolovka, USSR

Almost coherent "noises", generated by the moving charge density wave in a number of the Q1D compounds like the transition metal threechalcogenides, are interpreted in terms of the phase slip processes which take place at the conversion of the CDW current on the contact with a normal metal into the current of the ordinary carriers.

I. Introduction

Between many interesting results in the physics of the Q1D metals the one of the most importance is the experimental discovery of the so-called "Frohlich conductivity" - the transport of the electrical current by the moving CDW. This mechanism is now firmly established and have been examined rather narrowly. However, so far there is no exhaustive understanding of the phenomenon which is undoubtely related with the CDW motion. This is the generation of the almost coherent oscillations observed in a number of inorganic Q1D metals. While the "narrow band noises" generation is now observed in many of these materials, in most details it has been investigated for the transition metals threechalcogenides, especially, for $NbSe_3$. This compound is now available in the form of the quite perfect samples characterized by the record (for this group of materials) values of the low temperature conductivity. At the same time, the corresponding threshold electric fields, E_t, above which the Frohlich conductivity mechanism takes place, are also relatively low for $NbSe_3$. Probably this is somehow related with the fact that this compound does not display a strong conductivity anisotropy, i.e. the specific onedimensional features are not strongly pronounced in it. In what follows we shall basically appeal to the experimental results which have been obtained namely for this compound, besause the microscopic model equations suggested below, will have the quantitative meaning only if the 3D conductivity anisotropy is not too large.

Most of theories devoted to the generation phenomenon relate it, by some means or other, to the CDW impurity pinning in the bulk. Recently strong experimental arguments [4,5] have been obtained in favour of a local origin of the generation mechanism (see also the discussion in [6]). The explanation for these observations given in [5,7] , has been essentially based on the picture that due to the better contact material conductivity the effective electric field in the vicinity of contacts is lower than the threshold field. Hence, near a contact the CDW is fixed and it moves only far off contacts. The crossover between two regimes takes place in a "phase slip" process. The reasons why the "phase slip" occurs in a sort of the periodic regime remain unclear in this interpretation.

The idea of the "phase slip centers" (PSC) has been independently introduced in [8] in connection with the discussion of boundary conditions at the contact between the Peierls material and an ordinary metal where the conversion of the CDW current into the normal carriers current is to occur. Then it is possible to show that a nonlinear periodic regime appears near the boundary which, for instance, at the given current generates the periodic voltage component with the amplitude independent on the sample length and the volume impurity concentration. The mechanism is to prevail either in pure enough samples or when the applied electric field considerably exceeds the threshold field. The fields remain still quite low (the nonlinearity of the excitations current is negligable: $eE\xi_0/T_p \ll 1$). In [8] it was argued that the PSC position, x_0, is large, $x_0 \gg \xi_0$. Otherwise, the auto--oscillatory regime takes place far enough off the real border to smooth away its roughnesses and inhomogenities which could destroy the coherency of the oscillations and widen the spectrum lines.

II. Microscopic Model

One obviously needs some equations describing these nonstationary processes. To reproduce the typical properties of the Q1D conductors qualitetively, it is desirable to point out a simple model which, however, is able to describe the structural transition. The CDW, appearing at this transition, has to be characterized by a superstructure vector, Q, with at least one component incommensurate with the lattice period. All these conditions can be fulfilled in an anisotropic metal with the electron spectrum corresponding to the two open Fermisurface sheets positioned near the $\pm K_F$-planes in the zone, and satisfying the approximate "nesting" conditions

$$\varepsilon(p+Q) \simeq -\varepsilon(p) . \tag{1}$$

Eq.(1) can be a result of the quasi-onedimensionality of the conductor and $Q=(2K_F,0,0)$. The phase transition is provided by the three-dimensional phonon spectrum [9]. In the tight binding model Eq.(1) is exact [10], but $Q=(2K_F, 1/2,1/2)$. In what follows namely this last case is accepted because a substantial simplification is possible for this model. In fact, it is known that the structural transition temperature, T_p, is reduced by impurities. In the Q1D metals the large defect concentration introduces complications due to the specific 1D localization effects and also leads to the strong CDW pinning. However, if the electron spectrum (even if its Fermi surface consists of two open parts) reveals the 3D dispersion also in the transverse direction, this model would have a sort of the "gapless regime" when the presence of impurities makes the transition temperature very small: $T_p \tau \ll 1$. The derivation of the time-dependent equations for the order parameter $\Delta = |\Delta| exp(i\varphi)$ (Δ plays the role of a gap in the electron spectrum and is proportional to the Peierls distortion) is described for this limit in [8,11]. Here we shall write down the equations directly in their dimensionless form (for definitions see [11,12]). For purposes of this paper let us only mention that the length scale is $\xi_o \sim \hbar v/T_p$ and the field is measured as $eE\xi_o/T_p$. Thus, one has:

$$\dot{\Delta} + iE_x\Delta - \Delta - \nabla^2\Delta + |\Delta|^2 = 0, \tag{2}$$
$$j_x = E_x - \lambda|\Delta|^2 E_x - \bar{\lambda}|\Delta|^2\dot{\varphi}.$$

In the current expression $\lambda, \bar{\lambda} \ll 1$ $(T_p\tau \ll 1)$. This corresponds to the obvious fact that at small $T_p \ll 1/\tau$ the CDW contribution into the current transport is small. In the first approximation the electric field, E, is determined by the given current and therefore in the first of Eq.'s (2) for the gap it plays the role of an external parameter. If the solution for Δ (r,t) is found, the correction terms (of order of λ, $\bar{\lambda}$) in the expression for current will provide the voltage oscillations (at the fixed current). The first of Eq.'s (2) is also convenient to re-write in the different form:

$$|\dot{\Delta}| - (1 - Q^2 - |\Delta|^2)|\Delta| + \nabla^2|\Delta| = 0, \tag{3}$$
$$(\dot{\varphi} + E_x)|\Delta|^2 - div(|\Delta|^2 Q) = 0$$

(Here $Q = \nabla\varphi$).

Far off the contact $\varphi = -E_x t$ $(\Delta = exp(-iE_x t))$. This corresponds to the homogeneousely moving CDW. Near the contact the CDW is fixed (see somewhat below). Close to x=o in Eq.(3) the dependence on time can be omited and the solution would be

$$|\Delta|^2 = 1 - Q^2 \;; \quad 3Q - \ln\frac{1+Q}{1-Q} = E_x\, x. \tag{4}$$

This solution, as it is easy to verify, becomes absolutely unstable at $Q \gtrsim 1/\sqrt{3}$, i.e. at $X_0 \sim E^{-1}$. The physical meaning of such an instability is merely a manifestation of the fact that the finite $Q = \nabla\varphi$ corresponds to the local change of the superstructure vector, or to some deviation from the condition of the optimal Fermi surface nesting. The instability mentioned above would already appear at small perturbations. At the straight forward solution of the nonlinear equation for the gap the stability of the solution (4) is, of course, determined by the nonlinear terms. Therefore the position X_0 for the PSC cannot be pointed out from some general considerations. The solution (4) is written here in so far as it shows that X_0 could be large at low $E \ll 1$. The numerical integration [12] actually gives

$$X_0 \simeq 1{,}15\, E^{-0{,}28}. \tag{5}$$

In other words, while X_0 is large compared with the coherence length, the growth of X_0 with the field decrease is rather slow.

III. Boundary conditions

At the numerical calculations [12] the conditions on the boundary (x=0) have been taken in the form $\Delta(X=0,t) = \Delta_0\ (\Delta_0 \gg 1)$. Any real interface between two different metals always introduces atomic deformations near the contact (including the $2K_F$-deformation). The distortions fastly decrease inside each metal. However, near the surface they are not very sensitive to the fact whever the Peierls structural transition have already taken place or not. This means that at least in the applicability region for the Ginzburg-Landau approach a temperature independent "tail" of comparatively large deformations is always present near the boundary and any solution for the gap is to fit this "tail" near x=0. In the intermediate region, where the expansion over Δ is still applicable (but $|\Delta| \gg 1$), in the first of Eq.'s (2) or in Eq.'s(3) the small terms could be omited. As the result, one gets

$$|\Delta|^2 Q = q = const,$$
$$|\Delta(x)| = \frac{1}{\sqrt{2}\,(x - x^*)} \simeq \frac{1}{\sqrt{2}\,x} \tag{6}$$

Where q is determined by the solution for off x=0 and X^* is an effective shift of the boundary, which is dependent on the internal structure of the interface and is therefore negligably small compared with the Ginzburg-Landau scale. Eq.'s(6) suggest the effective boundary conditions for Eq.(2): $|\Delta(x)| = (\sqrt{2}\,x)^{-1}$ and Q=0 at $x \to 0$. This form is, however, not convenient for the numerical calculations and in [12] the growth of $|\Delta|$ at x=0 was simulated by taking $\Delta(x=0)=\Delta_0>1$. According to [12], the PS-position, x_0, is not too sensitive to the value of Δ_0. Nevertheless, it would be interesting to verify the sensibility of the solution with respect to the choice of the correct boundary conditions.

It is now relevant to stress the dynamical character of the PS-centers. In fact, let us assume that at t=0 $\Delta(x_0,t=0)=0$:

$$\Delta(x,t=0) = a|x-x_0|.$$

Linearizing Eq.(2) and using the standard solution for the parabolic equations, one obtains the gap behaviour at $t>0$:

$$\Delta(x,t) = \frac{\exp(1-iEx)t}{2\sqrt{\pi t}}\, a \int_{-\infty}^{+\infty} |x'-x_0|\, e^{-\frac{(x-x')^2}{4t}}\, dx'.$$

Therefore the zero value at x_0 in the spacial $\Delta(x,t)$ distribution disappears at larger t. The gap "grows over" till the next cycle beginning when it again passes across zero. This result completes the description of the PS process as the nonlinear autogenerating process.

IV. Threshold field for the finite size sample

The phase slippage mechanism described above exists even in pure samples due to the CDW stoppage near the contact boundary. The formulation of the problem can be inverted: in a finite size sample the CDW motion will be possible only if its size, L, is large, or the applied field is strong enough, since the domain near the contact, where the CDW is fixed, increases with the field decrease. At first sight an estimation of the threshold field, $E_t(L)$, could be found from Eq.(5) (substituting L for x_0 and then inverting this relation). The numerical calculations [12] give

$$E_t(L) \simeq 2{,}55L^{-1.23}. \qquad (7)$$

The substantial difference between Eq.(7) and Eq.(5) is obliged to the long range interaction of the PS-centers at the two ends of the sample, as it is discussed in [12].

V. "Vortexes"

In the homogeneous solution, discussed for simplicity so far, the gap passes across zero simultaneousely in the whole plane across the sample. Such a solution is hardly stable for a thick enough sample. In [5,7,8] it has been pointed out that a different regime is more plausible in this case, which would correspond to the chain of moving "vortexes". The phase changes by 2π after encircling each separate vortex (In other words, such a "vortex" is the dislocation in the Peierls superstructure). Therefore the phase transport due to the CDW motion can be compensated by the motion of these dislocations across the sample. Beside the computational complications obviousely appearing at the solution of this threedimensional problem, there are many additional questions. The main of them concerns the energetic barriers on the sample surface for the dislocations penetration into the sample. The same question is left for the so-called "strands" (longitudinal domains), which have been first observed in $NbSe_3$ [13] and are not understood so far.

VI. On the extention of the microscopic theory

The above theory suggests a simple but nontrivial scheme which explicitly describes the phase slippage. Unlike the phase hamiltonian or the model nonlinear pendulum equation, the PSC's are directly connected with the passage of the local order parameter across zero. Due to this the problem of the charge accumulation (which is always proportional to the phase growth at the nonzero gap) does not appear at all in the theory. However, the choice of the "gapless" (high impurity concentration) limit, while it seriousely simplifies the mathematical treatment, does omit some features important for the more realistic situations. These are various relaxation processes for the momentum and energy. The relaxation between different subsystems (electrons, phonons and the "condensate") becomes important (at low impurity concentration), especially at low temperatures, and is strongly dependent on the fact whever the system is of a semimetal type ($NbSe_3$), or it is of the dielectric type (TaS_3). The account of the relaxation processes would introduce some new characteristic scales for the magnitude and the phase variation of the gap. The phenomenological approach based on an generalization of Eq.'s(2,3), seems to be more adequate to the problem.

REFERENCES

1. R.M.Fleming, G.G.Grimes. Phys.Rev.Lett., **42**, 1423 (1979)
2. P.Monceau, J.Richard, M.Renard. Phys.Rev., **B25**, 931 (1982)
3. A.Zettl, G.Grüner, A.H.Thompson. Phys.Rev., **B26**, 5760 (1982); A.Zettl, C.M.Jackson, G.Grüner. ibid., p.5773

4. J.C.Gill. Solid State Comm., 44, 1041 (1982)
5. N.P.Ong, G.Verma. Phys.Rev., B27, 4495 (1983)
6. Proc. of Intern. Symp. on Nonlinear Transport in Inorganic Q1D Conductors, Hokkaido University, Sapporo, Japan, 1983
7. K.Maki. ibid., p.17
8. L.P.Gor'kov. Pis'ma Zh.Eksp.Theor.Fiz., 38, 76 (1983)
9. L.P.Gor'kov, E.N.Dolgov, A.G.Lebed'. Zh.Eksp.Theor.Fiz., 82, 613 (1982)
10. B.Horowitz, H.Gutfreund, M.Weger. Solid State Comm., 39, 541 (1981)
11. L.P.Gor'kov. Zh.Eksp.Theor.Fiz., 86, 1818 (1984)
12. I.Batistič, A.Bjelis, L.P.Gor'kov. J.Physique, 45, 1049(1984)
13. K.K.Fung, J.W.Steeds. Phys.Rev.Lett., 45, 1696 (1980)

Phase Vortices and CDW Conduction Noise*

Kazumi Maki
Department of Physics
University of Southern California
Los Angeles, CA 90089-0484 USA

We review a theoretical aspect of the phase vortex model proposed by Ong, Verma and Maki to account for the quasi-periodic voltage oscillation observed in the non-Ohmic regime of the CDW states in $NbSe_3$ and TaS_3. Phase conservation requires a train of phase vortices at the domain boundary whenever two adjacent CDW domains move with different sliding velocities. These phase vortices move along the boundary with the number of vortices which sweep the boundary per second being proportional to the difference in the sliding velocities. In addition to conserving phase these vortices convert the CDW current into normal current (i.e. the quasi-particle current). This implies that the sources of the normal current are localized and move with the phase vortices. Therefore a periodic array of phase vortices gives rise to voltage oscillations with a fundamental frequency proportional to the sliding velocity, when the CDW is pinned on one side of the array and moving on the other side.

*Supported by the National Science Foundation under grant number DMR-82-14525.

1. Introduction A series of quasi-one dimensional compounds exhibit characteristic CDW transport. For a small electric field the electric current is Ohmic. However when the electric field exceeds a small threshold field E_T (5 to 10^2mV/cm), the conductivity depends on E[1]. Simultaneously a voltage oscillation[2] is developed with the fundamental frequency proportional to J_{CDW} (i.e. the current in excess of the Ohmic current).

Most of the features of CDW transport are interpreted in terms of the CDW being pinned for $E<E_T$ and sliding for $E>E_T$. However the voltage oscillation appears to have eluded a simple interpretation until now. I shall describe here a theoretical aspect of the phase vortex model, which was proposed recently by Ong, Verma and Maki[3] in order to account for the voltage oscillations (or "narrow band noise".) Of the earlier theories the most popular is the classical particle model by Grüner et al.[4]. They assumed that the motion of the CDW is described by a simple ϕ, which is the phase of the CDW condensate. Furthermore the impurities are assumed to produce a periodic potential in ϕ. This model predicts a well defined threshold field E_T and the non-Ohmic behavior above the threshold field where the CDW starts sliding. The appearance of the oscillatory current above the threshold is also predicted. On the other hand, the model cannot describe the E dependence of the excess current qualitatively. However, this is not so surprising as the CDW condensate possesses many degrees of freedom[5] and it is certainly impossible to describe every aspect of the CDW transport in terms of a single ϕ. The classical model assumes implicitly that the noise is due to impurities and it is a bulk phenomenon. Recent expeirments[6,7] by Ong and coworkers indicate, on the contrary, that the noise source is highly localised. Furthermore the cleaner sample exhibits stronger voltage oscillation, indicating that the impurity is not important for the noise generation. Therefore we have to look for the origin of noise elsewhere.

2. Phase Conservation Let us recall first that the contacts introduce major perturbation in most of the quasi-one dimensional CDW systems. This is because the sample crystals are usually small (with transverse diameters of the order of 10 μm for $NbSe_3$) and the contacts (silver paint which completely envelops the sample) are better conductors than the sample. Thus even when the electric field in the bulk is much larger than the threshold E_T, the electric field under the contacts is in general much smaller than E_T. The advancing CDW in the bulk clashes with the stationary CDW under the contact. In other words, the phase carried by the bulk CDW accumulates at the boundary between moving and stationary CDW. Note that whenever the bulk CDW advances by $\lambda_{CDW} = 2\pi Q^{-1} = \pi k_F^{-1}$ the phase increases by 2π. The accumulated phase can be eliminated by boiling off all the CDW condensate at the phase boundary. However, there is a

much more economical way to get rid of the accumulated phase, namely by phase slippage. When one phase vortex sweeps along the phase boundary it carries away the excess phase of 2π. Therefore everytime the bulk CDW advances by λ_{CDW}, one phase vortex is required to sweep along the phase boundary. When a train of phase vortices is sweeping the phase boundary, the vortex velocity v_V is related to the sliding velocity of CDW v by

$$v_V/\ell = v/\lambda_{CDW} = f \qquad (1)$$

from the phase conservation, where ℓ is the distance between the phase vortices and f is the observed voltage oscillation frequency. Therefore phase conservation regulates the characteristic frequency of the vortex lattice.

More generally whenever two CDW domains with different sliding velocities v_1 and v_2 are adjacent to each other, the phase conservation requires a train of phase vortices moving transversally along the domain boundary. The characteristic frequency in this case is given by $f = (v_1-v_2)/\lambda_{CDW}$.

In the bulk the sliding velocity of the CDW is determined from the equations

$$I = \sigma_a E + \sigma_b(E)\,(E-E_T) \qquad (2)$$

$$= \sigma_a E + eQ^{-1} n_c v \qquad (3)$$

where $\sigma_b(E)$ is the nonlinear part of the conductivity ($\sigma_b(E)=0$ for $E<E_T$), I is

the total current, and n_c is the condensate density. Solving Eqs. (2) and (3), we obtain

$$v = 0 \text{ for } E<E_T$$

$$= Q(en_c(\sigma_a+\sigma_b))^{-1}\sigma_b(I-\sigma_a E_T) \text{ for } E>E_T. \qquad (4)$$

For a constant I, the sliding velocity v assumes a maximum value at T = 52K in $NbSe_3$, (approximately where $\sigma_a E_T$ assumes a minimum value.[7]) As already stated the sliding velocity of the CDW practically vanishes under the contacts. Therefore the frequency of the voltage oscillation is dictated by v given in Eq. (4) through Eq. (1).

When a large localized temperature gradient is applied[8], the temperature gradient can also create two domains of the CDW, one sliding with a different velocity from the other. Such a domain boundary generates the voltage oscillation with frequency $f = (v_1-v_2)/\lambda_{CDW}$, where v_1 and v_2 are sliding velocities of the adjacent domains.

When a weak temperature gradient is applied throughout the system, the response of the CDW is more complicated. For a small temperature gradient the bulk CDW is expected to slide as a monodomain with a sliding velocity given by Eq. (4) averaged over the sample. However, when the temperature gradient exceeds a critical value the bulk domain appears to split into many domains, where each domain slides with the sliding velocity corresponding to the local temperature[7]. Experimentally the threshold temperature gradient appears to be quite small[7]. Indeed this experiment establishes unequivocally the locality of the noise sources.

3. Phase Vortices as Current Convertor We now examine the role of the phase vortex in more detail[9]. A phase vortex moving with velocity v_V in the y direction is given by

$$\phi(\vec{r},t) = \tan^{-1}([y-v_V t]/x) \quad (5)$$

where ϕ is the phase of the CDW condensate. Here we consider an isotropic system, as the effect of anisotropy in the elastic coefficient is easily eliminated by a scale transformation[10]. The CDW current direction is taken along the x axis.

Substituting Eq. (5) into the expressions for expressions of the CDW charge density and CDW current density

$$\rho_{CDW} = en_s Q^{-1} \partial\phi/\partial x \quad (6)$$

and

$$J_{CDW} = -en_c Q^{-1} \partial\phi/\partial t \quad (7)$$

where n_s and n_c are the CDW charge density and the CDW condensate density, we obtain

$$\rho_{CDW} = en_s Q^{-1}(y-v_V t)/(x^2+(y-v_V t)^2) \quad (8)$$

and

$$J_{CDW} = en_c Q^{-1} v_V x/(x^2+(y-v_V t)^2) \quad (9)$$

where J_{CDW} is along the x direction. Equation (8) means that a phase vortex does not carry any charge but, instead, an electric dipole moment directed in the y direction. Equation (9) indicates that a phase vortex is a source or sink of the electric current with total current

$$2\pi e n_c Q^{-1} v_V = e n_c \lambda_{CDW} v_V.$$

Let us now consider the CDW near the contact pad area. For x>0, we assume that the CDW is sliding with velocity v towards the contact, while the CDW under the contact (x<0) is stationary. Then phase conservation requires that a train of phase vortices sweep the boundary interface in the yz plane at x=0, with velocity v_V and vortex density ℓ^{-1} as given in Eq. (1). Substituting the current associated with each vortex (recall that each vortex is a current source or sink) given by $e n_c \lambda_{CDW} v_V$, we find that the train of vortices not only guarantees phase conservation but also current conservation. Since there is no CDW current in the region x<0, the CDW current in the region x>0 carried by the sliding CDW is converted into the quasi-particle current.

However, since the current sources associated with individual vortices are rather localized as seen from Eq. (9), the normal current exhibits a periodic variation in the y direction in contrast to the CDW current, which gives rise to a periodic oscillation in the observed voltage. In order to make this argument more quantitative, we note first that Eq. (9) for the current associated with a phase vortex is not valid far away from the center of the vortex. Rather Eq. (9) should be replaced by

$$J_{CDW} = -2en_c Q^{-1} v_V \; \partial/\partial x \;\; K_o(r/L) \qquad (10)$$

where $K_o(z)$ is the modified Bessel function, $r = (x^2+(y-v_V t)^2)^{1/2}$ is the distance from the vortex and L is the Fukuyama-Lee-Rice length[10], which is the characteristic screening distance of the system. Then Eq. (10) may be approximated as

$$J_{CDW} = -(1/4L_t)(en_c \lambda_{CDW} v_V)\operatorname{sech}^2([y-v_V t]/L_t) \qquad (11)$$

for $x \leqq 0$. Therefore the amplitude of the voltage oscillation is given by

$$\Delta V = \begin{cases} 1/2(en_c \lambda_{CDW} v_V \rho)L^t/a & \text{for } L_t < a \\ 1/2(en_c \lambda_{CDW} v_V \rho) & \text{for } L^t > a \end{cases} \qquad (12)$$

where a is the transverse dimension of the sample (the sample thickness in the y direction) and L_t is the transverse Fukuyama-Lee-Rice length ($L_t \sim 0.1L$) and ρ is the effective resistance of the sample near the contact. The L_t dependence in Eq. (12) follows from the fact that the quasi-particle current source is localized within the width of $2L_t$. The predicted ρ and a^{-1} dependence of the voltage oscillation is observed in some studies on $NbSe_3$ and orthorhombic TaS_3 samples.[11]

4. <u>From Chaotic to Periodic Oscillation</u> So far we have considered only a regular array of phase vortices. However when the number of phase vortices

in the sample is rather small, their motion is rather chaotic. As the number of phase vortices increases, the distance between vortices ℓ decreases. When ℓ becomes less than L_t the transverse Fukuyama-Lee-Rice length, vortices feel a strong mutual repulsion. Therefore for $\ell<L_t$, the stable configuration should be a regular vortex lattice. This implies that there should be a transition from chaotic to periodic motion as the vortex density is increased. This transition is very analogous to the commensurate- to-incommensurate transition in one-dimensional systems in the presence of a pinning potential as discussed by Aubry.[12] In a CDW crystal the walls where phase vortices enter and exit provide pinning centers even in the absence of impurity scattering. Indeed a recent observation[11] in the CDW conduction noise in TaS_3 indicates such a transition, although the details of the transition is not clear.

5. Conclusion We have reviewed a theoretical aspect of the phase vortex model. We have shown that, whenever different parts of the CDW slide with different velocities, phase vortices are necessarily found in order to satisfy phase conservation. The local nature of the noise source and the length independence of the voltage oscillation are natural consequences of the model. The model enables us to interpret a number of recent experiments on $NbSe_3$ and TaS_3. However, unlike another model developed by Gor'kov,[13] the present model is still semi-phenomenological although we believe that our model is more realistic except for samples with thickness much less than 1 um. Therefore more work on the vortex dynamics is clearly desirable.

Acknowledgement I would like to thank Nai Phuan Ong for a number of discussions on experimental results in his laboratory. Most of the ideas developed here are the fruit of numerous discussions with him. I wish to thank also Pierre Monceau, M. Renard, M.C. Saint-Lager, and Z.Z. Wang for a stimulating discussion on their recent thermal gradient experiments.

† † †

References

1. R.M. Fleming and C.C. Grimes, Phys. Rev. Lett. 42 1423 (1979); R.M. Fleming, Phys. Rev. B22, 5606 (1980).
2. P. Monceau, J. Richard, and M. Renard, Phys. Rev. Lett. 45, 43 (1980), and Phys. Rev. B25, 931 (1982).
3. N.P. Ong, G. Verma, and K. Maki, Phys. Rev. Lett. 52, 663 (1984).
4. G. Grüner, A. Zawadowski and P.M. Chaikin, Phys. Rev. Lett. 46, 511 (1981).
5. See for example, L. Sneddon, M.C. Cross, and D.S. Fisher, Phys. Rev. Lett. 49, 292 (1982); D.S. Fisher, Phys. Rev. Lett. 50, 1486 (1983); L. Sneddon, Phys. Rev. Lett. 52, 65 (1984); J.B. Sokoloff and B. Horovitz, J. Phys. (Paris) Colloq. 44 C3-1667 (1983).
6. N.P. Ong and G. Verma, Phys. Rev. B27, 4495 (1983).
7. G. Verma and N.P. Ong, Phys. Rev. B 30, (1984).
8. This type of experiment was recently performed by P. Monceau, M. Renard, M.C. Saint-Lager and Z.Z. Wang (private communication).
9. A brief description is given in K. Maki, N.P. Ong and G. Verma, Proc. Int. Conf. on the Physics and Chemistry of Low Dimensional Synthetic Metals, Abano Terme 1984, to be published in Mol. Cryst. Liq. Cryst.
10. P.A. Lee and T.M. Rice, Phys. Reve. B19, 3970 (1979).
11. N.P. Ong, C.B. Kalem and J.C. Eckert, Phys. Rev. B 30 (1984).
12. S. Aubry, Physica 7D, 240 (1983); S. Aubry and P.Y. Le Daeron, Physica 8D, 381 (1983).
13. L.P. Gor'kov, Pis'ma Zh. Eksp. Teor. Fig. 38, 76 (1983).

DAMPING OF CDW-CONDENSATE MOTION BY INTERACTION WITH THERMAL PHASONS*

S. Takada[§], M. Wong and T. Holstein
University of California, Los Angeles, CA 90024

This paper contains a calculation of the damping of the q=0 phason (driven by an external microwave field of frequency ω) due to interaction with thermally ambient phasons. The obtained expression for the damping constant is frequency independent and varies as T^{-2} in the temperature domain relevant for experimental comparison. Its order of magnitude although smaller than that observed experimentally is still considered to be in satisfactory agreement, in view of uncertainties in knowledge of basic parameters and the preliminary character of the theory (neglect of interaction with thermal amplitons and the effects of long range Coulomb interactions.

Experiments on the microwave conductivity of incommensurate CDW-condensates ($\omega/2\pi$ = 10--35 GHz) have shown that, for relatively pure samples (of, e.g., $NbSe_3$ and TaS_3) $\sigma_{CDW}(\omega)$ is dominated by intrinsic damping, the effects of impurities being of subsidiary importance. This paper describes a calculation of the damping arising from the interaction of the microwave-driven, q=0, phason mode with thermally ambient phasons.

We start from a 3-d generalization of the Lee, Rice, Anderson Hamiltonian[1] in which the electron spectrum is still considered 1-d, whereas the phonon spectrum is 3-d. Concretely

$$H = \sum_{K_Z,\pm} \varepsilon_{K_Z\pm} C^+_{K_Z\pm Q/2}\, C_{K_Z\pm Q/2} + \sum_{\underline{q},\pm} \Omega_q (b^+_{q\pm Q}\, b_{q\pm Q})$$

$$+ \gamma N^{-1/2} \sum_{K_Z,q} C^+_{K_Z+q_Z+Q/2}\, C_{K_Z-Q/2} \left(b_{q+Q} + b^+_{\underline{q}-Q}\right) + \text{adj}$$

$$+ \sum_{K_Z} \Delta \left(C^+_{K_Z+Q/2}\, C_{K_Z-Q/2} + \text{adj}\right) \qquad (1)$$

The first three terms of (1) are basically the Fröhlich Hamiltonian, written in a two-band notation, with k_Z and $\underline{q}$ taken relative to $\pm k_F$ and $\pm Q\hat{z}$ ($Q=2k_F$), respectively ($|k_Z|<k_F$, $|q_Z|<Q$). The last term represents the self-consistent one-electron periodic potential associated with the CDW-condensate, with Δ, the gap parameter being given by

$$\Delta = \frac{2\gamma}{\sqrt{N}} \langle b_q \rangle \tag{2}$$

where $\langle b_q \rangle$ is the mean-field lattice distortion. In our treatment we have introduced the idealizations

$$\varepsilon_{K_Z \pm} = \pm V_F \, K_Z \tag{3a}$$

$$\Omega_q^2 = \Omega^2 + \Omega_\perp^2 \, (\underline{q}_\perp) \tag{3b}$$

(where $\underline{q}_\perp = (q_x, q_y, 0)$). These simplifications (together with the neglect of the $\underline{q}$-dependence of the EP coupling constant, γ) are appropriate for our problem in which, as it turns out, the relevant values of q_z are small ($q_z \sim \frac{2\Delta}{V_F} << Q$). In the detailed calculations, we have used for $\Omega_\perp^2 \, (q_\perp)$ the phenomenological expression

$$\Omega_\perp^2 \, (q_x, q_y) = \frac{\Omega_t^2}{2} \left[\sin^2 \frac{q_x b}{2} + \sin^2 \frac{q_y b}{2} \right] \tag{4}$$

(b = transverse lattice distance) with Ω_t describing the transverse frequency dispersion of the background phonon spectrum.

Following Kurihara[2], we introduce phason and ampliton operators according to the prescription

$$b_{\varphi q} = \frac{1}{\sqrt{2}i} \, (b_{q+Q} - b_{q-Q}) \tag{5a}$$

$$b_{\alpha q} = \frac{1}{\sqrt{2}} \, (b_{q+Q} + b_{q-Q}) \tag{5b}$$

(and similarly for $b^+_{\varphi q}$ and $b^+_{\alpha q}$). As shown by Kurihara[2], when these operators are introduced into the interaction term of (1), the phason and ampliton modes automatically get decoupled in lowest order in the sense that the "phason-ampliton" self-energy part

$\pi_{\alpha\varphi}(\underline{q}, \nu_\ell) \equiv$ 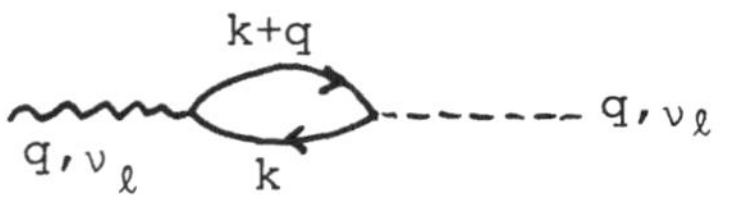 $= 0$ (6)

(Here, wavy solid and straight-dashed lines represent phason and ampliton Green's functions (given explicitly below), whereas solid straight lines represent zero'th order electron Green's functions which, in Nambu matrix formalism, take the form[2]

$$G^{(0)}(k,\zeta_n) = \frac{\zeta_n^2 - \varepsilon_K\tau_3 + \Delta\tau_1}{\zeta_n^2 - E_K^2} \qquad (\zeta_n = (2n+1)i\pi T) \tag{7}$$

where the τ_i (i=1,2,3) are Pauli matrices operating in the space of the two sub-bands, $k_z \pm Q$, and where

$$E_K = \left[\varepsilon_K^2 + \Delta^2\right]^{1/2} \tag{8}$$

The lowest-order ampliton and phason Green's functions are giben by[1,2]

$$D_{\varphi,\alpha}(q,\nu_1) = \frac{2\Omega_q}{\nu_\ell^2 - \omega_{\varphi,\alpha}^2(q)} \qquad (\nu_\ell = 2\pi i l\cdot T) \tag{9}$$

where

$$\omega_\varphi^2(q) = \Omega_q^2 - \Omega\Omega_q + \frac{\Omega_q}{\Omega}\,\omega_\alpha^2\,\frac{\xi\sinh^{-1}\xi}{\left(1+\xi^2\right)^{1/2}} \tag{10a}$$

and

$$\omega_\alpha^2(q) = \Omega_q^2 - \Omega\Omega_q + \frac{\Omega_q}{\Omega}\,\omega_\alpha^2\,(1+\xi^2)\,\frac{\sinh^{-1}\xi}{\xi\left(1+\xi^2\right)^{1/2}} \tag{10b}$$

with $\xi \equiv \frac{v_F q_2}{2\Delta}$ and $\omega_\alpha^2 = \lambda\,\Omega^2$, $\lambda = \frac{\gamma^2}{\Omega E_F}\,\frac{n_e}{N} = \sinh^{-1}\left(\frac{E_F}{\Delta}\right)$ being the dimensionless electron phonon coupling constant. For sufficiently small values of q_z and $q_\perp$, eqs.(10a,b) reduce to three dimensional generalizations of the more familiar expressions[1,2].

In order to calculate the damping of the q=0 "driven" phason it is necessary to take account of higher order contributions $\pi'_\varphi(0,\nu_\ell \to \omega)$ to its self-energy part. Thus

$\pi'_\varphi(0,\omega) =$

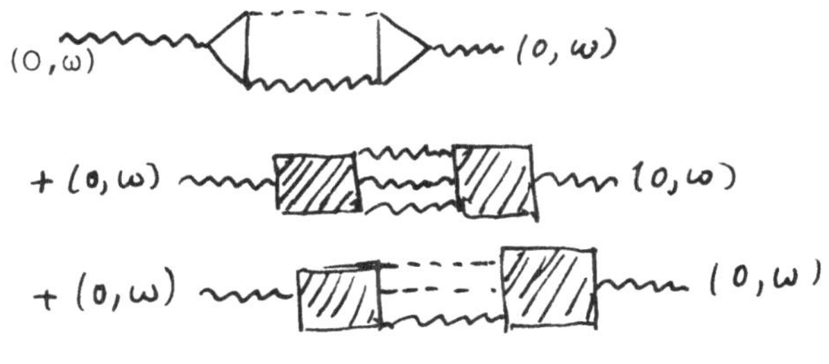

wherein the four-phason vertex

$$\hat{g}_4^{(4)}\left(\cdot\{q_i\},\ \{\nu_{1_i}\rightarrow\omega_\varphi(q_i)\}\right) \equiv$$

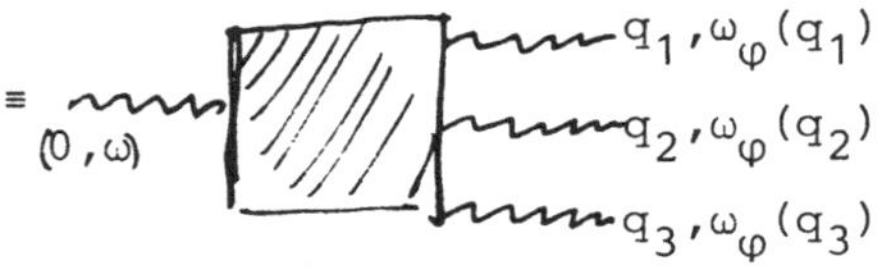

with $\underline{q}_1 \pm \underline{q}_2 \pm \underline{q}_3 = 0$ (12a)

and $\omega = \omega_\varphi(q_1) \pm \omega_\varphi(q_2) \pm \omega_\varphi(q_3)$ (12b)

is given by the diagrammatic equation

whereas the two-phason, two-ampliton vertex

$$g_4^{(3)}\left(\{q_i\}\{\nu_{1\,i}\}\right) \equiv$$

is given by

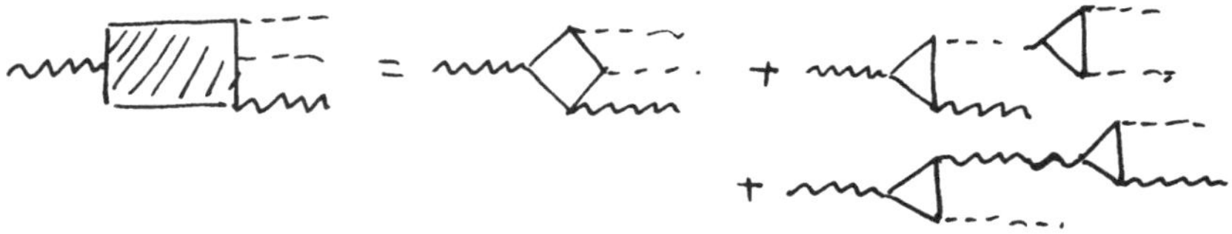

We remark that, as shown by Kurihara, vertices which contain an <u>odd</u> number of external <u>phason</u> lines give <u>vanishing</u> <u>contributions</u> in the idealized model described by (1). From this most useful selection rule (of which eq.(6) constitutes the simplest example) it also follows that the three-phason vertex

$$q_3\ (\{q_i\},\{\omega_{\varphi i}\}) =$$

$$= 0 \qquad (13)$$

For the case of a more realistic Hamiltonian in which (a) eq. (3a) is augmented by a term proportional to k_z^2, and (b) the constants Ω and γ are allowed to depend on q_z (as in the standard Fröhlich Hamiltonian), the vertex $g_3(\{q_i\}, \{\omega_{\varphi i}\})$ gives a non-vanishing contribution proportional to the product of the three external wave-vectors, q_{1z}, q_{2z}, q_{3z} (plus terms in higher powers of the q_{iz}). However, in our problem, one of the $\underline{q}$'s is zero, so that (13) is still valid.

We also observe that, as is easily seen, the first diagram of eq.(11) makes no contribution to damping since, with $\omega \ll |\omega_\alpha(q) - \omega_\varphi(q)|$, intermediate energy cannot be conserved. Deferring until later the consideration of the third diagram of (11) - i.e., the two-phason, two-ampliton process - we focus on the evaluation of the second diagram of (11) - i.e. the four-phason process. In particular we calculate its imaginary part, $\mathrm{Im}\,\pi_\varphi^{(2)}(0,\omega)$ which is related to the corresponding contribution to the damping coefficient $\Gamma_\varphi^{(2)}(\omega)$ via the general relation

$$\Gamma_\varphi(\omega) = -\frac{2\Omega}{\omega}\,\mathrm{Im}\,\pi_\varphi(0,\omega) \tag{14}$$

Summing over internal frequency variables and utilizing (14), we obtain

$$\Gamma_\varphi^{(2)}(\omega) = \frac{\pi\Omega^4}{N^2}\,\frac{1}{T}\sum_{\underline{q}_1,\underline{q}_1}\left[\tilde{g}_4^{(4)}(\underline{q}_1,\underline{q}_2)\right]^2 n_1 n_2 (n_3+1)/\omega_1\omega_2\omega_3$$

$$\times\left[\delta(\omega+\omega_3-\omega_1-\omega_2) + \delta(-\omega+\omega_3-\omega_1-\omega_2)\right] \tag{15}$$

where

$$n_i = \left[e^{\omega_i/T}-1\right]^{-1} \tag{16a}$$

$$\omega_i \equiv \omega_\varphi(q_i) \tag{16b}$$

with q_3 and ω_3 given by (12a,b) and where $\tilde{g}_4^{(4)}(q_1 q_2) \equiv \tilde{g}_4^{(4)}(\{q_i\}\{\omega_i\})$ for the special values of q_i and ω_i given by (12a,b).

(15) represents a process in which the q=0 "driven" phason interacts with an "incoming" thermal phason to produce two "outgoing" $q \neq 0$ phasons. As in standard lattice dynamics, the requirements of wave-vector and frequency conservation impose severe restrictions,

especially for small external frequency, ω. Indeed, utilizing (12a,b), we find that the conservation laws yield

$$\omega = [C(\hat{q}_2) - \eta\, V_\varphi(q_1)]q_2 \tag{17}$$

where $\underline{V}_\varphi(q) \equiv |\nabla_q \omega_\varphi(\underline{q})|$, $\eta = \cos\,[V_\varphi(q_1)\cdot q_2]$ and $C_\varphi(\hat{q})$ is the orientation-dependent phason velocity at $q \to 0$. Two cases present themselves. (1) If the equation

$$C(\hat{\underline{q}}_2) = \eta \cdot V_\varphi(\underline{q}_1) \tag{18}$$

can be satisfied for a 3-d domain of values of $\underline{q}_1$, then values of $\underline{q}_2$ of the order of BZ-boundary $\underline{q}$'s will play a dominant role in the $\underline{q}_2$-integration, leading to the appearance in $\Gamma_\varphi^{(2)}(\omega)$ of a factor $\sim \log(\Omega/\omega)$. (2) If, however, (18) cannot be satisfied, the permissible values of $\underline{q}_2$ will cover a range $\sim\left(\frac{\omega}{c_z}, \frac{\omega}{c_\perp}, \frac{\omega}{c_\perp}\right)$ where c_z and $c_\perp$ are obtainable from (10b) and (4). Using (10b) and (4), we can in fact show that, for $\Omega_t < \Omega$ (as is usually the case) only the second of these two possibilities obtains; thus, for sufficiently small ω, q_2 is also small. This circumstance permits us to use the approximation

$$\tilde{g}_4^{(4)}(\underline{q}_1,\underline{q}_2) \to \tilde{g}_4^{(4)}(\underline{q}_1,0) \equiv \tilde{g}_4^{(4)}(\underline{q}_1)$$

With this simplification, it was then found possible to evaluate $\tilde{g}_4^{(4)}(q)$ for all possible values of q_z and $\underline{q}_\perp$. The result is

$$\tilde{g}_4^{(4)}(\underline{q}) = (\pi v_F\lambda/2\mu a)\left[\sinh^{-1}\xi/\xi\left(1+\xi^2\right)^{1/2}\right]\left(1-\frac{\Omega}{\Omega_q}\right) \tag{19}$$

where $\mu \equiv 4\Delta^2/\lambda\Omega^2$ is the ratio of the CDW-mass (per electron) to the electron mass, and a the lattice spacing along the chain direction.

We note that, from (19) and the definition, $\xi \equiv q_z v_F/2\Delta$, given above, the domain of q_z-integration is indeed limited to $q_z \lesssim 2\Delta/v_F$, as claimed above.

In the final integration of (15) over $\underline{q}_1$, analytically simplified approximations to both (19) and (10a) were employed. For the domain of moderate temperatures ($T \gtrsim \Omega,\ \Omega_t$) we obtain

$$\Gamma_\varphi^{(2)}(\omega) \approx 2\pi\Omega\ (T/\Omega_t)^2\ (\lambda^{5/2}/\mu)\ (\pi/8)^2;\quad (2\Omega^2\lambda/\Omega_t^2 << 1) \tag{20a}$$

$$\Gamma_\varphi^{(2)}(\omega) \approx \pi^2\sqrt{2}\ (T/\Omega_t)\ (T\lambda^2/\mu)\ (\pi/8)^2\ ;\quad (2\Omega^2\lambda/\Omega_t << 1) \tag{20b}$$

We note that both of these expressions are independent of frequency and vary quadratically with temperature. For an order of magnitude estimate, we insert into (20b) the following representative values $\Omega \sim 2\,\Omega_t \sim T \sim 150^{\circ}K$, $\lambda \sim .7$, $\mu \sim 100$. The result is

$$\frac{\Gamma_{\varphi}^{(2)}}{2\pi} \sim 75 \text{ GHz} \tag{21}$$

We now discuss the contribution, $\Gamma_{\varphi}^{(3)}$, of the 2-phonon 2-ampliton process described by the third diagram of (11). An initial consideration had indicated that momentum and energy conservation restrictions would render this contribution negligible. However, as in the text preceeding eq.(17), we find that conservation requirements are satisfied provided that

$$\omega = [C(\hat{q}_2) - \eta\, V_{\alpha}(q_1)]\, q_2 \tag{22}$$

where $V_{\alpha}(q) \equiv |\nabla_q \omega_{\alpha}(q)|$. Comparison with (17) makes it quite apparent that the domain of $\underline{q}_1$, $\underline{q}_2$ space which satisfies (22) is quite comparable to that which satisfies (17). Since, as it turns out, the relevant interaction parameter $g_4^{(2)}(q_1)$ is also comparable to $\tilde{g}_4^{(4)}(q_1)$, we expect $\Gamma_{\varphi}^{(3)}$ to be of order of magnitude comparable to $\Gamma_{\varphi}^{(2)}$ (for $T \gtrsim \Omega$). A detailed calculation of $\Gamma_{\varphi}^{(3)}$ will be presented in our more complete report.

Finally, we consider the effect of long-range Coulomb interaction. As shown by Lee and Fukuyama[3] this interaction shifts the frequency of $q \approx 0$ phason modes up to the plasma frequency

$$\omega_{P\ell} = \left(\frac{4\pi n e^2}{m^* \varepsilon_o}\right)^{1/2} \tag{23}$$

where ε_o is the dielectric constant arising from virtual single-particle, over-the-gap, excitations. In the absence of any mitigating factors, this alteration basically suppresses the contributions of both the 4-phason and 2-phason 2-ampliton processes, due to the impossibility of satisfying energy conservation. However, realistically, at the temperatures of interest (T = 150--200 K), the dielectric constant is augmented by a contribution

$$\varepsilon_{qp} = \frac{4\pi\sigma_{qp}}{i\omega} = \frac{4\pi n_{qp} e^2 \tau_{qp}}{i\omega m} >> \varepsilon_o \tag{24}$$

associated with thermally excited electron-hole quasiparticles (n_{qp}, τ_{qp} = quasiparticle density and relaxation time). This modification has special relevance for the low-q phason discussed in the text following eq.(18). Indeed, upon introducing (23) and (24) into (9), one finds that the energy-conserving delta functions in (15) get replaced by the spectral functions

$$\frac{1/\tau^*}{\omega_2^4 + \frac{1}{\tau^{*2}}\left(\pm\omega + \omega_3 - \omega_1\right)^2}$$

where $\tau^* \equiv \frac{m^*}{m} \frac{n_{\sigma_0}}{n} \tau_{qp}$. The further evaluation leads to a result which, apart from a factor $\sim \log \frac{1}{\omega\tau^*}$, is numerically comparable to (20a,b).

We now discuss the relationship of our calculation to the experimental observations on the microwave conductivity, $\sigma_{CDW}(\omega)$, of the CDW-condensate. In the frequency range 10 - 35 GHz,

$$\sigma_{CDW}(\omega) = \frac{ne^2}{m^*\Gamma_\varphi} = \frac{ne^2}{m\tilde{\Gamma}_\varphi} \tag{25}$$

$$\text{where } \tilde{\Gamma}_\varphi \equiv \mu\, \Gamma_\varphi \tag{26}$$

is independent of the mass ratio, $\mu = m^*/m$. From the temperature dependence of Γ_φ as given by (20a,b), we have $\sigma_{CDW}(\omega) \sim 1/T^2$, which is in qualitative agreement with the observed T-dependence of $\sigma_{CDW}(\omega)$ (in the range 150 - 200 K; below the range, the observed T-dependence - decreasing sharply with diminishing T - is most likely associated with the onset of a number of additional factors (e.g., competition with impurity-pinning, long-range coulomb interactions, incipient lattice commensurability).

A second point of comparison is contained in the experimental observation that $\sigma_{CDW}(\omega)$ is comparable in magnitude to the d.c. (or microwave) conductivity in the metallic phase ($T > T_c$),

$$\sigma_m = \frac{ne^2}{m\Gamma_m} \tag{27}$$

where Γ_m is the transport relaxation rate due to the standard process of one-phonon emission and absorption by single electrons. The standard calculation of this process yields

$$\Gamma_m \simeq 4\pi\lambda T \tag{28}$$

From (20b), (26), and (28), one has (line above eq.(21))

$$\frac{\tilde{\Gamma}_\varphi^{(2)}}{\Gamma_m} \approx \frac{\pi^3\sqrt{2}}{256}\,\lambda(T/\Omega_t) \sim 0.23 \tag{29}$$

as compared with the experimental ratio

$$\Gamma_\varphi/\Gamma_m \sim 1 \tag{30}$$

Concerning this disagreement, we should note that, apart from the uncertainty in the value of Ω_t, (29) omits the contribution of the 2-ampliton, 2-phason process (i.e., $\Gamma_\varphi^{(3)}$) as well as modifications of both $\Gamma_\varphi^{(3)}$ and $\Gamma_\varphi^{(2)}$ by long-range Coulomb interactions. Under these circumstances, it is our opinion that the discrepancy between (29) and (30) should not be taken too seriously.

A somewhat more serious disagreement is exhibited in the comparison of the absolute magnitude of $\sigma_{CDW}(\omega)$, as predicted by (25) (together with (20a,b)) with the experimental value. For TaS_3, $n \approx 5\times10^{21}$ cm^{-1}; assuming m to be the free electron mass m_e, we have

$$\sigma \approx 2.7\times10^{16}\ sec^{-1} \approx 3\times10^4\ (ohm\text{-}cm)^{-1}$$

in contrast to the experimental value $\sim 2.5\times10^3$ $(ohm\text{-}cm)^{-1}$. Thus the theoretical value exceeds the experimental one by a factor of 12.

We note that in the metallic region, the theoretical value of σ_m exceeds the experimental value by a factor ~ 3. Nevertheless, in our opinion, even these discrepancies do not seem to warrant substantial revisions of the basic structure of the theory. We remark in this connection, that, apart from the above-noted uncertainties, the proper value for the "bare" electron-band mass, $m \sim m_e$ is not known.

* Supported by NSF Grant No. DMR 81-15542.
§ Permanently at: Institute of Physics, University of Tsukuba, Ibaraki 305, Japan

1 P.A. Lee, T.M. Rice, P.W. Anderson, Sol.St.Commun. 14, 703 (1974).
2 S. Kurihara, J.Phys.Soc.Japan 48, 1821 (1980).

"INDUCTIVE" RESPONSE FROM NONLINEAR MIXING IN CDW'S

S.N. Coppersmith, Brookhaven National Laboratory, Upton NY 11973 and
P.B. Littlewood, AT&T Bell Laboratories, Murray Hill, NJ 07974

We show that nonlinear inertial effects are consistent with highly dissipative linear response if one accounts for the CDW's internal degrees of freedom. Many apparently contradictory experimental results are reconciled.

The model of a CDW as an extended elastic medium pinned to impurities appears consistent with many features of its transport,[1-6] but controversy remains as to whether its internal modes must be considered in order to describe the motion or if an effective theory for one degree of freedom (representing the center of mass)[7] is adequate.

We consider the nonlinear dynamic response of a CDW and show that even if the linear response is purely relaxational, the nonlinear response (with both d.c. and a.c. driving fields) displays features characteristic of inertial systems. This apparent "inertial" or "inductive" behavior actually reflects extra dissipation occurring at frequencies which are harmonics or subharmonics of the internal "washboard frequency" of the CDW.

This result is used to resolve seemingly contradictory experimental data on sliding CDW systems. Measurements of the frequency-dependent linear conductivity $\sigma(\omega)$ of $NbSe_3$ indicate an unobservably small CDW inertia in the frequency range ($\lesssim 10$ GHz) of present experiments. Nevertheless, Tessema and Ong[8] observed inductive loops in the I-V characteristics of $NbSe_3$ at frequencies well within this range. More recently several authors have reported the appearance of subharmonic as well as harmonic steps in the d.c. I-V characteristics in the presence of an a.c. driving field.[9,10] While this behavior can be modelled by that of a massive point particle in a sinusoidal potential, the parameters are grossly inconsistent with those extracted from linear conductivity measurements.[11] We also predict "ringing" oscillations of the CDW in response to a sudden change in the driving field; such oscillations have recently been observed in $K_{0.3}MoO_3$ by Fleming et al.[12]

We take Fukuyama and Lee's[1] classical model of phase pinning of the CDW by random impurities. The equation of motion for the local distortions $u(\vec{r})$ of the CDW is[3]

$$
\begin{aligned}
&m\ddot{u}(\vec{r},t) + \lambda u(\vec{r},t) - K\nabla^2 u(\vec{r},t) \\
&= \rho_c E(t) + \rho(\vec{r}) V_z(\vec{r} + u(\vec{r},t))(1 + \partial u/\partial z)^{-1}
\end{aligned}
\qquad (1)
$$

Here m is the mass density of the CDW, λ a damping constant and K the elastic constant for the CDW medium.[13] The CDW couples to the electric field E via the full uniform collective density ρ_c, and to the random impurity potential $V(\vec{r})$ through the periodic CDW amplitude $\rho(\vec{r})(=\Sigma_{\vec{g}}\rho_{\vec{g}}\exp(i\vec{g}.\vec{r}))$. We assume that the deformation $u(\vec{r},t)$ lies only in the incommensurate z-direction (though its magnitude depends on x and y), and the subscript z denotes differentiation.

The characteristic frequency $\omega_0 = \lambda/m$ above which inertial response of the CDW becomes important has been estimated from the linear a.c. conductivity at high frequencies and is >> 10 MHz for all materials studied so far. Therefore we neglect the first term in eqn. (1) and use a purely dissipative equation of motion.

At d.c. bias E_0 much greater than the threshold field E_t, the response can be calculated in perturbation theory in powers of the impurity potential V.[3,4] We calculate the extra current induced by a small additional field $h(\vec{k},\omega)$ at frequency ω and wavevector k, $j(\vec{k},\omega) = \sigma(\vec{k},\omega)h(\vec{k},\omega)$. The conductivity can be written

$$\sigma(\vec{k},\omega) = i\omega\rho_c^{\,2}\left(G^{-1}(\vec{k},\omega) - \Sigma(\vec{k},\omega)\right)^{-1} . \tag{2}$$

G is the Green's function in the absence of impurities, and the "self-energy" $\Sigma(\vec{k},\omega)$ arises from the scattering of low frequency modes of the CDW by the random impurity potential $V(\vec{r})$. The poles of $\sigma(\vec{k},\omega)$ describe the renormalized frequencies of the elastic modes.

In the absence of pinning, all the modes are dissipative ($\omega(\vec{k})$ pure imaginary) but when the random potential is included, explicit evaluation of $\Sigma(\vec{k},\omega)$ to second order in perturbation theory in V reveals that the poles of eqn. (2) become complex for finite ω, and so the phonons describe damped oscillations. This has a crucial effect on the uniform response of the system to a time varying field.

We restrict our consideration to a purely sinusoidal CDW of wavevector $\vec{Q}$, and calculate the voltage response h(t) to a sudden change in the applied d.c. current by an amount j_o at time t = 0. In Figure 1, we show h(t) for $\beta = |\rho Q|^2 Q_z^{\,4} \times |V(Q)|^2(\lambda v Q_z)^{-1/2}(4\sqrt{2}\pi k)^{-1} = 0.2$ and several values of $\alpha = (\lambda v/kQ_z)^{1/2}$. As the CDW adjusts to its final velocity, the current undergoes damped "ringing" oscillations with a frequency close to the washboard frequency $Q_z v$. The overall magnitude of the oscillations is proportional to β, and the decay of the oscillations is controlled by the parameter α. These oscillations arise from <u>enhanced dissipation</u> of low frequency modes of the sliding CDW at frequencies close to the washboard frequency rather then from intrinsic inertia. The range of frequencies which are stongly damped is of order $\alpha Q_z v$ about the washboard frequency; Figure 1 shows the calculated conductivity $\sigma(\omega)$ which has well-defined structure near the washboard frequency. Qualitatively similar behavior is seen in numerical studies of incommensurate pinning models,[15] and is a clear feature in all the experimental data.[12,17] The ringing phenomenon in $K_{0.3}MoO_3$ has recently been observed by Fleming et al;[12] clearly it mimics an inductive response despite the negligible CDW inertia. Numerical simulations and qualitative arguments indicate that the oscillations become larger for fixed α and β as the current step j_o is increased, as is observed experimentally.[12] The ringing is not seen below threshold in contrast to the behavior of a model with inertia but only one degree of freedom.

Inductive loops in response to a large a.c. field $E_1\cos\omega t$ (but with no d.c. driving field) are also expected within this model. A perturbative (in V) calculation of the behavior of a CDW pinned by random impurities diverges because the

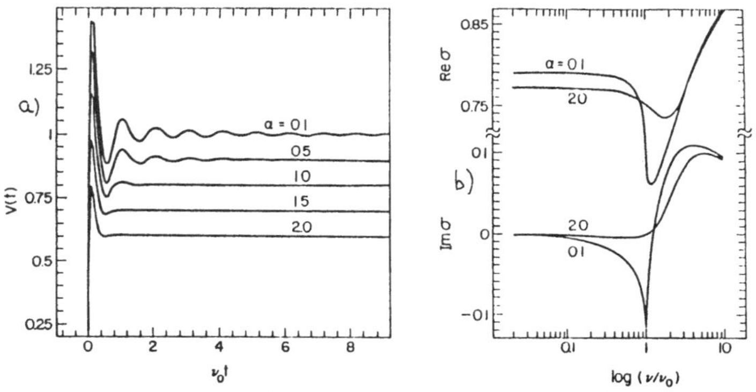

Fig. 1. (a) Ringing oscillations of the voltage versus time in response to a unit step in the CDW current for several values of the parameter α, and β = 0.2. Successive curves have been displaced for clarity, and $\nu_o = Q_z v/2\pi$ is the washboard frequency.
(b) Real and imaginary parts of the conductivity σ(ω), from eqn. (4), with parameters $\rho_c{}^2/\lambda = 1$, and β = 0.2.

oscillating field crosses threshold, so we study instead the one-dimensional incommensurate pinning model, with $V(q) = V/2\left(\delta(q + Q) + \delta(q - Q)\right)$, which has no threshold field for small V.[16] We expand about the zeroth order solution $u(t) = (\rho_c E_1/\lambda\omega)\sin\omega t$, and obtain the spatially-averaged velocity to second order in V. The response appears inductive if the current lags the voltage; to look for this we plot v(t) versus E(t) and examine the resulting Lissajous figures (Figure 2). Numerical simulation of the same model gives results similar to those obtained in perturbation theory. The results mimic qualitatively the inductive loops observed by Tessema and Ong[8] in $NbSe_3$, though a detailed comparison with experiment is not possible because of the nature of the potential we have used.

Finally, we comment on the subharmonic interference effects seen by Brown et al.,[9] and Sherwin and Zettl.[10] Harmonic steps (sharp peaks in the derivative dV/dI characteristic measured in the presence of a large a.c. field at frequency ω) had previously been observed corresponding to CDW velocities $Q_z v = p\omega$ for integer p.[17] The massless version of eqn. (1) leads to such anomalies at integral harmonics in second order of perturbation theory.[3,13] A tedious but straightforward calculation shows that anomalies in dV/dI occur at $Q_z v = (p + 1/2)\omega$ in fourth order.[18] In general, one expects anomalies in dV/dI at velocities $Q_z v = (p/q)\omega$ to appear in the $2q^{th}$ order of perturbation theory. The high order of perturbation theory necessary to see these steps does not imply that they are a small effect, since the perturbation series diverges as the CDW velocity approaches zero near threshold, and the field is driven below threshold in the experiments.

Brown et al.[9] model their results by neglecting the internal elastic degrees of freedom of the CDW but requiring the presence of a non-negligible mass in order to explain the observed subharmonics in terms of mode-locking.[19] Our model does not yield true mode-locking, and a detailed analysis of the shapes of the steps for small a.c. driving fields could be used to distinguish between the two models. We also note that there is evidence for both subharmonics and harmonics in the

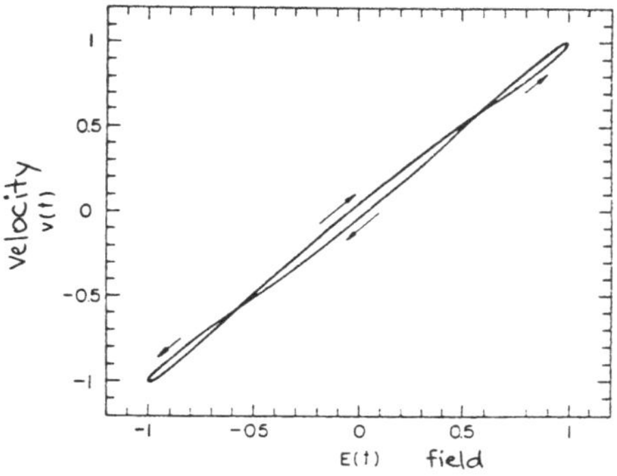

Fig. 2. Inductive loops obtained from perturbation theory of an incommensurate pinning model in one dimension.

"ringing" oscillations observed by Fleming et al. in $K_{0.3}MoO_3$, when the velocity is close to threshold,[12] which is consistent with the picture descibed above.

In conclusion, we have shown that an extended, classical and purely dissipative model has a number of properties usually considered as characteristic of one-particle inertial systems. This result reconciles seemingly incompatible experiments on charge-density-wave dynamics and provides strong evidence that the internal modes of a CDW crucially affect its transport properties. We also predict "ringing" oscillations, which have been recently observed by Fleming et al.[12]

We would like to thank R.J. Cava, D.S. Fisher, R.M. Fleming, S. Kivelson, G. Mozurkewich, N.P. Ong and L.F. Schneemeyer for helpful discussions. Work at Brookhaven was supported by the Division of Material Sciences, U.S. Department of Energy, contract number DE-AC02-76CH00016.

REFERENCES

1. H. Fukuyama and P.A. Lee, Phys. Rev. B17, 535 (1978).
2. P.A. Lee and T.M. Rice, Phys. Rev. B19, 3970 (1979).
3. L. Sneddon, M.C. Cross and D.S. Fisher, Phys. Rev. Lett. 49, 292 (1982).
4. D.S. Fisher, Phys. Rev. Lett. 50, 1486 (1983) and to be published.
5. L. Sneddon, Phys. Rev. B29, 719 and 725 (1984).
6. R.A. Klemm and J.R. Schrieffer, Phys. Rev. Lett. 51, 47 (1983).
7. G. Grüner, A. Zawadowski, and P.M. Chaikin, Phys. Rev. Lett. 46, 511 (1981).
8. G.X. Tessema and N.P. Ong, Phys. Rev. B23, 5607 (1981).
9. S.E. Brown, G. Mozurkewich, and G. Grüner, Phys. Rev. Lett. 52, 2277 (1984).
10. M. Sherwin and A. Zettl, to be published.
11. M. Azbel and P. Bak, to be published.
12. R. Fleming, L.F. Schneemeyer and R.J. Cava, to be published.
13. An anisotropic medium can be accounted for by using scaled variables (see Ref. 2 and Ref. 7).
14. A. Schmid and W. Hauger, J. Low Temp. Phys. 11, 667 (1973).
15. S.N. Coppersmith, unpublished and L. Sneddon, Phys. Rev. Lett., 52, 65 (1984).
16. See, for example, S.N. Coppersmith and D.S. Fisher, Phys. Rev. B28, 2566 (1983).
17. A. Zettl and G. Grüner, Solid State Commun. 46, 501 (1983), Phys. Rev. B29, 755 (1984).
18. We are grateful to Daniel Fisher for pointing this out to us.
19. Ref. 14, Appendix and Ref. 12.

MICROSCOPIC THEORY OF INTERACTION OF CDW WITH IMPURITIES

A. Zawadowski and I. Tüttő
Central Research Institute for Physics, H-1525 Budapest, POBox 49, Hungary

S.E. Barnes
Department of Physics, University of Miami, Coral Gables, Florida 33124 USA

P.F. Tua and J. Ruvalds
Department of Physics, University of Virginia, Charlottesville, Virginia 22901 USA

In the presence of CDW the perturbation in the electron density around an impurity is calculated in all orders of the perturbation theory considering backscattering only. In the vicinity of the impurity Friedel oscillations are superimposed on the CDW. Furthermore, an anharmonic effective potential is derived for the interaction between the CDW and the impurity. In the second order of the perturbation theory the previous results of Josephson type are reproduced. Finally the case of magnetic impurities is considered.

INTRODUCTION AND GENERAL DISCUSSION

A considerable portion of the papers in the present volume is dealing with sliding CDW and their interaction with impurities[1]. Other papers focus on the interaction of the CDW with electric contacts[2]. The question whether the contacts or the impurities play the determinant role in the pinning of the CDW and in generating the narrow band noise seems us to be far from being settled if any of these two does it at all. The theoretical approaches cover many different ideas which include quantum tunneling[3], solitons[4], etc.

The present work is dealing with the deformation of the CDW in the immediate vicinity of a single impurity and it is motivated by several theoretical and experimental problems.

Considering these theoretical problems the following may be mentioned:

(i) the classical rigid CDW model[5] does not take into account any deformation in the CDW;

(ii) the Fukuyama Lee Rice[6] theory for the deformation of the CDW is capable of describing long range deformations but ignores certainly any perturbation on a length scale smaller than the amplitude coherence length $\xi_o = v_F/\Delta_o$ where v_F is the Fermi velocity and Δ_o is the gap or the order parameter for the CDW.

(iii) microscopic processes have been recently considered by Barnes and Zawadowski[7] which take place in the immediate vicinity of the impurity and it has been shown that these processes in the second order of backward scattering on the impurity have a strong resemblance to the Josephson effect in superconducting junctions. The processes of higher orders, however, have not been considered, therefore the physical relevance of those calculations has remained somewhat in doubt.

(iv) Friedel oscillation must occur in the electron density around an impurity. As the periodicity distance of the Friedel oscillation and the wave length λ of the CDW is the same and is the inverse of twice the Fermi momentum k_F, therefore, a strong interaction and competition between the CDW and the local perturbation around the impurity is expected.

From experimental point of view some of the questions to be answered are the following:

(i) does the effective potential by which the interaction between the CDW and the impurity can be taken into account have a sinusoidal form or is there a strong deviation from that?

(ii) is λ or $\lambda/2$ the periodicity of that effective potential?

(iii) do the ratios of the harmonics in the narrow band noise depend on the temperature or not?

(iv) do the magnetic impurities with only magnetic interaction pin the CDW or not?

Concerning the theoretical motivations (iii) and (iv) further remarks will be made.

The calculations of the processes which show analogy to the Josephson effect are based on the following argument[7]. The CDW phase is characterized by electron-hole pairs with total momentum $\pm$ Q and with total spin S=0. There are two different types of pairs depending whether the electron or hole is on the right hand side of the electron dispersion curve (see Fig. 1). The second order term calculated by Barnes and Zawadowski takes into account a transition of two electrons from the same side of the dispersion curve to the opposite one by backward scatterings on the impurity. This process contributes to the transition between the pairs of different types. The macroscopic phase of the right and left electrons are denoted by φ_R and φ_L, respectively. The interference between the right going pairs with total momentum Q and left going ones with -Q results in the formation of the CDW and the relative phase $\varphi=\varphi_L-\varphi_R$ determines the position of the CDW. The Josephson like process characterized also by the phase φ. The third order in perturbation theory gives, however, finite amplitude in contrast to the Josephson effect. Namely, the electron moves in an effective periodic potential due to the CDW and the scattering on this potential results in a momentum transfer $\pm$ Q (this can be taken into account by the anomalous Green's function). Thus, if the excited electron is scattered by the impurity e.g. from right to left in the process depicted in Fig. 1 it may be scattered additionally both by the CDW and the impurity. Including these two extra scattering processes one type of the pairs can be scattered into the other type in third order of the impurity scattering as well, thus it gives a renormalization of the process calculated in the second order. Similar processes exist in higher orders and these are the subject of the present work. Furthermore, it will be seen that such microscopic processes describe the deformation of the CDW on a length scale much shorter than the amplitude coherence length ξ_o.

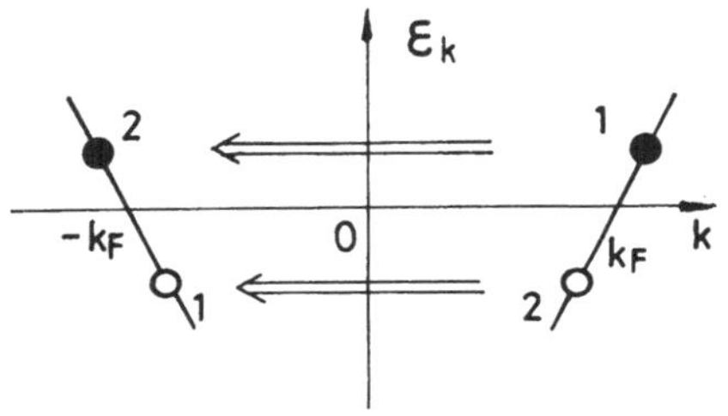

Fig.1. 1-d dispersion curve with the two types (labelled by 1 and 2) electron-hole pairs forming the CDW. The arrows are indicating the two back scatterings on the impurity which represent a transition beween the different pairs.

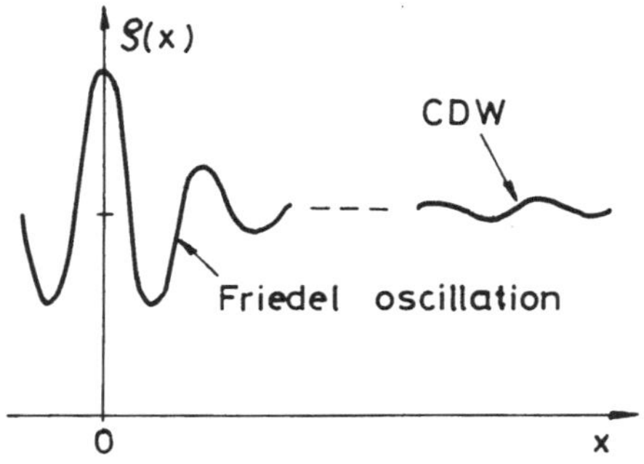

Fig.2. Schematic plot of the electron density around an impurity.

The motivation concerning the Friedel oscillation in the CDW needs also longer discussion. The electron density around an impurity is shown schematically for a realistic case in Fig. 2. In the vicinity of the impurity the electrons are affected by the impurity potential (0.1-1eV) and by the mean field due to the CDW ($\Delta_o \sim 0.01$ eV). As at the impurity site the effect of the impurity dominates over the CDW, Friedel oscillation is formed around the impurity. If the height of the impurity potential is comparable with the band width then the amplitude of the Friedel oscillation becomes comparable with the total electron density in the band. Because of the large damping of the Friedel oscillation at larger distances measured form the impurity, there exists a crossover distance x_o beyond which the CDW dominates the Friedel oscillation. The interesting feature to be studied is the following. Both oscillations have the same wave length, but the impurity tries to lock the phase of the oscillation to the impurity in order to have the maximal or minimal electron density at the impurity depending on the sign of the impurity potential. In general this locking phase is different from the phase φ of the CDW, thus the phases of these two regions must be adjusted in the crossover region around x_o.

In this picture the interaction between the sliding CDW and the impurity is very different from the case of the rigid CDW where the position of the CDW determines the electron density at the impurity site which interacts locally with the impurity. In the strong impurity potential case according to the present theory, however, the phase of the Friedel oscillation is locked by the impurity. This oscillation with large amplitude interacts with the CDW by interference and the interference energy is responsible for the force F exerted by the impurity. It must be emphasized that the Friedel oscillation dominates over a few atomic distances (a), while the CDW is coherent on the length scale of the amplitude coherence length ξ_o. In most of the cases $\xi_o \gg a$, thus the phase of the CDW can not be very different on the two different sides of the impurity just beyond the region dominated by the Friedel oscillation.

Turning to the case of a magnetic impurity[8] one can consider the impurity to be static for a very short time. In this case the Friedel oscillation locked to the impurity will have the opposite phase for up and down spin electrons thus the electron gas is magnetically polarized nearby the impurity. It is obvious from this that the CDW interacts with magnetic impurity in contrast to the rigid CDW. This interaction has been demonstrated in the second order of the exchange interaction between electrons and the impurity.

In the following a single impurity will be considered which interacts with the CDW whose phase is φ at large distances $x \to \pm \infty$, thus we ignore here the phase deformation of the CDW which occurs on a length scale much larger than ξ_o. The later deformation must be treated in the framework of a Ginsburg Landau theory (see e.g. Lee Rice[6]). For the sake of simplicity a strictly one dimensional model will be treated, but there is no reason to believe that the main features of our results do not hold for higher dimensions.

CALCULATIONS

The Hamiltonian H for the present problem is the sum of the Hamiltonian H_{el} of the interacting electron gas and of H_{imp} of the interaction between a single impurity and electrons, thus $H = H_{el} + H_{imp}$. The electron gas forming the CDW will be treated in mean field approximation and the effective Hamiltonian H_{CDW} is

$$H_{CDW} = \sum_{k\sigma} \varepsilon_k a^+_{k\sigma} a_{k\sigma} + (\Delta_o e^{i\varphi} \sum_{p,\sigma} a^+_{p+\frac{Q}{2},\sigma}\, a_{p-\frac{Q}{2},\sigma}) + (c.c) \qquad (1)$$

where $a^+_{k\sigma}$ are the free electron creation and annihilation operators, $Q = \frac{2\pi}{\lambda}$ is the wave vector of the CDW, Δ_o is the gap and φ is the phase of the order parameter. The dispersion ε_k is linearized at the Fermi energy as $\varepsilon_k = v_F(|k| - \frac{Q}{2})$ and a symmetrical momentum cut off p_o is applied around the Fermi levels, furthermore, $p_o v_F = D$ is the cutoff energy which is of the order of the bandwidth. The quasi particle energy is $E_p = E(k) = (\Delta_o^2 + (v_F p)^2)^{1/2}$ where $k = \pm Q + p$.

To describe the electron impurity interaction the electron field operator can be split into the left and right parts as $\psi(x) = \psi_R(x) + \psi_L(x)$ where $\psi_{R,L}(x) = L^{-1/2} \sum_{k \gtrless o} e^{ikx} a_k$ and L is the length of the one dimensional sample and x is the space coordinate. The electron density in the CDW is

$$\rho^{(o)}(x) = 2(\rho_o - \rho_1 \cos(Q\,x - \varphi)) \qquad (2)$$

Where the factor 2 is due to the electron spin. In the scattering of electrons on the impurity located at x=o only the back scattering is kept, thus

$$H_{imp} = T(\psi^+_R(o)\psi_L(o) + \psi^+_L(o)\psi_R(o)) \qquad (3)$$

where T is the scattering amplitude.

The thermodynamical Green's function technics will be applied and then zero temperature limit is taken. The renormalized Green's function is defined as $G_{\alpha\beta}(x,x';\tau-\tau') = -\langle T_\tau\{\psi_\alpha(x;\tau)\psi^+_\beta(x';\tau')\}\rangle$ and in the absence of the impurity $G^{(o)}$ can be written of

$$G^o_{\alpha\beta}(x,x';\tau-\tau') = e^{i(\frac{Q}{2}(\alpha x - \beta x') + \varphi_\alpha - \varphi_\beta)} \hat{G}^{(o)}_{\alpha\beta}(x - x';\tau-\tau') \qquad (4)$$

where $\alpha,\beta = \pm 1$ for R, L and a gauge transformation $\psi_\alpha(x) \to e^{i\varphi\alpha}\psi_\alpha(x)$ is applied $(\varphi_L - \varphi_R = \varphi)$ and $\hat{G}_{\alpha\beta}$ are the well known Green's functions in Gorkov's formulation.

In the following the electron density $\rho(x)$ the order parameter $\Delta(x)$ and the thermodynamical potentials are defined as

$$\rho(x) = 2\, G_{\alpha\beta}(x;x;\tau \to -o) \qquad (5)$$

$$\Delta(x) = 2g\, G_{RL}(x;x;\tau\to -o) \tag{6}$$

and

$$\delta\Omega = \int_o^T \frac{dT'}{T'} \langle H_{imp}(T')\rangle \tag{7}$$

where g is the effective electron-electron coupling constant. Eq. (7) for the thermodynamical potential correspond to the summation of the ring diagrams. All these quantities above are expressed in terms of the one electron Green's functions, for which the following Dyson Eq. holds

$$\begin{bmatrix} G_{RR} & G_{RL} \\ G_{LR} & G_{LL} \end{bmatrix} = \begin{bmatrix} G^{(o)}_{RR} & G^{(o)}_{RL} \\ G^{(o)}_{LR} & G^{(o)}_{LL} \end{bmatrix} \left[1 + \begin{bmatrix} O & T \\ T & O \end{bmatrix} \begin{bmatrix} G_{RR} & G_{RL} \\ G_{LR} & G_{LL} \end{bmatrix} \right] \tag{8}$$

where all space arguments are taken at x=x'=o and the energy variable is ω_n. The unrenormalized Green's function can be calculated directly and one obtains

$$G^{(o)}_{RR}(O;O;\omega_n) = -\frac{i\omega_n \frac{1}{\pi} \operatorname{arctg} \left(D/\sqrt{\Delta_o^2 + \omega_n^2}\right)}{v_F(\omega_n^2 + \Delta_o^2)^{1/2}} \tag{9}$$

furthermore $G_{LR} = G_{RR} \frac{\Delta_o}{i\omega_n} e^{i\varphi}$.

The electron density at the impurity can be split as

$$\rho(o) = \rho_n(o) + \rho_a(o) \tag{10}$$

where ρ_n is the part given by normal and ρ_a by the anomalous Green's function. Using Eqs. (8) and (i) one can show, that $\rho_n(o)$ is not renormalized, thus $\rho_n(o) = 2\rho_o$. The anomalous part, is however, strongly renormalized and the result is shown in Fig. 3, where $\rho(o)$ is plotted against the dimensionless coupling $T/2v_F$ for different CDW position given by φ.

The result can be summarized as:

- for T=O $\rho_a(o) = -\rho_1 \cos\varphi$ described the CDW (see Eq. 2)
- for $T/2v_F << 1$ the CDW is weakly perturbated

at the crossover coupling $T_{cr} = \pi^2 v_F^2 \Delta_o/(gD)$ the amplitude of the perturbation by the impurity is comparable with Δ_o. ($T_{cr}/2v_F \sim 2.10^{-1}$ for $g = -0.1(2\pi v_F)$ and D ~ 1eV and $\Delta_o \sim 10^{-2}$eV)

- for $T > T_{cr}$ the impurity is dominating the CDW

at $T \to +\infty$ $\rho_a = -2\rho_o$ thus for positive T the total density at the impurity site is zero and $\rho(o) = 4\rho_o$ holds for negative T (the band is completely filled by electrons).

For a realistic system $|T/2v_F| \sim 0.1-1$ is a good estimation.

The behaviour of the gap is similar since $\Delta(o) = 2g\, \rho_a(o)$.

The thermodynamical potential Ω has been calculated as a function of the position of the CDW (function of φ) for an intermediate strong coup-

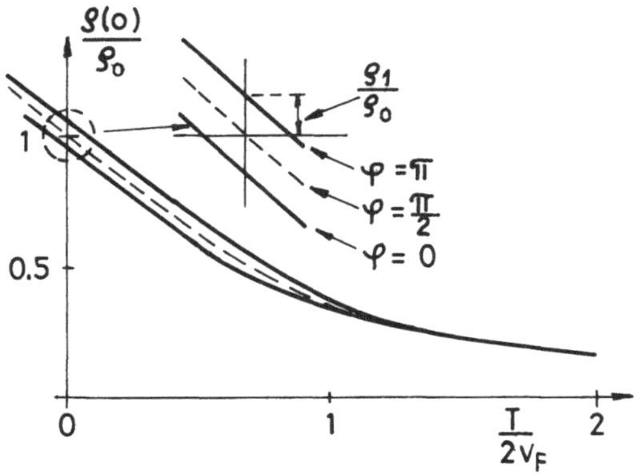

Fig.3. Electron density at the impurity site is plotted against the dimensionless scattering strength. The details near zero coupling are enlarged. The curve for $T < 0$ can be obtained by inversion through the point $T = 0$, $\rho(o)/\rho_o = 1$

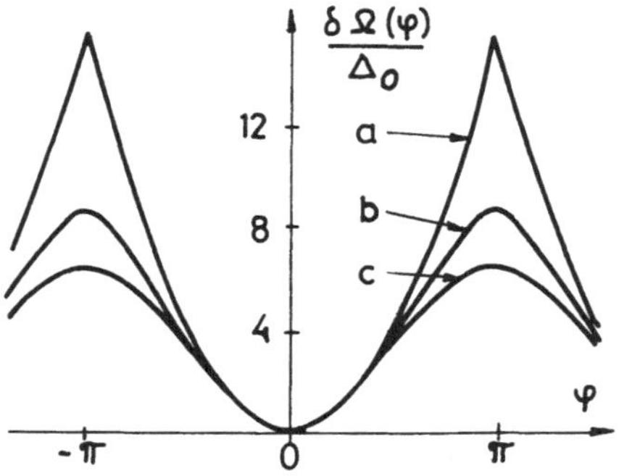

Fig.4. The effective potential (b) for the interaction between the impurity and the CDW as a function of the position φ of the CDW. For comparison a quadratic (a) and a sinusoidal (c) potential fitted to the effective potential at $\varphi = o$ are also shown

ling strength. Taking the zero temperature limit the interaction energy as a function of φ is shown in Fig. 4 which potential has a form between the sinusoidal and quadratic dependence.

The Friedel oscillation can be calculated for $x >> a$, where a is the atomic distance. The Green's function $G_{\alpha\beta}(x,x,i\omega_n)$ can be calculated from the generalization of the Dyson Eq.(8). The change in the electron density due to the impurity can be calculated for intermediate distances $a << x << \xi_o$ and the result is given as

$$\rho(x) - \rho^{(o)}(x) = \frac{2}{\pi} \frac{\frac{T}{2v_F}}{1 + (\frac{T}{2v_F})^2} \frac{\cos Qx}{x} \tag{11}$$

which is just the Friedel oscillation and it is independent of φ. The crossover distance x_o beyond which the CDW dominates is given as $x_o = -\xi_o \, g/(2\pi v_F) \cdot \frac{T}{2v_F} / (1 + (\frac{T}{2v_F})^2)$. Furthermore for $x >> \xi_o$ the Friedel oscillation can not be formed, because the energy of the electron-hole pairs necessary to build this oscillation is less than the CDW gap Δ_o, thus the Friedel oscillation can only tunnel into this region and therefore it decays exponentionally.

The order parameter is just the sum of the contribution of the CDW and of the impurity without CDW in a fairly good approximation, thus for $a << x < \xi_o$

$$\Delta(x) = (\Delta_o\, e^{i\varphi} - g\cos\varphi/4\pi \frac{\frac{T}{2v_F}}{1+(\frac{T}{2v_F})^2} \frac{1}{x})\, e^{-iQx} \tag{12}$$

The pinning force F is a function of the relative position of the CDW and the impurity. The force F can be obtained as the derivative of the interaction energy Ω with respect to the position of the CDW (φ/Q), thus

$$F = -Q\,\frac{\partial\Omega}{\partial\varphi} \tag{13}$$

For rigid CDW the force would be the same for $\varphi=\pi/6$ and $\varphi=5/6\pi$ but a significant difference is shown in Fig. 5 where the force is plotted against the coupling T at fixed φ. The force calculated shows a maximum at $T/2v_F$ where the height of the impurity potential is comparable with the band width. In the weak coupling region this result reproduces the previous ones obtained by the Josephson analogy. The asymmetry is the most significant for $T/2v_F \sim 1$ while the curve become sinusoidal for very weak and very strong couplings.

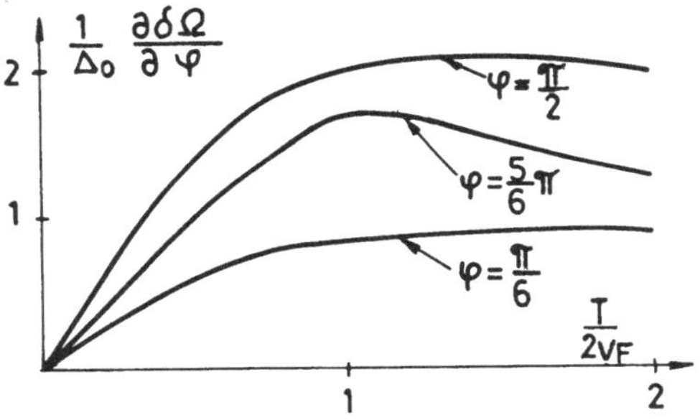

Fig.5. The force is plotted as a function of the dimensionless scattering strength at different CDW positions φ. The difference between curves for φ = = $\pi/6$ and φ = $5\pi/6$ demonstrate the anharmonic feature of the potential

There is another way to calculate the force F, namely as the time derivative of the total momentum P carried by the electrons. In a good approximation P is $P = 1/2 \quad (N_R - N_L)$ holds where $N_\alpha = \int \psi_\alpha^+(r)\psi_\alpha(r)\,dr$ for α = (R.L) and then

$$\frac{dP}{dt} = -i\,[PH] = QT(\psi_R^+(o)\psi_L(o) - \psi_L^+(o)\psi_R(o)) \tag{14}$$

The expectation value of this operator $\partial P/\partial t$ can be calculated as $\langle U^+(t)\,\frac{\partial P}{\partial t}\,U(t)\rangle$ where U(t) is the time development operator. The results obtained by these two different method have been compared in the second and third orders of the perturbation theory in T.

The classical equation of motion can be combined with the effective potential $V_{eff}(x) = \Omega(\varphi)$ shown in Fig. 4 and it can be written as

$$\gamma\,\frac{dX}{dt} = qE - \frac{\partial}{\partial X}\,V_{eff}(x) \tag{15}$$

where X is the center of mass coordinate of the CDW, γ is a damping factor, E is the electric field, q is the charge of the CDW considered and the inertia term is neglected. This equation of motion can be solved however, only by numerical methods. The solution is periodic in time with periodicity denoted by ω_o, but that is very anharmonic, thus the narrow band noise is generated. In the case of sinusoidal potential V_{eff} it is well known that ratio of the intensity of the subsequent harmonics I_{n+1}/I_n is independent of n. This behaviour is changed essentially in the intermediate strong coupling region, furthermore, the ratios I_{n+1}/I_n are essentially enhanced for larger couplings, but $I_1 > I_2 > I_3 > I_4$ holds.

Finally, the calculation of Tua, Zawadowski and Ruvalds[8] for magnetic impurities are commented. The interaction Hamiltonian of the usual exchange model is

$$H_{imp} = -J\vec{S}\ \{(\psi_R^+(o)\vec{\sigma}\psi_L(o)) + (\psi_L^+(o)\vec{\sigma}\psi_R(o))\} \tag{16}$$

where J is the exchange coupling, $\vec{S}$ and $\vec{\sigma}$ are the impurity spin and the Pauli operators, respectively. The calculation can be carried out in a way which is very close to the one just discussed above (see Eq. (14)), but the calculation is performed only in the lowest order proportional to J^2. The only new feature is that the impurity correlation function $\langle S^i(t)S^i(o)\rangle$ occurs, which is approximated by their free spin values. In this calculation the first nonvanishing order is the second one, and, therefore, the impurity potential has the periodicity $\lambda/2$ instead of λ. The magnetic field dependence of the effective potential is weak, that occurs on a scale of 100kG.

SUMMARY

We have presented a theory which provides the perturbation by a single impurity in the electron density of a 1-d electron gas. In the electron scattering by the impurity only the backward scattering has been considered, but it has been taken into account exactly. The only approximation has been made is the mean field treatment of the CDW Hamiltonain given by Eq. (1). One can show[9] that the renormalization of the mean field order parameter can be taken into account by the renormalization of the impurity scattering amplitude which is about 10% in realistic cases. The effect of higher order terms in the effective potential and in the narrow band noise has been briefly discussed and the details will be published elsewhere[9]. The inclusion of forward scattering and the treatment of clusters of impurities are beyond the scope of the present paper. The effective potential due to a single impurity has half periodicity $\lambda/2$

only if the impurity is magnetic. Finally, it can be remarked, that the higher order terms have always additional temperature dependence (see Ref.7.), thus these corrections can be ignored as the CDW transition temperature T_c is approached.

We are thank J. Sólyom for critical reading of the manuscript. This work was supported in part by National Foundation Grant No DMR-81--20827 and by the Department of Energy Grants DE-AS05-81-ER10959 and DE-FG05-84ER45113.

REFERENCES

(1) See e.g. the papers by J. Bardeen, D. Fischer, R.A. Klemm P.B. Littlewood, L. Sneddon in the present volume.

(2) See e.g. the papers by L.P. Gorkov, K. Maki and N.P. Ong in the present volume.

(3) J. Bardeen and J.R. Tucker in the present volume.

(4) B. Horovitz in the present volume.

(5) P. Monceau, J. Richard and J. Rerard, Phys. Rev. B25, 931 (1982) and G. Grüner, A. Zawadowski and P.M. Chaikin, Phys. Rev. Lett. 46, 511 (1981)

(6) See e.g. P.A. Lee and T.M. Rice, Phys. Rev. B19, 3970 (1979)

(7) S.E. Barnes and A. Zawadowski, Phys. Rev. Lett. 51, 1003 (1983)

(8) P.F. Tua, A. Zawadowski and J. Ruvalds, Phys. Rev. B29, 6525 (1984)

(9) I. Tüttő and A. Zawadowski, to be published.

QUANTUM EFFECTS IN THE JOSEPHSON APPROACH TO A CDW

S.E. Barnes
Department of Physics, University of Miami,
Coral Gables, Florida 33124, USA

The earlier work of Barnes and Zawadowski is generalized to a time dependent formalism. This permits the study of a.c. mixing, the equivalent of photon assisted tunneling and other quantum effects. Because the gap is so large such quantum effects and in particular the equivalent of photon assisted tunneling are not important for typical frequencies encountered in experiments to date. However a theory which describes the local dynamics by the Josephson modified classical equations but which includes fluctuations as a way of representing the flexibility of the CDW does lead to a scaling relation which might be compared that proposed by the Illinois group and based upon Bardeen's tunneling theory.

The Illinois group[1] has suggested harmonic mixing experiments and agreement with the following scaling relation as being strong evidence in favour of Bardeen's tunneling theory[2] for CDW conduction:

$$\sigma(\omega,V_o) = [(I'(V_o'+\alpha\omega) - I'(V_o'-\alpha\omega))/2\alpha\omega] \qquad (1)$$

where $I'(V_o') = I(V_o)$ the CDW current, $V_o' = \mathrm{sgn}(V_o)(|V_o|-V_T)$ and where V_o is the voltage. Within Bardeen's theory Eqn. (1) describes photon assisted tunneling. Related formulas to this scaling formula describe various ac mixing experiments. In connection with such experiments it is emphasized that there is no phase shift of the mixed signal. On the other hand there is no 'natural' explanation of why the modified voltage V_o' rather than just V_o enters the scaling law Eqn. (1).

The author and Zawadowski[3] have described a theory of CDW oscillations with a very strong parallel to the Josephson effect in SIS junctions (S=superconductor and I=insulator). It is perhaps natural to ask for a CDW, what is the role of the equivalent of photon assisted tunneling in an SIS junction ? Even without performing the calculation it is obvious that this cannot explain Eqn. (1). The energy scale for (1) is the pinning energy while for photon assisted 'tunneling' (really photon assisted scattering) the energy scale is the gap which is very much too large. Below is described the re-derivation of our earlier theory in a time dependent formalism which permits a detailed evaluation of such processes. It turns out photon assisted scattering leads to a term which damps the CDW motion, a contribution to the term usually denoted η in the 'classical' model[4].

As with our earlier approach only the lower order terms are evaluated. Omitted are terms which represent the elasticity of the CDW. The resulting theory can only describe CDW motion within a single Lee-Rice domain[5]. If fluctuations are included as a way of describing elasticity then a scaling law similar to (1) can be derived. That the ac current (or a mixing signal) has no phase shift for voltages below threshold has an explanation in terms of this approach.

The Hamiltonian is written as:

$$H = H_{CDW} + H_{int}(t) \;, \qquad H_{int}(t) = T(A(t)+A^{+}(t)) \tag{2}$$

where $A(t) = \sum_i \psi_R^{+}(R_i,t)\psi_L(R_i,t)$. Here H_{int} is a reduced interaction term which describes right going electrons being scattered, via impurities, into left going electrons. This is the equivalent of the oxide barrier in a SIS junction. In a quasi-one-dimensional material a steady current corresponds to a relative displacement of the Fermi surface for left and right going electrons, i.e. the CDW current: $J = ev_F(N_R-N_L)$, which, when inverted, can be used to define a 'voltage' or at least the equivalent of the potential difference across a SIS junction: $eV = (\varepsilon_F^R - \varepsilon_F^L) = (J/ev_F N(\varepsilon_F))$. For a dc current there is no time dependence to H_{CDW} and it will be assumed any effect of the rf current is negligible. The important time dependence is that of $A(t)$. Substituting $J = J_o+J_1\cos\omega t$ gives:

$$\psi_L(1) = \psi_L^{o}(t)\exp\{-i(e/\hbar)(V_o t+(V_1/\omega)\sin\omega t\} \tag{3}$$

The charge acceleration can be expressed in terms of $A(t)$:

$$(dJ/dt) = (2iev_F T/\hbar)(A(t) - A^{+}(t)) \tag{4}$$

which is very similar to the formula for the current in a SIS junction. The result for dJ/dt contains two types of contribution, (i) a normal current term corresponding to the correlation function $X(t) = i\theta(t) \langle[A(t),A^{+}(0)]\rangle$ and (ii) a Josephson current corresponding to the correlation function $\phi(t) = i\theta(t)\langle[A(t),A(0)]\rangle$. If $\omega \ll \Delta$, the gap, then:

$$\langle dJ/dt\rangle_N = - \eta J(t) \tag{5}$$

where

$$\eta = \text{const x } \{\text{Im}X(\omega)/\omega\} \tag{6}$$

The Josephson current corresponds to a formula of the form:

$$\langle dJ/dt\rangle_J = F(2eV_o/\hbar) + [(F(2eV_o/\hbar+\omega) - F(2eV_o/\hbar-\omega))/\omega](2eV_1/\hbar)\cos\omega t \,. \tag{7}$$

where with the present level of theory $F(x) = \text{const.}(1/x)$. Equation (5) represents a damping term while (7) clearly has a form related to (1).

Because flexibility of the CDW is not included, the above approach can only be interpreted as having meaning for a single Lee-Rice domain. In extending the theory to real systems it is envisaged that, because of this flexibility and thermal motion, the phase at a given impurity site makes a type of random walk. Such a walk corresponds to a random current, this corresponding in turn to some distribution in V_o denoted $P(V_o)$. The important point relative to the properties of the simple 'classical' model is that such a distribution implies each classical unit comprising an ill defined Lee-Rice domain is always above threshold, an assumption made in the derivation Eqn. (7). This also results in the ac current (or mixing signal) having no phase shift. With these assumptions Eqn. (7) is valid both above and below threshold and $F(x)$ is given in terms of $P(x)$ by:

$$F(x) = (J_1^2/\eta e^2 v_F N(\varepsilon_F))\int dx' P(x')(x+x')^{-1} \tag{8}$$

Writing $(2eV_o/\hbar) = \alpha J$ and including normal conduction, the ac conductivity is given by an equation of the form:

$$J_\omega = \sigma E_\omega + [(K(\alpha J+\omega) - K(\alpha J-\omega))/\omega]J_\omega \tag{9}$$

where $K(x)$ is determined by the dc response:

$$J = \sigma E_o + \sigma_1 K(\alpha J) \tag{10}$$

This relationship is very similar to Eqn. (1). In particular, if the crude approximation $J_{CDW} = \text{const}(V_o - V_T) = \beta V_o'$ is made, the two results are the same. This approximation is reasonable except close to threshold where it would appear (1) does not agree well with experiment.

Finally, does the frequency-voltage scale agree with experiment and the tunneling theory value ? The relation $\alpha J = \omega$ or $\alpha\beta V_o' \simeq \omega$ is that for Shapiro steps. Using Shapiro step data[6] for a different material (and presumably different dimensions) gives about 0.25mV/MHz to be compared with the Miller et al[1] value of 0.7mV/MHz for the significant parameter they denote as α. It seems possible the present approach might agree with the data as well as, or possibly better than, that based upon the tunneling model.

This work was supported in part by NSF Grant No. DMR81-20827.

1. J.H. Miller, Jr., J. Richard, R.E. Thorne, W.G. Lyons, J.R. Tucker and J. Bardeen, Phys. Rev. B29, 2328 (1984), and J.H. Miller, Jr., J. Richard, J.R. Tucker and J. Bardeen, Phys. Rev. Lett. 51 (1983).

2. J. Bardeen, Phys. Rev. Lett. 42, 1498 (1979), and 45, 1978 (1980).

3. S.E. Barnes and A. Zawadowski, Phys. Rev. Lett. 51, 1003 (1983).

4. G. Gruner, A. Zawadowski and P.M. Chaikin, Phys. Rev. Lett. 46, 511 (1981), P. Monceau, J. Richard and J. Renard, Phys. Rev. B25, 931 (1982).

5. P.A. Lee and T.M. Rice, Phys. Rev. B19, 3970 (1979).

6. A. Zettl and G. Gruner, Solid State Commun. 46, 501 (1983).

FOKKER PLANCK THEORY OF THE CLASSICAL CHARGE DENSITY WAVE MODEL WITH CURRENT NOISE

F. Gleisberg and W. Wonneberger
Department of Physics, University of Ulm,
D-7900 Ulm, W.-Germany

Theory and numerical results are given for ac conductivity under bias and dc conductivity under one or under two ("harmonic mixing") strong ac excitations for the classical charge density wave model with current noise.

The phenomenological classical model - equivalent to the RSJ model in the theory of the Josephson effect - can account for a number of observations on electrical transport by sliding charge density waves (CDW) in quasi one-dimensional conductors (cf. [1]). The model can be extended by adding noise forces to describe additional transport parameters. Thermal noise has been considered in [2-4]. However, nonequilibrium current noise seems to be of greater importance [5] since it is caused by impurities as well as by the contacts.

In the present work results are given for two nonlinear transport quantities taking into account current noise by solving the Fokker Planck equation via the method of matrix continued fraction (MCF) expansion. We consider:

a) Nonlinear ac response under a dc bias: $\sigma(\omega;\ E_\omega,\ E_0)$.

b) Nonlinear dc response under dc bias plus two strong ac fields at frequencies ω and 2ω: $I_0\ (E_0;\ E_\omega,\ E_{2\omega})$.

Both responses are superimposed on the oscillating background of the narrow band noise [6] and require a proper definition of the averaging procedures for the underlying instationary stochastic process.

Mathematical Framework

The stochastic classical model with current noise for a sinusoidal pinning potential and in a proper scaling is defined (cf. [5]) by the following Fokker Planck equation for the probability distribution $W(\Phi,t)$ for the phase Φ of the CDW in a phase domain:

$$\frac{\partial}{\partial t} W = \frac{\partial}{\partial \Phi} \{-\sin\Phi - c(t) + T_N \frac{\partial}{\partial \Phi}\} W \ . \tag{1}$$

In (1), time and frequencies are scaled to $\tau_c \equiv \omega_c^{-1} = \gamma/\omega_0^2$ [7,8], γ denotes the damping constant and $\omega_0 \ll \gamma$ the pinning frequency. The electric field $c(t)$ is measured in units of the bare threshold field strength $E_T = m_F\omega_0^2/e_0Q$ (m_F: Fröhlich mass, $Q = 2k_F$: CDW wave vector) and taken as

$$c(t) = c_0 + c_1 \cos\omega t + c_2 \cos 2\omega t. \tag{2}$$

T_N measures current noise: $T_N = \xi \,\overline{\langle\dot{\Phi}\rangle}^{0}$. (3)
corresponding to the shot noise limit. In (3), brackets denote a stochastic average while $\overline{\quad\quad}^{0}$ means a time average according to

$$\overline{f(t)}^{\,\omega} \equiv \lim_{T\to\infty} \frac{1}{T}\int_0^T dt\, e^{i\omega t} f(t). \tag{4}$$

Setting $c_0 = n\omega + \Delta c_0$, (5)
the instationary solution $W(\Phi,t)$ is expanded as

$$W(\Phi,t) = \sum_{q=-\infty}^{\infty} \exp[iq\Phi - iq(n\omega t + \frac{c_1}{\omega}\sin\omega t + \frac{c_2}{2\omega}\sin 2\omega t)]\ \alpha_q(t). \tag{6}$$

The rotating wave type transformation in (6) is suggested by related work in [9]. Defining a vector $\underset{\sim}{a}_q$ according to

$$(\underset{\sim}{a}_q)_l \equiv \overline{\alpha_q(t)}^{\,l\omega} \tag{7}$$

casts (1) into a tridiagonal vector recurrence relation

$$q\,\mathbb{A}\,\underset{\sim}{a}_{q+1} + \mathbb{B}_q\underset{\sim}{a}_q + q\,\mathbb{C}\underset{\sim}{a}_{q-1} = \underset{\sim}{0}. \tag{8}$$

The matrices in (8) are given in terms of Bessel functions by

$$(\mathbb{A})_{lk} = \frac{1}{2}\sum_p J_{l-n-k-2p}\left(\frac{c_1}{\omega}\right) J_p\left(\frac{c_2}{2\omega}\right) = -(\mathbb{C})^*_{-l-k}\,,$$
$$(\mathbb{B}_q)_{lk} = -(T_N q^2 + iq\Delta c_0 - il\omega)\delta_{lk}\,. \tag{9}$$

Eq. (8) can be solved numerically by MCF methods [10] as has been done for the harmonic mixing problem in [4]. To this order, mass operators are defined via

$$\underset{\sim}{a}_q = \mathbb{M}_q\,\underset{\sim}{a}_{q-1}\,. \tag{10}$$

Eq. (8) is then transformed into the MCF:

$$\mathbb{M}_q = -\frac{q}{\mathbb{B}_q + q\mathbb{A}\mathbb{M}_{q+1}}\,\mathbb{C}\,. \tag{11}$$

$\mathbb{M}_1$ determines the current response functions to all orders in the applied fields c_0, c_1 and c_2 as exemplified in two applications:

a) Nonlinear ac response under dc bias.

In this case we set $c_2 \equiv 0$ while c_1 is arbitrary. If j_ω is the current density at frequency ω, we define

$$\sigma(\omega;E_\omega,E_0) = \frac{j_\omega}{E_\omega} = \frac{2}{E_T}\frac{j_\omega}{c_1} = \frac{2}{c_1E_T}\frac{e_0}{\pi}\,\omega_c\,\overline{\langle\dot{\Phi}\rangle}^{\,\omega} \equiv \sigma_0\,\overline{\langle\dot{\Phi}\rangle}^{\,\omega}/c_1\,. \tag{12}$$

Here, $\sigma_0 = Qe_0^2\,/\,\pi m_F\gamma$ is the conductivity of the freely sliding CDW

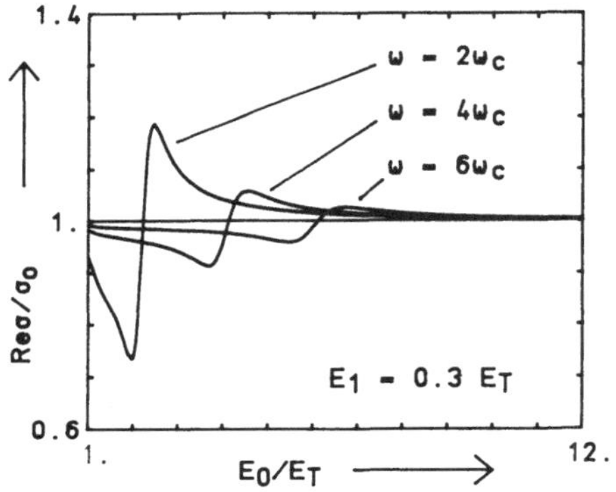

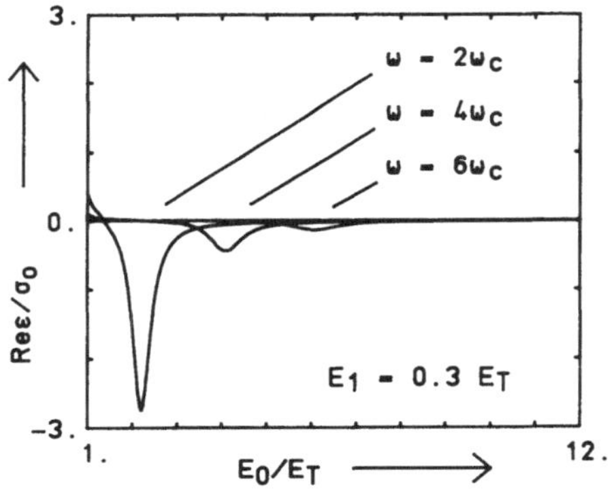

Fig. 1 and 2: Real parts of scaled conductivity and dielectric function vs. dc bias for several frequencies and almost small ac signal ($E_1 = 0.3E_T$). Finite widths of interference regions is due to current noise.

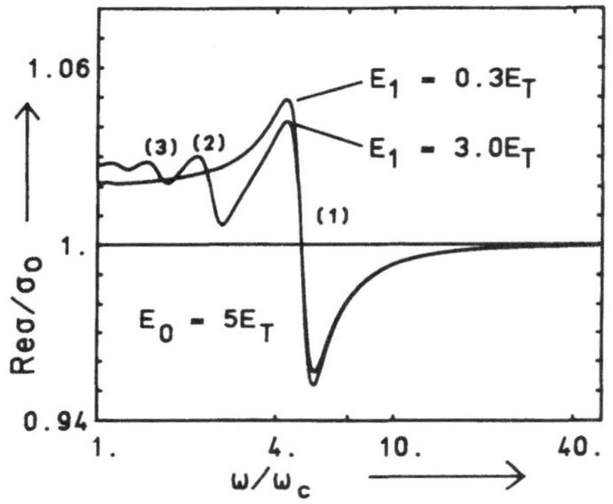

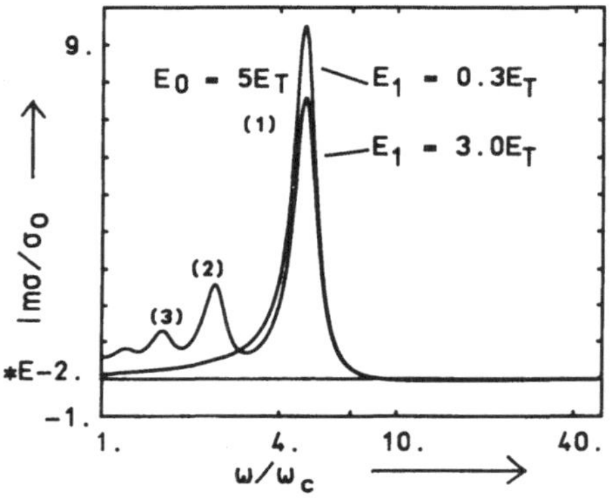

Figs. 3 and 4: Scaled complex conductivity vs. frequency for almost small ($E_1=0.3E_T$) and strong ($E_1=3E_T$) ac signal. Interference regions according to $\omega_{ph}=n\omega$ are indicated by integer (n).

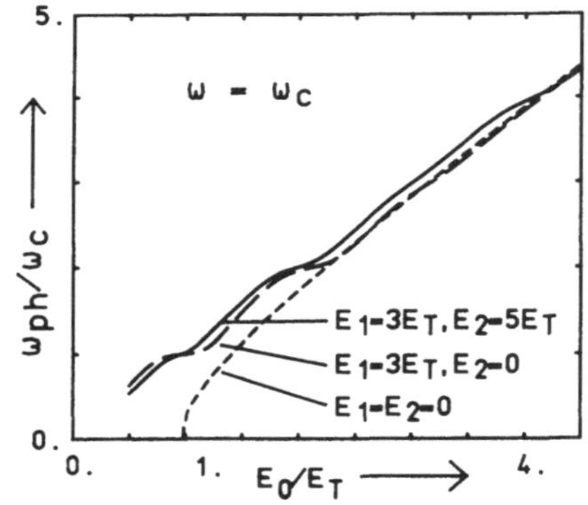

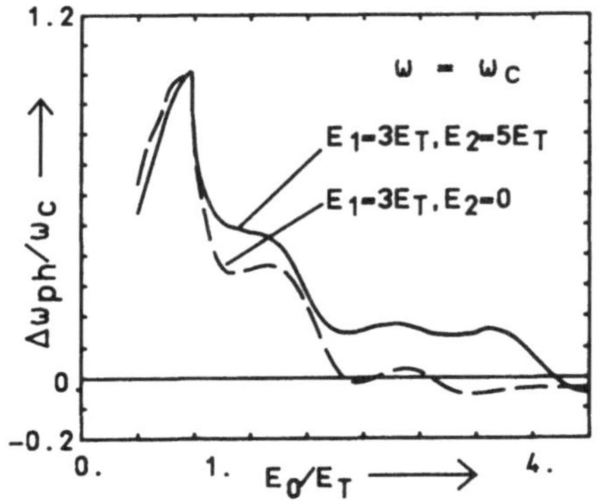

Figs. 5 and 6: Scaled basic frequency ω_{ph} of current oscillations vs. dc bias for two strong ac signals ($E_1=3E_T$, $E_2=5E_T$) at frequencies ω and 2 . Difference frequency $\Delta\omega_{ph}$ caused by ac signals is also shown.

and is about that of the normal electrons. With the corresponding average (4), we find

$$\tilde{\sigma}(\omega;c_1,c_0) \equiv \frac{\sigma}{\sigma_0} = 1 + \frac{i}{c_1}\sum_m J_m\left(\frac{c_1}{\omega}\right)\left[(M_1)_{-n-m+1,0} - (M_1)^*_{-n-m-1,0}\right]. \quad (13)$$

The linear ac response is the $c_1 \to 0$ limit of (13). Some results for this case have been given in [11] for purely thermal fluctuations. A number of relevant numerical results for the present case are plotted in Figs. 1 to 4 and should be compared with the measurements in [12].

b) Current voltage curve under ac excitations (CVC).

For a dc bias according to (5), $\overline{\langle\dot{\Phi}\rangle}^0 = c_0 - \overline{\langle\sin\Phi\rangle}^0 \propto I_{dc}$ is evaluated:

$$\overline{\langle\dot{\Phi}\rangle}^0_n = c_0 - \mathrm{Im}\left\{\sum_{k,l}(-1)^{k+n} J_{k+2l+n}\left(\frac{c_1}{\omega}\right) J_l\left(\frac{c_2}{2\omega}\right)(M_1)_{k,0}\right\}. \quad (14)$$

Corresponding results are shown in Figs. 5 and 6. The current steps at $n\omega = \langle\dot{\Phi}\rangle \equiv \omega_{ph}$, where ω_{ph} is the basic frequency of narrow band noise [13] are clearly seen.

Finally, we comment on the mechanism of the current noise underlying (3). For ideal shot noise of free electrons, the noise strength ξ in (3) equals $\pi/2$, giving a very strong effect. However, screening in the conversion region near the contacts where CDW electrons are converted into normal electrons, will reduce this value substantially. It is also conceivable [5] that pinning centers inside a phase domain cause a current noise with $\xi \cong \pi/2\,(\Delta q/e_0)^2(n_i/n_0)$ where Δq is the pinning charge and n_i/n_0 the concentration ratio of impurities and CDW electrons. Finally, multi-domain and boundary effects are supposed to contribute to the current noise. As a representative value, we have assumed $\xi = 0.1$ in the numerical calculation.

References:

[1] G. Grüner, Physica 8D (1983) 1
[2] W. Wonneberger, F. Gleisberg, Solid State Commun. 23 (1977) 665
[3] W. Wonneberger, H.-J. Breymayer, Z. Phys. B43 (1981) 329
[4] H.-J. Breymayer, H. Risken, H. D. Vollmer, W. Wonneberger, Appl. Phys. B28 (1982) 335
[5] W. Wonneberger, H.-J. Breymayer, Z. Phys. B56 (1984) 241
[6] R. M. Fleming, C. C. Grimes, Phys. Rev. Lett. 42 (1979) 1423
[7] G. Grüner, A. Zawadowski, P. M. Chaikin, Phys. Rev. Lett. 46 (1981) 511
[8] P. Monceau, J. Richard, M. Renard, Phys. Rev. B25 (1982) 931
[9] W. Schleich, C.-S. Cha, J. D. Cresser, Phys. Rev. A29 (1984) 230
[10] H. Risken, The Fokker-Planck Equation, Synergetics Ser. 18, Berlin-Heidelberg-New York: Springer 1984
[11] A. N. Vystavkin, V. N. Gubankov, L. S. Kuzmin, K. K. Likharev, V. V. Migulin, V. K. Semenov, Rev. Phys. Appl. 9 (1974) 79
[12] A. Zettl, G. Grüner, Phys. Rev. B29 (1984) 755
[13] P. Monceau, J. Richard, M. Renard, Phys. Rev. Lett. 45 (1980) 43

TRAVELLING CHARGE DENSITY WAVES : A MEAN FIELD TREATMENT

B.G.S. Doman
Department of Applied Mathematics and Theoretical Physics,
University of Liverpool, P.O. Box 147, Liverpool L69 3BX, U.K.

Travelling charge density waves are described using the thermodynamic Lagrangian formalisms developed by Fischbeck to derive the CDW phonon spectrum. It is shown that travelling charge density waves can exist at temperatures higher than those at which stationary CDWs exist.

We examine the properties of a slowly moving incommensurate charge density wave. Following Fischbeck[1] we shall describe the system using a thermodynamic Lagrangian. The kinetic energy comes from the slow motions of the ions as the CDW passes. The potential energy is the thermodynamic free energy of the electron lattice system for the given instantaneous distribution of ionic displacements.

To evaluate the Free Energy we start from the Fröhlich Hamiltonian

$$H = \sum_{k\sigma} \epsilon_k c^+_{k,\sigma} c_{k,\sigma} + \frac{1}{\sqrt{N}} \sum_{k,q,\sigma} g_q c^+_{k+q,\sigma} c_{k,\sigma} (b_q + b^+_{-q}) + \sum \omega_q b^+_q b_q \, , \quad (1)$$

in which $\epsilon_k = -W\cos ka \sim Wa(|k| - \pi/2a) = \hbar v_F(|k| - \pi/2a)$. In the mean field approximation we take

$$\Delta_q = g_q (\langle b_q \rangle + \langle b^+_{-q} \rangle)/\sqrt{N} \quad (2)$$

For a nearly half filled band, Δ_q will be large for $q \sim \pi/a$, and negligible for $q \sim 0$. For an incommensurate CDW with wave vector $Q = \frac{\pi}{a} + \kappa$, we only need sum over q = odd multiples of Q since $2nQ = 2n\kappa \bmod (2\pi/a) \sim 0$.

The free energy F is given by

$$F = -2k_B T \int \ln(1 + e^{-\beta(E-\mu)}) \rho(E)\, dE + \frac{1}{\lambda} \sum_q \Delta_q^2 \quad (3)$$

where ρ is the electron density of states in the mean field approximation. We can evaluate ρ from the electron Green's function.

Following closely the work of Mertsching and Fischbeck[2,3], we find that the difference between the free energy with the CDW and without is given by

$$F - F_N = \frac{\kappa}{\pi^2 W} \int_{-\pi/2\kappa}^{\pi/2\kappa} dx \left[c_1 \Delta^2 + c_2 \Delta^4 + c_3 \Delta^6 + (\hbar v_F)^2 (c_2 + 5c_3\Delta^2) \Delta'^2 + \tfrac{1}{2}(\hbar v_F)^4 c_3 \Delta''^2 \right] \quad (4)$$

where $\Delta(x) = \sum_{l=\text{odd}} \Delta_l e^{il\kappa x}$, which is a slowly varying function of x for small κ , and

$$c_1 \approx \ln \frac{4\pi}{\beta\Delta_0} + \mathrm{Re}\,\psi\left(\tfrac{1}{2} + \tfrac{i\beta\mu}{2\pi}\right),$$

$$c_2 \approx -\frac{1}{8}\left(\frac{\beta}{2\pi}\right)^2 \mathrm{Re}\,\psi''\left(\tfrac{1}{2} + \tfrac{i\beta\mu}{2\pi}\right), \tag{5}$$

$$c_3 \approx \frac{1}{192}\left(\frac{\beta}{2\pi}\right)^4 \mathrm{Re}\,\psi''''\left(\tfrac{1}{2} + \tfrac{i\beta\mu}{2\pi}\right)$$

The lattice displacement at na, $\delta(na) = (-1)^n \Delta(na)/g\sqrt{2\omega}$. The lattice KE is thus

$$\tfrac{1}{2} M \sum \left(\dot{\delta}(na)\right)^2 = \frac{M}{2g\sqrt{2\omega}} \sum \left(\dot{\Delta}(na)\right)^2 = \tilde{c}_0 \frac{\varkappa}{\pi} \int_{-\pi/2\varkappa}^{\pi/2\varkappa} \dot{\Delta}^2\, dx \tag{6}$$

We now change units so that $x \to aW/\sqrt{c_3/|c_2|}\, x$, $\varkappa x \to \varkappa x$, $\Delta \to \sqrt{|c_2|/c_3}\,\Delta$, $E \to (|c_2|^3/\pi W c_3^2) E$.

The Lagrangian $L = T - (F - F_N)$ becomes

$$L = \frac{\varkappa}{\pi} \int_{-\pi/2\varkappa}^{\pi/2\varkappa} dx \left[-z\Delta^2 + \Delta^4 - \Delta^6 + (1 - 5\Delta^2)\Delta'^2 - \tfrac{1}{2}\Delta''^2 + c_0' \dot{\Delta}^2 \right] \tag{7}$$

where $z = c_1 c_3 / c_2^2$ and $c_0' = (\tilde{c}_0 c_3 / c_2^2)\pi W$.

The Euler Lagrangian equation is

$$2c_0' \ddot{\Delta} + \Delta'''' + 2(1 - 5\Delta^2)\Delta'' - 10\Delta\Delta'^2 + 2z\Delta - 4\Delta^3 + 6\Delta^5 = 0 \tag{8}$$

This has a travelling wave solution $\Delta = \Delta(x - vt)$ with Δ satisfying

$$\Delta'''' + 2(1 + c_0 - 5\Delta^2)\Delta'' - 10\Delta\Delta'^2 + 2z\Delta - 4\Delta^3 + 6\Delta^5 = 0 \tag{9}$$

For $c_0 = 0$ and $5/27 < z < \frac{1}{2}$, Δ can be expressed in terms of the elliptic function sn. There is another unphysical elliptic function solution with positive $F - F_N$. For $z \sim 5/27$ the system is close to the tricritical point. As z approaches $\frac{1}{2}$ the amplitude of the CDW decreases and the elliptic function becomes close to a sine function.

Thus $z = \frac{1}{2}$ gives the incommensurate CDW - normal transition.

For $c_0 \neq 0$, the physical solution is no longer simply expressible in terms of sn although the non-physical solution still can be. For z close to $\frac{1}{2}$ we expect a solution is nearly a sine wave. Let us therefore suppose

$$\Delta = \sum_{n=\text{odd}} \Delta_n \cos n\varkappa x \tag{10}$$

We find a solution with small Δ_1, very small Δ_3, Δ_5 for which

$$(\varkappa^4 - 2(1+c_o)\varkappa^2 + 2z)\Delta_1 + (5\varkappa^2-3)\Delta_1^3 + \frac{15}{4}\Delta_1^5 + O(\Delta_3\Delta_1^2) + \ldots = 0 \tag{11}$$

If we neglect $O(\Delta_3\Delta_1^2) + \ldots$, we see that (11) has a non-trivial solution provided

$$\varkappa^4 - 2(1+c_o)\varkappa^2 + 2z < 0 \tag{12}$$

i.e. $2z < (1+c_o)^2$.

When $c_o = 0$ we have as before no CDW for $z > \frac{1}{2}$. For $c_o > 0$ i.e. for a moving CDW, we can have a solution for $z > \frac{1}{2}$. This means that these are travelling CDW solutions at higher temperatures than for stationary CDWs.

As it stands this theory predicts that the faster the CDW moves, the higher the temperature at which it can exist. This is clearly a deficiency of the theory. Fast moving CDWs would clearly not satisfy the initial assumptions of quasi-static, reversible thermodynamics, and the neglect of scattering/dissipative effects implicit in the use of a mean field approach.

References

1 H.J.Fischbeck Phys.Stat.Sol.(b)118,595,(1983)

2 J.Mertsching and H.J.Fischbeck Phys. Stat.Sol.(b)103,783,(1981)

3 J.Mertsching and H.J.Fischbeck Phys.Stat.Sol.(b)99,(1980)

CHARGE DENSITY WAVE TRANSPORT

COHERENT AND INCOHERENT EFFECTS IN CHARGE DENSITY WAVE TRANSPORT

G. Grüner
Department of Physics
University of California, Los Angeles, CA 90024

The interplay between coherent and incoherent effects associated with the dynamics of the charge density wave mode is discussed, by reviewing experiments concerning the response of the pinned and of the current carrying mode.

1. INTRODUCTION

A vastly oversimplified model of charge density wave (CDW) transport was proposed a few years ago by Zawadowski, Chaikin and myself[1] to account for the main experimental observations available at that time. The overall qualitative features of the dynamics of the CDW mode - ω and E dependent response and coherent current oscillations in the current carrying region - are accounted for by the model which neglects the internal dynamics of the collective mode. The model also served as guidance for experiments in which dc and ac fields are jointly applied and the dc and/or ac response of the nonlinear system is detected. In particular, the formal analogy of the equation of motion to that of the resistively shunted Josephson junction led to several experiments (and here the model is particularly successful). At present, the model is used to relate the experimental findings to models which treat the various aspects of the response of nonlinear driven systems. Although the fundamental difficulties of the model have been recognized for some time (in particular its conflict with observations that the collective mode is pinned by randomly positioned impurities), a large body of experimental data which point to the importance of internal modes in the dynamics have only recently emerged.

This short review summarizes the highlights and the failures of this so called classical particle description of charge density wave transport, along with possible reasons for the interplay between coherent and incoherent phenomena in these driven, random, nonlinear systems.

The author is well aware of other approaches which are successful in describing a broad range of experiments performed in the presence of joint ac and dc fields, the detailed form of the current-voltage characteristics and other features. In particular, quantummechanical descriptions of the dynamics of the collective mode are summarized by Professor Bardeen at this conference. The relation of our findings to these models will not be discussed here. It should be noted however, that once the ω and E dependent response is accounted for by various approaches, these often lead to similar predictions concerning more exotic phenomena, which in general follow from a nonlinear circuit theory. Attempts to distinguish between the various models on the basis of such experiments are summarized by J. Tucker.

2. THE CLASSICAL PARTICLE MODEL

The Hamiltonian of a one dimensional deformable charge density wave $\rho=\rho_o \cos(2k_F x+\phi(x))$ interacting with randomly positioned impurities is given in the presence of an applied electric field by[2]

$$H = \int dx\{\frac{\kappa}{2}(\frac{d\phi}{dx})^2 + \frac{M}{2}(\frac{d\phi}{dt})^2 + V_o\rho_o \sum_i \cos(2k_F x_i+\phi(x_i)) + \frac{eE_x\phi(x)}{2k_F}\} \qquad (1)$$

where the first term represents the elastic CDW, with stiffness constant κ, the second term is due to the inertia of CDW, the third term describes the interaction with impurities of strength V_o positioned at random, and the fourth term is the coupling to the external electric field.

Assuming that impurity pinning leads to an average periodic potential and choosing a sinusoidal form, the position dependent potential can be written as

$$V(\phi) = \omega_p^2(1-\cos\phi) \; . \qquad (2)$$

Neglecting the internal dynamics of the mode represented by the first term of Eq. (1), and adding a phenomenological damping term, the equation of motion of the average coordinate $x = \frac{\phi}{2k_F}$ is given by

$$\frac{dx^2}{dt^2} + \frac{1}{\tau}\frac{dx}{dt} + \frac{\omega_o^2}{2k_F}\sin 2k_F x = \frac{2k_F \; eE_x n_s}{m^*} \quad , \qquad (3)$$

where τ^{-1} is the damping coefficient, m^* the effective mass of the condensate, and n_s is the condensate density. By asserting that Eq. (3) represents the dynamics of the collective mode, it is assumed that the phase $\phi(x,t)$ can be expressed as

$$\phi = \phi_o(x) + \phi(t) \qquad (4)$$

where $\phi_o(x)$ represents the phase which adjusts the position of the CDW optimizing the interaction with impurities, and $\phi(t)$ leads to the time dependent part, which is related to the current by

$$j_{CDW} = \frac{n_s e}{\pi} \frac{d\phi}{dt} \quad . \tag{5}$$

Eq. (3) is that of a particle moving in a periodic potential. This simple nonlinear equation, which retains only the coherent aspects of CDW dynamics, will be contrasted with experiments first; incoherent effects, which point to the importance of internal dynamics of the mode will be summarized later. In all experiments discussed here, the electric field and current are applied and measured along the chain direction.

3. EXPERIMENTS: COHERENT EFFECTS

3.1. Frequency and electric field dependent conduction.

The ac response of Eq. (3) to small amplitude fields is that of a harmonic oscillator:

$$\begin{aligned} \mathrm{Re}\,\sigma_{ac}(\omega) &= \frac{n_s e^2 \tau}{m^*} \frac{\omega^2/\tau^2}{(\omega_o^2-\omega^2)^2+\omega^2/\tau^2} \\ \mathrm{Im}\,\sigma_{ac}(\omega) &= \frac{n_s e^2 \tau}{m^*} \frac{(\omega_o^2 - \omega^2)\ \omega/\tau}{(\omega_o^2-\omega^2)^2+\omega^2/\tau^2} \end{aligned} \tag{6}$$

with the low frequency dielectric constant

$$\varepsilon(\omega \to 0) = \frac{4\pi n_s e^2}{m^* \omega_o^2} \tag{7}$$

and maximum conductivity at $\omega = \omega_o$

$$\sigma(\omega = \omega_o) = \frac{n_s e^2 \tau}{m^*} \tag{8}$$

Experiments performed in the radiofrequency range on various materials ($NbSe_3$, TaS_3, $(TaSe_4)_2I$ and $K_{0.3}MoO_6$) when compared with Eq. (6) indicate an overdamped response $\omega_o^{-1} \gg \tau$. A fit of Eq. (6) to experiments performed on TaS_3 is shown in Fig. 1. Similar fits are obtained for other materials. The effect of inertia has recently been observed in

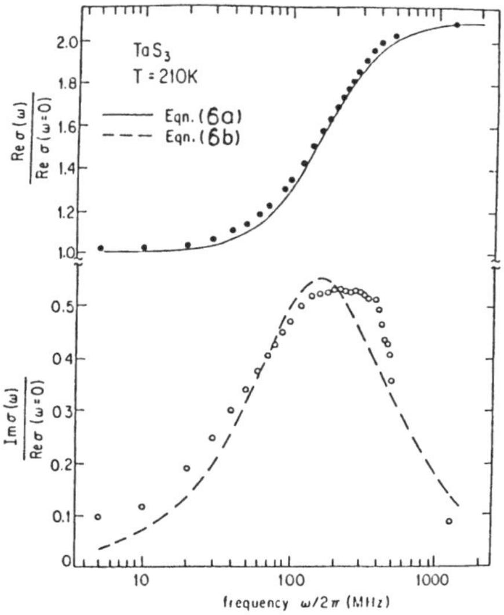

Fig. 1.

Frequency dependent conductivity in TaS_3. The full and dotted lines are fits to Eq. (6) with parameters given on the figure.

$NbSe_3$ and in TaS_3 at mm wave frequencies[4], leading to τ^{-1} and ω_o values in the GHz frequency region. In all cases, however, $\tau^{-1} > \omega_o$, thus the response is overdamped. Using Eq. (8), m^* can be evaluated, such procedure then gives good account for the dielectric constant given by Eq. (7).

Nonlinear conduction occurs when E exceeds a threshold field E_T, given by

$$E_T = \frac{\lambda}{2\pi} \frac{m^* \omega_o^2}{e} \qquad (9)$$

with $\lambda = \frac{\pi}{k_F}$ the period of the CDW. At high electric fields $\sigma(E \to \infty)$ saturates at a value given by Eq. (8).

Indeed, the measured high field and high frequency limits of the conductivity are the same in pure and in doped TaS_3 and also in $NbSe_3$[5]. Furthermore, the relation between the ω and E dependent response, as expressed by the product

$$\varepsilon(\omega \to 0) \cdot E_T = 2n_s e\lambda = 4n_\perp e \qquad (10)$$

where $n_\perp$ is the number of chains per unit area is confirmed experimentally in various materials (see Fig. 2). Here ε is the dielectric constant measured at frequencies around 1 MHz (much lower than $\omega_o^2\tau$), $n_\perp$ is evaluated from the known lattice structures. The full line leads to $\varepsilon E_T = 0.5e$, in an order of magnitude agreement with Eq. (10).

The model also leads to an oscillating current in the nonlinear region with a linear relation between the current per chain j and fundamental frequency f_o of the current oscillations. The relation is given by

$$j/f_o = 2e \frac{n(T)}{n(o)} \qquad (11)$$

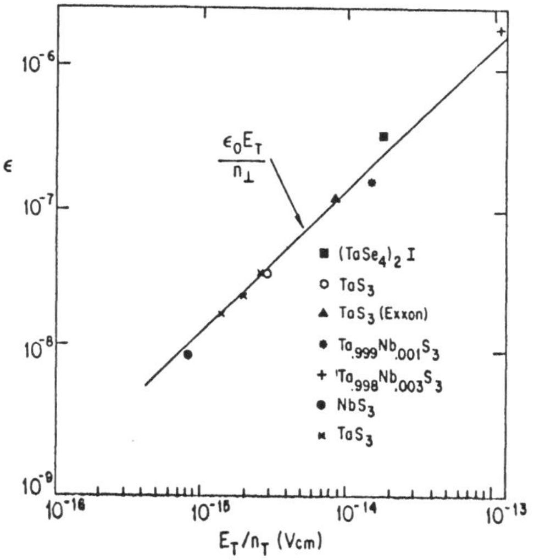

Fig. 2.

Increase dielectric constant versus threshold electric field. The full line corresponds to $\epsilon E_T = 0.5\ e$.

The linear relation between the current and fundamental frequency is observed in $NbSe_3$ over a broad range of driving fields (see Fig.3.) and also in good specimens of TaS_3 (both orthorhombic and monoclinic) and $(TaSe_4)_2I$, although the prefactor is under discussion at present[6].

Fourier transform of the time dependent current contains many harmonics with intensities (for an overdamped response)

$$I_n \sim (\alpha^2-1)^{\frac{1}{2}}[\alpha-(\alpha^2-1)^{\frac{1}{2}}]^n \cos[n(\omega_o t+\frac{\pi}{2}) + \sin^{-1}\alpha] \qquad (12)$$

where $\alpha = E/E_T$ and $\omega_o = 2\pi f_o$. Eq.(12) predicts a slow decay of the harmonics with increasing n, consistent with observations. Also the amplitude of the fundamental increases with increasing current and saturates in the high frequency limit. Such saturation has been observed in spectrum analysis studies[7], but experiments performed in the time domain indicate that I_o decreases in the high velocity limit[8]. This may be related to the importance of inertia effects, or to time dependent dephasing after an initial pulse.

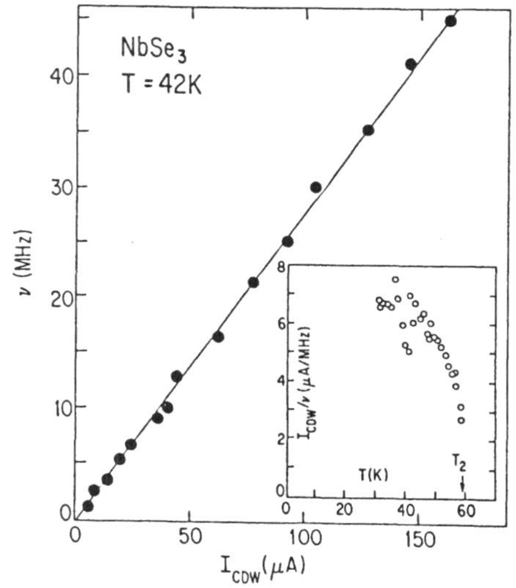

Fig. 3.

The relation between current I and frequency f in $NbSe_3$. *The insert shows the ratio I/f at various temperatures.*

3.2. Experiments in the presence of joint ac and dc fields.

With a highly nonlinear and frequency dependent conduction, the dynamics of the mode can further be explored by studying the response to joint ac and dc driving fields. The overall behavior of the response can also be described by Eq. (3), with $E_x = E_{dc} + E_{ac} \cos \omega t$. dc conduction can be induced with an application of large ac fields, and the effect is suppressed at high applied ac frequencies[9,10]. This simply follows from a strongly damped response. For parabolic (instead of sinusoidal) potentials the maximum displacement of the model driven by an ac field is given by

$$x_o = \frac{\frac{eE}{m^*} \omega_o^{-2}}{[1+(\frac{\omega}{\omega_o^2 \tau})^2]^{1/2}} \tag{13}$$

Assuming that a critical displacement is required to induce dc conduction, the ω dependent critical ac field is given by

$$E_{int} = E_{int}(\omega=0)\,[1-(\frac{\omega}{\omega_o^2 \tau})^2]^{1/2} \tag{14}$$

i.e. the ac amplitude required to induce dc conduction increases with increasing ω. Such an increase has been observed[9,10], and the experimental results can well be described by Eq. (13) with $\omega_o^2 \tau$ values close to those obtained from the ω dependent response. Similarly, arguments can be advanced to describe the modification of the ac response by the application of dc fields above threshold, and by the decrease of the low frequency dielectric constant with increasing dc field in the current carrying region[10].

3.3. Interference phenomena

In dimensionless form, Eq. (3) is given by

$$\frac{d^2\phi}{dt^2} + \Gamma \frac{d\phi}{dt} + \sin \phi = E/E_T \tag{15}$$

with $\phi = 2k_F x$, $\Gamma = \frac{1}{\omega_o^2 \tau}$ and E_T as given by Eq. (9). This equation is formally analogous to the resistively shunted Josephson junction model

$$\frac{d^2\theta}{dt^2} + G \frac{d\theta}{dt} + \sin \theta = I/I_j \tag{16}$$

where θ is the phase difference across the junction, I is the current through the junction, $G = (RC\,\omega_j)^{-1}$ and $\omega_j = 2eI_j/c\hbar$, with C the junction capacitance and I_j the critical current. This analogy (in which the current oscillations correspond to the ac Josephson effect) can extensively be used to search for interference and phase locking phenomena. In particular, steps in the dc I-V curves (called the Shapiro steps in the Josephson literature) can be observed whenever the frequency of the applied (or its harmonics) is the same as the fundamental oscillation frequency f_o of the current carrying mode[11]. Such behavior[12] is shown in Fig. 4. for $NbSe_3$. The height of the first step is given approximately by

$$\delta V = \beta V_t(\omega=0)\ J_1[V_1\omega_o^2\tau/\omega V_T(\omega=0)] \tag{17}$$

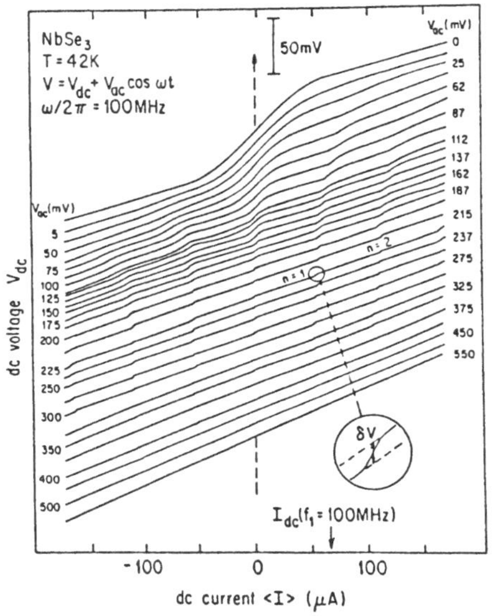

Fig. 4.
dc I-V traces in $NbSe_3$ in the presence of an applied ac field at frequency $\omega/2\pi=100$ MHz. The arrow indicates the dc current which leads to a noise frequency of 100 MHz.

where J_1 is the first order Bessel function and β is the fractional volume which responds coherently to the applied dc and ac driving fields. The Bessel function behavior has been observed in $NbSe_3$, with $\omega_o^2\tau$ similar to that observed for the frequency dependent conductivity. In pure and small specimens $0.1 < \beta < 1$ suggesting that a large fraction of the samples responds coherently to external perturbations.

The sensitivity of the method is greatly enhanced by taking the derivative of the I-V curve by using low frequency lock-in technique. Steps in the I-V curve correspond to peaks in the derivative. A typical experimental trace[13], obtained in $NbSe_3$, is shown in Fig. 5. The fundamental and higher harmonics are clearly visible, but one also observes a whole series of subharmonics, indicating mode locking whenever

$$p\omega_{ext} = q\ \omega_o\ . \tag{18}$$

Several of them are indicated on the Figure. Detailed experiments suggest that mode locking occurs for every integer value of p and q, indicating a Devil's staircase behavior. Up to approximately 70 steps have been

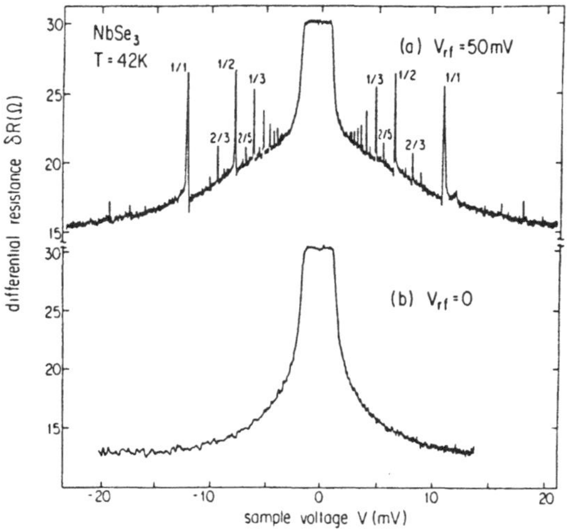

Fig. 5.

Differential resistance observed in NbSe$_3$ without and with an applied rf field. Some of the steps (see Eq. (18)) are indicated on the Figure.

detected in the interval up to the fundamental Shapiro steps, smaller steps are absorbed in the noise. Subharmonics have also been observed in orthorhombic TaS_3, with features similar to those shown in Fig. 5. The width of the steps in the derivative curve is however larger, and the amplitude smaller, indicating less coherence than in $NbSe_3$.

As both the intrinsic current oscillation, and the nonlinear response to an external sinusoidal applied field has many harmonics, intuitively one would expect that mode locking would occur between any harmonics of the two spectra, hence the condition Eq. (18). But calculations indicate that subharmonic steps do not occur for overdamped systems[15], but a small inertial term restores the mode locking. Mathematical models have been developed[15] to investigate the chaotic behavior of nonlinear systems, and one can compare the experiments to the predictions of these models by analyzing the step heights for different p and q. A fractal dimension can be extracted for the Devil's staircase structure, suggesting chaotic behavior, a complete Devil's staircase[14]. Other routes to chaos, such as period doubling bifurcations have also been observed and described in terms of Eq. (3)[15].

In spite of these developments, which demonstrate the usefulness of the phenomenological equation of motion, several important unresolved questions remain. The effect of normal electrons, the possible difference between the response to a constant current and constant voltage drive and the question of the effect of damping (as indicated by frequency dependent studies) has to be examined in detail. In particular, no subharmonic steps are expected to occur for an overdamped system, in apparent conflict with the observations. Here, as in other cases, the internal degrees of freedom of the mode may be of importance.

EXPERIMENTS: INCOHERENT EFFECTS

While the broad qualitative features of the observed nonlinear and frequency dependent response phenomena are reproduced by the model, several predictions are in clear conflict with the experimental observations. By neglecting the internal dynamics and the disorder associated with impurity pinning, the model leads to the following features:

- $\mathrm{Re}\sigma(\omega) \sim \omega^2$ and $\mathrm{Im}\sigma(\omega) \sim \omega$ in the low frequency ($\omega \to 0$) limit;
- divergent differential conductivity and differential dielectric constant at threshold

$$\frac{dI}{dE} \sim \frac{1}{E(1-E_T^2/E^2)^{\frac{1}{2}}} \qquad E > E_T \tag{19}$$

$$\frac{d\varepsilon}{dE} \sim \frac{d\varepsilon}{dE}(E=0)\,\frac{1}{(1-E_T^2/E^2)^{\frac{1}{2}}} \qquad E < E_T \tag{20}$$

  where $\frac{d\varepsilon}{dE}$ is defined as the (small ac field) dielectric constant in the presence of a dc field;
- no broad band noise and zero width for the Fourier components of the time dependent current;
- current oscillations in the thermodynamic limit;
- instantaneous transition between the pinned and the current carrying mode.

It is also clear that random impurity pinning leads, without internal deformations of the mode, to zero pinning energy in the thermodynamic limit. This is perhaps the most fundamental objection to the approach taken, and to the assumption which leads to Eq. (2).

In the following I will briefly survey the experiments relevant to the predictions listed above, these experiments all point to the importance of incoherent phenomena, and to the importance of the internal dynamics of the collective mode.

4.1. Pinned mode: frequency dependent conductivity.

Frequency dependent transport measurements extended to very low frequencies in materials which show CDW transport reveal serious discrepancies with the single oscillator response. In Fig. 6 $Re\sigma(\omega)$ and $Im\sigma(\omega)$ observed in TaS_3 is shown in a log log plot which emphasizes the low frequency response[17]. We observe that at frequencies below approximately 10 MHz

$$\sigma(\omega) = A(i\omega)^{\alpha} \qquad (21)$$

and similar behavior was found in $NbSe_3$, $(TaSe_4)_2I$ and $K_{0.3}MbO_6$ [18]. This particular power law is reminiscent to those found in glasses, spin glasses and strongly disordered solids. Phenomenologically, a broad distribution of relaxation times can be postulated to account for the ω dependent conductivity. Assuming a distribution for the crossover frequency $\omega_o^2\tau = \omega_{c.o.}$

$$P(\omega_{c.o.}) = \frac{1}{P_o}\,\frac{1}{1+(\omega_{c.o.}/b)^a} \qquad (22)$$

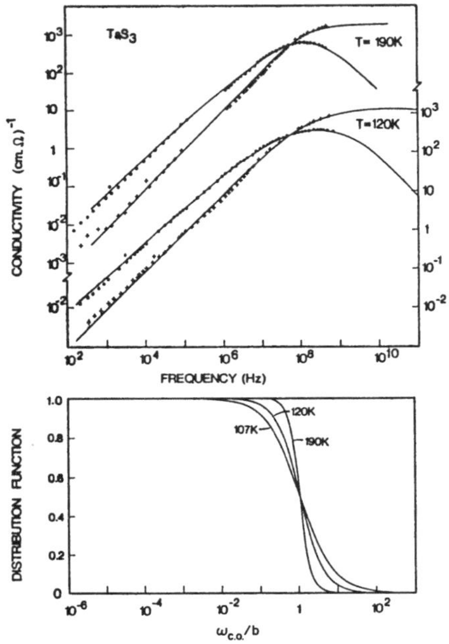

Fig. 6.

a) Frequency dependent conductivity, $Re\sigma(\omega)$ and $Im\sigma(\omega)$ observed in TaS_3 at two different temperatures,
b) The distribution function which is used to fit the data, see Eq. (22), at various temperatures.

an excellent fit is obtained as shown in the Figure. The distribution $P(\omega_{c.o.})$ obtained at various temperatures is also shown. While such phenomenological approach certainly does not represent a solution of the problem, it points in general to the importance of disorder in the dynamics of the pinned mode. This distribution of the relevant frequencies and/or relaxation times is also responsible for metastable states and the sensitivity of the system to small perturbations achieved by thermal and/or electric field pulses, and for the long time behavior of the decay of the remanent polarization (or in more fancy words left-right asymmetry) observed in these materials[19].

4.2. Current carrying mode: amplitude and spectral width of the current oscillations.

Early experiments[8,20] on the amplitude of the current oscillations (performed on high quality specimens) showed highly coherent response at dc fields near threshold, indicating that the specimens can be regarded as one coherent domain, and Eq. (4) is a reasonable approximation.

The volume dependence of the oscillation amplitude has been investigated recently both in $NbSe_3$[21] and in $(TaSe_4)_2I$[22]. The dependence of the amplitude of the fundamental $\Delta V/V$, measured at $f_o = 20$ MHz, on the length ℓ and on the cross section A of the $NbSe_3$ specimens is shown in Fig. 7.

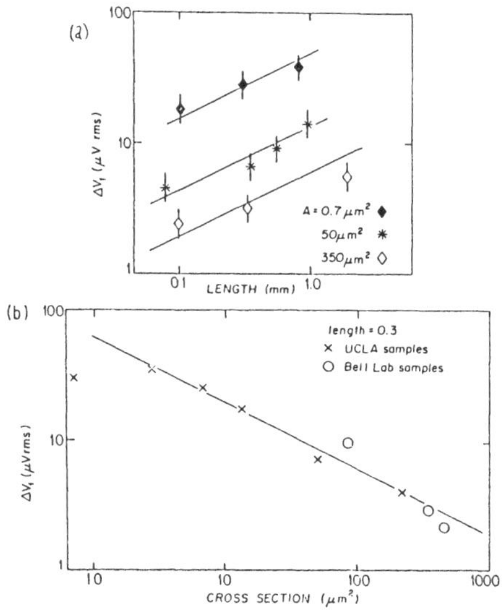

Fig. 7.

Oscillation amplitude ΔV_1 in $NbSe_3$ at 42.5 K, measured between 10 and 20 MHz, for samples of different dimensions. (a) ΔV_1 vs. length for samples of three cross sections A. The lines have slope ½. (b) ΔV_1 vs. A for samples of fixed length $\ell = 0.3$ mm. The line has slope − ½.

An $\ell^{-\frac{1}{2}}$ and $A^{-\frac{1}{2}}$ dependence observed leads to a volume dependent amplitude[23]

$$\frac{\Delta V}{V} = A\left(\frac{\Omega_c}{\Omega_s}\right)^{\frac{1}{2}}$$

This behavior, also observed in $(TaSe_4)_2I$, suggests that if the sample is broken into domains in which the phase can be taken as constant, then the current oscillates with the same frequency (leading to narrow spectral width) but with random phase. A can be evaluated for a single domain, and consequently the volume of coherent domains can be estimated leading to $\Omega_c \approx 1\mu m^3$.

While the Fourier transformed peaks are extremely narrow in $NbSe_3$, in more anisotropic materials they are broad, suggesting a distribution of collective mcde velocities within the specimens. This distribution is more evident for samples with larger cross sections as expected for the dynamics of a collection of weakly coupled chains or domains perpendicular to the chain direction.

A large amplitude broad band noise is also observed in the current carrying state. The low frequency spectral behavior observed in TaS_3 is shown in Fig. 8.[24] The functional dependence on the frequency is somewhat less than what would correspond to a 1/f noise, and the noise amplitude is orders of magnitude larger than what is observed in ordinary semiconductors. Similar to the large amplitude of the low frequency ac polarization, this suggests large coherent regions and collective fluctuations involving macroscopic length scales. In the absence of experiments performed over a broad frequency range a quantitative analysis of these fluctuation effects is not possible at present.

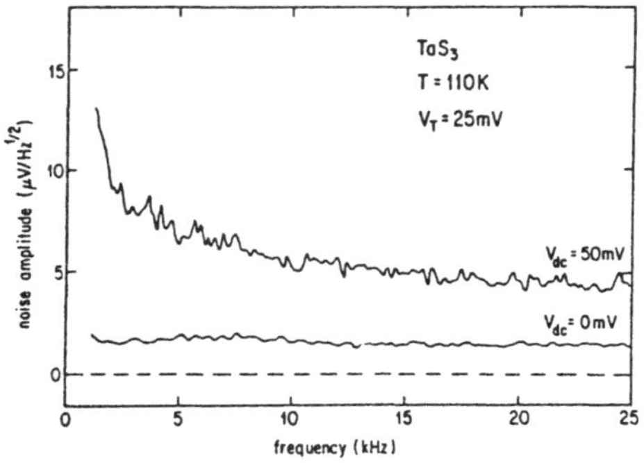

Fig. 8.
Low frequency broad band noise measured in TaS_3.

4.3. The transition between the pinned and the current carrying mode.

While early experiments showed a smooth onset of the sliding CDW mode, several recent measurements show hysteresis and switching phenomena associated with the transition between the two regions[25]. While inertia effects may, in principle, lead to such behavior, it is generally believed that the phenomenon is due to coupling between the various pinned domains and to the development of a coherent mode which executes the translational motion.

Hysteresis, first observed in $NbSe_3$, was shown to be associated with rather long times (of the order of milliseconds) required to establish the current carrying mode. Such switching between the pinned and current carrying mode is shown directly in Fig. 9,[26] where the time dependence of the voltage is displayed for a constant current. The large low frequency noise is associated with the current carrying mode, which is also represented by a lower resistance. The conditions under which such phenomena are observed are not clear at present, and it is likely that contact phenomena play an important role. It is important to recognize

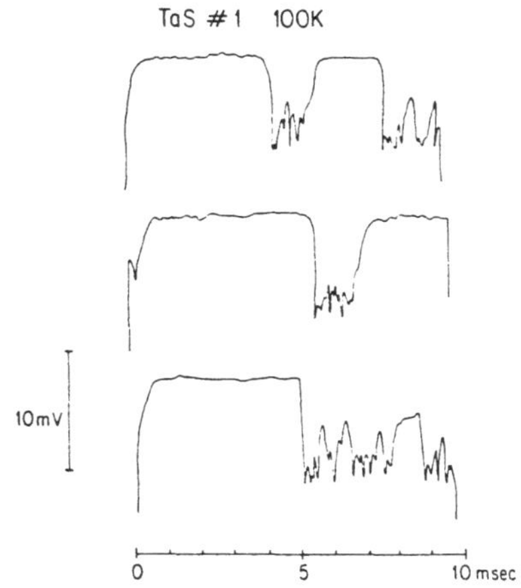

Fig. 9.
Several traces of the time dependent voltage observed in TaS_3, after application of a constant current pulse at $t = 0$ sec. The high voltage and low voltage states correspond to the pinned and current carrying modes.

that the phenomena is similar to those observed in coupled Josephson junctions and in granular superconductors. In both of these cases, weak coupling between elements which themselves display a nonlinear response plays an important role. The phenomenon is strongly reinforced by external passive circuit elements, and certainly has strong relations to long time, low frequency phenomena observed in $K_{0.3}MoO_6$ [27].

5. CONCLUSIONS

The experiments summarized above point to the interplay of coherent and incoherent effects associated with the dynamics of charge density waves, and suggest that a proper treatment of the Hamiltonian (Eq. (1), supplemented with damping effects) is required to account for the broad variety of phenomena observed.

Simple approaches, like involving a distribution of characteric parameters, or supplementing Eq. (3) with a noise term may lead to some of the coherent features seen experimentally. Neither of these however, are expected to account for effects associated with the internal dynamics of the collective mode. Computer simulations, which take into account the pinning by random impurities and are based on Eq. (2) reproduce many details of experiments, and so do the approaches taken by Fisher and by Klemm and Schrieffer, and by Sneddon[27].

It is also expected, that the interplay between coherent and incoherent effects is due to the fact that the relevant length scale, the Fukuyama-Lee-Rice phase-phase correlation length L, is comparable to the dimensions of the specimens. For a static CDW pinned by impurities

$$< \phi(x)\phi(o) > \sim \exp(-x/L)$$

with

$$L \sim K^2/V_o^2 \, \rho_o^2 \, c$$

where c is the impurity concentration. L as large as 10 μm to 1 mm, comparable to the lengths of specimens investigated. This length scale may play a fundamental role in the dynamics of charge density waves, with rather different behavior observed for rather small, clean specimens and in samples where L is much smaller than the sample dimensions. The dynamics of internal degrees of freedom may play an important role in the latter situation, but can perhaps be neglected in the former. Both cases appear to be interesting, but from different points of view. The dynamics of random, driven systems can be studied in the latter case, which also could be called a charge density wave glass. While the system is disordered, it is anticipated that coherence can be built up by application of large dc or ac fields. Such synchronization phenomena are discussed in detail in Ref. 13b. In the former case due to phase coherence which extends throughout the specimen experiments a'la Aharonov--Bohm could be envisioned, with possible searches for charge quantization, and other macroscopic quantum phenomena. Such experiments may also shed light on the relation between the classical and quantum descriptions of charge density wave transport.

ACKNOWLEDGEMENTS

The author gratefully acknowledges discussions and collaborations with many people. The experiments reported here were performed by S. Brown, G. Mozurkewich, L. Mihály, S. Sridhar, W.Y. Wu and A. Zettl. I have greatly benefitted from discussions with John Bardeen, S. Barnes, R. Klemm, A Jánossy, Leigh Sneddon, J.R. Schrieffer and A. Zawadowski. This research was supported by the National Science Foundation Grants DMR 84-06896 and DMR 81-21394.

REFERENCES

1. G. Grüner, A. Zawadowski and P.M. Chaikin: Phys. Rev. Lett. 49, 511 (1981) some of the details of the model have been worked out by P. Monceau et al: Phys. Rev. B25, 931 (1982)
2. H. Fukuyama and P.A. Lee: Phys. Rev. B17, 535 (1978)
   P.A. Lee and T.M. Rice: Phys. Rev. B19, 3970 (1979)
3. For a review of the experimental result see G. Grüner and A. Zettl: Physics Reports (to be published)
   A. Zettl and G. Grüner: Phys. Rev. B25, 2081 (1982)
4. S. Sridhar, D. Reagor and G. Grüner (this conference)
5. D. Reagor, A. Jánossy and G. Grüner (to be published)
6. See for example P. Monceau (this conference)
7. M. Weger, W.G. Clark and G. Grüner: Proceedings of the International Conference on Synthetic Metals, Les Arcs, 1982
8. John Bardeen, E. Ben Jacob, A. Zettl and G. Grüner: Phys. Rev. Lett. 51, 493 (1982)
9. G. Grüner, W.G. Clark and A.M. Portis, Phys. Rev. B24, 3641 (1981)
10. A. Zettl and G. Grüner: Phys. Rev. B25, 2081 (1982)
11. P. Monceau, J. Richard and M. Renard, Phys. Rev. Lett. 45, 43 (1980)
12. A. Zettl and G. Grüner, Phys. Rev. B29, 755 (1984)
13. a) S. Brown, G. Mozurkewich and G. Grüner, Phys. Rev. Lett. 52, 2277 (1984) and
    b) this conference
14. S. Brown, G. Mozurkewich and G. Grüner (to be published)
15. P. Bak and M. Hoegh Jensen, J. Phys. A15, 1983 (1982)
16. See A. Zettl (this conference)
17. W.Y. Wu, L. Mihály, G. Mozurkewich and G. Grüner, Phys. Rev. Lett. 52, 2382 (1984)
18. W.Y. Wu, L. Mihály, G. Mozurkewich and G. Grüner (this conference)
19. A. Jánossy, G. Mihály and L. Mihály (this conference)
    J.C. Gill (this conference) and references cited therein
20. R. Fleming, Solid State Comm.
21. G. Mozurkewich and G. Grüner, Phys. Rev. Lett. 51, 2206 (1983)
22. G. Mozurkewich, M. Maki and G. Grüner, Solid State Comm. 48, 5 (1983)
23. The detailed length dependence is controversial at present, see N.P. Ong et al: Phys. Rev. Lett. 52, 663 (1982)

24. A. Zettl and G. Grüner, Solid State Comm. 46, 29 (1982). Similar experiments on $NbSe_3$ were performed by J. Richard et al. J. Phys. C15, 7157 (1982), see also R. Klemm et al (this conference)

25. A. Zettl and G. Grüner, Phys. Rev. B26, 2298 (1982)

26. L. Mihály and G. Grüner, Solid State Comm. 50, 807 (1984)

27. For these approaches see the relevant papers in this proceedings.

THRESHOLD FIELD, ELECTRICAL CONDUCTIVITY AND TIME-DEPENDENT VOLTAGE IN TRANSITION METAL TRI- AND TETRACHALCOGENIDES

P. Monceau, M. Renard, J. Richard, M.C. Saint-Lager and Z.Z. Wang
Centre de Recherches sur les Très Basses Températures, CNRS, BP 166 X, 38042 Grenoble-Cédex, France

Non-linear transport properties have now been measured in numerous one-dimensional transition metal tri- and tetrachalcogenides. For all these compounds the threshold electric field above which the non-linear state appears goes through a minimum in the vicinity of the Peierls temperature transition. The value of this minimum is all the higher as the Peierls transition occurs at a higher temperature. The variation of the fundamental frequency of the time-dependent voltage is shown to follow a $(E-E_c)^{\gamma}$ law (with $\gamma \sim 1.5$) near the threshold indicating a collective pinning. Finally non-linear properties of a $NbSe_3$ sample have been studied when both contacts are above the Peierls temperature : in these conditions an a.c. voltage is still detected. This result is discussed in relation with theories on the origin of this voltage generation.

Many of the chains which form the transition metal tri- and tetrachalcogenides, namely $NbSe_3$[1], NbS_3 type II[2], TaS_3 with the orthorhombic and the monoclinic unit cell[3], $(TaSe_4)_2I$[4] and $(NbSe_4)_{10}I_3$[4] distort themselves below the Peierls transition temperature. The wave-lengths of the distortions and of the associated charge density waves (CDWs) of all these compounds are incommensurate with the main lattice, except for the orthorhombic TaS_3 for which the component along the chain axis locks to the commensurability of four atomic distances at $T_o' \sim 130$ K[3]. Among this family $NbSe_3$ is the unique compound undergoing a Peierls transition which remains metallic at low temperatures. For all the other compounds the Peierls distortion is associated to a metal-semiconducting transition. Fig. 1 shows the variation of the absolute resistivity normalized to its value at room temperature in a logarithmic scale as a function of $10^3/T$. In the insulating state the resistivity variation follows the activation law corresponding to single excitations through the gap $\Delta'(T)$ such as $\rho(T) = \rho_o \exp[-\Delta'(T)/2kT]$. The structure of the unit cell, the Peierls transition temperature, the ratio between the CDW gap 2Δ and the Peierls temperature, the components of the distortion and the absolute resistivity are summarized in Table I. It can be seen that the ratio $2\Delta/kT_c$ is much higher than 3.5 as found in the mean field theory which can be explained either by the unidimensionality of these compounds and/or by a strong coupling electron-phonon interaction[5]. The value $2\Delta \sim 700$ K for $NbSe_3$ for the lower CDW gap in $NbSe_3$ has been obtained from tunneling experiments[6].

Threshold electric field

All the compounds listed in Table I exhibit non-linear transport properties[7] at any

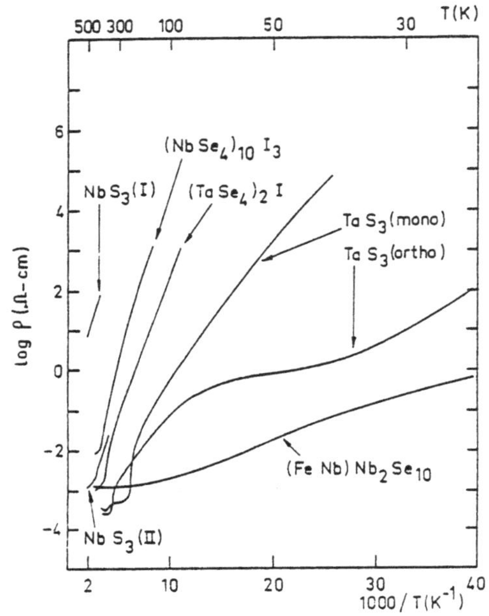

Fig. 1

Variation of the absolute resistivity (in a logarithmic scale) as a function of the inverse of temperature for transition metal tri and tetrachalcogenides undergoing a Peierls transition.

temperature below the Peierls transition when the applied electric field exceeds a threshold value, E_c. The temperature variation of E_c is drawn in Fig. 2. The data correspond to samples in each family which have shown the lowest threshold values.

For all the compounds, E_c decreases in a small temperature range in the vicinity of the Peierls transition and goes through a minimum.

The minimum value of E_c is in the range of 0.1-1 Vcm^{-1}. Its value is all the higher as the Peierls transition occurs at higher temperature. It can be roughly estimated that log E_c varies linearly with T_P.

E_c increases sharply at low temperatures. The slope of log E_c as a function of T is roughly the same for each compound except for the orthorhombic TaS_3 one. This behaviour of E_c at low temperatures is not understood but the exponential increase of E_c when T is reduced and the close relationship between the minimum value of E_c and the Peirls transition temperature indicates the importance of thermal activation in the CDW dynamics. Maki has shown that for the one-dimensional case, a soliton-antisoliton pair could be

TABLE I

	Symmetry	Peierls trans. temper. (K)	2Δ (K)	$\frac{2\Delta}{kT_c}$	Distortion components a^*	b^*	c^*	ρ(Ωcm) at T=300 K	chains per unit cell
NbS_3 type II	monoclinic	330	4400	13.3	0.5	0.352	0	8 10^{-2}	8
$NbSe_3$	monoclinic	{145 59	700	11.9	0 0.5	0.24117 0.26038	0 0.5	2.5 10^{-4}	2×3
TaS_3	orthorhombic	215	1600	7.44	? 0.5	0.1 0.125 (T<130 K)	0.255 0.250	3.2 10^{-4}	24
	monoclinic	{240 160	1900	11.9	0 0.5	0.253 0.247	0 0.5	3 10^{-4}	2×3
$(TaSe_4)_2I$	tetragonal	263	3000	11.4	1	0	0.943	1.5 10^{-3}	2×1
$(NbSe_4)_{10}I_3$	tetragonal	285	3900	13.7	0	0	0.487	1.5 10^{-2}	2×1

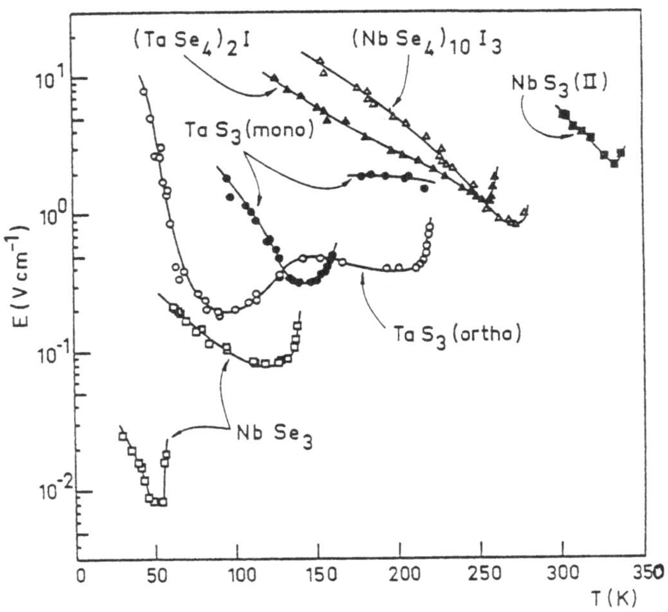

Fig. 2

Variation of the threshold electric field for compounds exhibiting non-linear transport properties as a function of temperature.

created by tunneling by the electric field[8]. In two or three dimensions domain walls or bubbles are created by thermal activation. The sliding CDW is achieved not by the motion of the whole CDW but only continuously creating and expanding domain walls. In this case Maki finds that the threshold varies like T^{-1} (ref. 9).

Collective pinning

The non-linearity has been ascribed to the motion of the CDW in the pinning potential created by the impurities[10,11]. The simplest models consider the phase of the CDW, rigid into a finite size domain. Above the threshold the CDW current is modulated with frequencies multiple of the fundamental one such as $v \sim (E^2-E_c{}^2)^{1/2}$. More realistic models take into account deformations of the CDW resulting from its interaction with impurities. When the local distortions of the CDW are small i.e. when the velocity of the CDW is large, the impurity pinning is only a perturbation and asymptotic laws can be derived[12]. But in the vicinity of E_c the CDW distortions are large. Fisher[13] considers the CDW depinning in the frame of critical phenomena. He finds that, above E_c, the velocity of the CDW follows the law :

$$v \sim (E - E_c)^{3/2}$$

When $E >> E_c$, v is naturally proportional to E. The power law coefficient, 3/2 is the consequence of collective pinning. If the number of impurities decreases and becomes a finite but small number, eventually one, the result for rigid CDW motion, $v \sim (E-E_c)^{1/2}$ is recovered. Numerical methods have been studied by Sokoloff[14], Pietronero and Strässler[15]. When the size of the system increases, it is found that the curvature of v(E) has a definite tendency to become concave upwards and that the singularity at E_c is confined to a very narrow region. Fig. 3 shows a schematic variation of the CDW velocity, v, as a function of E according to the different models mentioned above.
The CDW velocity can be written as the product of the fundamental frequency, ν, measu-

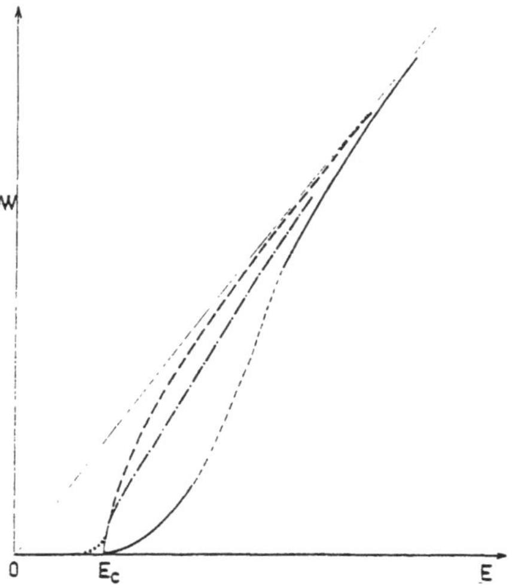

Fig. 3

Schematic variation of the CDW velocity as a function of the electric field :
--- for a classical rigid single domain motion,
··· for self-synchronized independent domains,
—— for a deformable motion of the CDW,
-·-· for a sample with a finite size.

red by Fourier analysis of the time dependent voltage generated above E_c and of a periodicity. Fig. 4 shows the variation of ν as a function of $(E-E_c)$ near the threshold for an orthorhombic TaS_3 sample at different temperatures. The curves are $(E-E_c)^{\gamma}$. For temperatures between 130 K and 205 K, γ is approximately equal to 1.5 : 1.33 at T = 205 K, 1.45 at T = 159 K, 1.44 at T = 141 K, 1.63 at T = 131 K. But γ increases at lower temperatures. Its value is 2 at T = 121 K and 2.3 at T = 107 K and 81 K. Similar results are obtained for $(TaSe_4)_2I$ and $(NbSe_4)_{10}I_3$: $\gamma \sim 1.5$ in the vicinity of T_p and increases at lower temperatures. For $NbSe_3$ at the higher CDW transition we found $\gamma \sim 1.1$ at 137 K and 129 K and 1.26 at T = 96 K. Therefore a collective pinning theory accounts

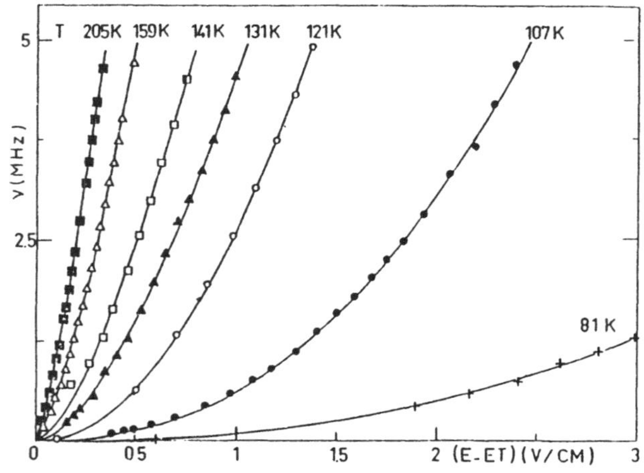

Fig 4

Variation of the fundamental frequency (or the CDW velocity) as a function of $E-E_c$ for orthorhombic TaS_3. The curves are $(E-E_c)^{\gamma}$ with γ in the text.

for the electric field variation of the CDW velocity, much better in any case than the classical rigid model.

Time-dependent voltage generation

Of prime importance is the nature of the time dependent voltage in the non-linear state : either a bulk effect or a local effect. Sneddon[16] has shown that, in the thermodynamic limit, there is no a.c. voltage generated in the sample. Mozurkewich and Grüner[17] have reported results showing a variation of the current oscillation in $\Omega^{-1/2}$, therefore vanishing in the infinite volume limit. But Ong et al.[18] have found that the a.c. voltage amplitude is independent of the length of the sample. This latter result strengthens the model independently proposed by Ong et al.[18] and Gork'ov[19] taking into account boundary conditions : the electrode is a local pinning center and the accumulation of charges in its vicinity is removed by the creation of phase-slip centers ; the a.c. voltage would be a local effect generated at the electrode. Experiments under thermal gradient have been performed by Ong et al.[18] and Zettl et al.[20] with contradictory results. The latter authors find that the change in the current oscillations is just this one expected from a change in average temperature. On the contrary, Ong et al.[18] have measured a splitting of the fundamental frequency, each fundamental being associate with each end of the sample. When the hot end of the sample is driven normal, the amplitude of the frequency associated to this electrode vanishes to zero.

We have performed non-linear studies with thermal gradient applied to $NbSe_3$ with a different temperature profile. The sample mounting configuration is schematically drawn in Fig. 5. On the two pyrex blocks is glued a strain gauge heater and a small isolated copper plate for homogeneizing the temperature. The contacts a, b, c, d are isolated from the cryogenic liquid (argon, freon) by a polymeric rubber. The tempera-

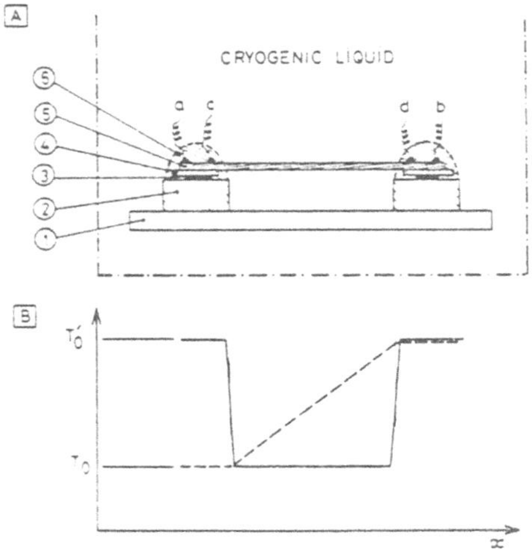

Fig. 5

A) Sample mounting configuration for thermal gradient experiment.
(1) sample holder
(2) pyrex block
(3) heater
(4) isolated copper plate
(5) $NbSe_3$ sample
(6) polymeric rubber

B) —— Temperature profile with the configuration shown in A.
--- Temperature profile in the experiment of Ong et al.

ture difference between each contact area (T_o') and the bath (T_o) is measured by a thermocouple. The temperature profile is shown in Fig. 5B ; the change in temperature takes place in a length comparable to the transverse dimension of the sample i.e. a few microns. The length of the sample in the cryogenic liquid can be varied from 2 mm to 2 cm. Temperature difference as large as 70 K can be achieved. We have studied the upper CDW in $NbSe_3$ setting the bath temperature, T_o, at 95 , 115 and 125 K. We have heated the contact area above T_2 = 145 K as measured by the thermocouple voltage and by monitoring the resistance of the sample between the contacts a,c and d,b. When $T_o' = T_o$ we first apply a current to the sample to drive it in the non-linear state and the Fourier spectrum of the voltage is recorded ; then, with the same d.c. current, we increase T_o' above T_2 on both contacts or on one of them. We find that the Fourier spectrum remains unchanged, without splitting as observed by Ong et al.[18]
It appears very unlikely that CDW wave lengths can evaporate to give electrons except somewhere when the order parameter $\Delta \to 0$. Therefore the CDW velocity is conservative which implies that along the sample the electric field is a function of the local temperature. Moreover the conservation of the total current leads along the sample to a different gradient of the Fermi level (which controls the excitation current). So that if T varies enough on a sufficient length L, the Fermi level crosses the gap. At this point $\Delta \to 0$ and the sample breaks in two parts (or more) with different velocities. So two important parameters appear in the non equilibrium CDW phenomena : one is the temperature variation and the other the length on which it is applied. If the temperature difference occurs on a relative large distance, the sample is able to break in two or more parts (the critical temperature difference depending of the length of the sample, the value of its gap, the temperature of the cold end, ...) which may be the case in the experiments of Ong et al.[18] while in the experiment described in Fig. 5 the temperature variation is too abrupt to allow this partition.

Conclusions

Although intense work as experimental as theoretical have been performed on the CDW dynamics of these inorganic one-dimensional compounds, many fundamental questions remain unsettled such as the explanation for the increase of the threshold at low temperature, the volume dependence of the current oscillation, the role of the electrodes for the conversion of the condensate into a normal current, ... which need further investigation in the future.

Acknowledgements - We would like to thank A. Meerschaut and F. Levy for providing us with the samples and L.P. Gork'ov and K. Maki for stimulating discussions.

References

1. R.M. Fleming, D.E. Moncton and D.B. McWhan, Phys. Rev. B 18 (1978) 5560.
2. C. Roucau, J. Phys. 44 (1983) C3-1725.
3. Z.Z. Wang, H. Salva, P. Monceau, M. Renard, C. Roucau, R. Ayrolles, F. Levy, L. Guemas and A. Meerschaut, J. Physique-Lettres 44 (1983) L-311.
4. C. Roucau, R. Ayrolles, P. Gressier and A. Meerschaut, J. Phys. C 17 (1984) 2993.
5. C.M. Varma and A.L. Simons, Phys. Rev. Lett. 51 (1983) 138.
6. A. Fournel, B. Oujia, and J.B. Sorbier, Proceedings on the International Conference on the Physics and Chemistry of Low-Dimensional Synthetic Metals, Abano Terme Italy , June 1984, Mol. Crystals, in press.
7. For a review see N.P. Ong, Can J. Phys. 60 (1981) 757.
   G. Grüner, Physica 8D (1983) 1, G. Grüner and A. Zettl, to be published,
   P. Monceau in Electronic Properties of Inorganic Quasi One-Dimensional Metals, Part II, edited by P. Monceau, D. Reidel Publishing Company Holland (1985).
8. K. Maki, Phys. Rev. Lett. 39 (1977) 46.
9. K. Maki, Phys. Lett. 70 A (1979) 449.
10. P. Monceau, J. Richard and M. Renard, Phys. Rev. Lett. 45 (1980) 43.
11. G. Grüner, A. Zawadowski and P.M. Chaikin, Phys. Rev. Lett. 46 (1981) 511.
12. L. Sneddon, M.C. Cross and D.S. Fisher, Phys. Rev. Lett. 49 (1982) 292.
13. D.S. Fisher, Phys. Rev. Lett. 50 (1983) 1486.
14. J.B. Sokoloff, Phys. Rev. B 23 (1981) 1992.
15. L. Pietronero and S. Strässler, Phys. Rev. B 28 (1983) 5863.
16. L. Sneddon, Phys. Rev. B 29 (1984) 719.
17. G. Mozurkewich and G. Grüner, Phys. Rev. Lett. 51 (1983) 2206.
18. N.P. Ong, G. Verma and K. Maki, Phys. Rev. Lett. 52 (1984) 663.
19. L.P. Gork'ov, Pis'ma Zh Eksp. Teor. Fiz. 38 (1983) 76, J.E.T.P. Letters 38 (1983) 87.
20. A. Zettl, M. Kaiser and G. Grüner, preprint.

SOLITONS IN TaS_3. EXPERIMENT

F.Ya.Nad
Institute of Radioengineering and Electronics, USSR
Academy of Sciences, 103907 Moscow, Marx Avenue 18

The paper presents the results of experimental study of optical properties, dc and ac conductivity, and dielectric constant of orthorhombic TaS_3 in a wide range of temperatures and electric fields. The results are interpreted in terms of an essential role played by solitons in determining the properties of TaS_3.

A good deal of experimental and theoretical results related to unusual properties of quasi-one-dimensional materials are available by now (see e.g./1,2/). Nevertheless, many aspects of the physical phenomena in question remain vague. It has become clear recently that solitons play an important role in these phenomena. The paper reports on the experimental results on the electric and optical properties of one of the most interesting quasi-one-dimensional materials TaS_3. The results were obtained in the Institute of Radioengineering and Electronics USSR Acad.Sci. by F.Ya.Nad`, S.K.Zhilinskii, and M.E.Itkis. It is demonstrated that the interpretation of the results necessarily involves the concept of nonlinear excitation of charge density waves (CDW), i.e., solitons.

Peierls transition. Temperature range 220-110 K

The study was conducted on samples of orthorhombic TaS_3 with different degree of perfection taken from a number of batches /3-6/. The temperature of three-dimensional ordering T_p in high-quality samples /5,6/ reached 220 K, and the minimal threshold field was 0.3 V/cm. The experimental techniques were described in detail elsewhere /4,5/. Fig.1 plots the temperature dependence of conductivity $\sigma(T)$ normalized to its room temperature value, σ_0. The temperature dependence of the derivative d $(\log\sigma)$/d(1/T) had in the interval 300-100 K a single minimum at T_p=217 K. Four regions can be singled out roughly in the range 350-4 K /4/ : 1 - $T > T_p$, 2 - $T_p > T > 100$ K, 3 - $100 > T > 20$ K, and 4 - $30 > T > 4$ K. In this paper we deal with regions 2, 3, and 4.

After conductivity sharply decreases at $T \simeq T_p$, we observe on the $\log(\sigma/\sigma_0)$ vs 1/T curve a nearly linear region from, roughly 200 K to 100 K; it corresponds to activation conduction with energy gap Δ = 770 K (in this particular sample). As a rule, Δ varies slightly among samples from different batches. Its values lie in the interval 750-840 K, the lowest being close to those reported in /7/. The slope of log (σ/σ_0) begins to diminish around 100 K (fig.1). We also measured transverse conductivity $\sigma_\perp$. The activation energy of $\sigma_\perp$ in

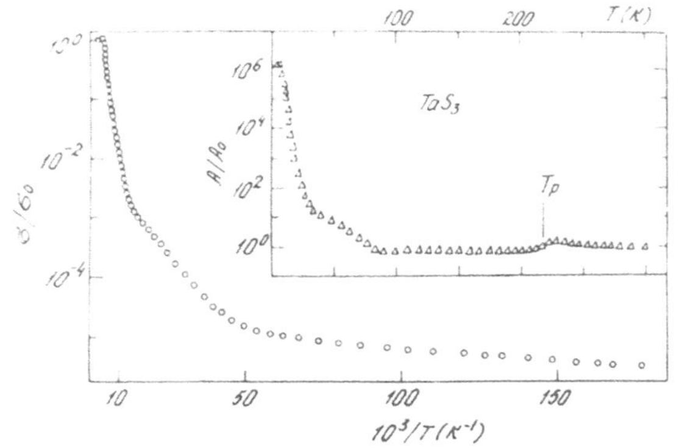

Fig.1. Temperature dependence of conductivity of TaS_3 normalized to its room temperature value σ_0 . The inset shows temperature dependence of the anisotropy $A= \sigma_{\parallel} / \sigma_{\perp}$ normallized to its room temperature value A_0 .

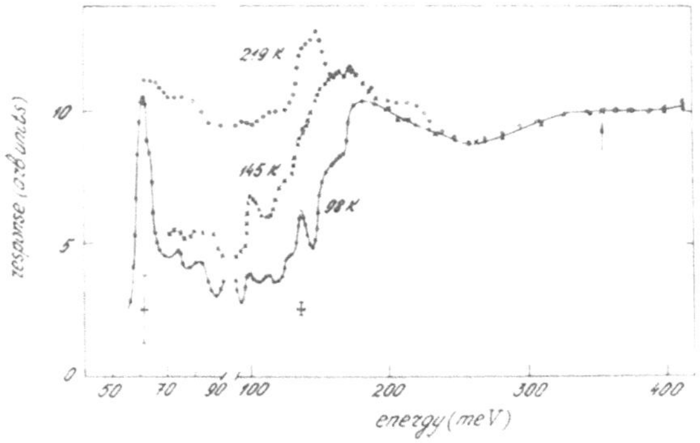

Fig.2. Spectral responce for sample A at temperatures indicated on the curves. The curves were scaled to coincide at the point indicated by arrow.

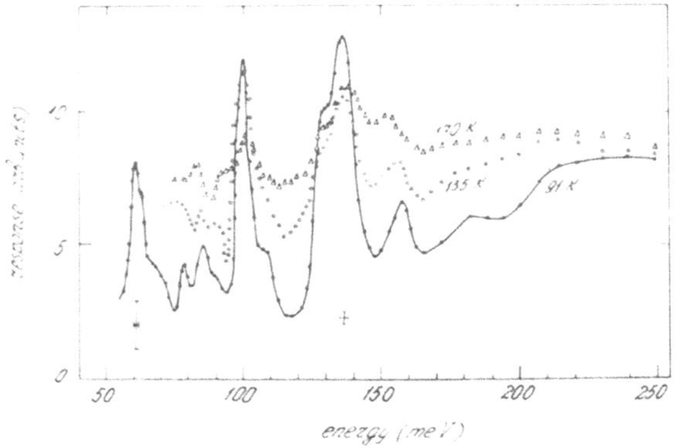

Fig.3. Spectral responce for sample B at temperatures indicated on the curves.

the range 200-100 K was the same as for $\sigma_{\parallel}$. The inset in fig.1 shows the temperature dependence (300-4.2 K) of the anisotropy $A=\sigma_{\parallel}/\sigma_{\perp}$ normalised to its room temperature value A_o. As temperature tends to T_p, anisotropy slightly increases /4/. Below T_p anisotropy drops below that in the metallic state (at 300 K). The variation of A between 200 and 90 K is quite small. At still lower temperatures anisotropy rises sharply, with knees at 90 and at 30 K. In high-quality samples conductivity anisotropy at 5 K exceeded 10^8. The I-V curves of the samples were recorded in the range 300-4.2 K. They were linear at $T<T_p$ up to a certain threshold electric field E_T, but strongly nonlinear at $E>E_T$ /3,4,7/. As a rule, E_T varied only moderately in the region 200-100 K. The minimum $E_T \simeq 0.4$ V/cm was reached in this sample at T=100 K. As temperature decreased further, E_T began to grow sharply /4,8/.

A study of spectral behavior of optical properties can yield important information on the energy band structure and on the types of excitations in a Peierls dielectric. We have analysed the response of orthorhombic TaS_3 in the range 3-22 μm at temperatures from 90 to 320 K /6/. The response, namely, the relative change in resistance $\Delta R/R$ due to irradiation, was measured in electric fields below the threshold field. The response in the whole investigated temperature and spectral range, including the microwave (9.5 GHz), was, first, proportional to d log R/dT and, second, was phase-shifted by $\approx 90^o$ relative to the incident radiation modulated at a low frequency (40-200 Hz). The response decreased in inverse proportion to modulation frequency. However, we failed to separate the response component in phase with the modulated radiation, that could correspond to such fast processes as e.g. photoconductivity. The time constant of the microwave response was 0.1 s. All these results show, together with the data of /9/, that the response of TaS_3 crystals to both IR and microwave radiation is mostly bolometric.

Fig.2 plots the spectral response per unit incident power for a sample A with E_T =0.8 V/cm at T=100 K. It was shown in /6/ that the response in our thin (≈ 3 μm) samples was determined by absorption in optical transitions. Fig.1 shows that the response in the range 400-250 meV and at all temperatures depends only slightly on the energy of radiation quanta. Two basic features are found on the spectral response curves at low temperatures (curve at 98 K). First, the maximum at $h\nu_1$ =185 meV is followed by a considerable drop in absorption, with the edge at $h\nu_2$ =125 meV (recently a similar result was independently obtained in /10/). Second, a narrow absorption peak is observed at $h\nu_3$ =62 meV, its fine structure revealed at higher spectral resolution. Note that $\nu_2 \approx 2\nu_3$. As temperature is increased (fig.2), the maximum at ν_1 shifts towards lower energies and survives up to the highest temperatures reached in our experiment (320 K). This shift is accompanied by a steady rise of absorption in the range from ν_2 to ν_3.

Fig.3 is the spectral response curve of sample B with a lower E_T =0.3 V/cm, at three temperatures $T < T_p$. As in sample A, we find a drop in absorption, with the edge at 120 meV (the structure being negltcted). A clearly defined structure is observed on this drop: sharp peaks at 157,136, and 100 meV, and weaker peaks in the range 70-90 meV. A similar structure was observed in all investigated samples, including A (fig.2), although in a less pronounced form. At higher temperatures peaks remain at almost unchanged positions up to 170 K, but become lower; the structure gets blurred. The observed structure is not a result of diffraction or interference because the position of peaks is the same in samples with substantially different transverse dimensions.

The explanation of the obtained results will be given in terms of different types of charge carriers existing in quasi-one-dimensional Peierls conductors. Thermal activation keeps a certain fraction of electrons free in the range of $T < T_p$ after the Peierls gap has formed. But the main fraction condenses into CDWs. Furthermore, nonlinear excitations - amplitude and phase solitons - can exist in a CDW. In some cases experimental data in the range 200-100 K can be explained by taking into account only free electrons and uniform CDW. The temperature dependence of both longitudinal and transverse conductivities in a weak electric field ($E < E_T$) is caused by thermal activation of electrons across the Peierls gap, with CDW pinned. The nonlinear conduction at $E > E_T$ is usually attributed to CDW motion. However, the characteristics of this motion and of the processes involving CDW remain largerly unclear. Thus, complete understanding of CDW formation and generation effects has not been reached yet. A number of conclusive arguments in favor of a local (at a contact) generation mechanism, by formation of phase slip centers in CDW /11-13/ or of a vortex chain /14/ were recently suggested. It is also likely that solitons may be generated at $E > E_T$ close to a contact (or to another strong nonuniformity).

The following interpretation of the obtained spectral response can be suggested. The rise in optical absorption (fig.2) at $h\nu_2$=125 meV (see also /10/) is caused, as we think, by the optical excitation of electron-hole pairs form the ground state of CDW across the Peierls gap, and corresponds to the absorption edge of orthorhombic TaS_3. The excitation energy that we have measured is in agreement with the thermal activation energy $\Delta \approx 60-70$ meV (700-800 K). The growth of absorption in the forbidden energy range at higher temperatures is indicative of a continuous transition from a gap to a pseudogap. The maximum on the spectral curve at the absorption edge survives up to 320 K which is much higher than the temperature of three-dimensional ordering T_p. This seems to indicate that the specific feature of state density, corresponding to the pseudogap, still exists at $T > T_p$. This result agrees with the observation of X-ray diffraction maxima corresponding to a superlattice, at $T > T_p$ /15/. The fact that at $T < T_p$ we observe a decrease in anisotropy (fig.1) supports T_p as the temperature of three-dimensional ordering. The absorption edge is far from abrupt, which is typical for quasi-one-dimensional Peierls dielectrics /16/. The interaction that is specifically strong in these materials is that between excitations and phonons corresponding to the Peierls deformation of the lattice. Estimates based on the data of /17/ give for the blurring of the absorption edge a value that within an order of magnitude is close to the experimental result.

A narrow peak appearing at 62 meV (fig.2), i.e., at energy Δ , is caused in all likelihood by optically excited bound electronic states in the middle of the Peierls gap. It can be suggested that this bound state corresponds to an amplitude soliton with zero charge (the electron charge being completely screened) and spin-1/2 /17/. Such soliton states can be formed as a result of self-localization of free electrons and holes or can be localized at crystal defects. The shape of the spectral line at $h\nu = \Delta$ (fig.2) is in qualitative agreement with the theory of absorption due to solitons /19/. This interpretation is also supported by the fact that we could not separate appreciable photoconductivity from the bolometric signal. This points to a very short lifetime of optically excited electrons and holes, due to their fast relaxation (selflocalisation) into soliton states over a time 10^{-13} s /16/.

The peak 61 meV is found in sample B with lower threshold field (E_T =0.3 V/cm, fig.3) but its height is smaller than in sample A with larger E_T . In contrast, it reveals a much better defined structure of the absorption edge. One of the possible explanations of this structure is the effect of discommensurations on absorption edge /20, 6/. These discommensurations are formed when the CDW is nearly commensurate.

The presence of amplitude solitons does not affect appreciably the electric conductivity of the samples, but must be evidenced by their magnetic properties. The activation temperature dependence of the spin magnetic susceptibility, with activation energy ≈ 600 K, was observed in orthorhombic TaS_3 in /18/. This activation energy is considerably lower than that of conduction due to free electrons (≈ 800 K). Moreover, the effective mass of the spin carrier is found to be 5 times that of a free electron or hole. Consequently, it would be rather difficult to ascribe the spin susceptibility to free electrons or holes; the data are in much better agreement with the theory describing solitons in such systems /17/: the soliton activation energy is less than Δ, and must be slightly higher than $2\Delta/\pi$ (in the present case $2\Delta/\pi = 510$ K), and the effective mass calculated in /17/ coincides with the experimental value.

Temperature range 100-4 K

Many properties of orhtorhombic TaS_3 change drastically in the vicinity of 100 K. A fairly sharp bend is observed on $\sigma(T)$ curves in a narrow temperature range, after which the curves become less steep (fig.1). This temperature range can be put in correspondence with activation energy that in different samples varies from 150 to 230 K, thus being less than above 100 K by a factor of 4-5. The transverse conductivity $\sigma_\perp$ continues to decrease at $T < 90$ K, with activation energy close to the value in the range 200-100 K where it is identical for $\sigma_\parallel$ and $\sigma_\perp$ /4,21/. The difference in activation energies of $\sigma_\parallel$ and $\sigma_\perp$ appearing at $T < 100$ enhances the crystal anisotropy (fig.1). Threshold field also starts to rise sharply at about 100 K /4,8/. In pure samples E_T increases by a factor of 10-15 at 50 K. In the same temperature range the nonlinear part of conductivity $\sigma_n \exp(-E/E_o)$ also changes drastically ($E \gg E_T$). E_o is a characteristic field /7/. In a strong field $E \gg E_o$ (usually $E_o \lesssim 5E_T$) the nonlinear part of conductivity is simply σ_n and must be independent of both temperature and field. Fig.4 plots σ_n as a function of 1/T for different field strengths $E \gg E_T$ ($E_T = 0.4$ V/cm at T=100). The curves show that $\sigma_n \approx \sigma_o$ in the region from 200 to ≈ 120 K and indeed σ_n is practically independent of temperature and field, in accordance with the results reported in /7/. However, at still lower temperatures σ_n is found to reduce sharply, in activation manner, with activation energy depending on electric field.

The slope of the $\sigma(T)$ curve smoothly changes at the beginning of the next temperature region 30-4 K, the curve becoming again less steep (fig.1). At the same time, the slope of the temperature dependence of anisotropy increases still further (fig.1). The $\sigma(T)$ curve can be linearized in the range $20 > T > 4$ K by a fit $\log[\sigma(T)/\sigma_o]$ vs $[-(T_o/T)^{1/2}]$ /4/. The exponents 1 and 1/4 in this function result

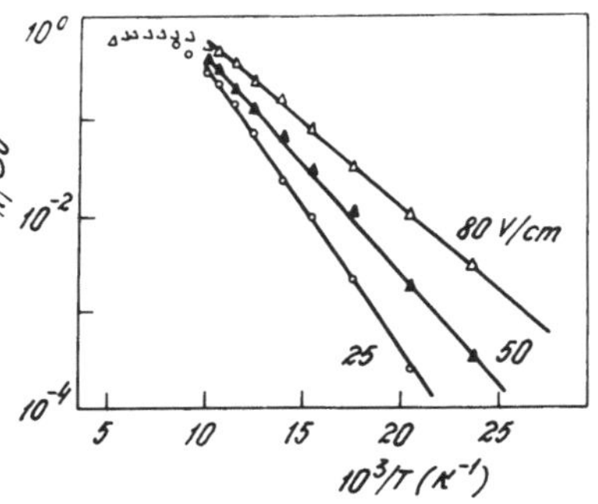

Fig.4. Temperature dependence of nonlinear part of conductivity at electric fields indicated on the curves.

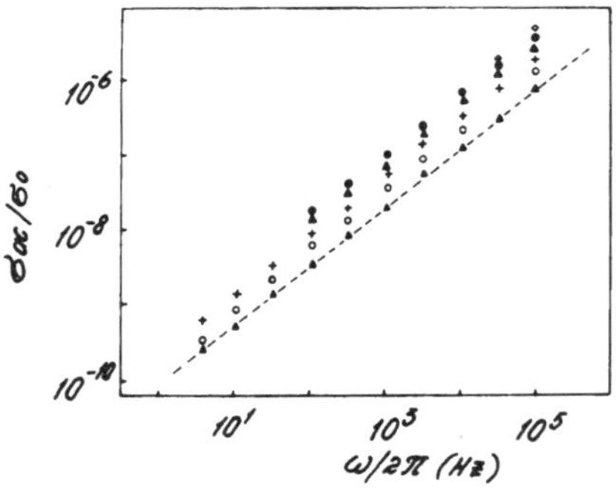

Fig.5. Frequence dependence of σ_{ac}/σ_0 at temperatures: T= ▲ 4.22, ○ 6.15, + 9.95, △ 14.0, ● 18.3, ◇ 35.4 K. Dashed line represent the low $\sigma_{ac} \sim \omega^s$, s=0.77.

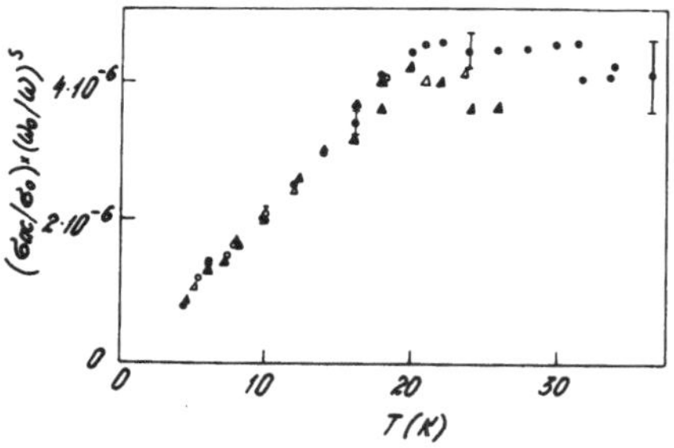

Fig.6. Temperature dependence of $(\sigma_{ac}/\sigma_0)(\omega_0/\omega)^s$ at frequencies: ● $\omega/2\pi$ = 99.1, ▲ 11.1, △ 1.11, ○ 0.111 kHz, $\omega_0/2\pi$=99.1 kHz, s=0.77.

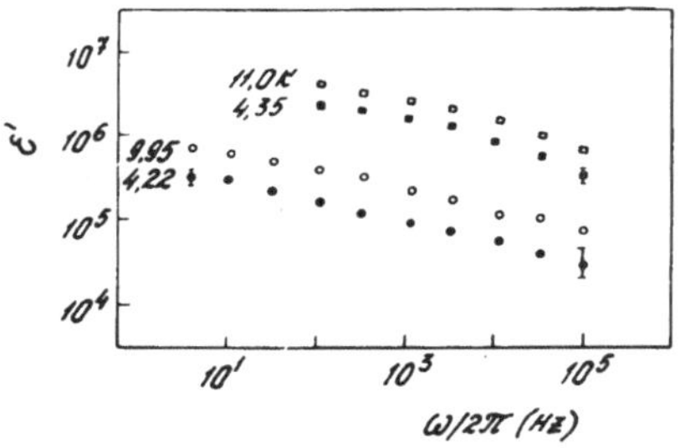

Fig.7. Frequency dependence of dielectric constant ε' for two samples at several temperatures.

in greater deviations from linearity. This means that electric conduction obeys a Mott-type law for variable range hopping conduction in a quasi-one-dimensional conductor /22/. This law holds in the range 20-4 K for all investigated samples.

We have conducted measurements of frequency dependence of the real and imaginary parts of conductivity in five samples from different batches, in the frequency range 3-10^5 Hz and in the temperature range 4.2-40 K. The detailes of the measurement technique can be found in /5/. The dependence of the real part on frequency is best pronounced at low temperatures becomes weaker at higher temperatures, and vanishes at $T \lesssim 40$ K. Fig.5 plots the frequency-dependent component of conductivity, σ_{ac} /5/. It shows that log (σ_{ac}/σ_o) is a linear function of log $(\omega/2\pi)$, giving a law $\sigma_{ac} \sim \omega^s$ typical for a wide class of disordered compounds /23/. In the sample shown in fig.5, s=0.77, and in other samples s varies from 0.77 to 0.85, being temperature-independent within experimental error. Fig.6 plots conductivity σ_{ac} as a function of temperature for fixed frequencies. In order to emphasize the temperature dependence of σ_{ac} the curves representing different frequencies ω were normalized to $(\omega/\omega_o)^s$. It was found that σ_{ac} was a linear function of T in the range 4-20 K for all frequencies, and tended to saturation at T 20 K.

The reported above experimental dependences of the real part of σ_{ac} on temperature and frequency show a behavior typical for the mechanism of hopping conduction in disordered systems /23/. Furthermore, a comparison with the appropriate theory for quasi-one-dimensional systems /24/ indicates that hopping proceeds through states in a restricted energy range ΔE about the Fermi level /5/. If $kT < \Delta E$, σ_{ac} increases linearily with temperature, and if $kT \gtrsim \Delta E$, σ_{ac} tends to saturation (fig.6).

The obtained imaginary part of conductivity yielded the value of the real part of complex dielectric constant ε'. Fig.7. shows ε' as a function of frequency in two samples. At helium temperatures ε' is sensitive to the degree of perfection of the samples, being higher in higher-quality samples. In all samples ε' was increasing almost linearly with temperature in the range 4-20 K /5/. Note that an appreciable frequency dispersion of ε' is observed even at the lowest frequencies. We also find that at $T < 20$ K the value of ε' was not altered when a dc bias was applied to a sample, while the real part of conductivity increased severalfold. An analysis of the frequency dependence of transverse conductivity has demonstrated that ε' was essentially anisotropic and that $\varepsilon'_\perp$ did not exceed 10^2, which is the minimal detectable quantity determined by the absolute error in sample capacitance. These data can be explained on the basis of the assumption that the main contribution to the imaginary part is made by CDW oscillations accompanied by substantial polarization effects. Low values of $\varepsilon'_{\perp}$ result from low transverse polarizability of CDW.

As follows from the above data, a large number of properties of orthohombic TaS_3 undergo changes at $T \lesssim 100$ K: activation energy of longitudinal conductivity changes, conductivity anisotropy sharply increases, threshold field increases, and the dependence of conductivity on electric field is modified. Electron diffraction data show that the incommensurate-commensurate transition is also completed in this temperature range, so that CDW is now totally commensurate with the pristine lattice /8/. The hysteresis loop on the conductivity-temperature curve diminishes below 100 K and completely disappears at about 60 K /25/. The temperature boundaries of all these effects presumably depend on the degree of perfection of samples, and their positions have to be elaborated. Nevertheless, a certain relation between these phenomena can already be suggested. In all likelihood most of the specific features observed in the temperature and field dependences of conductivity around 100 K are caused by CDW pinning on commensurations ($E < E_T$) and by the disappearance of discommensurations. It seems that this change increases the total pinning force which sharply enhances E_T. CDW as a whole becomes less mobile, and the nonlinear CDW excitations (amplitude and phase solitons) begin to contribute heavily to conduction /2,4,16,17,26,27/.

It seems very likely that phase solitons, and most of all the topological solitons with charge 2e corresponding to a local change in CDW phase by 2π /27/, play a predominant role in the conduction of TaS_3 at $T < 100$ K. This cannot appreciably increase the energy of excitations, despite the interchain coupling. These solitons have lower activation energy than amplitude solitons; within an order of magnitude it equals T_p, provided $T_p < \Delta$. Solitons cannot transfer from one chain to another, which explains the differences in activation energy and the enhanced anisotropy appearing at 100 K. As at high temperatures, here again transverse conduction goes through free electrons. The number of solitons decreases with decreasing temperature, and the degree of their localization increases, considerably reducing conductivity due to activation. The variable range hopping conduction becomes the predominant mechanism. As was shown in /23/, the behavior of temperature and frequency dependences of conduction is not affected, at least qualitatively, on whether the charge carriers are electrons or, say, polarons. It can be presumed that the ideology developed in the work on hopping conduction is applicable to solitons /28/. Quite likely, a finite number of localized solitons exist in our case at low temperatures, because of the presence of defects and impurities that fix CDW phase. The states that correspond to these solitons lie close to the chemical potential level, within an interval about 20 K wide. It is along these states that phase solitons move jumpwise from one pinning site to another, with CDW at rest. No final conclusions can be drawn so far from the described results on the mechanisms of electric conduction of the quasi-one-dimensional conductor TaS_3 in the whole range of 350-4 K. Nevertheless, the reported data point to an important role of soliton excitations in determining the properties of TaS_3.

I am grateful to V.B.Preobrazhenskii and Ya.S.Savitskaya for providing the samples, and to S.A.Brazovskii, N.N.Kirova, and M.E.Itkis for useful discussions.

1.Physics in One-Dimension, Springer Verlag, N.Y.1981
2.Proc. of the International Conference of the Physics and Chemistry of Polymer Conductors, J.Physique 44, Colloque C3 (1983)
3.S.K.Zhilinskii, M.E.Itkis, F.Ya.Nad`, I.Ya.Kal`nova and V.B.Preobrazhenskii, Inst. Rad. Electron. Preprint No 21 (1981)
4.S.K.Zhilinskii, M.E.Itkis, F.Ya.Nad`, I.Ya.Kal`nova and V.B.Preobrazhenskii, Sov. Phys. JETP 58, 211 (1983)
5.S.K.Zhilinskii, M.E.Itkis, F.Ya.Nad`, Phys.Sat.Sol.(a)81,367 (1984)
6.M.E.Itkis, F.Ya.Nad`, Pis`ma Zh. Eksp. Teor. Fiz. 39, 373 (1984)
7.A.H.Thompson, A.Zettl, G.Gruner, Phys. Rev. Lett. 47, 64 (1981)
8.Z.Z.Wang, H.Salva, P.Salva, P.Monseau, M.Rehard, C.Roucau, R.Ayroles, F.Levi, L.Guemas and A.Meershaut, J.Physique 44, L311 (1983)
9.J.W.Brill, S.L.Herr, Phys. Rev. B27, 3916 (1983)
10.J.W.Brill, S.L.Herr, Solid State Commun. 49, 265 (1984)
11.N.P.Ong, G.Verma, Phys. Rev. B27, 4495 (1983)
12.L.P.Gor`kov, JETP Lett. 38, 87 (1983)
13.L.P.Gor`kov, Zh. Eksp. Teor. Fiz. 86, 1818 (1984)
14.N.P.Ong, G.Verma, K.Maki, Phys. Rev. Lett. 52, 663 (1984)
15.C.Roucau, R.Ayroles, P.Monseau et al.Phys.Stat.Sol.(a)62,483 (1980)
16.S.A.Brazovskii JETP Lett. 28, 606 (1978)
17.S.A.Brazovskii Sov. Phys. JETP 51, 342 (1980)
18.D.C.Johnston, J.P.Stokes, Pei-Ling Hsieh, G.Grimes, J.Physique 44, Colloque C3, 1749 (1983)
19.A.H.Heeger, A.G.Mac Diarmid, Physics in One-Dimension, Springer Verlag, N.Y. 1981, p.179
20.S.A.Brazovskii, S.I.Matveenko, Sov. Phys. JETP 54, 818 (1981)
21.T.Takoshima, M.Ido, K.Tsutsumi, T.Sambongi, S.Honma, K.Yamaya, Y.Abe, Solid State Commun. 35, 911 (1980)
22.V.K.Shante, S.M.Varma, and A.N.Bloch, Phys. Rev. B8, 4885 (1973)
23.N.F.Mott and E.A.Davis, Electron Processes in Non-Crystalline materials, Clarendon Press, Oxford (1979)
24.V.V.Bryskin, Fiz. Tverd. Tela 22, 2048 (1980)
25.A.W.Higgs, J.C.Gill, Solid State Commun. 47, 737 (1983)
26.S.N.Artemenko, A.F.Volkov, Sov. Phys. JETP 54, 992 (1981)
27.B.Horovitz, J.A.Krumhanal, Solid State Commun. 26, 81 (1978)
28.A.I.Larkin, P.A.Lee, Phys. Rev. B17, 1596 (1978).

THERMAL GRADIENT EXPERIMENTS ON THE CHARGE-DENSITY-WAVE CONDUCTION NOISE SPECTRUM†.

N.P. Ong, G. Verma, and X.J. Zhang*, Department of Physics, University of Southern California, Los Angeles, Ca 90089-0484.
†Supported by U.S. National Science Foundation (DMR 81-09971).
*Permanent address: Dept. of Physics, Zhejiang University, China.

The firmest evidence supporting theories[1,2] which ascribe the voltage oscillations (narrow-band noise) to phase-slippage at the ends of the sample come from experiments performed with the sample in a thermal gradient[3,4]. In this short note we will review three classes of experiments which clarify different aspects of the problem. All experiments are done on high purity (RRR > 100) $NbSe_3$ in two-probe configuration, except otherwise indicated. Where appropriate we will stress experimental details and difficulties. See K. Maki, these proceedings, for theoretical details.

1.) MONOTONIC GRADIENT ON LONG SAMPLES. In these studies[3] one of the sample ends A is held at a fixed temperature T_A while the other B is incrementally scanned from T_A to a higher value, usually exceeding the transition temperature $T_c \sim 59$ K. The sample lengths exceed 2 mm. When $\Delta T = T_B - T_A$ increases from zero the single fundamental frequency splits into two frequencies f_1 and f_2. If T_A is well regulated then f_1 remains stationary while f_2 moves as T_B is raised (Fig.1.) Eventually as T_B crosses T_c f_2 decreases rapidly to zero, leaving f_1 as the sole frequency (Fig. 2.) In our studies a run was abandoned if the sample displays more than one fundamental frequency at zero ΔT. In samples with poor contacts (or ones that have undergone numerous thermal cycling) several fundamentals are often seen even in nominally zero ΔT. These are likely due to weakly-connected current paths near the contacts. (Recall that the sample frays easily at the cut ends and that sightly different electric fields E can induce different frequencies in independent fibers[5].) The different fundamentals clearly have nothing to do with behavior in a gradient.

Very often the spectrum can be cleaned up by applying fresh paint and cooling down again. Some clean samples will show a single fundamental at low current I which then splits into two closely spaced frequencies at higher I (in zero gradient.) These, again, are unrelated to the gradient results. The important point is that in a gradient <u>one set of frequencies is static while a second set moves according to the hot end T.</u> In particular, the second set decreases to zero when T_B exceeds T_c. No frequencies with behavior different from these two sets are observed.

2) MONOTONIC GRADIENT ON SHORT SAMPLES. In these studies the samples (between 0.8 and 0.3 mm in length) were attached to copper wires which were anchored to sapphire substrates. Separate diode sensors and heaters on the two substrates enabled T_A and T_B to be independetly regulated so that the sign of ΔT as well as its magnitude could be changed. (This was desirable because of the observation[4] of <u>frequency-locking</u> in short samples.) Furthermore, because of the surprisingly large thermal conductance of short samples we had to attach secondary sensors made of $NbSe_3$ samples to the copper support wires. We found that a large fraction of the imposed ΔT occurs along the support wires so that ΔT across the sample itself is greatly overestimated without using the secondary sensors. Figure 3 shows an example of frequency-locking in a 0.6 mm sample. With end A clamped at 40 K the two frequencies f_1 and f_2 merge continuously as T_B warms towards T_A. They stay locked until ΔT exceeds 2 K. Unlocking proceeds by a first-order jump. Such abrupt jumps are rare. We observed them in 2 out of 10 short samples examined. The linewidth narrows noticeably during the locking interval ΔT_ℓ (Fig.4.) A sample displaying locking over a 10 K range is shown in Fig. 5. (Note that in this run in which secondary sensors were not utilised the hot end T is badly overestimated.) In all 10 samples unlocking invariably occurs when the hot end exceeds 50 K, because the

order parameter at the hot end begins to diminish rapidly with increasing temperature. Unpublished reports that short samples were observed to oscillate at the average T when one end is high above T_c while the other is below T_c were not confirmed in our experiments.

3) RE-ENTRANT GRADIENT ON LONG SAMPLES. An interesting conclusion from the data in Fig. 2 is that if both ends of the sample are kept above T_c while the middle is cooled one can induce CDW conduction without observing narrow-band noise. To carry out this experiment we[4] anchored the ends of a 3 mm sample to a sapphire substrate with indium. The middle of the sample was kept in contact with a thick insulated copper wire by tension. (Varnish was also used in some samples.) Keeping the middle temperature T_M at 50K we warmed up the ends to above T_c in successive steps. The results are shown in Fig. 6, which directly verifies that CDW conduction will occur without noise if the condensate is kept away from the ends so that phase slippage is absent. If the thermal anchor in the middle of the sample is a source of phase-slippage (caused for e.g. by using a bare copper wire) then a stationary frequency may appear and persist even when the ends are heated above T_c. This can be readily verified to come from the middle contact by changing T_M while holding the end T fixed.

DISCUSSION. The results here show quite clearly that voltage oscillations arise at the ends of the sample rather than from an ac current in the bulk. Refs. 3 and 4 discuss in some detail the incompatibility of the bulk-origin theories. (In particular phase-slip models predict that frequency coherence in a gradient - i.e. locking - is enhanced when the sample ends are close together whereas in bulk theories the key parameter is the magnitude of the thermal gradient. The observation of locking in short samples, but not in long ones, clearly favors phase-slip models.)

† † †

References
1. N.P. Ong, G. Verma, and K. Maki, Phys. Rev. Lett. 52, 663 (1984); Kazumi Maki and N.P. Ong, to be published.
2. L.P. Gor'kov, Pis'ms Zh. Eksp. Teor. Fiz. 38, 76 (1983) [JETP Lett. 38, 87 (1983)].
3. G. Verma and N.P. Ong, Phys. Rev. B 30 (1984), in print.
4. X.J. Zhang and N.P. Ong, Phys. Rev., submitted; and unpublished.
5. N.P. Ong and G. Verma, Phys. Rev. B 27, 4495 (1983).

Figure Captions

Figure 1. Observed frequencies vs. temperature in $NbSe_3$ in a thermal gradient (open) and in zero gradient (solid symbols.) In left panel open triangles and circles represent two frequencies when the hot end is heated from 45 K to 52 K while the cold end is held at 45 K. In the right panel the cold end is held at 49 K while the hot end is heated to 55 K. Note the parallel trajectories when sample is uniformly heated (solid symbols, both panels.)

Figure 2. Frequency spectra of $NbSe_3$ when the hot end is heated from 52 K to 59 K (ascending order.) The cold end is held at 52 K. Note that one frequency vanishes while the other remains undiminished and unshifted.

Figure 3. Frequency locking in a 0.6 mm sample in a thermal gradient. As T_B is warmed towards T_A the two frequencies merge and stay locked for an interval of 4 K. When T_B-T_A exceeds 2K unlocking proceeds by a first order jump. The inset shows the sample mounting and the sites of secondary thermal sensors.

Figure 4. The full spectra of the data shown in Fig. 3. The linewidths narrow distincly during locking. No other fundamental frequencies are observed.

Figure 5. Frequency over a 10 K gradient range in a 0.6 mm sample. When T_B exceeds 45 K the two frequencies separate and move in accordance with the local T of each end. Because of the large gradient in the copper wires T_B is overestimated by 8 K. (No secondary sensors were used in this run.) The broken line is the single frequency in zero gradient.

Figure 6. Voltage noise spectra when the middle of the sample is kept cold at 50K while the ends are heated from 44 to 60 K. All fundamental frequencies vanish when the ends exceed 60 K. (Inset) Sample mounting showing cooling by thermal contact with insulated copper wire.

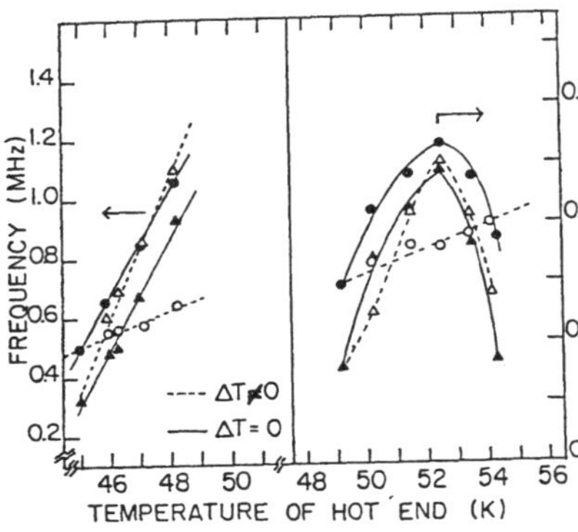

Figure 1

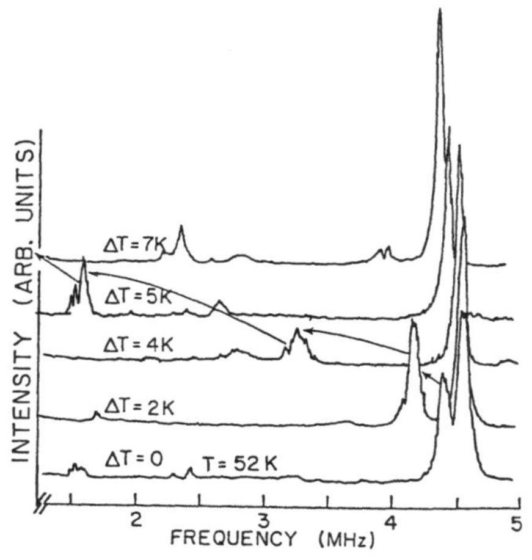

Figure 2

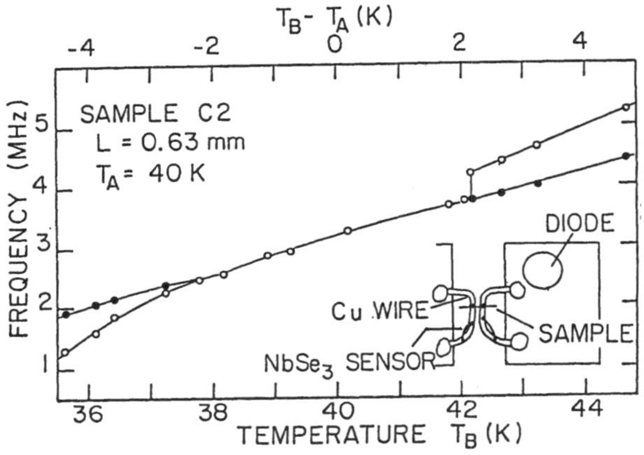

Figure 3

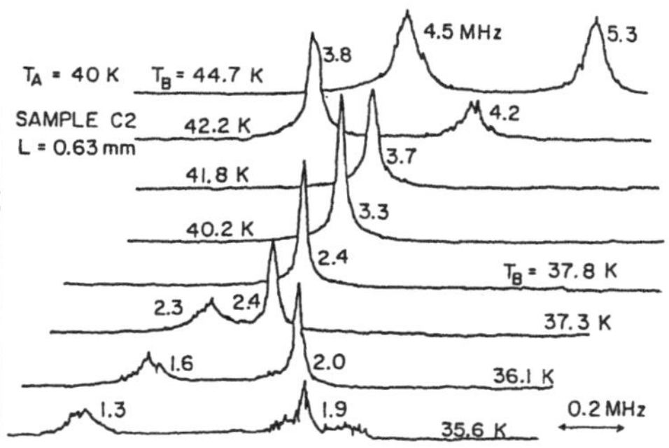

Figure 4

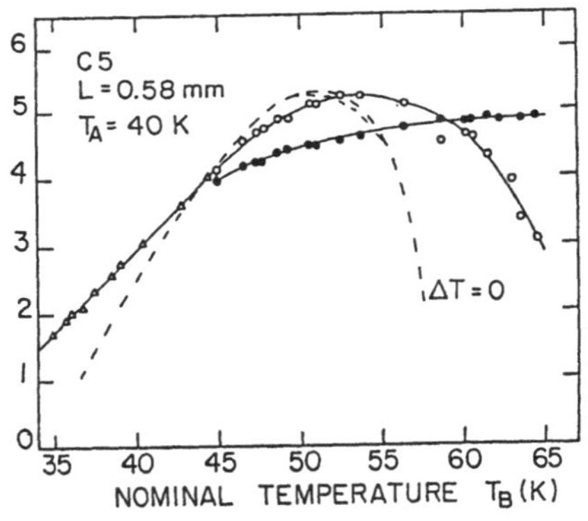

Figure 5

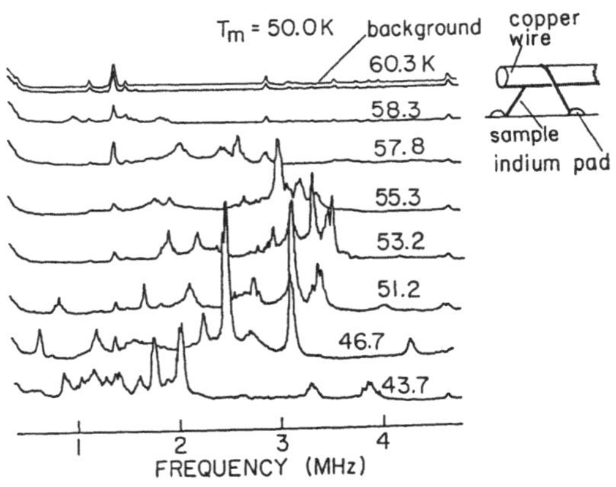

Figure 6

BROADBAND NOISE IN ORTHORHOMBIC TaS_3

J. P. STOKES, MARK O. ROBBINS, S. BHATTACHARYA and R. A. KLEMM
Corporate Research Science Laboratories, Exxon Research and Engineering Company, Annandale, New Jersey 08801

Abstract

We report experimental results on the broadband noise in sliding charge density wave conductor orthorhombic TaS_3. We propose that the noise has its origin in the interaction of a deformable condensate with random impurities. The experimental results are in excellent agreement with a phenomenological model based on fluctuations in threshold field due to deformations of the sliding condensate. The amplitude of the noise is directly related to the dynamic coherence volume of these fuctuations.

The onset of nonlinear electrical conduction beyond a threshold electric field E_T in charge density wave (CDW) conductors, such as $NbSe_3$ and TaS_3, is known to be caused by the sliding of the CDW which is pinned below E_T by the impurities. The appearance of noise, both narrow-band and broadband, in the nonlinear conduction regime has been studied extensively in recent years.[1,2,3] In this paper we report measurements of the broadband noise in orthorhombic TaS_3.

The measured broadband noise has the following characteristics. (1) Field dependence - The onset of noise is sharp and coincident with the onset of nonlinear conduction as evidenced by a comparison with the differential resistance measurement. (2) Frequency dependence - The noise power has a $f^{-\alpha}$ spectrum with $\alpha = 0.95 \pm .05$ (for $10 \text{ Hz} \leq f \leq 10^5 \text{ Hz}$) at 160 K and the spectrum is field independent except very close to the threshold voltage V_T.[4] (3) Sample size dependence - The r.m.s. noise voltage δV scales as $[\ell/A]^{1/2}$ where ℓ is the length and A is the cross-sectional area or, equivalently, $\delta V^2/V^2$ scales as the inverse volume. These results establish that the noise is a bulk (finite size) phenomenon and not associated with contacts.

In order to quantitatively study the behavior we propose the following model. At constant total current, fluctuations in the effective pinning force or V_T, cause fluctuations in the chordal resistance $R(= V/I)$ which is explicitly threshold voltage dependent. Within this model, therefore, the mean squared noise voltage is given by

$$\langle \delta V^2 \rangle = I^2 \langle \delta R^2 \rangle = I^2 \left(\frac{\partial R}{\partial V_T}\right)^2 \langle \delta V_T^2 \rangle \qquad (1)$$

Direct measurement of $\partial R/\partial V_T$ is not possible; so we assume that R is a function of $(V-V_T)$ only, i.e., $\partial R/\partial V_T = -\partial R/\partial V$. The latter is evaluated numerically from the I-V characteristics. Since $\partial R/\partial V$ is only weakly frequency dependent below 100 kHz, the frequency dependence is entirely contained in $\langle \delta V_T{}^2 \rangle$, i.e., $\delta V^2(\omega) = I^2(\partial R/\partial V_T)^2 \delta V_T{}^2(\omega)$.

Figures 1(a) and 1(b) show plots of the field dependence at two temperatures of the noise voltage measured at one frequency (ω = 300Hz, Q = 10) and of the numerically evaluated value of $I(\partial R/\partial V_T)$. Clearly, except very near V_T, they track each other accurately.[4]

It is now known from various experiments[4] that metastable states exist in CDW systems corresponding to long wavelength deformations of the phase of the CDW condensate. Such deformations alter the distribution of phases at impurity sites and therefore the pinning force exerted by the impurities on the sliding condensate. We suggest that this is the source of the threshold field fluctuations. If the transition between metastable states is thermally activated, then a distribution of barrier heights leads to a distribution of relaxation times. Indeed, such a distribution of barrier heights and a thermally activated behavior have been inferred for the CDW conductor $K_{.3}MoO_3$.[5] Even if the barrier height distribution is sharply peaked at an energy $E_p >> k_BT$, one still obtains a $f^{-\alpha}$ power spectrum for the noise, so long as the distribution function is slowly varying within k_BT. This is the analog of the Dutta-Horn model for $f^{-\alpha}$ noise in metals,[6] and explains the observed frequency spectrum.

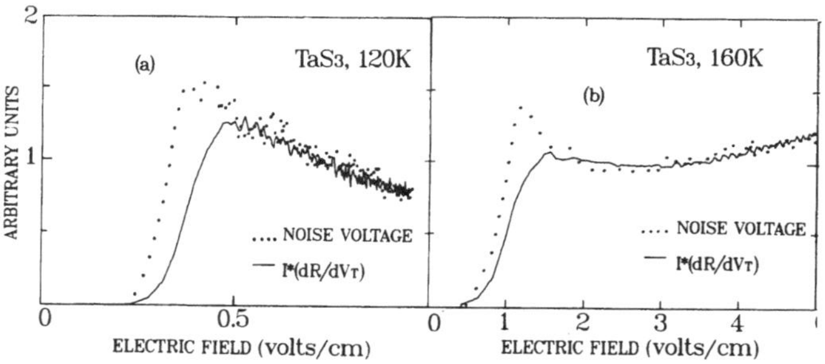

FIGURE 1 Field dependence of the broadband noise $<\delta V^2>$ measured at 300 Hz and $I^2(\partial R/\partial V_T)^2$.

We propose that the fluctuation in $E_T = V_T/\ell$ is coherent within a "coherent volume," λ^3. The net threshold voltage fluctuation across the entire sample of length ℓ and cross-sectional area A is the incoherent addition of these fluctuations. Assuming that the fluctuation in the pinning field, $<\delta E_T{}^2>$, is proportional to $E_T{}^2$, we obtain

$$<\delta V^2(\omega)> = I^2\left(\frac{\partial R}{\partial V_T}\right)^2 \cdot E_T{}^2 \cdot \lambda^3 \cdot \frac{\ell}{A}\, S(\omega,T) \qquad (2)$$

where $S(\omega,T)$ is the spectral weight function. Equation (2) produces the experimentally observed sample dimension dependence. This is analogous to the 1/N dependence of the noise power $[\delta V^2/V^2]$ in normal metals where N is the number of electrons.[6] In CDW systems, $N = \ell A/\lambda^3$ is the number of independent entities generating the noise. This number is small compared to the number of electrons in a metal; this factor, in in part, responsible for the large magnitude of the noise. Equation 3 allows information about the usually inaccessible quantity λ^3 to be obtained from measurements of the broad band noise amplitude.

In Fig. 2(a) we plot the temperature dependence, for $V = 2V_T$, of $\delta V^2(\omega)$ measured at 300 Hz. It grows rapidly below T_c and has a pronounced peak near 150K where an incommensurate-commensurate transition is thought to occur.[7] Figure 2(b) shows the temperature dependence of $I^2(\partial R/\partial V_T)^2 V_T{}^2$ measured directly. This quantity also grows rapidly below T_c and shows a pronounced peak at 150 K.

In Fig. 2(c) we plot the temperature dependence of the ratio of these two quantities i.e., $\delta V^2(\omega)/[I^2(\partial R/\partial V_T)^2 V_T^2]$. This, according to equation (2), reflects the temperature dependence of $\lambda^3 S(\omega,T)$. This grows below T_c and saturates gradually. The peak disappears. Several issues remain unresolved. First, we do not know what relation λ bears to the Lee-Rice length, nor to the dynamic coherence length ξ in ref. 8. A more microscopic theory is desirable. Second, a model for the temperature dependence of $S(\omega,T)$ is needed.

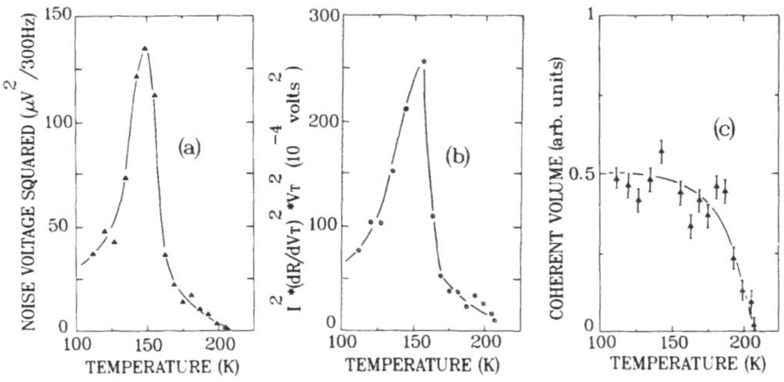

FIGURE 2 Temperature dependence of (a) $<\delta V^2>$ (300 Hz) (b) $I^2(\partial R/\partial V_T)^2 V_T{}^2$ and (c) $\lambda^3 S(\omega,T)$.

To conclude, we have demonstrated that a phenomenological model of threshold field fluctuations can accurately describe the broadband noise in sliding CDW conductors. It will be interesting to see if analogous models can be constructed for noise generation in other systems, such as charge transfer salts.

We acknowledge helpful discussions with A. N. Bloch, P. M. Chaikin, P. Dutta, D. C. Johnston, R. Koch and J. R. Schrieffer.

REFERENCES

1. See G. Grüner, Comments in Sol. State Phys. 10, 183 (1983).
2. A. Zettl and G. Gruner, Solid State Commun. 46, 29 (1983).
3. A. Maeda, M. Naito and S. Tanaka, Solid State Commun. 47, 1001 (1983).
4. G. Mihaly, GY. Hutiray and L. Mihaly, Solid State Commun. 48, 203 (1983). Also see J. C. Gill, ibid. 39, 1203 (1981); R. M. Fleming, ibid. 43, 167 (1982).
5. R. J. Cava, R. M. Fleming, P. Littlewood, E. A. Rietman, L. F. Schneemeyer and R. G. Dunn, Phys. Rev. B (to be published).
6. See P. Dutta and P. M. Horn, Rev. Mod. Phys. 53, 497 (1981).
7. P. Monceau, H. Salva and Z. Z.Wang, J. Phys. (Paris) 44, 1639 (1983).
8. D. Fisher, Phys. Rev. Lett. 50, 1486 (1983).

HIGH FIELD I-V CHARACTERISTICS OF ORTHORHOMBIC TaS_3

Zhang Dian-lin, Duan Hong-min, Lin Shu-yuan and Wu Pei-jun
Institute of Physics, Chinese Academy of Sciences, Beijing, China

The dc I-V characteristics of orthorhombic TaS_3 have been measured up to the field as high as 7000 V/cm in the temperature range of 77 K - 300 K. Some new features have been revealed. The results support the single-particle excitation picture of transport in quasi-1D conductors.

Some years ago there was an argument on whether a collective mode or the single-particle excitations are responsible for the properties characteristic of some organic conductors undergoing Peierls phase transition[1]. Things have been changed since a great deal of various experimental results on some transition metal trichalcogenides seems undoubtful to support the collective CDW motion in understanding of the dynamical properties of these compounds[2,3]. More and more people now believe the existence of such a collective CDW excitation. however, there is something in the model intrinsically not self-consistent in spite of its great success[4].

In a previous work[5] one of the authors proposed a transverse tunneling model to account for the preliminary results in high dc electric field for TaS_3. Now we have been able to extend the field to as high as 7000 V/cm, using the same experimental technique as was used in the previous work except for that in the present measurements we kept the electrical field constant and followed the variation of resistance with temperature. Several samples were measured but not all of these measurements were successful. Some samples were broken in the midway of measurements. The best curves were obtained for sample 7# as shown in Fig. 1.

Several new features have been revealed in our measurements:
1. There is a well-defined activation energy even in the highest field we reached, that is, for this compound we have the typical single-particle thermally activated conductivity of semiconductors;
2. At higher field a distinct change in activation energy around 130 K can be seen, which is fairly consistent with the incommensurate-commensurate transition found by Roucau recently[6]. The activation energies for sample 7# are 755 K and 585 K below and above the transition at 130 K, respectively. For another sample, 5#, the

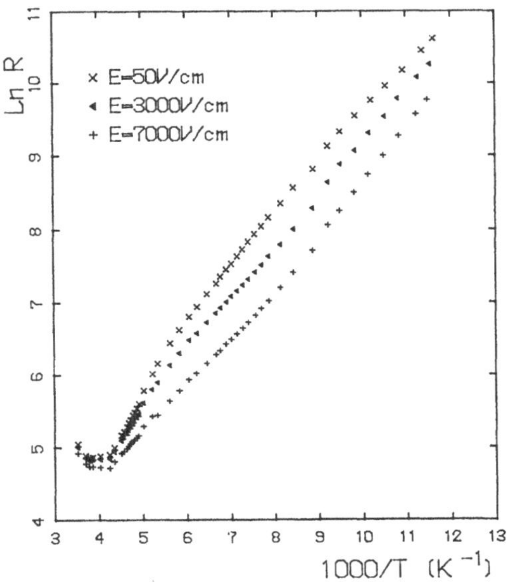

Fig.1 Relations of ln R versus 1/T for orthorhombic TaS_3 in different electrical fields.

corresponding values are 534 K and 420 K, somewhat lower than those of sample 7#. But the ratio of the activation energy below the transition to that above is almost the same for these two samples, being equal to 1.29;

3. In weak field one always finds some tendency to saturation in the ln R versus 1/T curve when temperature goes down and often attributes it to impurity conductivity. From Fig.1 we see that the saturation disappears in high fields. This means that the electrical field affects only the prefactor of the exponential expression for the conductivity of semiconductors and this prefactor causes the tendency to saturation;

4. In weak field there is a sharp resistance increase with decreasing temperature just below the CDW phase transition which is usually accounted for by the development of Peierls energy gap. But this sharp resistance increase vanishes in high field, which seems to indicate that the energy gap has been already formed above 215 K. This is not consistent with the mean field theory or with the temperature dependence of the energy gap determined by structure analysis in weak field;

5. From Fig.1 we see that there is a weak nonlinear field effect even above the Peierls transition temperature. In the previous work[5] we attributed this nonlinearity to heating effects. It seems that this is not so as seen from Fig.1 which shows the resistance decreases with increasing field even at room temperature, well above the resistance minimum, where heating effects should lead to resistance increase;

6. The room temperature resistances of the samples are around $10^2\ \Omega$. Using the resistivity data in reference 2, we get a much smaller cross section for the sample, about $0.06\ \mu^2$, than the actual geometrical cross section. This means that the anisotropy plays quite remarkable role in the measurements, which supports the model[5] of a nonuniform current distribution in the samples.

To our knowledge, there have been a lot of experiments examining the nonlinear field effect of quasi-1D conductors (in much lower field than our measurements and using pulse signal), keeping temperature constant. But none of them tried to trace the temperature dependence of resistance, keeping the field constant. The present work is the first of such measurements and the results are quite unexpected for the models which suppose that in the nonlinear region of field the conductance is the combined contribution of two quite different parts: the normal single-particle transport and a CDW collective motion. We should have expected quite different temperature dependence for these two kinds of current carriers. For example, the Bardeen expression gives $\sigma = \sigma_a + \sigma_b(1-E_T/E)\exp(-kE_T/E)$. There are no data available about how σ_b and k vary with temperature, but with the available E_T data it is unimaginable that this expression could give a temperature dependence similar to that of single-particle thermally activated conductivity. In other words, the temperature behaviour of CDW transport can not be the same as that of normal electrons.

If we admit that the single-particle excitations are responsible for the conductivity in both weak and strong fields and allow the prefactor of the exponential expression to vary with field, then our present results could be easily understood. But how could the prefactor depend on electrical field? There are two ways which may affect this prefactor: the field dependence of mobility and the variation of the "effective" carrier concentration with field. It is possible that both of them play roles in the field dependence of conductance. We notice that in weak field the ln R versus 1/T curve is not a straight line in the Peierls semiconductor state but is straightened in high field, which may be explained by that in weak field the mobility is temperature dependent and becomes independent of temperature as the field is increased. This is the case if we suppose that mobility is determined in weak field and higher temperature by phonon scattering and in strong field by the scattering of nonionic impurities. The carrier concentration for a semiconductor with one type of carriers is equal to $N_0\exp(-E_g/kT)$, where E_g is activation energy. To understand why the carrier concentration should vary with field, or why should $N_0 = N_0(E)$, we must take into account of the limited dimension of the CDW domains and that even in their metallic state the electrons are strongly localized in the perpendicular to chain directions for TaS_3. These two factors affect the transport properties by adding some kind of tunneling process in series with the normal thermally activated conductivity. Every tunneling will change N_o by a factor $A\exp(-t\phi^{\frac{1}{2}})$, where t and ϕ are the width and height of the potential barrier being tunneled, respectively. This factor, $A\exp(-t\phi^{\frac{1}{2}})$, is obviously field-dependent and thus gives an "effective" carrier concentration increasing with the field. For a given electrical field we have a well-defined carrier concentration and consequently, a typical thermally activated conductivity as shown in our measurements. This model is further supported by the facts that the apparent room temperature resistivity for our very short samples is higher than that for long samples and that nonlinear effects exist above Peierls temperature.

Besides nonlinear I-V relation, there are many dynamical properties including narrow-band noise and various metastable phenomena which seem to support the collective CDW motion. However, as shown above, the transport process which involves some kind of single-particle tunneling is a nonlinear process, whereas any nonlinear process will give rise to a kind of noise spectrum. The metastable phenomena are typical for materials having domain structures.

In conclusion, our high-field resistance measurements support the model of thermally activated single-particle transport and do not confirm the collective CDW motion.

Acknowledgement: the authors want to show their thanks to Prof. Yu Lu an C.W.Chu for their help.

References

1. J. Bardeen, Highly Conducting One-dimensional Solids, edited by J.T.Devreese, R.Evrard and V.E.van Doren (London: Plenum Press, 1980) p.374.
2. H.Salva, Z.Z.Wang, P.Monceau and M.Renard, Phil. Mag.,B49, 385 (1984) and references therein.
3. P.Monceau, J.Richard and M.Renard, Phys. Rev., B25, 931 (1982) and J.Richard, P.Monceau and M.Renard, Phys. Rev., B25, 948 (1982) and references therein.
4. Duan Hong-min and et al, to be published.
5. Zhang Dian-lin, Solid State Commun. 48, 369 (1983); Zhang Dian-lin, Acta Physica Sinica, 33, 779 (1984).
6. C.Roucau, J. Phys., Paris, 44, C3-1725 (1984).

INERTIAL DYNAMICS OF CDW TRANSPORT IN $NbSe_3$

D. Reagor, S. Sridhar and G. Grüner
Department of Physics, University of California
Los Angeles, California 90024, U.S.A.

We report the observation of effects due to the inertial mass of the CDW condensate in the high frequency conductivity of $NbSe_3$. Our measurements yield experimental values for the damping and effective mass and are in agreement with classical models of CDW transport.

The response of CDW materials to time-varying electromagnetic fields reveals unique information regarding the dynamics of the CDW. We have measured the frequency-dependent complex conductivity of $NbSe_3$ at very high frequencies (9 GHz to 95 GHz) in the microwave and millimeter wave range. The experiments reveal that the real part reaches a maximum at 9 GHz and then decreases with frequency, and that the dielectric constant is negative. We associate these features with the finite inertial mass of the CDW condensate. The measurements yield direct experimental values for the damping and the effective mass. Similar results have also been obtained for TaS_3.

Two types of experimental techniques have been employed. At microwave frequencies, the well-known cavity perturbation method was used with cavities constructed at 9 and 35 GHz.[1] At frequencies greater than 26 GHz, we have developed a new bridge method of measuring complex conductivity. At the overlap frequency of 35 GHz, the bridge and cavity measurements on the CDW materials discussed below are in excellent agreement.

The bridge method measures the complex impedance of a sample holder which is a shorted waveguide section in which the sample is placed parallel to the electric field. Treating the sample as a lossy dielectric with complex dielectric coefficient $\varepsilon' - j\varepsilon''$ $\sigma = \omega\varepsilon_0\varepsilon''$ we extract σ and ε' from the measured complex impedance using well-known expressions.[2] In order to satisfy the above assumption we use very thin and long ($\sim 10^{-8} cm^2 \times \sim 2$ mm) samples. The metallic region above the CDW transition (Fig. 1) serves to calibrate the technique.

We have carefully verified that the measurements of σ and ε' reported here are independent of power and that spurious effects such as heating are minimal. The results are reproducible for samples of the same preparation batch and also for a different batch.

Figure1-

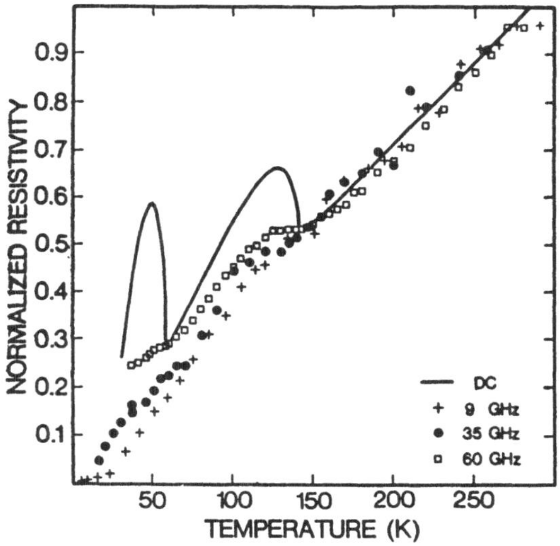

Figure 1 represents the temperature dependence of the normalized resistivity $\rho(\omega,T)/\rho_{RT}$ of normally pure $NbSe_3$ (RRR ~ 150, E_T ~ 50 mV/cm at 45 K) measured at frequencies 0 (dc), 9, 35 and 60 GHz. The absolute room temperature DC resistivity was 250 μΩ-cm. Above the upper phase transition, ρ is independent of frequency as expected for a metal. Below ~ 150 K, it will be observed that ρ increases (and hence σ decreases) with frequency for $\omega/2\pi > 9$GHz. The full frequency dependence of the CDW contribution $\sigma_{CDW} = \sigma(\omega) - \sigma_{DC}$, including data between dc and 500 MHz, is illustrated in Fig. 2(a) for a representative temperature T = 45 K. As Fig. 2(a) reveals, $\sigma_{CDW}(\omega)$ reaches a maximum at ~ 9 GHz. We call this maximum value, σ(9 GHz), the "high frequency limit" of the CDW conductivity - it represents the full unpinned dynamical response of the CDW.

Figure 2a- 2b) 2c)

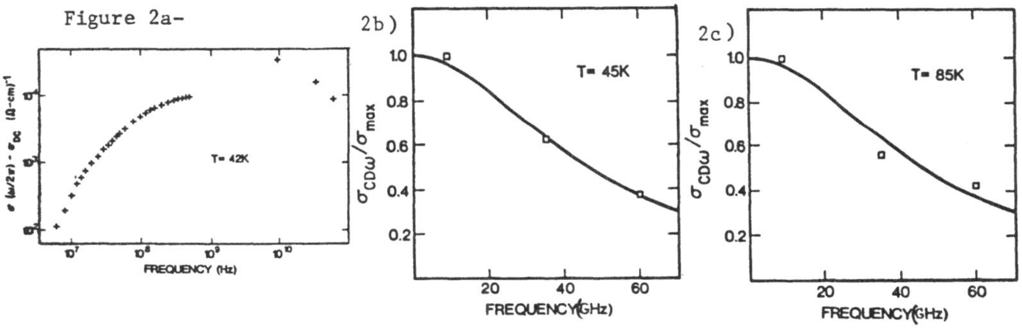

To quantitatively analyze our results we use the conductivity predicted by the classical model[4]

$$\sigma_{CDW}(\omega) = \sigma_{max} \frac{\omega^2/\tau^2}{(\omega^2 - \omega_o^2)2 + \omega^2/\tau^2}$$

where ω_o is the pinning frequency $1/2\pi\tau$ is the damping frequency and σ_{max} is the conductivity of the freely moving CDW. Confirming ourselves to the high frequency regime ($\omega^2 > \omega_o^2$) of the classical model, we have $\sigma_{CDW}(\omega) = \sigma_{max}/(1 + \omega^2\tau^2)$. Figures 2(b) and 2(c) demonstrate that the expression describes the data quite well. From a best fit we extract the damping frequency $1/2\pi\tau$ (see table).

TABLE. Experimental values for $1/2\pi\tau$ (GHz) and M_F/m_e for $NbSe_3$.

T	$1/2\pi\tau$ (GHz)	M_F/m_e
40K-45K	46	70
85K-110K	45	440

Using these experimentally determined values of $1/2\pi\tau$ and (9 GHz) = σ_{max} = $ne^2\tau/M_F$ are able to determine the effective or Frohlich mass, M_F, of the CDW condensate. We obtain M_F/m_e = 70 at 40K - 45K and 440 at 85K - 110K.

In calculating the effective mass in the lower phase we have assumed that the condensate from the upper phase transition contributes negigably to $\sigma - \sigma_{DC}$ at the lower temperatures. The total carrier density can be estimated from band filling and is approximately 5×10^{21} electrons/cc.[5] Since each phase transition involves only one of the three distinct chains we take the CDW carrier density, n, to be $1/3(5 \times 10^{21})$, for both phases. The absolute values agree with estimates of 10^2 and $6 = 10^2$ based on the LRA expression:[6]

$$\frac{M_F}{m_e} = 1 + \frac{4\Delta^2(0)}{\lambda(\hbar\,\omega_{2K_F})^2}$$

with the Peierl's gap, $\Delta(0)$, equal to 1.74 kTp, the electron-phonon coupling constant, λ, equal to 0.3 and the $2K_F$ phonon energy $\hbar\,\omega_{2K_F}$, equal to 40K. The relative values of the effective masses in the upper and lower phases are in remarkable agreement, experimentally M_u/M_ℓ= 6.3 and theoretically $M_u/M_\ell \simeq (T_p)^2_{upper}/(T_P)^2_{lower}$= 6.2. This indicates that λ and ω_{2K_F} are the same in the two phases.

We have also carried out measurements on orthorhombic TaS_3. For 130K < T < 210K we again find σ decreases with frequency for $\omega/2\pi >$ 9 GHz and the dielectric constant is negative. Again the classical model describes the data quite well. Using a carrier density of 5×10^{21}, we get $1/2\pi\tau \sim 92 - 130$ GHz and $M_F \sim 900$. At other temperatures, σ has a more complicated behavior but is in agreement with the known behaviour of TaS_3.

In conclusion, our experiments reveal the presence of inertial mass and damping effects in CDW dynamics. The results are in agreement with classical models of CDW transport.

We thank L. Mihaly and Wei-Yu Wu for useful discussions and for providing the low frequency data. This work was supported by NSF (DMR 84-06896), UCLA and equipment contributions from Hughes Aircraft Company.

References

1. L. I. Buravov and I. F. Schegolev, Prob. Tekh. Eksp. Instrum. Exp. Tech. (USSR) 14(2), 171 (1971).
2. N. Marcuvitz, "Waveguide Handbook," MIT Radiation Lab Series, McGraw Hill (1951).
3. W. Wu, L. Mihaly, and G. Grüner, submitted for publication.
4. G. Grüner, A. Zawadowski, and P. M. Chaikin, Phys. Rev. Lett. 46, 511 (1981).
5. G. Grüner and Z. Zettl, to be published.
6. P. A. Lee, T. M. Rice, and P. W. Anderson, Solid State Comm. 14, 703 (1974).

FREQUENCY DEPENDENT CONDUCTIVITY OF CDW COMPOUNDS

Wei-Yu Wu, L. Mihaly, G. Mozurkewich and G. Grüner
Physics Dept., University of California, Los Angeles, CA 90024, USA

Complex conductivity measurements on TaS_3, $NbSe_3$ and $(TaSe_4)_2I$ are reported in the frequency range of 100 Hz - 500 MHz. The low frequency response of the CDW system can be described by a power law frequency dependence with an exponent less than one. At higher frequencies a crossover to overdamped harmonic oscillators is observed and the data is fitted with a distribution of crossover frequencies.

The investigation of the frequency dependent conductivity $\sigma_{CDW}=\sigma'(\omega) + i\sigma''(\omega)$ of the CDW compounds is a powerful tool in exploring the CDW dynamics. The pinned electron condensate responds to the external field with a large polarization P. At high frequencies the time variation of the polariztion, i.e. the current associated with the CDWs, $\dot{P}=j_{CDW}$, is comparable to or greater than the current carried by the normal electrons. As the CDWs are pinned and the normal current is independent of the frequency ω (at least in the frequency range investigated in our studies) we may decompose the total conductivity in the form of $\sigma_{tot}(\omega) = \sigma_{CDW}(\omega) + \sigma_{dc}$, where σ_{dc} is the conductivity measured at the zero frequency limit.

In this study we investigated the charge density wave conductivity of TaS_3, $(TaSe_4)_2I$ and $NbSe_3$ in the frequency range of 100 Hz-500 MHz. At low frequenices (100 Hz $<$ f $<$ 100 kHz) we applied a bridge circuit connected to a lock-in amplifier in order to measure the real and imaginary part of $\sigma_{CDW} = \sigma_{tot} - \sigma_{dc}$. At high frequency (3 MHz - 500 MHz) we used an HP network analyzer.

Fig. 1 shows low frequency measurements on different CDW compounds.[1] We observed a universal power law frequency dependence $\sigma_{CDW} = \sigma_o(i\omega/\omega_o)^\alpha$ with $\alpha < 1$. The experimental values are $\alpha=0.96$ for $NbSe_3$ at 46.5 K, and 0.95-0.93 for $(TaSe_4)_2I$ at temperatures between 180 to 215 K. For TaS_3, α is almost independent of temperature ($\alpha=0.90 \pm 0.03$) between 70 K and 150 K while at temperatures closer to the transition temperature (T_P = 226 K) the exponent is increasing $\alpha=0.96$ at 193 K.

Power law frequency dependence of the conductivity was also observed earlier in impure and amorphous semiconductors and it is considered a characteristic property

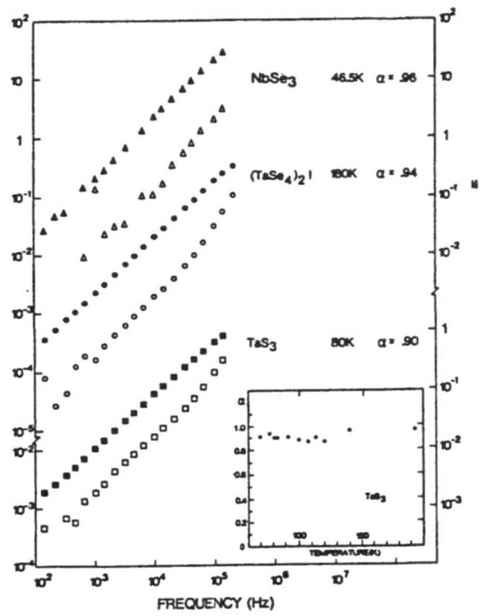

Fig. 1. Real (empty dots) and imaginary (full dots) part of the low frequency conductivity of different CDW compounds. The exponent obtained from the slope of the curves and from the ratio between σ' and σ'' using $t_g\,(\alpha\pi/2)=\sigma''/\sigma'$ is within the experimental error supporting $\sigma=\sigma_o(i\omega/\omega_o)^{\alpha}$. The insert shows the temperature dependence of α for TaS_3.

of glassy materials.[2] We want to emphasize that for CDW materials the magnitude of the effect (σ_o) is about 10 orders of magnitude higher than that observed in glassy semiconductors. None of the existing CDW theories predict this frequency dependence.[1]

Fig. 2 shows the real and imaginary part of the CDW conductivity for TaS_3 and $NbSe_3$ at higher frequencies, where deviations from the power law behavior begins to develop. The real part of the conductivity exhibits a sharp increase at higher frequencies while the imaginary part has a maximum. This overall behavior is expected for an overdamped harmonic oscillator.[3] However, a single harmonic oscillator with a response of

$$\sigma = \sigma_\infty \left(\frac{i\omega/\omega_{c.o.}}{1 + i\omega/\omega_{c.o.}} \right)$$

does not describe the details of the frequency dependence.[1] For TaS_3 we performed a detailed calculation and found that the three parameter fit

$$\sigma = \sigma_\infty \int \frac{i\omega/\omega_{c.o.}}{1 + i\omega/\omega_{c.o.}} \cdot \frac{1}{1 + (\omega_{c.o.}/b)^a}\, d\omega_{c.o.}$$

gives excellent results for temperatures above 100 K (Fig. 2). Using this formula we implied a distribution $P(\omega_{c.o.})$ of the crossover frequency $\omega_{c.o.}$; $P(\omega_{c.o.})$ is steplike at temperatures close to the phase transition and broadens at lower T. In a similar study a distribution of relaxation times was used to describe the frequency dependent response of $K_{0.3}MoO_3$.[4]

In conclusion, we demonstrated that the low frequency response of the CDW system follows a power law with an exponent less than one. We believe that this response is closely related to the metastable states observed in other experiments.[5] On the other hand the high frequency conductivity shows a relatively simple overdamped behavior. Although the relationship between the deformable CDW Hamiltonian and the phenomenological oscillators applied in this study has not been cleared yet, it is likely that at high frequencies the response is determined by small amplitude, local deformations at the pinning centers, while the low frequency response reflects collective effects involving many pinning centers.

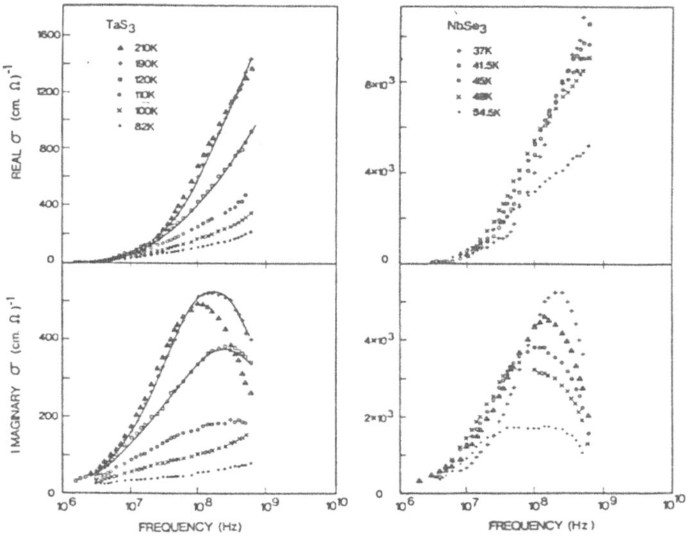

Fig. 2. Real and imaginary part of the CDW conductivity at higher frequencies. The solid lines fit the TaS_3 data at 120K and 190 K by distribution of crossover frequencies, discussed in the text.

This work was supported by NSF grant DMR 84-06896.

References

1. For the earlier measurements on TaS_3 see Wei-Yu Wu, L. Mihaly, G. Mozurkewich and G. Gruner, Phys. Rev. Lett. Vol. 52, p. 2382 (1984).
2. A.K. Jonscher, J. Material Science 16, 2037 (1981).
3. G. Gruner, A. Zawadowski and P.M. Chaikin, Phys. Rev. Lett. 46, 511 (1981).
4. R.J. Caver, R.M. Fleming, P. Littlewood, E.A. Rietwam, L.F. Schneemeger and R.G. Dunn, Phys. Rev. B, (to be published).
5. G. Mihaly and L. Mihaly, Phys. Rev. Lett. 52, 149 (1984).

AC CONDUCTIVITY OF THE BLUE BRONZE $K_{0.3}MoO_3$

R. P. Hall, M. Sherwin and A. Zettl
Department of Physics, University of California
Berkeley, California 94720 U.S.A.

Abstract: We have measured the low field ac conductivity of the blue bronze $K_{0.3}MoO_3$ in the charge density wave (CDW) state. For temperatures above 70K and over an extended frequency range, our results are consistent with the model proposed by Cava et. al. Below 70K, however, there appears at high frequencies an additional relaxation mechanism. Associated with this new mode are unusual hysteresis effects and an ac conductivity similar to that found in TaS_3 .

The low field ac conductivity of $K_{0.3}MoO_3$ is currently the subject of much interest. We have performed careful low field ac conductivity measurements in the frequency range 10 Hz to 2.3 GHz. Figure 1 shows the ac conductivity as function of frequency for several samples of $K_{0.3}MoO_3$ at two temperatures. At the higher temperature, 77K, the conductivity is well described by $\sigma(\omega) = i\omega\varepsilon(\omega)$, where[1]

(1) $$\varepsilon(\omega) = \varepsilon_\infty + (\varepsilon_0 - \varepsilon_\infty)[1 + (i\omega\tau_0)^{1-\alpha}]^{-\beta}$$

In figure 1a, a fit to this expression is indicated by the solid and dashed lines for Re $\sigma(\omega)$ and Im $\sigma(\omega)$ respectively. Note that the characteristic relaxation time, τ_0 , is unusually high. This value of τ_0 corresponds to a characteristic pinning frequency $\omega_1 = 1/\tau_0$, several orders of magnitude lower than characteristic frequencies seen in $NbSe_3$[2] and TaS_3[3].

As the temperature of the material is lowered, ω_1 moves to still lower frequencies. Figure 1b shows the ac conductivity at 42K. As indicated by the low frequency crossing of Re σ and Im σ, the characteristic frequency is now below 1kHz. Because of the

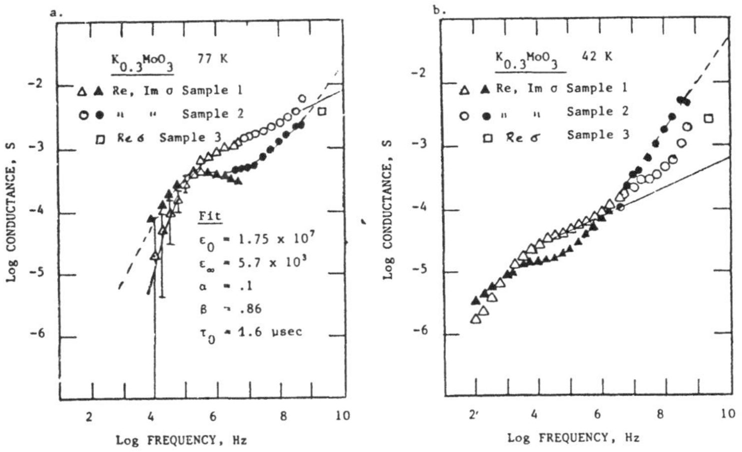

Figure 1: The impedance of $K_{0.3}MoO_3$ at 77K and 42K.

uncertainty in the data below 1kHz, only the high frequency limit of equation 1 is fitted. For $\omega\tau_0 \gg 1$, the real and imaginary components of $\sigma(\omega)$ obey simple power laws:

$$\text{(2)} \quad \text{Re}\,\sigma(\omega) \sim \omega^{1-\beta(1-\alpha)}$$

$$\text{Im}\,\sigma(\omega) \sim \omega$$

There are clear departures from equation 2 in the 42K data. The real component of $\sigma(\omega)$ goes as $\omega^{.26}$ between 10kHz and 1MHz, but then curves away from this behavior at higher frequencies. Im $\sigma(\omega)$ does not go as $\omega^{1.0}$, but rather as $\omega^{0.9}$ at high frequency. To explain these departures from equation 2, we look at $\sigma_{HF}(\omega) = \sigma_{CDW}(\omega) - \sigma_{LF}(\omega)$, where $\sigma_{LF}(\omega)$ is given by equation 1. When this is done, we find that Re $\sigma_{HF}(\omega)$ and Im$\sigma_{HF}(\omega)$ both obey the same rough power law, $\sigma(\omega) \sim \omega^{0.9}$. This behavior is identical to the low frequency behavior recently observed in the ac conductivity of TaS_3[4].

Preliminary work by Gruner[5] at UCLA suggests that both Reσ and Im σ continue to climb in the low microwave frequency range. Eventually Imσ must turn over at some crossover frequency ω_c. This second characteristic frequency would correspond to the more familiar

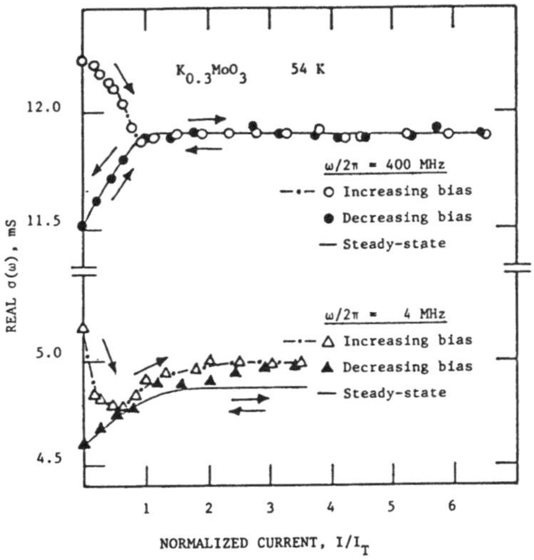

Figure 2: The effect of a dc bias on the high frequency ac conductivity of $K_{0.3}MoO_3$.

pinning frequency encountered in $NbSe_3$ and TaS_3. It is likely that the high frequency response is present at all temperatures, not just below 70K. At high temperatures (70K - 100K) the effect of the high frequency resonance is to a large extent masked by the low frequency relaxation process, which displays an Arrhenius behavior in both its characteristic frequency and its magnitude.[1]

Figure 2 shows the effect of an applied dc bias on the ac conductivity. The measurements were made according to the following procedure. First the sample was warmed above 54K and then cooled back down. Next the bias was swept out so that $I > I_T$, and then reduced back to $I = 0$. Subsequent sweeps of the bias were made until the curve of $\sigma(\omega)$ versus I became repeatable. As shown in the top of figure 2, a "steady-state" was reached as soon as I became greater than I_T when $\omega/2\pi$

was 400 MHz. At $\omega/2\pi$ = 4MHz, two or three sweeps of the bias were required to achieve a repeatable curve.

There are several unusual features to the data presented in figure 2. First is the existence of a metastable, enhanced ac conductivity state which can only be reached by temperature cycling. A related temperature hysteresis effect has been observed in the dc conductivity by Tsutsumi et. al.[6] This ac effect, however, can not be explained solely in terms of a changing dc conductivity offset; the magnitude and bias dependence of the ac effect is enhanced at 400 MHz compared to 4 MHz. Second, at 400 MHz the conductivity is independent of bias past $I = I_T$. Increasing the bias past E_T does not increase $\sigma(\omega)$ even though $\sigma(\omega)$ has not saturated at some high-frequency, high-field limit. The scaling relation between $\sigma(\omega)$ and $\sigma(E)$ observed in $NbSe_3$ and TaS_3 can apply to $K_{0.3}MoO_3$ only under limited conditions.

We thank R. M. Fleming and G. Gruner for useful discussions. This research was supported in part by NSF grant DMR-8400041.

References:

1. R. J. Cava, R. M. Fleming, P. Littlewood, E. A. Rietman, L. F. Schneemeyer and R. G. Dunn, to be published.
2. G. Gruner, L. C. Tippie, J. Sanny, W. G. Clark, and N. P. Ong, Phys. Rev. Lett. 45, 935 (1980).
3. A. Zettl and G. Gruner, Phys. Rev. B 25, Rap. Comm., 2081 (1982).
4. Wei-yu Wu, L. Mihaly, George Mozurkewich, and G. Gruner, Phys. Rev. Lett. 52, 2382 (1984).
5. G. Gruner, private communication.
6. K. Tsutsumi, T. Tamegai, S. Kagoshima, H. Tomozawa, and M. Sato, to be published.

SUBHARMONIC SHAPIRO STEPS, DEVIL'S STAIRCASE, AND SYNCHRONIZATION IN RF-DRIVEN CDW CONDUCTORS

Stuart E. Brown, George Mozurkewich,* and George Grüner
Physics Dept., Univ. of Calif., Los Angeles, CA 90024, USA

Explanation of recent Shapiro steps studies in CDW conductors requires consideration of internal degrees of freedom and of associated finite velocity correlation lengths. The synchronization of different regions of the specimens with increasing rf is demonstrated through coalescence of noise peaks, reduction of fluctuations in noise peaks, and narrowing of steps.

The ac-dc interference phenomenon known in Josephson junctions as Shapiro steps was discovered in CDW conductors by Monceau, Richard, and Renard,[1] who used it to monitor narrow band noise in $NbSe_3$. Later work by Zettl and Grüner[2] stretched the Josephson junction analogy further, and the authors' most recent studies of steps shed new light on the microscopics of CDW motion.[3] Here we outline: (1) the restrictions which can be placed on the CDW equation of motion; and (2) the extent to which CDW response can be synchronized throughout the specimen.

I. Equation of Motion

Perhaps the most intriguing aspect of CDW steps studies so far is the presence of subharmonic steps. "Principal" steps occur whenever the internal (narrow band "noise") frequency f_{int} is near an integer multiple p of the applied rf driving frequency f_{ext}. Subharmonic steps are not restricted by this rule; they occur whenever $pf_{ext} = qf_{int}$ for q an integer not equal to 1. Figure 1 shows Shapiro steps in TaS_3, represented as peaks in the differential resistance dV/dI. This representation emphasizes small, sharp details which would be missed in direct I-V curves. When huge rf is applied, the usual linear region vanishes and steps appear. In addition to the principal steps (p/q = 1/1, 2/1), one can see three small subharmonics (1/2, 1/3, 2/3). These have been found in several TaS_3 specimens, while a few showed fewer subharmonics and others showed more. The differential

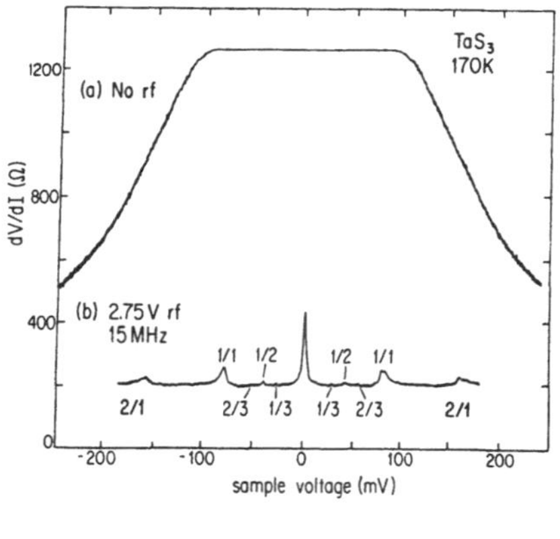

Fig. 1. dV/dI in TaS_3. Peaks in the lower curve correspond to principal and subharmonic Shapiro steps.

resistance at the peaks is much smaller than the linear value, R_o = 1270 Ω. Figure 2 shows part of a spectacular array of steps in $NbSe_3$. All subharmonic steps for $q \leq 14$ have been observed, plus some for larger q, for a total of some 80 steps. Not only are there more steps in $NbSe_3$, but dV/dI comes closer to the linear value R_o = 28 Ω.

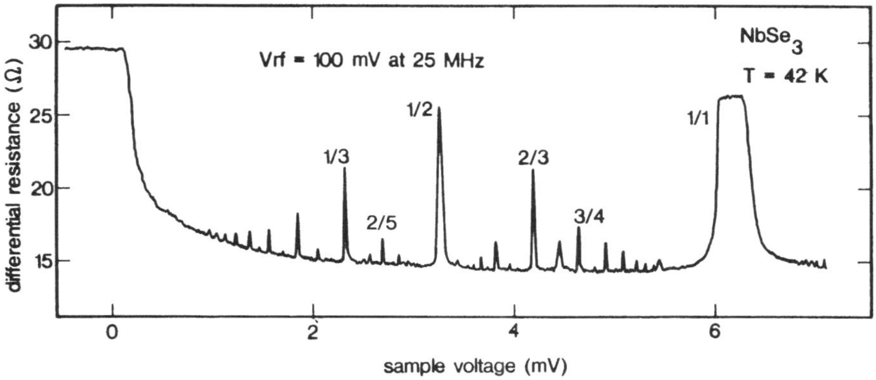

Fig. 2. Segment of a Shapiro step trace in $NbSe_3$.

The steps in $NbSe_3$ exhibit strong similarities to the devil's staircase behavior predicted theoretically[4,5] for the RSJ model of a Josephson junction. The steps generally are shorter and narrower for larger q, and there are so many steps that one begins to suspect they would fill the entire horizontal axis, if only one had adequate experimental resolution. The similarity is, in fact, more than superficial. By taking an expanded version of data like that which appears in figure 2, it is possible to extract a fractal dimension D = 0.91 ± 0.03 which agrees very closely with numerical prediction (D = 0.87)[5] and analogue simulation (D = 0.91 ± 0.04).[6] Reference 3 describes the analysis. Any value of D < 1 implies that the entire horizontal axis would be occupied by steps, if vanishingly small steps could be included. It is important to note, however, that the same value of D was found for several values of V_{rf} and within different intervals along the x-axis, in conspicuous disagreement with Refs. 5 and 6.

The presence of the subharmonic steps places restrictions on the underlying equation of motion for the CDW. The simplest equation which describes nonlinear and frequency dependent conduction, narrow band noise, and Shapiro steps treats the entire CDW as a single, damped classical particle in a periodic potential.[7] Supplemented by an inertial term it takes the form:

$$m\ddot{x} + \frac{1}{\tau}\dot{x} + \frac{m\omega_o^2}{Q}\sin(Qx) = eE \qquad (1)$$

where E is the applied (dc + rf) electric field. This equation is identical to the

RSJ equation for Josephson junctions, allowing results from the literature to be taken over directly. We therefore expect that the inertial term is essential for Eq. (1) to exhibit subharmonic steps.[4,8] However the observed frequency dependent conductivity is consistent with Eq. (1) only if the inertial term is negligible to at least several hundred MHz.[9] Hence Eq. (1) appears to be inconsistent with Figs. 1 and 2.

What is clearly lacking from Eq. (1) is allowance for internal degrees of freedom of the CDW, which are needed to account for pulse memory effects,[10] long time decays,[11] etc. The extra freedom may be modeled by assigning equations of form (1) with no inertial term to velocity-coherent regions in the sample, then adding coupling terms between the regions.[1] Such systems of coupled first order nonlinear equations can be expected[12] to exhibit mode locking and other behaviors reminiscent of the RSJ equation, and so it is possible that subharmonic steps might result. A distribution of coupling strengths between the regions might be able to explain the lack of strong dependence of D on V_{rf} and on V_{dc}.

II. Synchronization

The steps are most readily understood as regions of locking between the internal and applied frequencies when $pf_{ext} - qf_{int}$ is sufficiently small. If the locking within such regions is complete, the CDW velocity becomes fixed by f_{ext} and does not respond to changes in the applied dc voltage. Hence dV/dI rises to the linear resistance R_o attributable to uncondensed electrons alone. This situation pertains if the CDW velocity is coherent throughout the specimen. In reality the velocity coherence length must be finite, and f_{int} may vary spatially. If the variation of f_{int} is greater than the width of the region over which locking can occur, locking will be incomplete, and dV/dI will rise to a level less than R_o. Hence the height of dV/dI is expected to correlate with the degree of synchronization across the sample.

Doubling or tripling of narrow band noise peaks is frequently noted in power spectra of $NbSe_3$ current oscillations, graphically illustrating variation of f_{int} within a single specimen. If the splitting is small enough ($\lesssim$ 100 kHz), the locking effect of large V_{rf} can make the noise peaks coalesce, with corresponding signatures in the steps. An example is shown in fig. 3, for the $p/q = 1/2$ step. Within this step, dV/dI has two plateau levels. Within the lower level, the spectrum contains two peaks (fig. 3a), one of which is locked to $f_{ext}/2$, while the other moves with V_{dc}. Within the higher plateau, the two peaks have combined into a single peak at $f_{ext}/2$ which is independent of V_{dc} (Fig. 3b). Thus the higher dV/dI indicates greater synchronization throughout the sample.

Another common characteristic of narrow band noise in $NbSe_3$ is its fluctuations: the

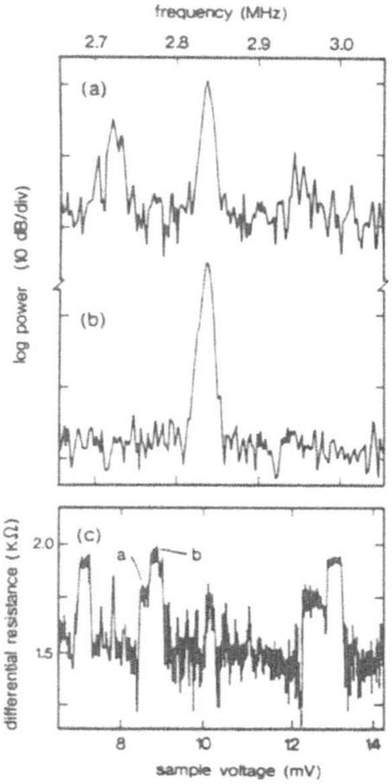

Fig. 3. Demonstration of synchronization in $NbSe_3$ at T = 45 K. (a) and (b) show noise peaks corresponding to labeled plateaus in the p/q = 1/2 step.

amplitude of the noise peak varies widely on a time scale of less than a second. These fluctuations can be reduced with application of large V_{rf}, provided V_{dc} is biased within a step. In samples for which dV/dI = R_o exactly, the time behavior of the fluctuations becomes entirely different, the amplitude becoming stable over the time frame of a minute or more. Figure 4 presents histograms of the noise amplitude, taken in the presence of large V_{rf} over two successive 15 minute periods. One period corresponds to locking on the p/q=1/2 step; for the other f_{ext} was detuned from $2f_{int}$. The 'locked' case exhibits several narrow peaks, each of which were formed by more or less successive measurements of the noise amplitude. The behavior suggests that in the presence of a large rf drive, most fluctuations are unable to disturb the relative phases of the CDW between regions of the sample. The larger fluctuations may allow the CDW to "realign" the phases to a state not much different in energy than before, and the amplitude of the noise takes on the value characteristic of that new state.

Finally, the more anisotropic CDW conductors such as TaS_3 and $(TaSe_4)_2I$ tend to have extremely broad noise spectra,[13] unless they are very pure.[14] This indicates a short velocity correlation length, which probably explains the small, wide steps of Fig. 1. Nevertheless, application of large V_{rf} sharpens the steps (Fig. 5), demonstrating that the phenomena of synchronization are relevant to these materials also.

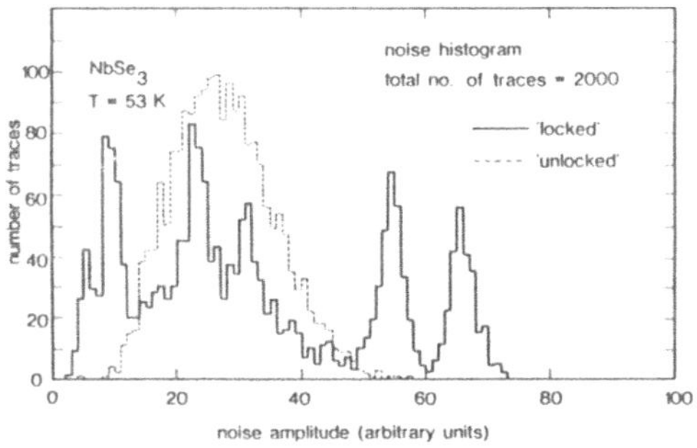

Fig. 4. Histograms of noise amplitude in $NbSe_3$ with with f_{int} tuned to, and detuned from locking criterion f_{ext} = 2 f_{int}. Each histogram contains 2000 points.

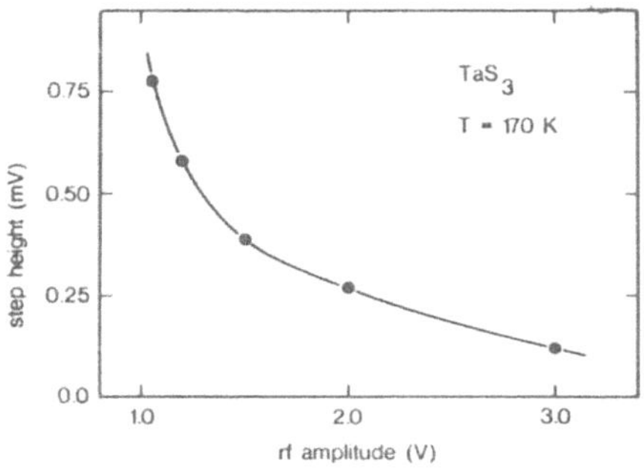

Fig. 5. Rf amplitude dependence of full width at half maximum of the p=1, q=1 step in TaS_3.

We thank L. Mihaly for discussions and suggestions. This work was supported by NSF Grant DMR 84-06896.

*Current address: Physics Dept., Univ. of Illinois, Urbana, IL 61801, USA

References

1. P. Monceau, J. Richard and M. Renard, Phys. Rev. Lett. 45, 43 (1980); Phys. Rev. B 25, 931, 948 (1982).
2. A. Zettl and G. Gruner, Phys. Rev. B 29, 755 (1984).
3. S.E. Brown, G. Mozurkewich and G. Gruner, Phys. Rev. Lett. 52, 2277 (1984).
4. E. Ben-Jacob, Y. Braiman, R. Shainsky and Y. Imry, Appl. Phys. Lett. 38, 822 (1981).
5. M.H. Jenson, P. Bak and T. Bohr, Phys. Rev. Lett. 50, 1637 (1983).
6. W.J. Yeh, D.-R. He and Y.H. Kao, Phys. Rev. Lett. 52, 480 (1984).
7. G. Gruner, A. Zawadowski and P.M. Chaikin, Phys. Rev. Lett. 46, 511 (1982).
8. M.J. Renne and D. Polder, Rev. Phys. Appl. 9, 25 (1974); J.R. Waldrum and P.H. Wu, J. Low Temp. Phys. 47, 363 (1982).
9. G. Gruner, A. Zettl and W.G. Clark, Phys. Rev. B 24, 7247 (1981).
10. J.C. Gill, Solid State Commun. 39, 1203 (1981).
11. G. Mihaly and L. Mihaly, Phys. Rev. Lett. 52, 149 (1984).
12. F.C. Hoppensteadt, Nonlinear Oscillations in Biology, (Am. Mathematical Soc., Providence, RI, 1979).
13. G. Mozurkewich, M. Maki, and G. Grüner, Solid State Comm. 48, 453 (1983).
14. H. Salva, Z. Z. Wang, P. Monceau, J. Richard, and M. Renard, Phil. Mag. B 49, 385 (1984).

MODE LOCKING AND CHAOS IN SLIDING CHARGE-DENSITY-WAVE SYSTEMS

P. Bak
Department of Physics
Brookhaven National Laboratory
Upton, NY 11973, U.S.A.

Sliding CDWs in ac electric fields may serve as model systems for the study of mode-locking phenomena and the transition to chaos in dissipative dynamical systems with competing frequencies. The mode-locking structure at the transition is expected to form a complete devil's staircase with fractal dimension D ~ 0.87. Indeed, Brown, Mozurkewich and Grüner have observed a multitude of steps in the I-V characteristics of $NbSe_3$ with an apparent fractal dimension D = 0.91 ± 0.03.

1. Introduction

Single crystals of the charge-density wave system $NbSe_3$ in dc electric fields exhibit current oscillations with a fundamental frequency ω_{int} proportional to the CDW current I[1]. If in addition to the dc field the crystal is subjected to an ac field with frequency ω_{ext}, the two frequencies will couple because of non-linearities, and the CDW frequency may lock into the external frequency, $\omega_{int} = (p/q)\omega_{ext}$, with p and q integers.[2]

The study of this phenomenon is interesting for two reasons. First, a quantitative investigation of the subharmonic structure may lead to new insight into the microscopic mechanisms for the sliding conductivity in a given compound. One may call this the "materials science" point of view. A second philosophy, which will be adopted here, is that sliding CDWs in combined ac and dc electric fields may serve as model systems for the transition to chaos in dynamical dissipative systems with competing frequencies. Experiments on CDW systems may thus serve to throw light on a much more general problem in physics. In particular it has been predicted, that a "universal" transition to chaotic behavior caused by interacion and overlap of resonances may occur at sufficiently strong coupling[3-5]. Below the transition to chaos the two frequencies are either locked or, with finite probability unlocked or quasiperiodic. At the transition to chaos the two frequencies are always locked, and the ratio between the frequencies assumes all rational values in an interval at the transition line, so the subharmonic step structure forms a "complete devil's staircase". The staircase defines a Cantor set with fractal dimension D ~ 0.87. This number is expected to be universal, i.e. the number does not depend on the underlying microscopic physics, as represented for instance by a differential equation. Hence, the prediction of a universal fractal dimension, and other universal indices can in principle be checked by studying the mode-locking structure at and near the transition in a large class of systems with competing frequencies exhibiting a transition to chaos.

In a sence the philosophy is the same as for critical behavior near second order thermodynamic phase transitions: for instance, in order to predict "Ising" critical behavior it is sufficient to analyze the symmetry of a given system. There is no need to establish a one-to-one correspondence between the "real" Ising model and the microscopic interactions. The "Ising model" for the transition to chaos to be discussed here is the so-called "circle map", and the predictions of universal critical properties stem mostly from investigations of a class of circle maps[4]. In addition, the circle map critical behavior has been confirmed by numerical studies of a differential equation representing charge-density-waves[3,6].

Indeed, quite recently Brown, Mozurkewich, and Grüner have observed scaling of the subharmonic structure in $NbSe_3$. They found $D = 0.91 \pm 0.03$ in fair agreement with our predictions[7]. The small apparent disagreement might be because the measurements were not performed sufficiently close to the transition line. There is no reason to believe that the disagreement is related to the nature of the underlying differential equation.

In the following we shall first briefly review the theoretical evidence for universal scaling behavior. In particular, it will be argued that a transition to chaos, represented by the circle map, might indeed take place in CDW systems subjected to ac and dc electric fields. Next, the recent results by Brown et al[7], and by others, will be discussed in the light of the theoretical predictions.

2. Charge density waves and circle maps

It has been suggested that the motion of CDWs in electric fields may be described by a simple differential equation[8], which in dimensionless form reads:

$$\alpha\ddot{\theta} + G\dot{\theta} + \gamma\sin 2\pi\theta = E_{dc} + E_{ac}\cos\omega t. \qquad (1)$$

This is the equation for a damped pendulum, driven by a constant torque E_{dc} and an oscillating force E_{ac}. The equation also describes the resistively shunted Josephson function in a microwave field[9]. For the CDW system, θ is the position of the CDW, G is the damping, α is the inertia "mass" term, and γ is the amplitude of a periodic potential which might well be a contact potential for a crystal containing only one domain. Figure 1 shows schematically a system obeying an

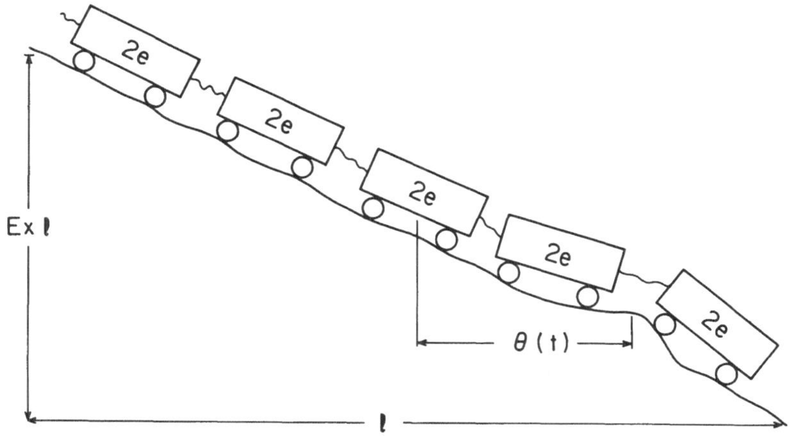

Figure 1. Artist's view of a sliding CDW. If the current carriers are solitons rather than CDW-periods the charges might be e/2 rather than 2e.

equation such as (1). The current carried by the CDW is proportional to the average velocity, $I = c \langle\dot{\theta}\rangle$. For small values of E_{dc} the CDW is pinned by the potential γ and the current is zero; for E_{dc} greater than a threshold value the CDW is depinned and slides with a positive average velocity, and the CDW carries a current. This motion corresponds to the rotating modes of the driven pendulum.

Assume now that we watch the system with "stroboscopic light" at regularly spaced intervals, $t_n = n\tau$, using the external force as a clock, $\tau = 2\pi/\omega$. The values of θ and $\dot{\theta}$ at t_{n+1} must be related to their values at t_n through equations of the form

$$\theta_{n+1} = g_1(\theta_n, \dot{\theta}_n) \qquad (2)$$

$$\dot{\theta}_{n+1} = g_2(\theta_n, \dot{\theta}_n).$$

The system can be described by a two dimensional return map, since the differential equation is of second order.

Because of dissipation it might well be that after a transient period θ_n becomes a "slave" of $\dot{\theta}_n$:

$$\dot{\theta}_n = h(\theta_n) \qquad (3a)$$

so $$\theta_{n+1} = g_1\big(\theta_n, h(\theta_n)\big) = f'(\theta_n) = \theta_n + f(\theta_n) \qquad (3b)$$

where $f(\theta_n) = f(\theta_n + 1)(\text{mod } 1)$ because of the translational symmetry of equation (1). Equation (3a) defines the so-called invariant circle, and (3b) is

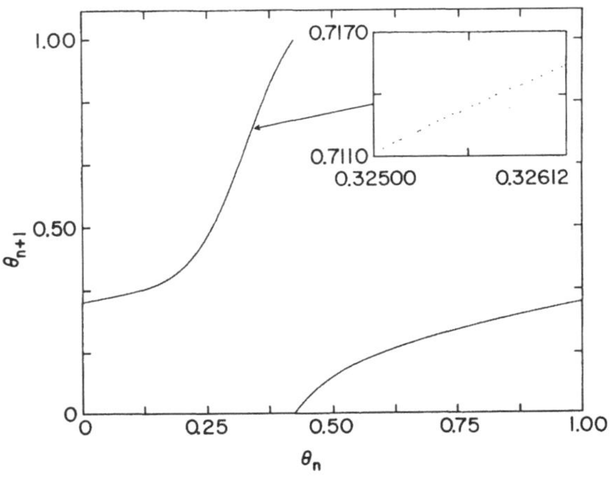

Figure 2. Return map calculated by integrating eq. (1) numerically. The smooth line is made up entirely of points. The motion for the particular choice of parameters ($\alpha = \gamma = E_{ac} = 1$; $E_{dc} = 1.4$, $\omega = 1.76$) is quasiperiodic (incommensurate). The map is monotonically increasing so we are below the transition to chaos. The inset is a magnification stressing the one-dimensionality (after Bohr et al, ref. 3).

called a circle map since it maps the circle $0 < \theta < 1$ onto itself. Whether or not the "dimensional reduction" from two to one as expressed by eqs (3a) and (3b) actually takes place depends on the specific system.

It has been shown numerically that for a wide range of parameters including a transition to chaos, the return map of equation (1) is indeed a one-dimensional circle map[3]. Figure 2 shows θ_{n+1}(mod 1) vs. θ_n(mod 1) for values of the parameters below the transition to chaos where the map is monotonic; at the transition the map tends to acquire zero slope at some point, and becomes non-invertible. Once the equivalence with the circle map has been established one can forget about the differential equation and simply study iterates of the map. Theoretical results derived from the study of discrete maps can be taken over and directly applied to the physical system.

Most theoretical work has been performed on the "sine circle maps":

$$\theta_{n+1} = \theta_n + \Omega + \frac{K}{2\pi}\sin 2\pi\theta_n \qquad (4)$$

This circle-map has a critical line, $K = 1$, where the derivative is zero for $\theta = 0$. Figure 3 shows the phase diagram for the circle map. Note the "Arnold tongues" where mode locking occurs. Think of K as the amplitude and Ω the dc driving voltage in the CDW system. Figure 4 shows the frequency (in units of ω),

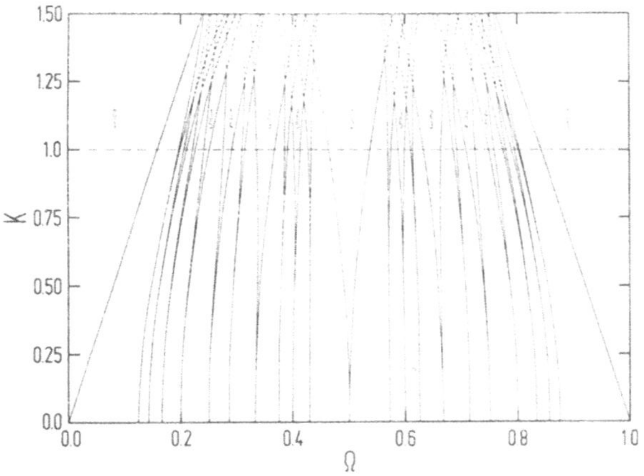

Figure 3. Phase diagram for the circle map. At the transition line the Arnold-tongues fill-up everything (after Jensen et al, ref. 4).

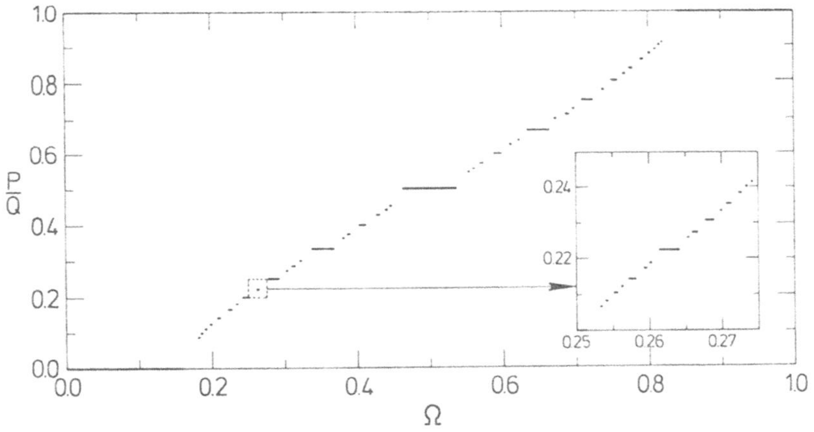

Figure. 4. Frequency locking at the transition to chaos at the critical line for the circle map (Jensen et al, ref. 4)

$$W = \lim_{n\to\infty} \frac{\theta_n - \theta_0}{n}$$

vs. Ω at the critical line $K = 1$[4]. The steps indicate the intervals where the frequency is locked at rational values $W = p/q$. The completenes of the staircase can be shown as follows[4]. Choosing a scale r one adds up the total length S(r) of steps which are larger than r in an Ω interval of length ℓ. Defining the function $N(r) = [\ell - S(r)]/r$ one finds

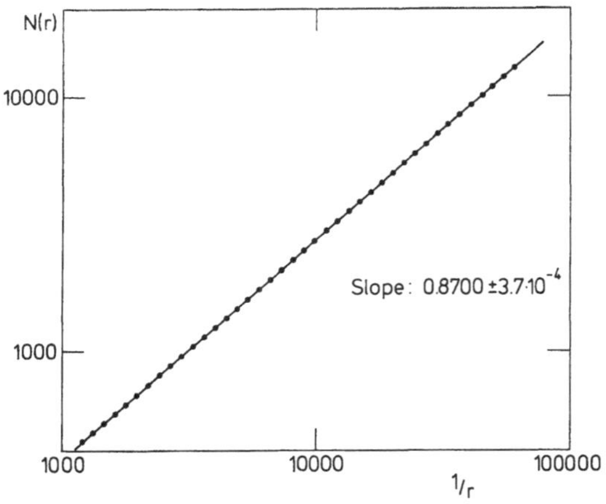

Figure 5. N(r) vs. r. The slope of the straight line gives $D = 0.8700 \pm 4\times10^{-4}$

$$N(r) = r^{-D}, \quad D = 0.8700 \pm 4 \times 10^{-4} \qquad (5)$$

as seen from the log-log plot, Figure 5. The total length of "holes", $\ell - S(r)$, vanishes as $r \to 0$, so the subharmonics fill-up the critical line. The exponent D is the fractal dimension of the Cantor set (of measure zero) which is the complementary set on the Ω axis to the intervals for which the frequencies are locked.

In general, of course, the periodic function in (3b) is not a pure sine-function. The return map depicted in Figure 2 contains ~10% higher harmonics. It has been shown that the addition of higher harmonics do not change the scaling dimension D: the claim of universality rests precisely on this observation[4].

In addition to the calculation on the circle map, the scaling behavior of mode locking at the transition to chaos has been verified by analog computer simulations on the differential equation (1)[6] and by a study of a two dimensional dissipative map of the form (2)[3]. Both calculations gave $D \sim 0.87$.

There appears to be a problem in interpreting experiments on CDWs in terms of an equation (1). A previous analysis of the frequency dependent conductivity indicates that the mass α/G is rather small (< 1/10).[2] It has been shown by Waldram and Wu, and by Renne and Polder[10] that for $\alpha = 0$ there are no subharmonic steps with $q > 1$. For small values of the mass α it can be shown analytically that the return map is a circle map[11]

$$\theta_{n+1} = \theta_n + \alpha h(\theta_n) \qquad (6)$$

where h is a periodic function. Hence, for small α the map (6) is monotonic and there can be no transition to chaos. Our original estimates for the possibility of observing the transition to chaos were therefore quite pessimistic[11-12].

However, as we shall see,[7,13] experiments have revealed a multitude of mode-locked steps, so the transition to chaos is well within reach. This means that the effective mass entering an appropriate differential equation simply can not be small. Because of the universality we expect the general scaling picture to apply irrespective of the underlying mechanisms leading to a significant mass and inductive behavior.

3. Experiments on sliding conductivity of $NbSe_3$ in combined ac and dc electric fields

In the experiments by Brown et al[7] the steps in the I-V curve signaling lock-in of the CDW velocity were determined by measuring the differential resistance dV/dI. They performed measurements with and without applied rf voltage at 25 MHz. The first thing to observe is that significant subharmonic steps of high order exist. There seems to be complete locking since dV/dI = constant (= the resistivity of normal electrons)[13] at the plateaus. As the amplitude of the rf field becomes larger and larger the steps become wider and wider as expected (see Figure 3). Indeed, it seems that the system can be brought to a state where the resonances overlap and chaos is expected. More experimental work is needed to accurately determine the transition line. Figure 6 shows a plot of N(r), defined in the discussion preceeding eq. (5), vs. r as determined experimentally by Brown

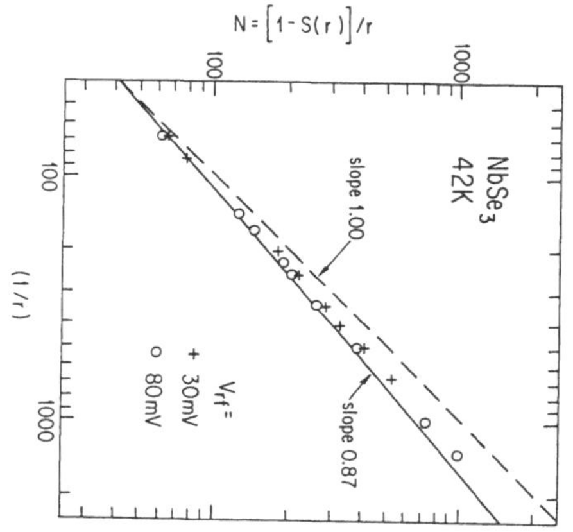

Figure 6. N(r) vs. (r) for $NbSe_3$ at T = 42K for two rf levels. The data fits D = 0.91 ±0.03 while the solid line corresponds to the circle map[4] (Brown et al, ref.7).

et al[7]. The slope of the straight line is D = 0.91 ± 0.03. We consider this to be in good agreement with the theoretical value D = 0.87.

How can it be that mode-locking occurs even if the effective mass estimated by the frequency dependent conductivity is small[2]? This is probably because the coefficient α is not related to the dielectric response in a simple way: There is little reason to believe that the linear response of a resting CDW, pinned, for instance, by frictional impurity forces, has anything to do with the inertial mass of the sliding CDW! To calculate the inertial term for the the sliding conductivity one must take into account processes at the contacts, as for instance vortex formation etc.[14]. These processes are not important for the linear response to small ac fields. More specifically, although the relation

$$I = c\,\langle\dot{\theta}\rangle$$

is valid, since infinite charges can not build up at the contacts, the equation

$$I(t) = c\dot{\theta}(t)$$

is not valid, i.e. there is no simple proportionality between the instant current and the instant velocity of the CDW as assumed in the linear response theory, and experiments on the frequency dependent conductivity can not give information on α. This has consequences also for the oscillating "noise" observed in CDW systems. Even if the motion of the CDW is described by a simple well-defined equation such as (1), the frequency dependent response $I_{ac}(t)$ depends on the nature of the contacts etc. In the model of Ong et al[14], even a CDW moving with constant velocity leads to an oscillating current: Indeed, Hall and Zettl[13] find that the mode locking occurs at well defined values of V_{rf}/V_t, where V_t is the threshold voltage, but the amplitude of the "noise" signal is spurious and fluctuates wildly from sample to sample.

It has beens suggested that a strongly dissipative model with zero inertial terms and with internal elastic degrees of freedom can account for some of the experiments[15]. Such models, however, fail to provide true mode-locking in sharp contrast with experiments.

Preferably, one should study the mode-locking phenomena in single-domain samples in order to avoid additional impurity related volume effects. For large enough samples the phase coherence must break down. The CDWs in different domains may slide either with different velocity, in which case several frequencies should be observable simultaneously, or with the same velocity but out of phase. In the latter case, although the CDWs independently follow the same equation of motion the resulting noise signal will more or less average out. Consider, for instance a number of single domain CDWs connected in parallel. The mode locking in each domain is expected to take place as described here. The frequencies of all domains will couple to the common rf signal, and phase coherence will be restored. Again, the noise signal will be spurious, but the mode-locking well-defined.

The circle map exhibits a wide range of chaos-related phenomena which might be studied experimentally on CDW systems. In addition to the mode-locking discussed here, there are infinite series of Feigenbaum bifurcations, intermittent chaos, and universal behavior associated with the transition to chaos for specific relative frequencies such as the "Golden mean"[16]. We suggest that further experiments be performed to explore these possibilities and test predictions of current theories.

References

1. R.M. Fleming and C.C. Grimes, Phys. Rev. Lett. 42, 1423 (1979); P. Monceau, J. Richard and M. Renard, Phys. Rev. Lett. 45, 43 (1980).
2. A. Zettl and G. Grüner, Solid State Commun. 46, 501 (1983); Phys. Rev. B29, 755 (1984).
3. P. Bak, T. Bohr, M.H. Jensen, and P.V. Christiansen, Solid State Commun. 51, 231 (1984); T. Bohr, P. Bak and M.H. Jensen, Phys. Rev. A (to be published).
4. M.H. Jensen, P. Bak and T. Bohr, Phys. Rev. Lett. 50, 1637 (1983); Phys. Rev. A (to be published).
5. For reviews see P. Bak in "Statics and dynamics of nonlinear systems" Edited by G. Benedek, H. Bilz, and R. Zeyher (Springer, Berlin, 1983) p. 160; P. Bak, M.H. Jensen, and T. Bohr in "Procedings of the 59th Nobel Symposium", Physica Scripta, to be published.
6. P. Alstrom, M.H. Jensen, and M.T. Levinsen, Phys. Lett. 103A, 171 (1984).
7. S.E. Brown, G. Mozurkewich and G. Grüner, Phys. Rev. Lett. 52, 2277 (1984).
8. G. Grüner, A. Zawadowski and P.M. Chaikin, Phys. Rev. Lett. 46, 511 (1981).
9. W.L. Stewart, Appl. Phys. Lett. 12, 277 (1968). D.E. McCumber, J. Appl. Phys. 39, 3113 (1968).
10. J.R. Waldram and P.H. Wu, J. Low Temp. Phys. 47, 363 (1982); M.J. Renne and D. Poulder, Rev. Phys. Appliqué 9, 25 (1974).
11. M. Ya. Azbel and P. Bak, Phys. Rev. B, to be published.
12. P. Bak, Proceedings of the International Symposium on Nonlinear Transport in Quasi-one-dimensional Conductors, Sapporo, Japan (Hokkaido University, Sapporo, 1984) p 13; R.P. Hall and A. Zettl, Phys. Rev. B30, 2279 (1984).

14. N.P. Ong, G. Verma and K. Maki, Phys. Rev. Lett. 52, 2419 (1984).

15. S.N. Coppersmith and P.B. Littlewood, preprint.

16. M.J. Feigenbaum, L.P. Kadanoff and S.J. Shenker, Physica 5D 370 (1982); S. Ostlund, D. Rand, J. Sethna, and E. Siggia, Physica 5D, 303 (1983).

CHAOS IN CHARGE DENSITY WAVE SYSTEMS

A. Zettl, M. Sherwin, and R.P. Hall

Department of Physics
University of California, Berkeley
Berkeley, California 94720 U.S.A.

We investigate chaotic response in the charge density wave (CDW) condensates of $(TaSe_4)_2I$ and $NbSe_3$. In $(TaSe_4)_2I$, non bifurcative routes to chaos occur when the pinned CDW is excited by an external ac electric field. The behavior is interpreted as that of a driven anharmonic oscillator. In $NbSe_3$ in the switching regime, a period doubling route to chaos occurs for combined ac + dc fields. The route to chaos is characteristic of instabilities in phase lock for systems of competing periodicities. Intermittent chaos is also observed in dc biased $NbSe_3$ with negative differential resistance. We interpret the chaotic behavior in terms of simple models with restricted numbers of degrees of freedom, and return maps appropriate to these models.

Introduction

There has been much study on turbulent or chaotic behavior in systems which have macroscopic numbers of degrees of freedom. Of particular interest is the existance of universality classes describing the onset of chaos, which provides a direct connection between highly complex real systems, and simplified models representing only a small number of degrees of freedom. Well known examples of universality are the period doubling route to chaos[1] and the onset of intermittent chaos[2].

We shall here be interested in the association of chaos with the dynamics of charge density wave (CDW) condensates. A number of phenomena are investigated in $(TaSe_4)_2I$ and $NbSe_3$ which can, to a surprising degree, be well explained in terms of simple deterministic equations of motion possessing one dimensional return maps. Our purpose at present is to gain insight into CDW dynamics by analyzing the particular route to chaos involved.

The anharmonic oscillator: application to $(TaSe_4)_2I$

As was first discussed by Lee, Rice, and Anderson[3], the low field ac response of a pinned CDW condensate might be expected to follow a damped harmonic oscillator behavior. For large ac drive fields, nonlinear anharmonic potential effects are inevitable. Under appropriate conditions, such anharmonic terms may lead to chaotic structure. Huberman and Crutchfield[4] were the first to demonstrate for a simple anharmonic oscillator a bifurcation cascade to the chaotic state, with a response characterized by a strange attractor in phase space. A signature of the chaos is a dramatic rise in broadband noise in the response spectral density. Subsequent studies of the damped pendulum[5] have again demonstrated routes to chaos for sufficiently large ac drive amplitude. In dimensionless form, the damped pendulum equation of motion reads

$$\beta \frac{d^2\theta}{dt^2} + \frac{d\theta}{dt} + \sin\theta = e_{dc} + e_{ac} \sin \Omega t \quad , \tag{1}$$

Eq.(1) also describes a resistively shunted Josephson junction[6], and it has been suggested to describe CDW dynamics in an approximate classical limit[7]. A critical parameter in Eq.(1) is β, which reflects system inertia. In the limit $\beta \to 0$, Eq.(1) does not predict dynamical chaos.

The low field ac conductivities of $NbSe_3$, TaS_3 , and $(TaSe_4)_2I$ all appear to represent "overdamped" response, for which β is vanishingly small. Indeed, experiments aimed at achieving chaos in the pinned CDW states of these materials, by simply driving the condensate with an ac field, have not been successful.

We introduce finite inertia into the CDW system by addition of a real inductor in series with the CDW crystal. By appropriate choice of circuit parameters and sample, a high Q resonance circuit can be created, as demonstrated in Fig. 1a. The low field response of this hybrid $(TaSe_4)_2I$ circuit is underdamped, i.e. β is substantially greater than zero. For this circuit, as the ac drive amplitude is increased, chaotic response results for the pinned CDW condensate, yielding high-level broadband noise in the current response spectrum.

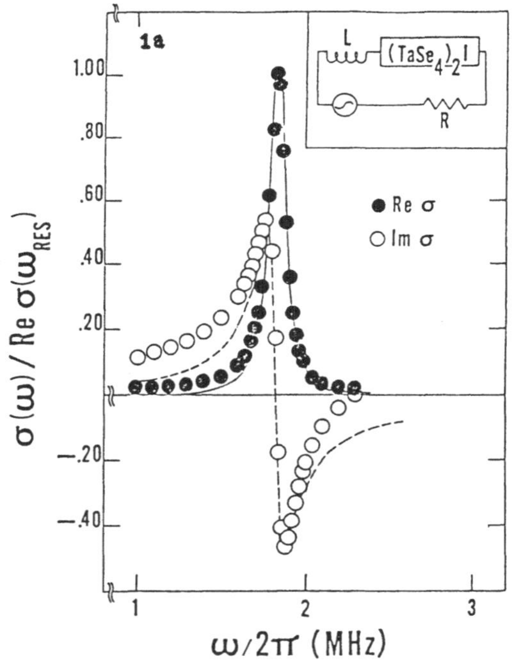

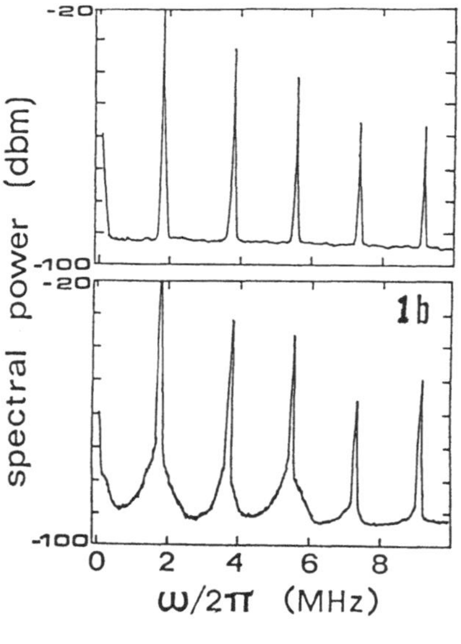

The onset of chaos is suden and non-bifurcative. Such non-bifurcative transitions
The onset of chaos is suden and non-bifurcative. Such non-bifurcative transitions
are quite possible in terms of Eq.(1), as has been discussed by various authors[5,8]. Eq.(1) is able to account well for the chaotic response observed in this hybrid CDW circuit.

<u>Phase lock and Shapiro steps: period doubling routes to chaos in NbSe.</u>

In NbSe., dramatic Shapiro steps[9] result for combined ac and dc drive fields, i.e. for finite e_{dc} and e_{ac} in Eq.(1). The steps are a manifestation of phase lock between the CDW condensate and the externally applied ac field. In the limit β = 0, subharmonic steps are not predicted. However, in $NbSe_3$ at T=42K, substantial subharmonic structure has been observed[10], as demonstrated in Fig. 2a. This behavior would indicate a finite β. A detailed analysis[11] of such subharmonic structure in terms of the Devil's staircase has suggested that at this temperature $NbSe_3$ is close to, but not at, chaos.

At temperatures below 42K, $NbSe_3$ may show switching behavior[12]. In a phenomenological sense switching drastically enhances β, leading to hysteresis

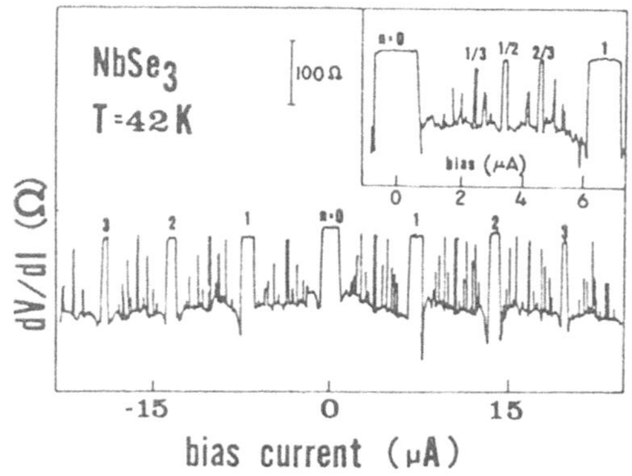

FIG. 2a. dV/dI vs dc bias current for $NbSe_3$. The rf frequency is 5 MHz and the rf amplitude is 7 mV. A rich spectrum of harmonic and subharmonic steps is observed. The inset shows the subharmonics in greater detail.

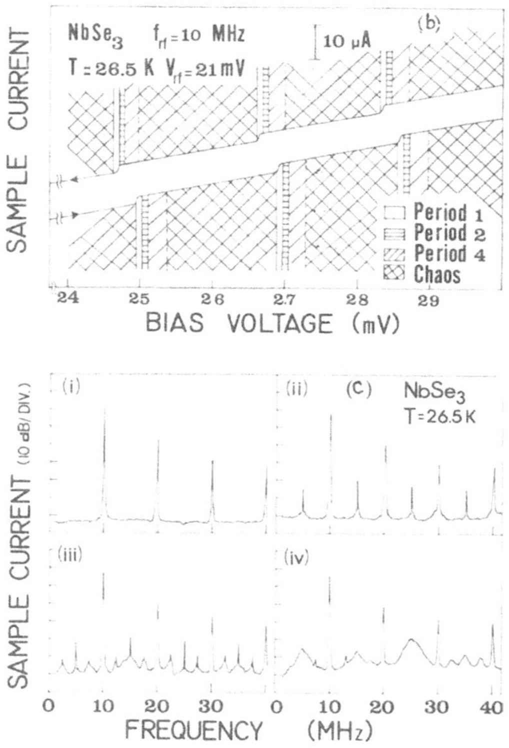

FIG. 2. (b) Schematic representation of current response in Shapiro step region, for forward- and reverse-bias voltage sweeps. (c) Frequency spectrum of current response in Shapiro step region. External rf drive frequency and amplitude as in (a). (i) V_{dc} = 25 mV, period 1; (ii) V_{dc} = 25.1 mV, period 2; (iii) V_{dc} = 25.2 mV, period 4; (iv) V_{dc} = 25.5 mV, chaos.

effects in the dc I-V characteristics, and also to modified Shapiro step structure. On each Shapiro step, an increase in dc bias field can lead to a well-defined period doubling route to chaos, as illustrated in Figs. 2b and 2c. The repetitive nature of the response is most easily interpreted in terms of the one dimensional return map of Eq.(1), i.e. the circle map[13]

$$\theta_{i+1}=\theta_i+\Gamma+C\sin 2\pi\theta_i \quad , \qquad (2)$$

where Γ represents the ratio of external drive frequency to internal (narrow-band noise) frequency. In the parameter range $C>1/2\pi$, the circle map displays stable mode-locked solutions, which bifurcate successively to chaos as Γ is smoothly increased. Since . is a modulo 1 variable, changing Γ to $\Gamma+n$, with n an integer, will not change Eq.(2). Thus the patterns of bifurcations to chaos will repeat as Γ is increased monotonically. The bifurcation sequence in dc bias observed in $NbSe_3$ is consistent with the periodicity of the behavior predicted by the return map of Eq.(1).

System bistability: intermittency and 1/f noise

The initial onset of CDW dc conduction is in general quite smooth with no evidence for divergent behavior. In switching samples, however, CDW depinning is dramatically sharp. "Intermediate" switching may result in negative differential resistance (NDR) just beyond the depinning threshold, as has been observed[14] in $NbSe_3$. The NDR region is associated with dramatic broadband noise response and additional random structure in the frequency spectrum, indicative of temporally intermittent chaos. Fig. 3a shows the voltage response spectrum of a $NbSe_3$ sample biased into the NDR region. The 1/f noise level is approximately 4 orders of magnitude larger than that associated with conventional CDW motion. Fig. 3b indicates the additional intermittent structure. Here data represent fast Fourier transforms of the voltage response, taken sequentially under identical experimental conditions. This type of chaotic behavior could arise from system bistability, as might accur for fluctuating current paths in the NbSe. sample (reflecting possibly macroscopic CDW domain structure). In the context of a domain model by Joos and Murray[15], there can exist an instability between having n and n+1 channels open and conducting CDW current. The system may then hop between the n+1 and n open channel states, effectively representing hopping between valleys of a bistable system. The model is equivalent to that considered by Ben-Jacob et al[16] for intermittent chaos in Josephson junctions. There the random-like hopping between states leads to intermittent chaos, with response

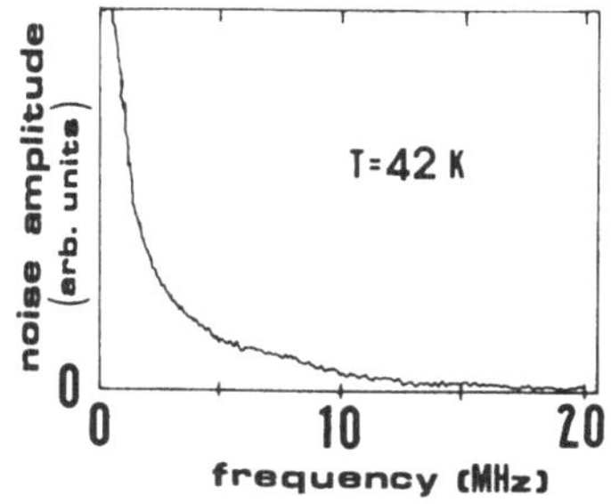

3a.

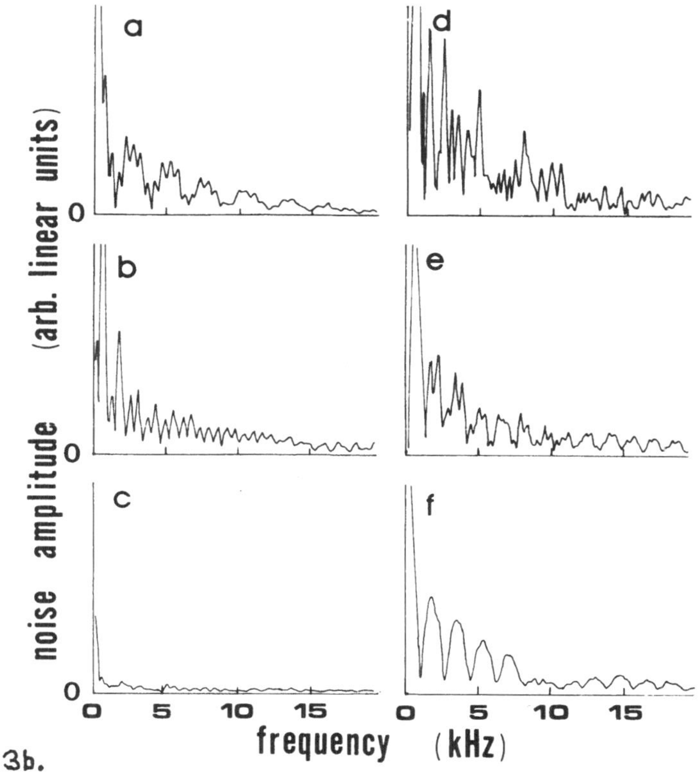

3b.

power spectra not unlike those of Fig. 3b. In $NbSe_3$, the 1/f noise and intermittent structure could well represent quite similar processes, with of course different characteristic attempt frequencies.

We thank B. Huberman, P. Bak, J. Bardeen, and G. Gruner for helpful interactions. This research was supported in part by a grant from the UCB Campus Committee on Research. One of us (AZ) is an NSF Presidential Young Investigator, and recipient of an Alfred P. Sloan Foundation Fellowship.

References

1. M.J. Fiegenbaum, J. Stat. Phys. 19, 25 (1978) and 21, 669 (1979)
2. P. Manneville and Y. Pomeau, Phys. Lett. 75A, 1 (1979); Y. Pomeau and P. Manneville, Commun. Math. Phys. 74, 189 (1980), see also J. Hirsh, B. Huberman, and D. Scalapino, Phys. Rev. A25, 519 (1982)
3. P.A. Lee, T.M. Rice, and P.W. Anderson, Solid State Commun. 14, 703 (1974)
4. B.A. Huberman and J.P. Crutchfield, Phys. Rev. Lett. 43, 1743 (1979)
5. D. D'Humieres, M.R. Beasley, B.A. Huberman, an A. Libchaber, Phys. Rev. A26, 3483 (1982)

6. W.C. Stewart, Appl. Phys. Lett. 12, 277 (1968); D.E. McCumber, J. Appl. Phys. 39, 3113 (1968)
7. G. Gruner, A. Zawadowski, and P.M. Chaikin, Phys. Rev. Lett. 46, 511 (1981)
8. E. Ben-Jacob, Y. Braiman, R. Shainsky, and Y. Imry, Appl. Phys. Lett. 38, 822 (1981)
9. A. Zettl and G. Gruner, Solid State Commun. 46, 501 (1983)
10. R.P. Hall and A. Zettl, Phys. Rev. B30, (1984)
11. S.E. Brown, G. Mozurkewich, and G. Gruner, Phys. Rev. Lett.
12. A. Zettl and G. Gruner, Phys. Rev. B26, 2298 (1982)
13. T. Geisel and J. Nierwetberg, Phys. Rev. Lett. 48, 7 (1982); H. Jensen, T. Bohr, P. Christiansen, and P. Bak, Brookhaven National Laboratory Report No. 33495 (1983)
14. R.P. Hall, M. Sherwin, and A. Zettl, Phys. Rev. Lett. 52, 2293 (1984)
15. B. Joos and D. Murray, Phys. Rev. B29, 1094 (1984)
16. E. Ben-Jacob, I. Goldhirsh, Y. Imry, and S. Fishman, Phys. Rev. Lett. 49, 1599 (1982)

CONTRIBUTION OF CDW MOTION TO THE HALL EFFECT AND TO THE TRANSVERSE CONDUCTIVITY IN TaS_3. EXPERIMENT.

Yu.I.Latyshev, Ya.S.Savitskaja, V.V.Frolov
Institute of Radioengineering and Electronics
of the Academy of Sciences of the USSR,
103907 Moscow, Marx av. 18, USSR

The contribution of sliding charge density wave (CDW) to the Hall voltage V_H and to the transverse conductivity $\sigma_\perp$ is observed in orthorombic TaS_3. The results are interpreted on the basis of a microscopic theory which takes into consideration the variation of the quasiparticle distribution function in the process of CDW motion.

In the last few years the considerable attention was paid to investigations of the properties of the CDW and phenomena accompanying its motion. It was found that CDW motion leads to the appearance of narrow band noise[1], to the decrease of the dielectric constant[2], to the extension of phase coherence to the macroscopic distances[2,3] and so on. In this paper we report on the experimental results concerning the influence of CDW motion on V_H and $\sigma_\perp$.

At first sight it seems that CDW motion should not induce any change in transverse characteristics (V_H, $\sigma_\perp$) because of one-dimensional character of CDW motion along the chains. This point of view has been supported earlier in experiments on non-linear Hall effect in $NbSe_3$[4,5]. In this papers no additional contribution to the linear dependence $V_H(E)$ was observed at fields E exceeding the threshold field E_T for non-linear longitudinal conductivity. Nevertheless, our measurements on o-TaS_3 (material with more pronounced properties of the Peierls transition and the Fröhlich conductivity than $NbSe_3$) prove that such an influence of sliding CDW on transverse characteristics takes place. This influence may be explained by a variation of the quasiparticle distribution function due to impurity scattering of the quasiparticles moving with the CDW. It leads to an additional contribution to the quasiparticle current (and, therefore, to V_H) proportional to the CDW velocity $\sim \dot{\chi}$[6-8] (χ is the phase of the CDW). On the other hand, if the motion of the CDW conserves phase correlation between different chains in the whole volume of the crystal, and the CDW wave vector Q is not parallel to conducting chains (axis c

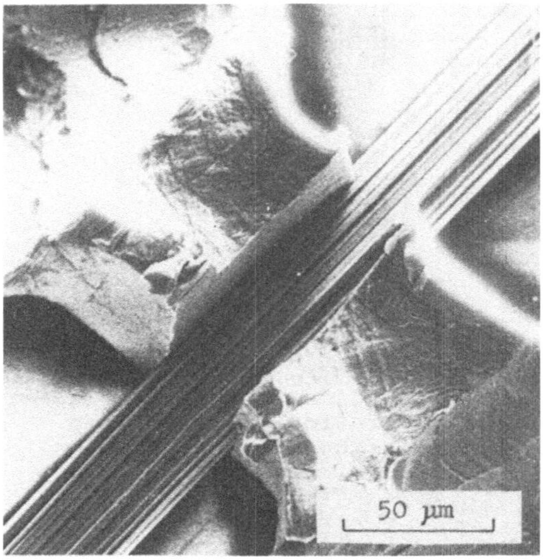

Fig. 1. Microphotograph of o-TaS_3 sample with transverse contacts.

in o-TaS_3), then superlattice motion may change the CDW phase in transverse direction. As a result, it induces the change in transverse conductivity without real motion of the CDW in this direction.

o-TaS_3 crystals of high quality (T_p= 220 K, E_T(at 110 K) = 0.2 ÷ 0.4 V/cm) were prepared by direct reaction of Ta and S at 650°C and crystallisation of the reaction products by thermal transport (temperature gradient 2 degrees per cm) in evacuated sealed quartz tube for two weeks. The samples used are single crystals of size 4 mm × ×50 µm × 10 µm. Transverse contacts were attached by In (see Fig. 1). They had a low contact resistance (0.1 Ohm at 300 K) and good reproducibility after thermocycling. The Hall voltage V_H was measured in b direction by nanovoltmeter Keithley-180 (H ∥ a, j ∥ c) and was symmetric when H was reversed. The dependence $V_H(H)$ was also linear over the all temperature region (400 - 77 K) and magnetic fields up to 8 kOe.

Earlier we investigated low-electric field-Hall effect ($E \ll E_T$) in o-TaS_3[9]. It was found that at room temperature the Hall constant R_H had a positive sign and was equal to 3.5×10^{-3} cm^3/C ($p = 2 \times 10^{21}$ cm^{-3}). At temperatures $T > T_p$ (region of metallic conductivity) R_H was temperature independent, and at $T < T_p$ $R_H \backsim \exp(\Delta/T)$ due to the formation of the CDW gap Δ ($\Delta = 1000$ K). We consider now our results on high-electric field-Hall measurements ($E > E_T$). The dependence $V_H(E)$ for two samples with different E_T is shown in Fig. 2. To avoid Joule heating up to 8 v/cm these measurements were carried out in liquid methane. The characteristic feature is a deviation of $V_H(E)$ downwards from the linearity[6]. This deviation was observed on three samples with various E_T. It was not observed at $E < E_T$, but was observed only at $E \gtrsim E_T$. Nonlinearity in $V_H(E)$ weakens with the temperature increase and disappears at $T > T_p$. All these facts allow

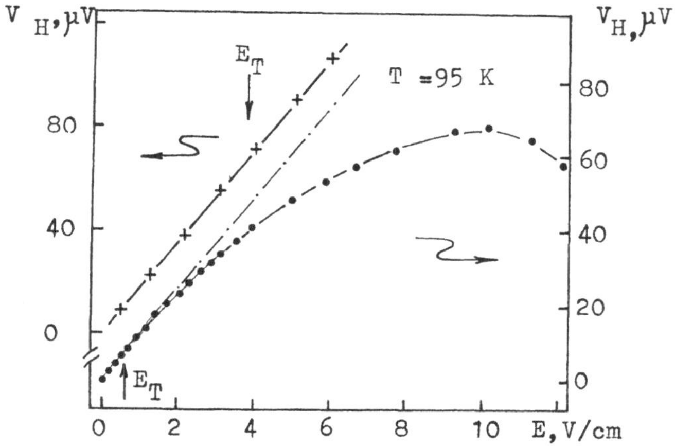

Fig. 2. Electric field dependence of the Hall voltage for o-TaS_3.

us to consider that the nature of the observed deviation is caused by the CDW motion.

Microscopic theory of Hall effect[7,8] also gives a decrease of V_H in the presence of CDW motion. This effect occures due to the decrease of the quasiparticle current[6,7]: $j = \sigma_1 E - a\frac{\sigma_1}{l}\dot{\chi}$ (σ_1 is a quasiparticle conductivity without CDW motion, l is a quasiparticle mean free path, $a \sim 1$) when the CDW moves ($\dot{\chi} \neq 0$). the last term appears because of additional impurity scattering of the quasiparticles moving with the CDW. According to the theory[6,7] $\frac{V_H(\dot{\chi} \neq 0) - V_H(\dot{\chi} = 0)}{V_H(\dot{\chi} = 0)} \backsim -\dot{\chi}/E$. As the estimates show[6], ratio $\dot{\chi}/E$ in TaS_3 grows with the decrease of a temperature, being in qualitative agreement with the experiment.

Transverse conductivity was measured at weak a.c. current at various values of d.c. fields E in longitudinal direction. Typical dependence of $\rho_\perp$ (in b direction) on E is shown in Fig.3 together with $\rho_{\parallel}(E)$ measured simultaneously at the same experimental conditions. As one can see from Fig 3, the change in $\rho_\perp(E)$ appears when $E > E_T$, i.e. when CDW moves. This correlation was observed at all temperatures in the interval ($77\ K < T < T_p$) and in the samples with larger values of E_T. Furthemore, measurements of $\rho_\perp(E_\perp)$ showed that $\rho_\perp$ was not changed up to the fields $E_\perp$, two orders of magnitude large than fields $E_{\parallel} \approx E_T$. Measurements reported here show the effect of CDW

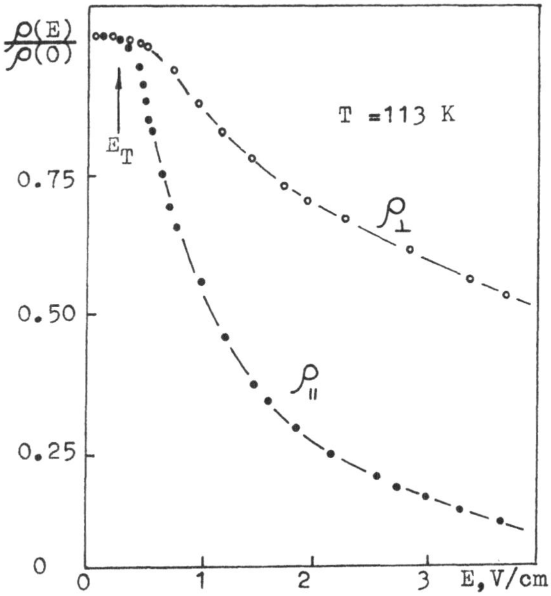

Fig. 3. Transverse (⊥) and longitudinal (∥) normalized resistivities vs. longitudinal field E for o-TaS_3.

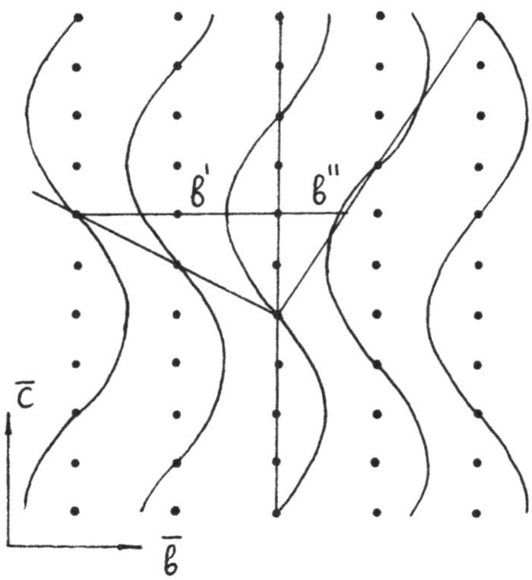

Fig. 4. Schematic view of charge distribution along distorted chains in o-TaS_3

motion on transverse conductivity. The explanation of this effect was proposed by S.N.Artemenko and A.N.Kruglov (see these Proceedings). It was shown that the effect of CDW motion on $\sigma_\perp$ may occur due to non-parallelism of t the CDW wave vector and the direction of chains in o-TaS_3. $\vec{Q} = (\frac{a^*}{2}, \frac{b^*}{8}, \frac{c^*}{4})$. We attempt to give graphic illustration of this idea in Fig.4. In this picture the charge distribution along distorted chains (axis c) in TaS_3 is drawn schematically. The phases of adjacent chains are shifted by $\Delta\chi = \frac{2\pi}{8}$. Assuming the phase correlation between the chaires in moving CDW we describe the motion of the CDW as the displacement of the whole picture along c-axis. Then it occures that the velocity of the phase change in transverse direction $\dot{\chi}_\perp$ to the left $\dot{\chi}_\perp^l$ and to the right $\dot{\chi}_\perp^r$ from an arbitrary chain are not equal. $\dot{\chi}_\perp^l$ and $\dot{\chi}_\perp^r$ in Fig.4 are proportional to the projections of constant phase lines to the b axis - b' and b''. Thus, the total change of phase in transverse direction in average is not zero ($b' \neq b''$). As a result, it leads to the change of $\sigma_\perp$. It can be easily understand, that no phase change in transverse direction occures when $\Delta\chi = 0, \Delta\chi = \frac{\pi}{2}$. The first condition corresponds to $Q \parallel c$, and the second one to the doubling of period in transverse direction. It should be noted that the effect of the CDW motion on $\sigma_\perp$ is caused by the deviation of thequasiparticle distribution function from its equilibrium value. It is interesting to study influence of CDW motion on $\sigma_\perp$ in monoclinic TaS_3 and $NbSe_3$ in which Q is known to be parallel to the chain axis below the first Peierls transition.

Acknowledgements. - We would like to thank S.N.Artemenko, E.N.Dolgov and A.N.Kruglov for stimulating discussions.

References

1. Fleming R.M., Grimes C.C., Phys. Rev. Lett. 1979, 42, 1423.
2. Gruner G., Zettl A. J. Physique. 1983,44, C3-1631.
3. Ong N.P., Verma G., Maki K. Phys. Rev. Lett. 1984, 52, 663.
4. Tessema G.X., Ong N.P. Phys. Rev. 1981, B23, 5607.
5. Kawabata K., Ido M., Sambongi T. J.Phys.Soc.Jap. 1981, 50, 1992.
6. Artemenko S.N.,Dolgov E.N.,Kruglov A.N., Latyshev Yu.I.,Savitskaya Ya.S.,Frolov V.V. Pis'ma Zh.Eksp.Teor.Fiz. 1984, 39, 258.
7. Artemenko S.N., Kruglov A.N. Fiz. Tverd. Tela 1984, 26, 2391.
8. Dolgov E,N. Sol. St. Commun. 1984, 50, 405.
9. Latyshev Yu.I., Savitskaja Ya.S., Frolov V.V. Pis'ma Zh.Eksp.Teor. Fiz. 1983, 38, 446.

CONTRIBUTION OF CDW MOTION TO THE HALL EFFECT AND TO THE TRANSVERSE CONDUCTIVITY. THEORY.

S.N.Artemenko and A.N.Kruglov
Institute of Radioengineering and Electronics of the Academy of Sciences of the USSR, 103907 Moscow, Marx av. 18, USSR

The contribution of sliding charge density wave (CDW) to the transport phenomena related to perpendicular to conducting chains charge transfer is investigated theoretically. The Hall effect and the transverse conductivity are calculated on the basis of a microscopic theory, neglecting the spatial deformation of the CDW.

The problem of the Hall effect in a quasi-one-dimensional (Q1D) conductor with CDW is of particular interest, because it was stated, that CDW motion cannot contribute to the Hall current. Besides, such a contribution was not observed experimentally in $NbSe_3$[1,2]. We have shown, that CDW motion changes the spectrum and the distribution of quasiparticles, thus contributing to the Hall current, this contribution being of the sign opposite to the Hall current in a crystal with the pinned CDW[3]. At low temperatures and strong electric fields the motion of CDW may change the sign of the Hall current.

To calculate the Hall current one can use the kinetic equations[4,5] for the Green functions. But to avoid cumbersome calculations, it is more convenient in this case to use a kinetic equation for the distribution function of quasiparticles, which we derive from the equations for the Green functions, following Keldysh technique. An equation of this type was derived formerly for superconductors[6].

We neglect 1D fluctuations assuming that the phonon spectrum is 3D enough and that the adiabatic condition $m/m^* \ll 1$ is fulfilled. The momentum relaxation time is considered to be large ($\Delta\tau \gg 1$, Δ is the order parameter). The kinetic equation is valid in the quasiclassical limit $qv_F \ll \Delta$, $\omega \ll \Delta$ (where ω, q are the characteristic frequency and the wave vector). We take into account only the impurity scattering, but the results can be easily generalised to the case of the elastic phonon scattering.

There are two types of quasiparticles, their spectrum being dependent on the CDW velocity $u=\dot{\chi}/Q$ (χ is the phase of the CDW, Q - its wave vector):

$$\varepsilon_p^{(1,2)}=\varphi(\vec{r},t)+\eta(\vec{p})-\tfrac{1}{2}(\vec{Q}\cdot\vec{\nabla}\chi)\pm\left[(\xi(\vec{p})+\dot{\chi}/2-(\vec{p}\cdot\vec{\nabla}\chi))^2+\Delta^2\right]^{1/2} \quad (1)$$

Where $\varepsilon(\vec{p})=p_1^2/2m+\varphi(\vec{p}_{tr})-\varepsilon_F$ is the 3D spectrum of the Q1D metal ($|\varphi| \ll \varepsilon_F$); $\eta(\vec{p})=\frac{1}{2}\left[\varepsilon(\vec{p}+\vec{Q}/2)+\varepsilon(\vec{p}-\vec{Q}/2)\right]$, $\xi(\vec{p})=\frac{1}{2}\left[\varepsilon(\vec{p}+\vec{Q}/2)-\varepsilon(\vec{p}-\vec{Q}/2)\right]$

Calculating the Hall current, we assume that Q is parallel to the chains, in this case $\eta = p_1^2/2m + \varphi(\vec{p}_{tr})$, $\xi = (Q/2m)p_1 = v_F p_1$, where p_1 and $\vec{p}_{tr}$ are longitudinal and transverse momenta.

Thus, we get the kinetic equation (the detailed derivation is in Ref.3):

$$\frac{\partial n_i}{\partial t} + \frac{\partial \varepsilon_p^{(i)}}{\partial \vec{p}}\frac{\partial n_i}{\partial \vec{r}} - \frac{\partial \varepsilon_p^{(i)}}{\partial \vec{r}}\frac{\partial n_i}{\partial \vec{p}} + \vec{E}\frac{\partial n_i}{\partial \vec{p}} - \frac{1}{c}\left(\left[\frac{\partial n_i}{\partial \vec{p}} \times \vec{H}\right]\cdot\frac{\partial \varepsilon_p^{(i)}}{\partial \vec{p}}\right) =$$

$$= \int \frac{d^2\vec{p}\,'_{tr}}{S} \int d\xi' (n_i(\xi,\vec{p}_{tr}) - n_i(\xi',\vec{p}\,'_{tr}))\Big[\nu_1(u_\xi u_{\xi'} + v_\xi v_{\xi'})^2 \delta(\varepsilon_p^{(i)} - \varepsilon_{p'}^{(i)}) +$$
$$+ \nu_2 u_{\xi'}^2 v_\xi^2 \delta(\varepsilon_p^{(i)} - \varepsilon_{p'}^{(i)} - (-1)^i \dot{\varkappa}) + \nu_2 u_\xi^2 v_{\xi'}^2 \delta(\varepsilon_p^{(i)} - \varepsilon_{p'}^{(i)} + (-1)^i \dot{\varkappa})\Big] + \tag{2}$$
$$+ \int \frac{d^2\vec{p}\,'_{tr}}{S} \int d\xi' (n_i(\xi,\vec{p}_{tr}) - n_j(\xi',\vec{p}\,'_{tr}))\Big[\nu_1(u_\xi v_{\xi'} - u_{\xi'} v_\xi)^2 \delta(\varepsilon_p^{(i)} - \varepsilon_{p'}^{(j)}) +$$
$$+ \nu_2 v_\xi^2 v_{\xi'}^2 \delta(\varepsilon_p^{(i)} - \varepsilon_{p'}^{(j)} - (-1)^i \dot{\varkappa}) + \nu_2 u_\xi^2 u_{\xi'}^2 \delta(\varepsilon_p^{(i)} - \varepsilon_{p'}^{(j)} + (-1)^i \dot{\varkappa})\Big] \,.$$

Here $i=1,2$; $j=3-i$; $u_\xi = \sqrt{\frac{1}{2}(1+\tilde{\xi}/\tilde{\varepsilon})}$; $v_\xi = \sqrt{\frac{1}{2}(1-\tilde{\xi}/\tilde{\varepsilon})}$; $\tilde{\xi} = \xi + \dot{\varkappa}/2 - (\vec{p}\cdot\vec{\nabla}\chi)$ $\tilde{\varepsilon} = (\tilde{\xi} + \Delta^2)^{1/2}$; S is the Brillouin zone cross section at $p_1=0$, ν_1 and ν_2 are the scattering frequencies without(with) the change of longitudinal momentum by $\vec{Q}$.

Longitudinal and transverse currents are expressed in terms of n_1 and n_2:

$$j_1 = -\int \frac{d^2\vec{p}_{tr}}{(2\pi)^3} \int d\xi \frac{\tilde{\xi}}{\tilde{\varepsilon}} (n_1 - n_2) + \frac{\dot{\varkappa} S}{(2\pi)^3} \,; \tag{3}$$

$$j_{tr} = -\int \frac{d^2\vec{p}_{tr}}{4\pi^3} \vec{v}_{tr} \int dp_1 (n_1 + n_2) \,; \quad \vec{v}_{tr} = \partial\eta/\partial\vec{p}_{tr} \,; \tag{4}$$

To calculate the Hall current we solve the Eq.2, the external fields E and H being taken into account by the perturbation theory. We consider a Q1D conductor with the small bending of the Fermi surface ($\eta \ll \Delta$), this case is realized in TaS_3. We consider the low temperature limit $T \ll \eta$. In this case we need to know the form of the function $\eta(\vec{p}_{tr})$ near its main extrema only. Let $\eta = \eta_1 - (p_{tr}-p_1)^2/2m_1$ in the vicinity of its highest maximum, and $\eta = -\eta_2 + (p_{tr}-p_2)^2/2m_2$ near its lowest minimum. These regions give the main contribution to the current due to the factors $\exp(\pm\varepsilon_p^{(1,2)}/T)$ in the distribution functions. Taking into account that in the low temperature limit the characteristic value of $\xi \sim (\Delta T)^{1/2} \ll \Delta$, we may conclude that the dispersion law (1) near the extrema is quadratic and the problem becomes close to the calculation of the Hall current in the semiconductor with the ellipsoidal spectrum. The resulting expression for the

Hall current in the case of pinned CDW is of the form:

$$j_{EH} = j_{EH}^{(1)} + j_{EH}^{(2)}; \quad j_{EH}^{(1,2)} = \pm A_{1,2}\, \sigma_{tr}^{(1,2)} lEH/\Delta ; \tag{5}$$

Contribution from the moving CDW has the form

$$j_{\dot{\chi}H} = j_{\dot{\chi}H}^{(1)} + j_{\dot{\chi}H}^{(2)}; \quad j_{\dot{\chi}H}^{(1,2)} = \mp B\, \sigma_{tr}^{(1,2)} \dot{\chi} H/\Delta ; \tag{6}$$

Where $\sigma_{tr}^{(1,2)} = (ST/6\pi^3 v_F \nu_0 m_{1,2})\exp(\frac{\eta_{1,2}-\Delta}{T})$; $\nu_0 = \nu_1 + \nu_2/2$; $\sigma_{tr} = \sigma_{tr}^{(1)} + \sigma_{tr}^{(2)}$ is the transverse conductivity in Peierls state; $l = v_F/\nu_2$ is the mean free path; $A_{1,2} = v_F \nu_2 S/8c(2\pi\Delta T)^{1/2} \nu_0 m_{1,2} \sim v_F \eta/c(\Delta T)^{1/2}$; $B = v_F \nu_2/2c\nu_0 \sim v_F/c$; indices 1 and 2 are related to the electron and hole components.

Let us discuss these results. If one neglects the effect of the electric field and of the scattering on the distribution function, then quasiparticles must be in equilibrium with the CDW and move together with it. But this equilibrium distribution should not contribute to the Hall current. The deviation of the distribution function from the equilibrium in the first order perturbation theory is caused firstly by the drift of the quasiparticles in the field E and, secondly, by the impurity scattering of moving together with CDW quasiparticles. These deviations are of opposite sign. In the second order of the perturbation theory the mechanisms mentioned above result in the appearance of the components (5) and (6) in the Hall current.

The contribution of the sliding CDW to the Hall current depends on the relation between $\dot{\chi}$ and E. Neglecting pinning, we get $\gamma\dot{\chi} = El$, where the factor γ is determined by the mechanism of friction of the CDW. In Ref.5 only friction mechanism connected with the quasiparticles at energies $|\varepsilon| > \Delta$ was taken into account. The underestimate value $\gamma = (\Delta/T)\exp(-\Delta/T)\ln(T/\nu) \ll 1$ was obtained. In fact, factor γ may be larger (see S.N.Artemenko and A.F.Volkov, paper in these Proceedings) due to localized electron states in the middle of the Peierls gap. The order-of-magnitude estimate based upon experimental data on the longitudinal conductivity gives, that the contribution of the sliding CDW to the Hall current is of the order of the Hall current in a crystal with motionless CDW, but has the opposite sign. The recent experiment on TaS_3[7] qualitatively confirms these predictions. The case of $\Delta \ll T$ was studied by Dolgov[8].

As it was stated above, we have studied the Hall effect assuming that the wave vector of the CDW $\vec{Q}$ is parallel to the conducting chains. If the vector $\vec{Q}$ has a component perpendicular to the chains, then the motion of the CDW contributes not only to the Hall effect, but to the transverse conductivity, too. To calculate this contribution, we use the kinetic equations for the Green function[5]. From the calculations similar to that in Ref.5, neglecting the dependence of $\nu_{1,2}$ on $\vec{p}_{tr}$, we obtain for the transverse current at $T \ll \Delta$:

$$j_i = \sigma^{(1)}_{i,k} E_k + \dot{\chi}\sigma^{(2)}_i / l \qquad (7)$$

where $\sigma^{(1)}_{i,k} = S(4\pi^3 \nu_0 v_F)^{-1} \langle u_i^+ u_k^+ + 2\sqrt{\pi} u_i^- u_k^- T/\Delta \rangle \exp(-\Delta/T)$; $\sigma^{(2)}_i = S(T/\Delta)^{1/2}(\sqrt{2}\pi^{5/2}\nu_0)^{-1}\langle u_i^- \mathrm{ch}(\eta/T)\rangle \exp(-\Delta/T)$; $u_i^+ = \partial\eta/\partial p_i^{tr}$; $u_i^- = \partial\xi/\partial p_i^{tr}$ $\langle\,\rangle = \int d^2\vec{p}_{tr}/S$. It follows from the expression for ξ, that if $Q \parallel p_1$, then the velocity $u_i^- = 0$, and the last term in Eq.7 vanishes.

We calculate the dependence of $\dot{\chi}$ on the electric field from the self-consistency condition for the phase. Neglecting pinning we get $\gamma\dot{\chi} = E_l l + (\vec{E}_{tr} \cdot \vec{l}_{tr})$, where $l_{tr\ i} = 2\pi^3\sigma^{(2)}_i / S$. Taking into account the friction of CDW associated with quasiparticles at $|\varepsilon| > \Delta$ only and supposing for simplicity the isotropic transverse electron spectrum, we obtain the relative change of the transverse conductivity

$$\frac{\delta\sigma_{tr}}{\sigma_{tr}} = 4\pi\left(\ln\frac{T}{\gamma}\right)^{-1} \frac{\nu_2}{\nu_0} \frac{\langle u^- \mathrm{ch}\,\eta/T\rangle^2}{\langle u^{+2}\rangle} (T/\Delta)^2 \qquad (8)$$

It is clear that in this case the relative contribution of the sliding CDW to σ_{tr} is less than its relative contribution to σ_l[5]. However, allowance for the friction associated with the other mechanisms results in the increase of γ and $\sigma^{(2)}$. This leads to increase of $\delta\sigma_{tr}/\sigma_{tr}$ and to decrease of $\delta\sigma_l/\sigma_l$, so that their values can be of the same order of magnitude. Our study gives the qualitative explanation of the transverse conductivity measurements on TaS_3 by Latyshev et al. (see these Proceedings).

We are grateful to A.F.Volkov for helpful advices and also to Yu.I.Latyshev, Ya.S.Savitskaya and V.V.Frolov for the access to their experimental results and for discussion.

1. Tessema G.X., Ong N.P. - Phys. Rev. B23, 5607 (1981).
2. Kawabata K., Ido M., Sambongi T. - J.Phys.Soc.Jpn.50,1992(1981)
3. Artemenko S.N., Kruglov A.N. - Fiz. Tverd. Tela 26, 2391(1984).
4. Gor'kov L.P., Dolgov E.N. - Sov. Phys. JETP 50, 203(1979).
5. Artemenko S.N., Volkov A.F. - Sov. Phys. JETP 53, 1050(1981).
6. Aronov A.G., Gurevich V.L. - Fiz. Tverd. Tela 16, 2656(1974).
7. Artemenko S.N., Dolgov E.N., Kruglov A.N., Latyshev Yu.I., Savitskaya Ya.S., Frolov V.V. - Pis'ma Zh.Eksp.Teor.Fiz. 39, 258(1984).
8. Dolgov E.N. - Solid State Comm. 50, 405(1984).

DEPENDENCE OF THE ELASTIC MODULUS OF TaS_3 ON THE CDW CURRENT

J.W. Brill
Department of Physics & Astronomy
University of Kentucky
Lexington, KY 40506-0055, USA

We report on the dependence of the Young's modulus of orthorhombic TaS_3 on current and voltage near the threshold (I_T,V_T) for non-Ohmic conduction. When the charge density wave current exceeds ~0.3% of the normal current, the relative modulus decrease $-\Delta E/E = A\sqrt{I_{CDW}/I_T}$, where $A \sim 7\times10^{-3}$ for all samples. Closer to threshold, $-\Delta E/E \propto (V/V_T - 1)^2$, with a sample dependent slope, suggesting that the modulus change is dominated by domain wall motion as $V \to V_T$, but that it is predominantly intrinsic to the depinned CDW at larger voltages.

Most investigations of the depinning of a charge density wave (CDW) have dealt with the unusual electronic transport properties observed when voltages greater than the depinning threshold, V_T, are applied. It has recently been discovered[1,2] that there are also anomalous lattice properties for $V > V_T$; i.e. the Young's modulus decreases by ~1% and the internal friction increases ($\Delta(1/Q)$~1%). These experiments, which have used the vibrating reed technique[3] of exciting flexural acoustic resonances (f_0 ~1 kHz) which are detected with very frequency sensitive phase-lock electronics, have concentrated on providing general descriptions of the effects, chiefly in orthorhombic TaS_3 (o-TaS_3). In this paper, we briefly review our earlier work[1] and discuss the results of our present research on the more detailed behavior of Young's modulus of o-TaS_3 as $V \to V_T$. It is hoped that such experiments will complement transport measurements in determining the correct model of CDW depinning.

The fundamental flexural resonant frequency of a fiber of length ℓ, thickness t, density ρ, and Young's modulus E is given by

$$f_0 = a_0 t/\ell^2 \sqrt{E/\rho} \qquad (1)$$

where a_0 is a constant of order unity which depends on the boundary conditions. The change in internal friction (e.g. as a function of voltage) is given by $\Delta(1/Q)$, where Q is the quality factor. As discussed in Reference (1), Eq. (1) does not apply if uniaxial stress is applied to the sample, (e.g. by the current leads) which is difficult to avoid in o-TaS_3 due to the thinness (~3μm) of the samples. In our early work[1], the stress was kept low by gluing one end of the sample to a rigid rod and using a 50 μm constantan wire for the second current lead, thereby roughly clamping the second end. The magnitude and temperature dependence of the resonant frequency indicated that the effective modulus as determined from Eq. (1) was largely that of the TaS_3 sample, with a contribution of <30% resulting from uniaxial stress and the finite stiffness of the wire.

The results for such a sample, grown at UCLA, are shown in Figure 1; the data is corrected for temperature changes due to Joule heating as discussed in Reference (1). Current, voltage, resonant frequency, and quality factor (i.e. amplitude) were measured simultaneously. The voltage at which the internal friction and modulus change is the same as that at which the CDW becomes depinnned and the resistance decreases. The main observations were that:

1) The internal friction saturates at a voltage $V < 2V_T$.

2) The change in internal friction decreases with increasing temperature.

3) The modulus decrease $-\Delta E/E$ showed no sign of saturating at the highest voltages measured ($\sim 3V_T$). (Mozurkewich <u>et al.</u>, however, did observe saturation in this voltage range.[2])

4) For "high voltages" ($V > 1.2\ V_T$), $-\Delta E/E \underset{\sim}{\alpha} I_{CDW}{}^p$, where $1/3<p<1/2$. (I_{CDW} is the CDW current: $I_{CDW} = I - V/R_0$, where R_0 is the normal resistance.)

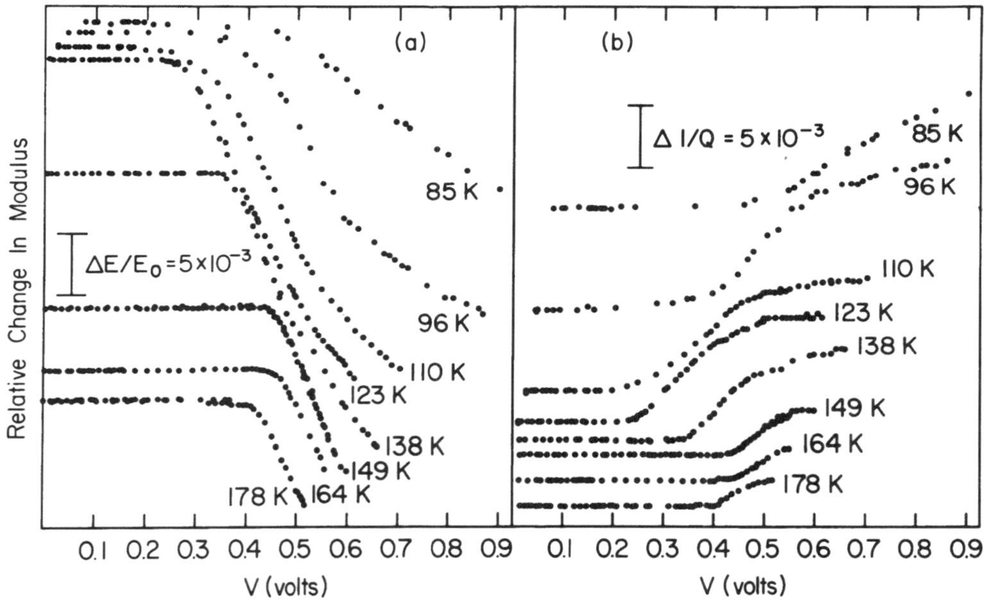

Figure 1 - (a) Relative change in Young's modulus and (b) change in 1/Q vs. voltage for sample #1 (constantan clamp) at several temperatures. Vertical offsets are arbitrary. $Q^{-1}(V=0)\sim 1.4\times 10^{-3}$ at all temperatures. (From Ref. 1).

We have further reduced the effects of uniaxial stress by replacing the constantan wire with a crystal of $TaSe_3$ (which is metallic with no CDW transition) of thickness comparable to the o-TaS_3 sample. The $TaSe_3$ fiber was bent for strain relief and mounted perpendicularly to the sample, so that differential thermal contraction would at most flex the sample. The boundary condition, and therefore the constant a_0 (in Eq. (1)), is now ambiguous. In general, several resonant modes were observed, one of which had a temperature dependence similar to that of a "pure" flex mode of o-TaS_3 (i.e., with one end free).[3] Figure 2 shows the variation of "effective modulus" for a TaS_3 sample before and after adding the $TaSe_3$ "clamp". Although detailed data were

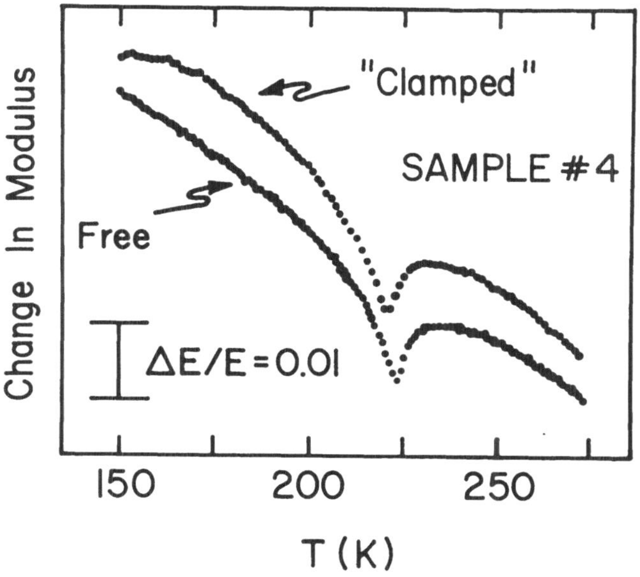

Figure 2 - Relative change in Young's modulus vs. temperature before and after adding a $TaSe_3$ "clamp". The vertical offset is arbitrary.

only taken for such "pure" flexural modes, other modes with very different temperature dependences (e.g. shear modes which are coupled to the electrostatic transducers through the $TaSe_3$ fiber) often showed stronger dependences on applied voltage; i.e. other elastic constants seem to depend more strongly on voltage than the Young's modulus. Data were taken on samples grown in different tubes at the University of Kentucky.

The voltage dependence of the modulus near threshold for a few samples at different temperatures is shown in Figure 3. It is seen that as $V \to V_T$, the modulus decreases quadratically with voltage:

$$-\Delta E/E = B(V/V_T - 1)^2 \qquad (2)$$

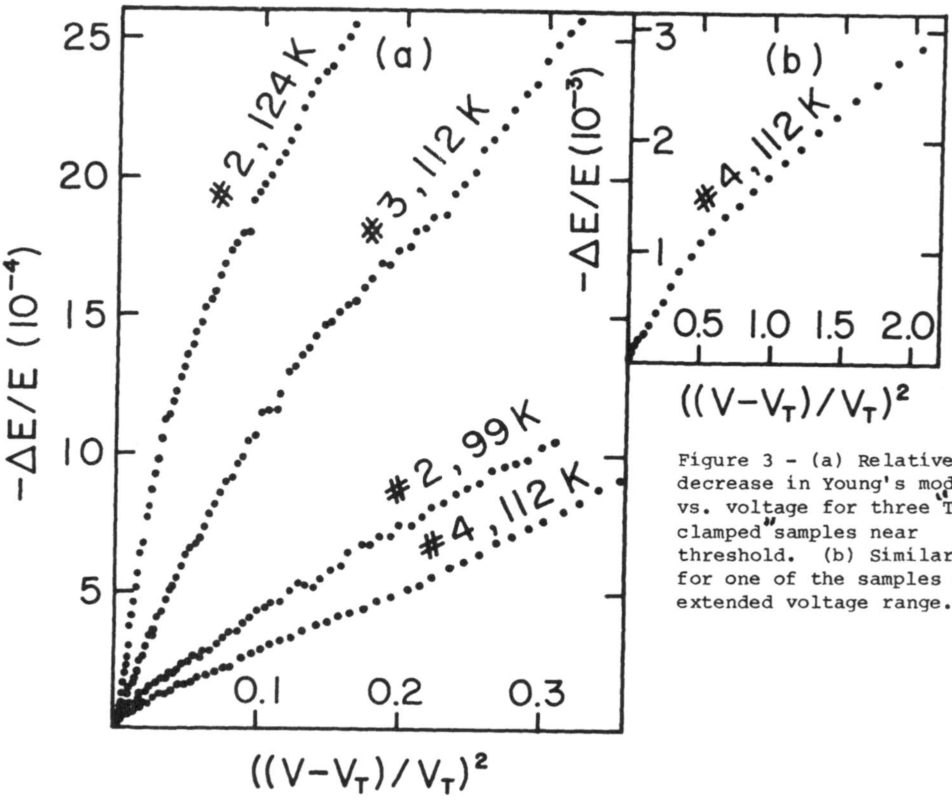

Figure 3 - (a) Relative decrease in Young's modulus vs. voltage for three "$TaSe_3$ clamped" samples near threshold. (b) Similar plot for one of the samples over extended voltage range.

The constant B is very temperature (as is also clear from Figure 1) and sample dependent, suggesting that its size depends on defects. Since the CDW domain walls become depinned as the voltage exceeds threshold,[4] this initial modulus decrease is possibly due to the relaxation of domain walls under the oscillating strain. The CDW transition temperature, and therefore the density of condensed electrons is very stress dependent,[5] and strain, like temperature,[4] will therefore affect the equilibrium domain configuration. When the CDW is pinned, the domains cannot respond to the oscillating strain, so that the measured modulus is unrelaxed, while for $V > V_T$ it is relaxed and therefore smaller.[1] As discussed in Reference (1), an upper limit to the relaxation strength can be estimated by assuming that the entire sample is a single relaxing domain wall in which case

$$-\Delta E/E\Big)_{\substack{MAX\\ RELAX}} \sim hv_F/(\Omega E\lambda^2) \sim 0.02 \qquad (3)$$

where λ is the wavelength and Ω the area per chain.

The modulus change remains quadratic in voltage until $-\Delta E/E \sim 1.5\text{x}10^{-3}$, above which

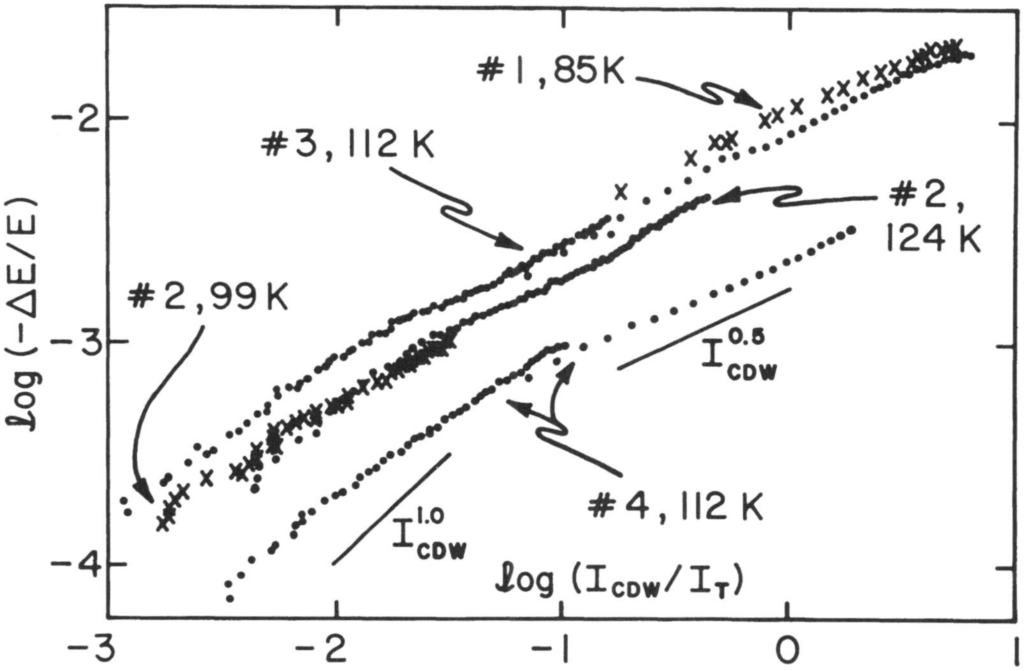

Figure 4 - Relative decrease in Young's modulus vs. I_{CDW}/I_T for several samples at different temperatures.

the slope decreases slightly. In Figure 4, we have plotted $-\Delta E/E$ vs. the CDW current, normalized to the normal current at threshold, $I_T = V_T/R_0$, in a logarithmic plot. It is seen that for $-\Delta E/E \gtrsim 10^{-3}$, $-\Delta E/E = A\,(I_{CDW}/I_T)^p$, with $p = 1/2$. For smaller voltages, the slopes increase toward p=1, although this identification remains tenuous due to scatter in the data and the critical dependence on the choice of R_0. However, it is clear that the modulus varies as $\sqrt{I_{CDW}}$ for at least an order of magnitude in ΔE, corresponding (for some samples) to a range in CDW current of from 1% of I_T to more than 6 I_T. Very surprising is the weak dependence of A on sample and temperature; we have $A = (7\pm3)\times10^{-3}$. (These samples have internal friction anomalies varying by more than an order of magnitude!) This suggests that the square root dependence is intrinsic to the depinned CDW state in o-TaS_3, and might provide a convenient test of models of CDW motion. Further work is needed to see how this behavior changes (saturates) at larger currents and whether this behavior holds for other materials, as well as to determine whether the modulus variation with voltage can more clearly be separated into a dynamic relaxation component (varying as $(V-V_T)^2$ as $V\to V_T$) and a static component dominating at larger voltages. Work on the latter question will require more complete measurements of the internal friction as well as Young's modulus.

"Acknowledgements"--Some crystals were kindly provided by Dr. G. Grüner of UCLA. This work was supported in part by the U.S. National Science Foundation under Grant No. DMR-8318189.

References:

1) J.W. Brill and W. Roark, Phys. Rev. Lett., (20 August, 1984).
2) George Mozurkewich, P.M. Chaikin, W.G. Clark, and G. Gruner, to be published in Solid State Commun.
3) J.W. Brill, Solid State Commun. 41, 925 (1982).
4) A.W. Higgs and J.C. Gill, Solid State Commun. 47, 737 (1983).
5) R.S. Lear, M.J. Skove, E.P. Stillwell, and J.W. Brill, Phys. Rev. B29, 5656 (1984).

LOW FREQUENCY ELASTIC PROPERTIES OF MATERIALS CONTAINING A SLIDING CDW

George Mozurkewich,* P. M. Chaikin,**, W. G. Clark, and G. Grüner
Physics Department, U.C.L.A., Los Angeles, CA 90024 USA

The Young's moduli of TaS_3, $(TaSe_4)_2I$, and $NbSe_3$ are found to soften when the samples are biased in excess of the threshold voltage for nonlinear conduction. The sliding state also shows increased internal friction. The softening may be explained by coupling between the phonon and phason modes.

Sliding charge density waves (CDWs) have been extensively studied by electrical techniques, such as nonlinear conductivity for voltages $V > V_T$, current oscillations, ac-dc interference effects, etc. Nonelectrical properties including heat capacity, elastic moduli, and magnetic susceptibility have also been measured, but only in the zero bias limit in which the CDW does not move. In particular, Brill and Org[1,2] showed that the Young's modulus E undergoes a dip near T_p, which in TaS_3 is quite large ($\approx$ 1%). Because the CDW is a coupled electron-phonon system, one might expect that these nonelectrical properties will be modified in the sliding state. We have measured the elastic properties of three CDW materials for $V > V_T$, finding that E is significantly smaller and the internal friction greater than in the pinned state.[3,4]

Because of the fibrous samples, we used the vibrating reed technique.[2,5] The samples were mounted at both ends to allow passage of current, and they were surrounded by helium gas to minimize Joule heating. Details are given elsewhere.[5,3] A feedback loop locked the driving frequency to the center f of the mechanical resonance response. Frequency shifts correspond to charges of E ($\Delta E/E = 2\, \Delta f/f$ for small shifts), and increased internal friction reduces the magnitude A' of the forced mechanical vibrations.

Figure 1 shows f and A' vs applied voltage V for $(TaSe_4)_2I$. The arrows locate

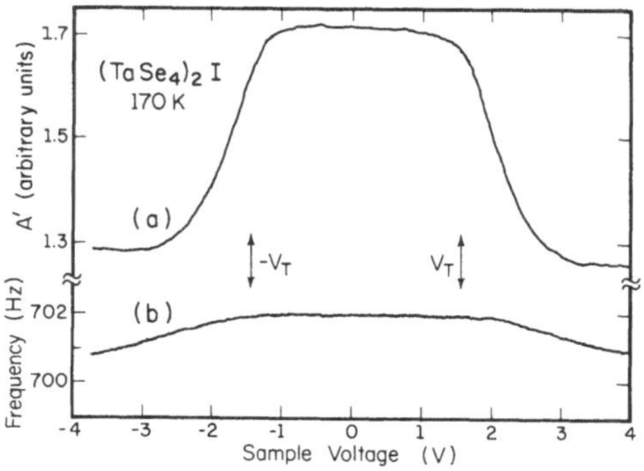

Fig. 1: Dependence of (a) mechanical resonance amplitude and (b) frequency on voltage bias in $(TaSe_4)_2I$. The vertical arrows show $\pm V_T$ determined from I-V curves.

$V = \pm V_T$. The frequency (and therefore E) is independent of V up to V_T, then E softens as V increases further. While in this figure E continues to decrease at the largest V applied, in some samples E clearly attains a constant E_∞ for V greater than a few times V_T. Figure 1 also indicates a dramatic increase of internal friction in the sliding state, as A' decreases suddenly above V_T. The difference $\Delta E = E_0 - E_\infty$ between the pinned value of modulus, E_0, and the limit E_∞ is zero at T_p = 261 K and was found to increase as T is reduced below T_p.

Figure 2 shows the bias dependence of E in TaS_3. A constant E_∞ for large V is clearly attained, and the growth of ΔE as the temperature is lowered below T_p = 222 K can also be seen. Figure 3 shows the threshold behavior of A' in a different TaS_3 sample at 77 K.

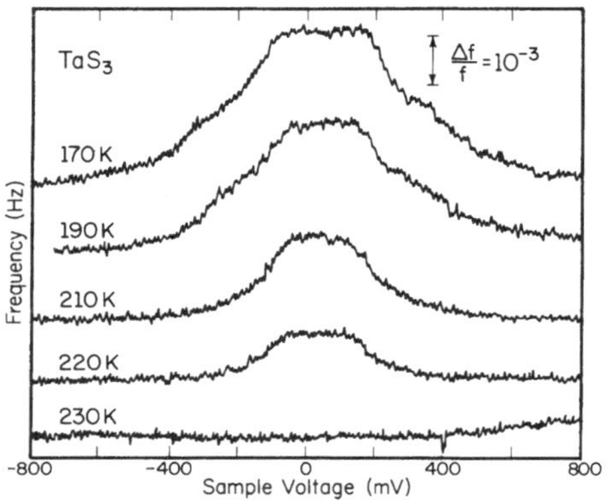

Fig. 2: Bias dependence of resonant frequency in TaS_3 at four temperatures below and one temperature above T_p = 222K. f≈350 Hz. The curves are offset vertically for clarity.

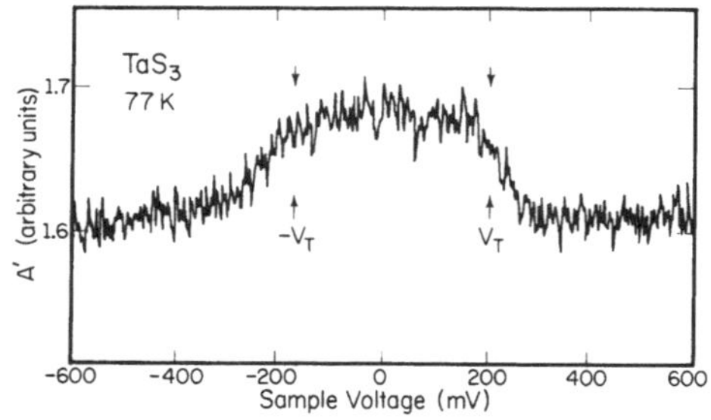

Fig. 3 Bias dependence of resonance amplitude in TaS_3 at 77 K. f ≈ 3000 Hz.

Finally, Fig. 4 shows the bias dependence of the modulus in $NbSe_3$. Below the upper transition (T_p = 142 K), $\Delta E/E$ is about an order of magnitude smaller than in either TaS_3 or $(TaSe_4)_2I$, while no change in E is discernible to a level of $<10^{-4}E_o$

below the lower transition (T_p = 59 K). It appears that $\Delta E/E$ is a strongly increasing function of T_p, although the role of the remnant normal electrons in $NbSe_3$ is not yet clear.

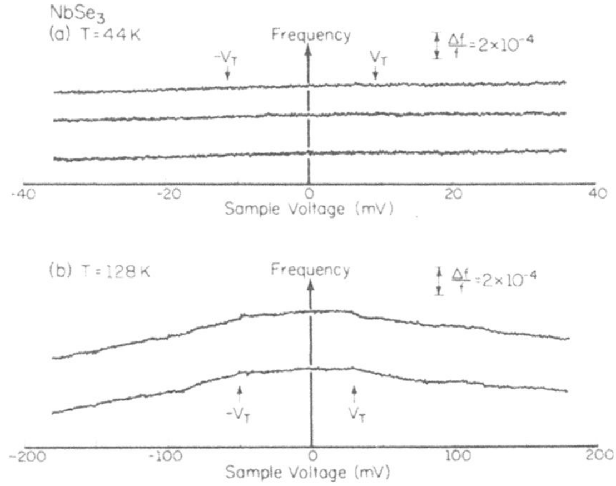

Fig. 4 Bias dependence of resonance frequency in $NbSe_3$ (a) below T_2 = 59 K, and (b) below T_1 = 142 K. The vertical arrows indicate the threshold for nonlinear conduction. f ≈ 1650 Hz. Successive voltage sweeps are offset for clarity.

In all three materials, the beginning of the softening is correlated with the threshold for nonlinear conduction, thereby proving that the effect is associated with motion of the CDW. It is clear that a lattice containing a moving CDW should be softer than one containing a pinned CDW because, by overcoming the pinning, the former situation has an additional degree of freedom. A detailed explanation of the effect has not yet been given. Here we present a mode coupling approach,[6] in which the stiffness change reflects mixing of phason character into the phonon modes.

The phason and phonon modes will be regarded initially as independent. Let u_p represent the displacement field of the positive lattice (with wavevector $q \approx 0$) and u_c the displacement field of the negative CDW condensate (with wavevector $2k_F + q$). Neglecting pinning, the potential energy will contain one elastic term for each mode, plus a Coulomb interaction:

$$W = \frac{1}{2} E \left(\frac{\partial u_p}{\partial x}\right)^2 + \frac{1}{2} K \left(\frac{\partial u_c}{\partial x}\right)^2 + \frac{2\pi e^2}{q^2 + k_{TF}^2} \left[\bar{p} \frac{\partial u_p}{\partial x} - \bar{c} \frac{\partial u_c}{\partial x}\right]^2 \qquad (1)$$

Here E and K are the lattice and CDW stiffness respectively, and $\bar{p}$ and $\bar{c}$ are the ion and condensed electron number densities. The uncondensed electrons are represented through their screening wavevector k_{TF} in the Coulomb term. If M_+ and m_F are the ion and Fröhlich masses, the Langrangian density is

$$L = \frac{1}{2} M_+ \bar{p} \left(\frac{\partial u_p}{\partial t}\right)^2 + \frac{1}{2} m_F \bar{c} \left(\frac{\partial u_c}{\partial t}\right)^2 - W. \qquad (2)$$

By applying the Lagrangian equations of motion to u_p and u_c and diagonalizing, the normal mode frequencies are found to be

$$\omega_+^2 = q^2 \left[\frac{K + (4\pi e^2/k_{TF}^2)\overline{c}^2}{m_F\overline{c}}\right] = q^2 v_\phi^2 \tag{3a}$$

$$\omega_-^2 = q^2 \left[\frac{E + (4\pi e^2/k_{TF}^2)\overline{p}^2}{M_+\overline{p}}\right] = q^2 v_S^2 \tag{3b}$$

provided the phason velocity v_ϕ is much greater than the phonon velocity v_S.

In the pinned case, translational symmetry is broken by a restoring force of the form

$$W \to W + \frac{1}{2} P \, (u_p - u_c)^2 \tag{4}$$

for small $(u_p - u_c)$. This term mixes the modes so as to introduce the famous pinning gap into the phason branch. Provided P is the dominant term in W, the gap is found at $\omega_+^2 \approx P/m_F\overline{c}$, and the phonon branch shifts to

$$\omega_-^2 = q^2 \, \frac{E + K + (4\pi e^2/k_{TF}^2)\,(\overline{p} - \overline{c})^2}{M_+\overline{p} + m_F\overline{c}} \quad . \tag{5}$$

Now the pinning term in (4) is really an expansion of a periodic potential, so we argue that it averages to zero for a rapidly sliding CDW, and the result (3) is regained. Comparing (3b) to (5), the frequency shift becomes

$$\frac{\Delta f_-}{f_-} \approx \frac{1}{2} \, \frac{K + (4\pi e^2/k_{TF}^2)\,(\overline{c}^2 - 2\overline{pc})}{E} \tag{6}$$

Numerically, $E \approx 10^{12}$ dyn/cm^2 (Ref. 2), $K \approx 10^9$ dyn/cm^2 (Ref. 3), $k_{TF} \approx 2\pi$ (1 to 10 Å)$^{-1}$, and $(4\pi e^2/k_{TF}{}^2)\overline{p}^2 \approx 10^8$ dyn/cm^2. Therefore $\Delta f/f \approx 10^{-3}$, which is of the right order of magnitude.

We are grateful to H.-B. Schüttler for helpful discussions. This work was supported by NSF grants DMR84-06896 (G.M. and G.G.), DMR82-05810 (P.M.C.), and DMR81-03085 (W.G.C.).

* Current address: Physics Department, University of Illinois at Urbana-Champaign 1110 W. Green St., Urbana, IL 61801 USA

** Current address: Physics Department, University of Pennsylvania, Philadelphia, PA 19104 USA

1. J. W. Brill and N. P. Org, Solid State Comm. 25, 1075 (1978).
2. J. W. Brill, Solid State Comm. 41, 925 (1982).
3. G. Mozurkewich, P. M. Chaikin, W. G. Clark, and G. Grüner, submitted to Solid State Comm.
4. Similar data have been obtained in TaS_3 by J. W. Brill and W. Roark (preprint).
5. T. Tiedje, R. R. Haering, and W. N. Hardy, J. Acoust. Soc. Am. 65, 1171 (1979).
6. M. B. Walker, Can. J. Phys. 56, 127 (1978).

THE CONDUCTIVITY OF ORTHORHOMBIC TaS_3 UNDER UNIAXIAL STRAIN

Preobrazhensky V.B., Taldenkov A.N., Kalnova I.Ju.
Kurchatov Institute of Atomic Energy, 123182, Moskow, USSR

The low field ("linear") conductivity in a Peierls state of orthorhombic TaS_3 was found to be strongly dependent on uniaxial strain S along the chain direction, increasing linearly against S at low deformations with a subsequent abrupt fall at large S. The origin of such a behavior is discussed.

There are two highly unexplored aspects in physics of 1-D conductivity which demand a systematic study of electronic structure and transport properties of linear chain compounds while the lattice parameters are changed. The variance of intrachain distances appears to be of importance due to commensurate-uncommensurate transition, while the changes of interchain spacings-due to its influence on transport properties (we would like to emphasize, that in pure 1-D system the conductivity is impossible).

The availability of TaS_3 in a form of very perfect wiskers offers a unique opportunity to study the properties of this compound under extremely high elastic deformations (in excess of 1,5 %) unavailable for massive single crystals.

It was found recently [1] that the uniaxial stretch deformation along the chain direction has a strong influence on the small field conductivity of orthorhombic TaS_3. A further study of the effect of uniaxial strain on $\sigma_{||}$ of TaS_3 is presented.

The samples (typically 3x0,01x0,003 mm^3) were prepared from elements by heating stoichiometric ratios of Ta and S. The strain was applied by special device known as a strain transformer [2].

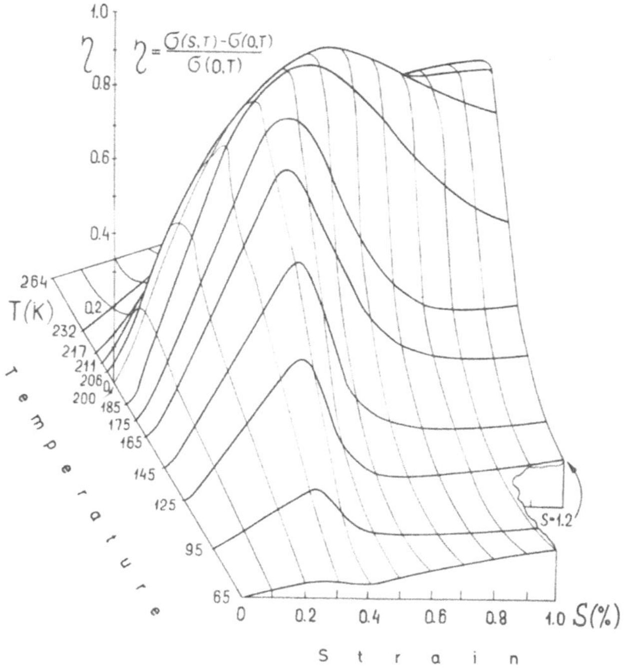

Fig. 1. The strain dependence of normalized excess conductivity

The samples were mounted on a stretcher by silver paint or clamped between a sapphire substrate and an annealed gold wire, served as electrode. Special experiments show, that if gold electrodes used, the contact resistance is extremely low, allowing a two probe configuration to be used. The usual ac syncronous detection technique was used with $E < 100$ m V/cm.

The following characteristic features are observed in all samples studied. At low strain σ increases linearly against S (Fig.1), the tensoconductivity $\beta = \frac{1}{\sigma(0,T)}\frac{\partial\sigma}{\partial S}$ being strongly temperature dependent (Fig.2). At large S an abrupt fall of σ is found. The temperature dependence of the strain S, at which a steepest drop of is observed shown in insert Fig.3. All σ changes are reversible at $S < 1{,}5$ %.

First we discuss whether the increase of conductivity is determined by the rise of carrier concentration, or by change of their mobility. Assuming, that the decrease of $\mathcal{E}_F$ caused by the increase of intrachain spacing is a dominant factor, one may find from $\exp(-1/\lambda)$ (1) (where $\lambda = g^2/\mathcal{E}_F\,\omega_Q$ - electron-phonon coupling constant), that $\frac{1}{\Delta}\frac{d\Delta}{d\mathcal{E}_F} = -\frac{1}{\mathcal{E}_F}\left(\frac{1-\lambda}{\lambda}\right)$ is always negative. Hence an increase of Δ and a decrease of both the carrier concentration and σ should be expected. This is in an obvious contradiction to the experiment: i) σ grows rapidly at low strain, ii) no rise of T_p is observed; instead a substantial broadening of the Pierls transition especially pronounced at low temperature side is found, which simulates

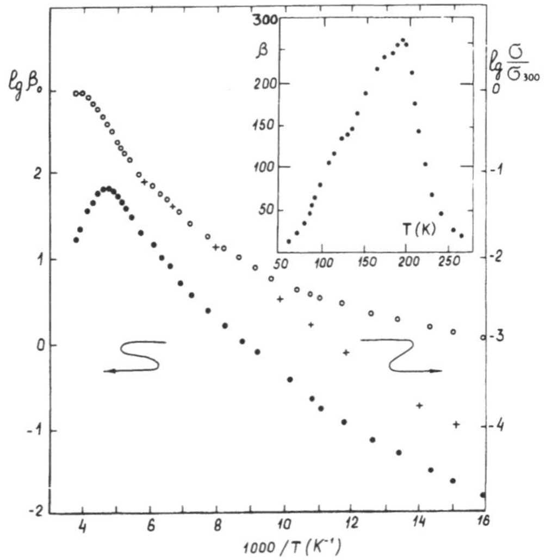

Fig.2. The strain derivative and conductivity VS 1/T dependence.

• - $\beta_0 = \frac{1}{\sigma_{300}}\frac{\partial\sigma(S,T)}{\partial S}$;

o - $\sigma^{\parallel}/\sigma^{\parallel}_{300}$;

+ - $\sigma^{\perp}/\sigma^{\perp}_{300}$ $\sigma^{\perp}$data from [5]

In the insert; the temperature dependence of tensoconductivity β .

a lowering of T_p, iii) The activation energy is almost unchanged (Note: to account for an observed enlargement of σ to approximately a double value a 15% decrease of Δ is needed). Thus the change of Δ fails to explain the conductivity rise at least at small S, and the observed increase of is attributed mainly to the variance of carrier mobility, while the Pierls gap remains almost unaffected.

The change of normalized conductivity against S may be expressed as:

$\eta(S)=\alpha_1 S_{\parallel} + \alpha_2 S_{\perp} + \alpha_3 S_{\parallel}^2 + \alpha_4 S_{\perp}^2 + \ldots$ (2) where $S_{\perp}$ is related to $S_{\parallel}$ by Poisson relation. Ido et at [3] found that at hydrostatic pressure of 13 kbar the conductivity of TaS_3 also rises to approximately a double value. From the linear dependence of σ vs S in our experiments and a positive sign of the pressure derivative of in [3] one may conclude that the second term in (2) associated with transversal contraction of the sample is responsible for conductivity changes.

The role of transversal contraction is confirmed by the fact that the stress derivative $\beta_0 = \frac{1}{\sigma_{300}} \frac{\partial \sigma}{\partial S}$ follows an activation law with $E_a \approx 800$ K troughout the whole temperature range just as the transversal conductivity do (Fig.2), while $E_{\sigma}^{\parallel}$ drops to 1/3 of its original value at T < 100 K.

Let now discuss some mechanisms which can lead to a strong dependence of mobility on transversal strain. If the longitudional mobility is controled by defects in conducting chain, the decrease of interchain spacing will favour the transfer of electrons from one chain to another "bypassing" the defect region.

Another, - and more attractive, - explanation of this effect is

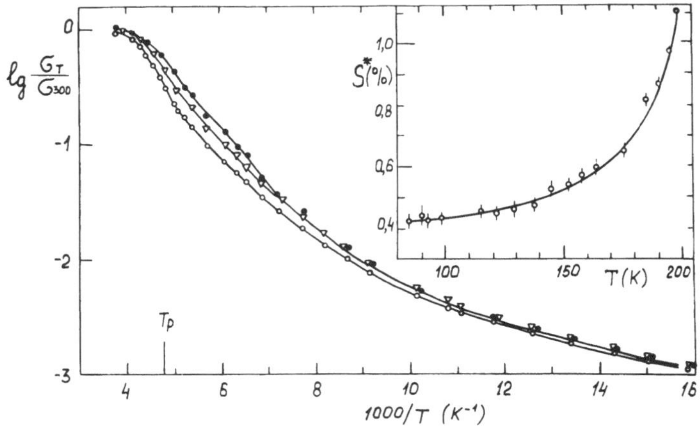

Fig.3. The normalized conductivity vs 1/T dependence at fixed strain
o - S = 0
▽ - S = 0.4 %
● - S = 1.2 %
In the insert: the temperature dependence of S*
Solid line corresponds to
$S^*(T)=\frac{A}{(T_p-T)^{1.04}}$

based on the assumption that in unstretched TaS_3 the carrier mobility is low because the one particle excitations are unstable against soliton or polaron formation [4] , while the enbanced interchain interaction in a strained sample suppresses this unstability.

The strain has strong influence on high temperature conductivity, which is determined by one particle excitations, but it fails at low T, where σ is attributed to collective excitations. This observation is in accordance with both our proposals.

An abrupt fall of σ at S=S* is most likely a result of some kind of stress enhanced electronic transition (parhaps another CDW transition).

1 Preobrazhensky V.B., Taldenkov A.N., Kalnova I.Yu. - Proc. of conference "Inhomogeneous electronic states", Novosibirsk, March 1984, p.126; Preobrazhensky V.B., Taldenkov A. N., Kalnova I.Yu. - Pis'ma JETP. 40, N 5, 182 (1984).

2 Gaidukov Ju.P., Danilova N.P., Tscherbina-Samoilova M.B. - Pribori i technicka experim. n 1, 250 (1979); for a short description of strain device see a companion paper. This conference.

3 Ido M., Tsutsumi K., Sambongi T., Mori N. - Sol.State Commun. 29, 399 (1979).

4 Brasovskii S., Kirova N., Yakovenko V. - Journ. de Physique 44, Suppl C-3- 1525 (1983)

5 Takoshima T., Ido M., Tsutsumi K., Sambongi T. - Sol.State Commun. 35, 911 (1980).

OHMIC AND NONLINEAR TRANSPORT OF $(TaSe_4)_2I$ UNDER PRESSURE

L. Forró*, H. Mutka**, S. Bouffard, J. Morillo
S.E.S.I., Cenfar, B.P. N°6, 92260 Fontenay Aux Roses, France

A. Jánossy
Central Research Institute for Physics, H-1525 Budapest, P.O.B.49, Hungary

We present the effect of hydrostatic pressure on the resistivity of $(TaSe_4)_2I$ in both ohmic and nonlinear regimes. The phase transition temperature of 262 K at ambiant pressure initially increases at the rute of 1 K/kbar, has a maximum and above 12 kbars it starts to decrease. The semiconducting energy gap in the Peierls state decreases by 50 K/kbar, i.e. 1.7%/kbar. The effect of pressure on the threshold field is also drastic. At 190 K it decreases from 2.0 V/cm at ambiant pressure to 0.8 V/cm at 12 kbars, but it continous to show the exponential temperature dependence. For the explanation of these data we propose a model whereby the temperature dependence of the threshold field depends on the normal carrier density.

$(TaSe_4)_2I$ is a quasionedimensional metal which undergoes Peierls transition[1] at T_p = 262 K. In the low temperature semiconducting phase the charge density waves (CDW) are pinned by defects. At low electric field the current is carried by the normal electrons excited over the semiconducting energy gap(Δ). Above a threshold electric field (E_T) the CDW is depinned, the current carried by the system becomes nonlinear with the applied voltage. From somewhat below T_p, decreasing the temperature E_T increases exponentially[1,2].

The pressure dependence of sliding CDW conductors has been little investigated. The published papers on $NbSe_3$ and orthorombic TaS_3 deal with the pressure dependence of the conductivity in the ohmic regime. Applying pressure to $NbSe_3$ the resistivity anomalies at 144 K and 59 K shift down in temperature and disappear rapidly. In TaS_3 T_p and Δ decrease at a rate of 1.3 K/kbar and 4 K/kbar respectively.

In this paper we report the pressure dependence of electrical transport in $(TaSe_4)_2I$ in both the ohmic and nonlinear regimes.

EXPERIMENTAL

The crystals were grown at the University of California Los Angeles. On samples of typical dimensions 3.0x0.05x0.05 mm^3 gold contacts were evaporated, and four electrodes were attached by silver paint. Up to 12 kbars a copper-beryllium chamber was used coupled to a gas compressor. The pressure transmitting medium was helium gas. This setup enabled us to make measurements either at constant pressure varying the temperature or at constant temperature varying the pressure. At 21 kbars kerozene was the pressure transmitting medium with a sealed piston rod. In both cases the pressure was monitored inside the chamber with InSb resistance sensor.

RESULTS

On figure 1 the resistance normalized to R.T. is plotted versus inverse temperature at several pressures. The insert shows the phase transition region on a larger scale. The phase transition temperature T_p was determined by the peak in the logarithmic derivative of the resistance. It is clearly seen on the insert of Fig. 1 that T_p first increases with pressure but at 21 kbars it is decreasing, values are plotted on Fig. 2a.

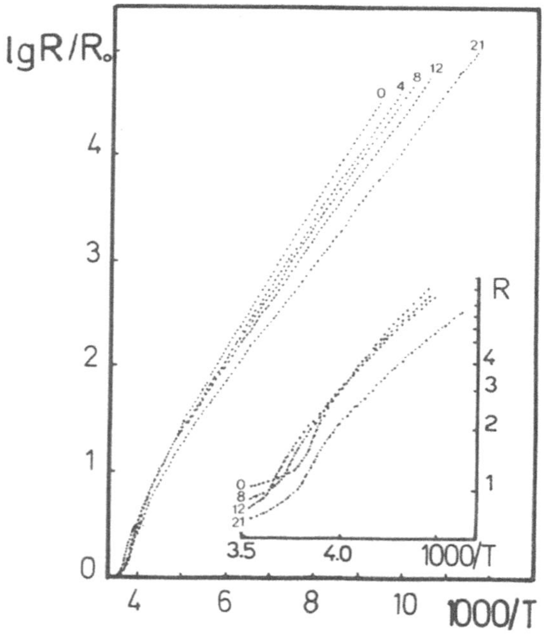

Fig. 1.

Resistance normalized to room temperature versus 1/T. The insert shows the phase transition regime on a larger scale. The numbers on the curves denote pressures in kbar.

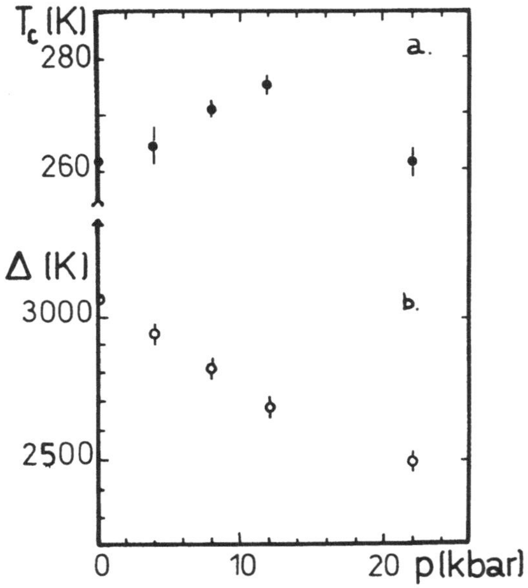

Fig. 2.

a) Peierls transition temperature versus pressure; b) Semiconducting energy gap versus pressure.

The resistance follows an activated behaviour $R = R_o \exp(\Delta/2kT)$; the energy gap for the normal excitations Δ was deduced from Fig. 1 in the 90-140 K temperature range. Δ decreases with pressure smoothly in the range of 0 to 21 kbars with the rate of 50 K/kbar (figure 2b).

The threshold field E_T for depinning CDW conductivity is somewhat smeared and we adapted arbitrarily the definition of linearly extrapolating from dV(I)/dI curves. This method overestimates E_T somewhat. As a function of pressure at fixed temperature E_T decreases. The decrease with pressure of the threshold normalized to ambiant pressure is somewhat faster than that of the normalized resistance (figure 3). At ambiant pressure the tem-

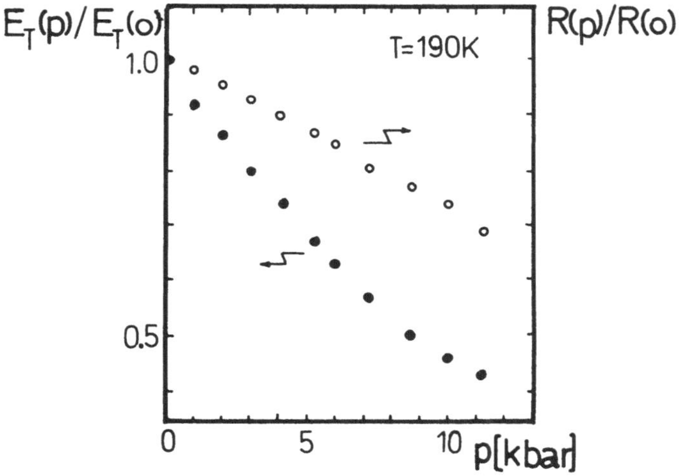

Fig. 3.

Threshold field and resitance normalized to ambiant pressure versus pressure at 190 K.

perature dependence of E_T agrees with that of reference 1 and 2. At 12 and 21 kbars E_T rises as the temperature is lowered less rapidly (figure 4).

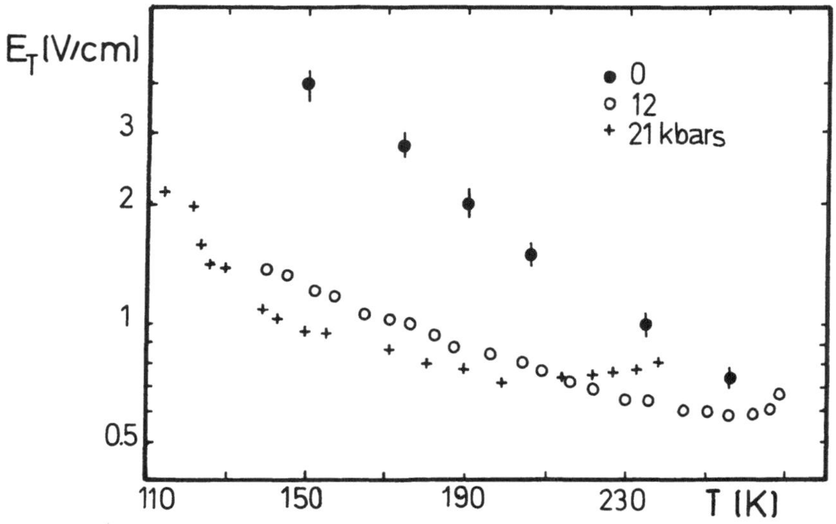

Fig. 4.

Temperature variation of the threshold field E_T for normal pressure, 12 kbars and 21 kbars.

DISCUSSION

The most remarkable effect in the ohmic region is the pressure dependence of the Peierls transition temperature. It has been already suggested[2] that T_p is probably greatly reduced from T_{MF} (mean-field transition temperature) in $(TaSe_4)_2I$ by large 1d fluctuations. Initially as interachain coupling increases by pressure 1d fluctuations are suppressed and T_p increases as approaches T_{MF}. Increasing the pressure further an other effect appears reversing the pressure dependence of T_p. Namely, as the system becomes more and more three dimensional (3d) the instability occurs at lower temperatures since in the limit of a 3d material no Peierls transition occurs. We belive this is the reason for the decrease in T_p observed at 21 kbar.

Below T_p, in the ordered phase the mean-field theory gives $\Delta \sim \exp(-1/\lambda)$, where λ is the electron-phonon coupling constant. Stiffening the lattice by pressure λ increases, Δ decreases as seen on Fig. 2b.

The pressure dependence of E_T and Δ are correlated (Fig. 2b and Fig. 3). This can give an idea for the pressure and temperature variation of E_T. Charged defects pin the CDW electrostatically. If the normal electron density is high enough, the defects are screened in a short distance and pinning is less effective. By lowering the temperature screening length increases, the CDW is more and more pinned and E_T increases. Pressure acts in the opposite way. Δ decreases with pressure so the normal electron density increases and E_T is reduced (Fig. 4).

We greatfully acknowledge the samples provided by M. Maki from UCLA, and useful discussions with J.R. Cooper and G. Grüner.

*Present address: Institute of Physics of the University, P.O.B.304, 41001 Zagreb, Jugoslavia

** Present address: Technical Research Centre of Finland, SF-02150 Espoo 15, Finland

REFERENCE

1. Z.Z. Wang, M.C. Saint-Lager, P. Moncean, M. Renard, P. Gressier, A. Meerschaut, L. Guemas, J. Rouxel, Solide State Commun. 46, 325 (1983)
2. M. Maki, M. Kaiser, A. Zettl, G. Grüner, Solid State Commun. 46, 497 (1983)
3. A. Briggs, P. Moncean, M. Nunez-Regueiro, J. Peyrard, M. Riboult, J. Richard, J. Phys. C13, 2117 (1980)
4. M. Ido, K. Tsutsumi, T. Sambongi, N. Mori, Solid State Commun. 29, 399 (1979)

HYSTERESIS AND METASTABILITY

PINNING, METASTABILITY AND SLIDING OF CHARGE-DENSITY-WAVES

P. B. Littlewood

AT&T Bell Laboratories
Murray Hill, New Jersey 07974

Incommensurate charge density waves pinned by random impurities will exhibit metastable behavior in the pinned phase for applied d.c. bias voltages below threshold. Numerical simulations of a one-dimensional system and analytic calculations within a mode-coupling approximation are presented.

The phenomenon of sliding charge-density-waves (CDW) is by now exceedingly well-documented, and an extraordinarily-rich catalogue of observations has been made.[1] It is convenient to separate the experimental observations into three broad classes, which may overlap somewhat in practice but are distinguishable in principle. I define the classes as: bulk, equilibrium properties in an infinite system; bulk, nonequilibrium properties; and finite-size or contact effects.

This is of course a theorist's distinction made for theorists' purposes, and presupposes a particular model for CDW current transport. The distinction between equilibrium and nonequilibrium properties is essentially between low-frequency measurements made at d.c. biases exceeding threshold and measurements made on the CDW at biases below threshold. In principle, the sliding CDW will reach an equilibrium configuration at long times (although the characteristic time for equilibration will be very long for applied fields close to the threshold field E_T); below threshold, the pinned CDW will always be in a metastable, nonequilibrium configuration, at least at zero temperature. As bulk equilibrium properties we have therefore the nonlinear (d.c.) I-V characteristics and the response to both an a.c. field and a d.c. field exceeding threshold (including the Shapiro steps, at both harmonics and subharmonics).[2] Nonequilibrium phenomena are much more widespread and include: hysteretic a.c. I-V characteristics; the response to trains of pulses crossing threshold (one of the so-called "memory" effects, whereby the response depends on the polarity of the most recent pulse);[3] "freezing" of the q-vector at low temperatures;[4] changes in the low-field conductivity in response to thermal pulses;[5] and the frequency-dependent dielectric response at bias fields below threshold.[6] The most famous phenomenon observed in sliding CDW materials is probably the narrow-band "noise" observed at the washboard frequency for a CDW sliding in a pure d.c. field,[7] which I have not included as a bulk property of an infinite system. A periodic response at the washboard frequency is in fact expected for a random pinning model of CDW's, in *equilibrium* but on a *finite* although perhaps very long) length scale.[8] A different model for the origin of the narrow-band noise involves consideration of the conversion of the collective CDW charge density into normal carriers near the contacts.[9] The contacts certainly make a very large perturbation on the CDW system and may also be the origin of some of the broad band noise, or intermittent behavior.

The bulk equilibrium properties of the sliding CDW at fields exceeding threshold are now largely understood,[8,10] although the details of the critical behavior beyond mean field theory close to threshold are not yet completely understood. I shall principally be concerned here with nonequilibrium properties of the metastable states below threshold.

The behavior of the CDW on long length scales (>> CDW period) and low frequencies (<< CDW optic phonon frequency) is well-described by the Fukuyama-Lee[11] phase pinning model of an elastic CDW of incommensurate wavevector Q, where the periodic part of the charge density is given by

$$\rho(r) = \rho_0 \cos(\mathbf{Q}\cdot\mathbf{r} + \phi(\mathbf{r})) \ . \tag{1}$$

At temperatures well below the CDW onset, fluctuations in the amplitude ρ_0 of the CDW are very unfavorable energetically, and the pinning of the CDW is produced principally by fluctuations in the phase $\phi(\mathbf{r})$, or local position of the CDW. The energy of the CDW interacting with short ranged impurity potentials $V_i(\mathbf{r}-\mathbf{R}_i)$ at position $\mathbf{R}_i$ and in the presence of an electric field $\mathbf{E}$ is (in d dimensions)

$$\mathcal{H} = \int d^d x \left\{ \frac{1}{2} K |\nabla\phi|^2 + \sum_i V_i(\mathbf{r}-\mathbf{R}_i)\, \rho(\mathbf{r}) - \rho_c E \phi(\mathbf{r}) \right\} \tag{2}$$

Here the electric field couples to the full collective charge density ρ_c of the CDW, and the spatial dimensions have been scaled so as to make the elastic constant K isotropic. The last term in Eq. (2) will cause the CDW to slide in a large enough electric field, leading to a CDW current $j(r) = \rho_c \dot{\phi}(\mathbf{r})$. In order to describe the dynamics we use a purely relaxational equation of motion $\lambda_0 \dot{\phi} = -\delta\mathcal{H}/\delta\phi$. For short-ranged impurity potentials $V_i \delta(\mathbf{r}-\mathbf{R}_i)$, we have then

$$\lambda_0 \dot{\phi} = K\nabla^2\phi + \rho_c E + \textstyle\sum_i V_i \delta(\mathbf{r}-\mathbf{R}_i)\, \rho_0 \sin(\mathbf{Q}\cdot\mathbf{R}_i + \phi(\mathbf{R}_i)) \ . \tag{3}$$

Because $\mathbf{Q}\cdot\mathbf{R}_i$ (modulo 2π) is a random number, the effect of the impurities will be to destroy long-range phase-coherence of the CDW on some length scale L. We briefly review the results of scaling arguments of Lee and Rice.[12] The strength of the pinning is characterized by the dimensionless parameter $\delta = \rho_0 V / K c^{(2-d)/2}$, where c is the impurity concentration and $V = \langle V_i^2 \rangle^{1/2}$. For $\delta >> 1$, the impurity pinning dominates over the elastic energy so that $\phi(\mathbf{R}_i) \simeq (2n+1)\pi - \mathbf{Q}\cdot\mathbf{R}_i$, and $L \sim c^{-1/d}$. For $\delta << 1$, the phase distortions take place over a much longer length scale in order to take advantage of fluctuations in the impurity pinning in a volume L^d containing a large number of impurities. Thus the characteristic pinning energy per unit volume is

$$E_{pin} \sim KL^{-2} - \rho_0 V (cL^d)^{1/2} L^{-d} \ . \tag{4}$$

In less than four dimensions, equation (4) is minimized for finite L, given by $Lc^{1/d} = \delta^{2/(d-4)} >> 1$. In the absence of an electric field, there is a trivial degeneracy of the pinned solutions under a translation $\phi(r) \rightarrow \phi(r) + 2n\pi$, thus the threshold field E_T is estimated to be of order $(Kc^{2/d}/\rho_c)\delta^{4/(4-d)}$.

Numerical studies of equation (3) in one dimension provide considerable insight into the properties of the system.[13,14] Static pinned solutions are most easily found by taking an initial guess for $\phi(r)$ and by following the time evolution of Eq. (3) toward a static solution. A typical example is shown in figure 1 for a chain of 200 δ-function impurities placed at random, in zero electric field with the parameters $\delta = 0.1$, $Qc^{-1} = 100$, and using free boundary conditions. Clearly the correlation length of the phase is longer than the average impurity spacing (here we set $c = 1$), as one expects for small δ. The pinned state is not unique, and also shown in fig. 1 is a different state obtained from a different initial configuration. The relationship between the two pinned states is made clear if we plot the difference in phase between them, also shown in fig. 1. We see that the two states are nearly identical

Fig. 1

Two metastable solutions (solid lines) of the CDW phase $\phi(x)$ for a 1D chain of 200 impurities placed at random, and their difference (dashed line).

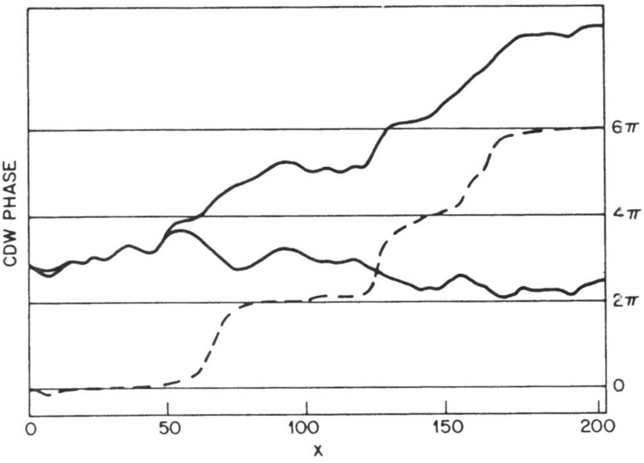

(modulo 2π) over long regions, separated by well-defined phase slips of 2π. The thickness of the phase slip region is characteristically of order the Lee-Rice length $L = \delta^{-1/2}$, which now sets the fundamental microscopic length scale for the problem. The similarity of these phase slips to discommensurations in a CDW system pinned by a periodic potential is no accident, and is a direct consequence of the periodic nature of the CDW. However, in the random case the "discommensurations" are defined only relative to a reference pinned state, and these phase slips are not free to move.

We also note that the two pinned states in figure 1 have different values of $\langle \nabla\phi \rangle$, averaged over the sample. Correspondingly the averaged *measured* value of the q-vector is $Q + \langle \nabla\phi \rangle$. In most CDW systems, the equilibrium q-vector is temperature-dependent; however changes in the q-vector can be accomplished only by continuous deformations of the phase, which will be hindered by the random pinning. Thus at low enough temperatures such that thermal fluctuations can be neglected, the q-vector may be expected to "freeze" at a nonequilibrium value; the maximum departure from equilibrium which the system can sustain will be $\delta q \sim 1/L$.[13]

The above remarks apply to a purely sinusoidal incommensurate CDW. Close to an incommensurate-commensurate transition, the CDW is far from sinusoidal, and is better described as commensurate regions pinned to the lattice separated by discommensurations (DC) of thickness ζ and spacing R with $R >> \zeta$. Then the q-vector is $Q_C + 1/Rp$, where Q_C is the commensurate value and p the order of commensurability. The transition is then described by a chemical potential for the DC which becomes zero at the C-I transition temperature, plus a repulsive interaction between neighboring DC proportional to $\exp(-R/\zeta)$. Once R/ζ becomes large (i.e. $(Q-Q_C)\zeta << 1$), the repulsive interactions between the DC will no longer be sufficient to overcome the pinning of individual DC to impurities or disorder (the pinning itself will be accomplished by fluctuations in the position of the DC in order to accommodate itself to the random potential) and the Q-vector will again "freeze" at a value close, but not equal, to Q_C.[15] This seems to be a rather commonly observed phenomenon.

Now we consider the effects of increasing the electric field E. We start from a low energy pinned state at zero field, and increase the electric field in small increments allowing the phase $\phi(x)$ to continuously deform to minimize its energy. The strength of the electric field is conveniently measured in terms of the scaled variable $\epsilon = \rho_c E/K$, and we apply periodic boundary conditions. In figure 2, we plot the equilibrium solutions at increasing field ϵ measured relative to the initial equilibrium state $\phi_0(x)$ at zero field. At low fields $\epsilon \lesssim 0.1$, the CDW remains within a single valley of the potential, but as the field is increased further, local pieces of the CDW "run over" into the next well. This process continues as the field is increased, generating fluctuations in the CDW phase on longer and longer length scales. Eventually, as ϵ approaches the threshold field ϵ_T (for this configuration $0.24 < \epsilon_T < 0.25$), these fluctuations grow to approach the system size and the CDW breaks free and begins to slide. The uppermost curve in figure 2 is in fact a snapshot of the moving solution at a field $\epsilon = 0.25$ just exceeding threshold. We see that the very long length scale persists; because of random fluctuations in the pinning potential, more strongly pinned regions of the CDW lag the average phase.

Fig. 2

Evolution of the CDW phase measured in units of 2π for increasing field ϵ, with $\delta = 0.1$. The curve for $\epsilon = 0.25$ is a snapshot of a moving solution, and the dashed line is the configuration taken up when the field is turned off.

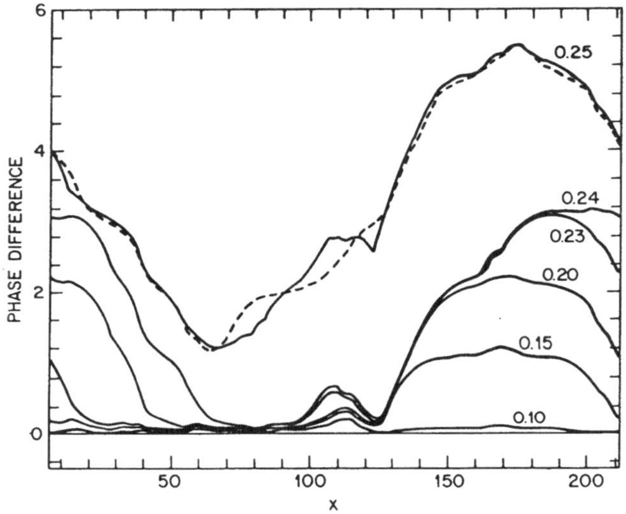

If the electric field is switched off suddenly, the CDW does not relax back to the initial configuration, but rather into a nearby metastable state with disorder built in on the long length scale characteristic of the sliding state at that particular field as shown in the dashed curve in fig. 2. Thus it is possible to move the CDW from one pinned metastable state to another by application of the field, and it is clear that the field need not exceed threshold to accomplish this. Such behavior provides a plausible interpretation of the hysteresis, and "memory" effects of the CDW in response to current pulses.[3] If the initial state at zero field is well annealed into a low energy metastable state with no long-range fluctuations in the phase, the response to a pulse just exceeding threshold will be slow as the necessary long range phase correlations must be built up. However, if the field is turned off, the long-range fluctuations will be frozen in, and the response to a second pulse in the *same* direction will be fast. if a pulse is applied in the *opposite* direction, the slow response reappears because the long range correlations have to be first "unwound" and then "rewound" in the opposite direction (the strongly pinned regions will always lag).

Figure 3 shows the results obtained from numerical simulation of the response to a double pulse, starting from a low energy initial state. The parameters are the same as before, except that we have now chosen $\delta = 10.0$. It is more convenient to work in the strong pinning limit in order to reduce the

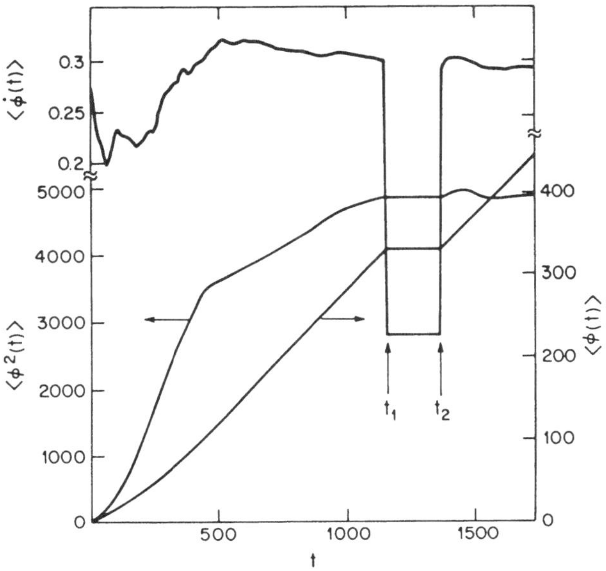

Fig. 3 The current response to a double pulse of field $\epsilon = 0.7 > \epsilon_T$, from a low energy pinned state, with parameters $\delta = 10, \lambda = 1$. The lower curves show the time evolution of the average phase and the fluctuations $<(\phi - <\phi>)^2>$. The current has been time-averaged over short times to remove the narrow-band noise.

sensitivity to finite size effects, although the qualitative behavior is the same. The lower part of figure 3 shows the time evolution of $\langle\phi(x,t)\rangle$, and $\langle(\phi - <\phi>)^2\rangle$ (averaging over position), which confirms the qualitative picture above.

We have also studied the linear response of individual metastable states to a small a.c. field $h(\omega)$, which defines the dielectric response function $\epsilon(\omega) = \langle d\phi(\omega)\rangle/dh(\omega)$ at finite d.c. bias. While the properties of different metastable states at the same bias field will not be identical, there are consistent changes in $\epsilon(\omega)$ observed when we follow the evolution of a single low energy state as the d.c. bias is increased (fig. 4). The response expected from a *uniformly* pinned state would be $\epsilon(\omega) = a/(\omega^2\lambda^2 + D)$. This is a reasonable description of the original low-energy state at zero bias, but at higher fields we see the development of a well defined cusp in Re $\epsilon(\omega)$ at low frequency. This cusp is a notable feature of experiments on pinned CDW's,[6] and arises in this model because of the presence of low energy localized eigenmodes, corresponding to the instability of local pieces of the CDW rolling over into a neighboring valley as we see in fig. 2. We caution that we do not have sufficient numerical data to fit the frequency-dependence at low ω, owing to the graininess of the eigenvalue distribution in our finite sample. Moreover, the results at a given field are not unique; also shown in figure 4 is the response of a high energy pinned state at zero field (in fact the state obtained by "annealing" at a field $\epsilon = 7.0$ at time t_1 in figure 3). The inset to figure 4 shows that $\epsilon(0)$ diverges as $\epsilon \rightarrow \epsilon_T$, so that the conductivity will become finite at threshold.

In order to perform analytic work, it is convenient to study a continuum version of Eq. (3), when we study the properties only on length scales much longer than the impurity separation. Changing length

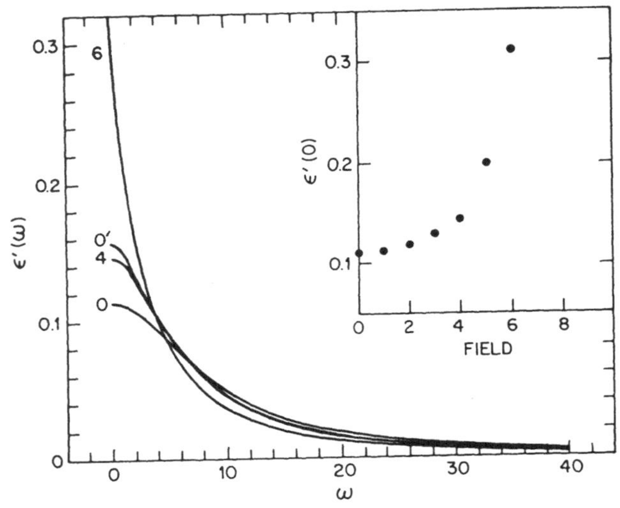

Fig. 4 Real part $\epsilon'(\omega)$ of the frequency dependent dielectric response function for different bias fields below threshold. The inset shows the field-dependence of $\epsilon'(0)$ for a *single* run as a function of increasing field.

scales by $x \rightarrow xc^{1/d}$, the equation of motion can be written

$$\lambda\dot{\phi}(\mathbf{x},t) = \nabla^2\phi(\mathbf{x},t) + V(\mathbf{x}) \sin(\theta(\mathbf{x}) + \phi(\mathbf{x},t)) + \epsilon . \tag{5}$$

Here $\lambda = \lambda_0 K^{-1} c^{(d-2)/2}$, and $V(\mathbf{x})$ and $\theta(\mathbf{x})$ are random fields with $\langle V(\mathbf{x})^2\rangle^{1/2} = \delta$. We assume that there exists a static solution of Eq. (5) $\phi_\epsilon(\mathbf{x})$ at some field $\epsilon < \epsilon_T$, in which case the linear response $\psi(\mathbf{x},t)$ to a small additional applied field $h(\mathbf{x},t)$ is given by

$$\lambda\dot{\psi} = \nabla^2\psi - [r + f(\mathbf{x})]\psi + h(\mathbf{x},t) . \tag{6}$$

The quantities r and $f(\mathbf{x})$ are field dependent, in the sense that they depend on the solution $\phi_\epsilon(\mathbf{x})$

$$r + f(\mathbf{x}) = - V(\mathbf{x}) \cos(\theta(\mathbf{x}) + \phi_\epsilon(\mathbf{x})) . \tag{7}$$

We have separated the two terms so that $f(\mathbf{x})$ contains only fluctuations in the pinning potential, $\langle f(\mathbf{x})\rangle = 0$, and $r(>0)$ is the average pinning energy. We also make the assumption that $f(\mathbf{x})$ can be treated as a Gaussian random variable, characterized only by its variance $\langle f(\mathbf{x})f(\mathbf{x}^1)\rangle = \Delta\delta(\mathbf{x}-\mathbf{x}^1)$; and solve equation (6) within the self-consistent Born approximation (which is strictly valid only in the limit of a large number of dimensions, and therefore we neglect all q-dependence). The average response to a field at finite frequency is given by $G(\omega) = \langle\psi(\omega)\rangle/h(\omega)$

$$G(\omega)^{-1} = G_0(\omega)^{-1} - \Delta(G^{-1}(\omega) - 1)^{-1} \tag{8}$$

where $G_0^{-1} = i\omega\lambda + r$. The solution to equation (8) can be written in the form $G(\omega)^{-1} = \bar{r} + z(\omega)$, where

$$\bar{r} + 1 = \frac{1}{2}(r+1)\left\{1 + \sqrt{[1-4\Delta/(r+1)^2]}\right\} \tag{9}$$

and as $\omega \to 0$,

$$z(\omega) = i\omega\bar{\lambda} = i\omega\lambda\ /[1-\Delta/(\bar{r}+1)^2]\ . \tag{10}$$

The fluctuations in the pinning potential lead to a reduction in the effective mass $\bar{r}$, and an enhancement of the damping constant $\bar{\lambda}$. We note that there are real solutions to Eq. (9) only provided that $4\Delta/(r+1)^2 < 1$. As this quantity approaches unity, the solution of equation (9) becomes complex (but $\bar{r}$ remains always non-zero) while $\bar{\lambda}$ diverges. The origin of this instability (which does not correspond to threshold) can be easily seen if we calculate a second order response function $C(\omega) = \langle\psi(\omega)\psi(\omega)\rangle/h(\omega)^2$; within the same approximations as before we find for low frequencies

$$C(\omega) = -\ (\omega\lambda)^{-1}\text{Im}\ G(\omega) = (\bar{\lambda}/\lambda)\ \bar{r}^{-2}\ . \tag{11}$$

The divergence of $\bar{\lambda}$ is thus a signature that the response of the fluctuations to a small external low-frequency field is diverging, whereas the average linear response remains finite. Equation (10) is an example of a pseudo (i.e. zero-temperature) fluctuation-dissipation theorem whereby the fluctuations induced by an applied field are related to the damping.[16]

Small disorder Δ (measured relative to the initial pinned state $\phi_\epsilon(\mathbf{x})$, which is itself disordered) corresponds to the situation of a low energy pinned state well below threshold. By integrating the solutions of Eq. (6) and (7) to obtain the dependence of r and Δ on electric field (assuming an adiabatic evolution of $\phi_\epsilon(\mathbf{x})$), one finds that r decreases with increasing field, while Δ increases. This is intuitively obvious from the numerical studies we showed before. For each metastable state, there is a characteristic field $\epsilon_0 < \epsilon_T$ where $\bar{\lambda}$ diverges and the metastable state vanishes; this corresponds to the "rolling-over" of domains into a nearby well, i.e. a new metastable state, as we saw in figure 2. As ϵ approaches ϵ_0, we find $\bar{r} = \frac{1}{2}\ (r(\epsilon_0)\ -1)\ +\ 0((\epsilon_0-\epsilon)^{1/2})$ and $\bar{\lambda} \sim (\epsilon_0-\epsilon)^{-1/2}$. Precisely at ϵ_0, $\bar{\lambda}$ diverges but $z(\omega)$ is finite; $z(\omega) \sim (i\omega)^{1/2}$, leading to a cusp in Re $G(\omega) \equiv$ Re $\epsilon(\omega)$ with a square-root frequency dependence. The new metastable state so obtained will have similar properties to the previous one as it only differs locally; the characteristic $|\omega|^{1/2}$ cusp will persist in this picture right up to threshold, when the final metastable state disappears and the CDW begins to slide.[17]

Finally, a few comments on thermal effects. Because the typical pinning energy of a domain (αL^{d-2}) is quite large compared to kT, (primarily because of the very long length scales involved in weak pinning), it might be imagined that thermal effects would play no role. However, we have seen that the electric field can force the CDW into very high energy pinned states, with very small barriers separating them from nearby states. The cusps observed in $\epsilon(\omega)$ at low frequency[6] arise from the presence of distributions of barrier heights extending to zero, and thermal activation over low barriers will be likely. Thermal activation over barriers will relax the CDW toward lowest energy pinned states. Thus as the CDW relaxes, activation over higher and higher barriers has to take place; quite generally this will lead to relaxations proceeding as a power of log (time).[3,5]

Acknowledgements

Some of this work was performed in collaboration with T. M. Rice and C. M. Varma. The author has profited greatly from conversations with S. N. Coppersmith and D. S. Fisher on points of theory and also with R. J. Cava, R. M. Fleming, G. Grüner, N. P. Ong and L. F. Schneemeyer on the experimental situation.

REFERENCES

1. For reviews see G. Grüner, Physica 8D, 1 (1983) and N. P. Ong Can. J. Phys. *60,* 757 (1982).
2. S. E. Brown, G. Mozurkewich and G. Grüner, Phys. Rev. Lett. *52,* 2277 (1984); M. Sherwin and A. Zettl, to be published.
3. J. C. Gill, Solid State Commun. *39,* 1203 (1981); R. M. Fleming and L. F. Schneemeyer, Phys. Rev. *B28,* 6996 (1983).
4. D. W. Ruesink, J. M. Perz and I. M. Templeton, Phys. Rev. Lett. *45,* 734 (1980); E. Fawcett, R. Griessen, and C. Vettier, in *Transition Metals 1977* ed. M. J. G. Lee, J. M. Perz and E. Fawcett, IOP Conf. Proc. 39, (Inst. of Physics, London, 1978) p. 592.
5. G. Mihály and L. Mihály, Phys. Rev. Lett. 52, 109 (1984)
6. G. Grüner, in *Proc. of Int. Symposium on Non-Linear Transport and Related Phenomena in Inorganic Quasi One-Dimensional Conductors,* Sapporo, Japan (1983), p. 77; R. J. Cava, R. M. Fleming, P. B. Littlewood, E. A. Rietman, L. F. Schneemeyer and R. G. Dunn, to be published; W. Wu, G. Mozurkewich and G. Grüner to be published.
7. R. M. Fleming and C. C. Grimes, Phys. Rev. Lett. *42,* 1423 (1979)
8. D. S. Fisher, to be published, and Phys. Rev. Lett. *50,* 1486 (1983).
9. N. P. Ong, G. Verma and K. Maki, Phys. Rev. Lett. *52,* 663 (1984).
10. L. Sneddon, M. C. Cross, and D. S. Fisher, Phys. Rev. Lett. *49,* 292 (1982); L. Sneddon, Phys. Rev. *B29,* 719 and 725 (1984).
11. H. Fukuyama and P. A. Lee, Phys. Rev. *B17,* 535 (1978).
12. P. A. Lee and T. M. Rice, Phys. Rev. *B19,* 3970 (1979).
13. P. B. Littlewood and T. M. Rice, Phys. Rev. Lett. *48,* 44 (1984).
14. H. Matsukawa and H. Takayama, Solid State Commun. *50,* 283 (1984); N. Teranishi and R. Kubo, J. Phys. Soc. Japan *47,* 720 (1979); J. B. Sokoloff, Phys. Rev. *B23,* 1992 (1981).
15. T. M. Rice, S. Whitehouse and P. B. Littlewood, Phys. Rev. *B24,* 2751 (1981).
16. This picture bears a strong resemblance to models of dynamics of spin-glasses, originally due to S.-K. Ma and J. Rudnick, Phys. Rev. Lett. *40,* 589 (1978) and considerably expanded by H. Sompolinsky and A. Zippelius, Phys. Rev. *B25,* 6860 (1982).
17. The mean-field solution of this model by D. S. Fisher (ref. 8) reads to a $|\omega|$ cusp below threshold (as long as the system exhibits hysteresis) and D. S. Fisher has given general arguments as to why the *same* power law should be seen in low dimensions. Experimentally, cusps in $\epsilon(\omega)$ are observed (ref. 6) but the exact ω-dependence is not yet clear.

DISTORTION, METASTABILITY AND BREAKING IN CHARGE-DENSITY WAVE TRANSPORT: RECENT EXPERIMENTS ON NIOBIUM TRISELENIDE, SUGGESTING A NEW MEAN-FIELD APPROACH

J. C. Gill

H. H. Wills Physics Laboratory, University of Bristol,

Tyndall Avenue, Bristol BS8 1TL, England.

Some quantities relevant to mean-field models of charge-density wave (CDW) motion have been measured in $NbSe_3$ between 144K and 60K. Data are presented on the threshold field E_T, including the contribution from breaking (phase-slip) at the current terminals; on the increase of Frohlich current with field E, confirming the predicted variation as $(E-E_T)^{3/2}$ near threshold; and on the conduction due to transitions between metastable distorted states, and thus on the elastic modulus of the CDW. It is tentatively concluded that, at least between 60K and 90K, the conduction near threshold is restricted mainly by the need to maintain phase-slip at the boundaries of regions, perhaps pinned at surfaces, which do not join in the general motion. A new mean-field model, mathematically equivalent to those suggested by Fisher and by Sneddon, but related also to the phenomenological model of Tua and Zawadowski, is proposed to describe this.

Electrical conduction through the motion of incommensurate charge-density waves, first suggested by Frohlich[1], has now been seen in $NbSe_3$ and several other materials[2]. The experiments show inertial effects to be negligible, so that the applied field $\underline{E}$ is opposed by the forces provided by motional damping, limiting the conductivity in high fields, and by the 'pinning' of the CDW to the imperfect crystal lattice, which prevents motion in a steady field until E exceeds a threshold value E_T. Although the motion of a CDW under the influence of these forces is not yet understood in detail, none of the experimental results seems beyond description in semi-classical terms. The possibility that the motion is a macroscopic manifestation of quantum-mechanical tunneling[3] now seems remote[4].

Of semi-classical models, the simplest liken the CDW to a particle moving in a periodic potential. They reproduce some features of the conduction, notably the periodic modulation of the current observed in response to a steady field[5], and the interference phenomena seen when steady and alternating fields are superposed[6], but are unrealistic in assuming the CDW to be in effect rigid, for if it were it could not, if incommensurate, be pinned at all in the infinite-volume limit.

Ample evidence that CDWs are not rigid is provided by electrical memory effects[7], by electron microscope studies[8], and recently by the direct observation, in X-ray diffraction, of current-induced changes in wavevector[9]. While their deformability enables CDWs to be pinned, it does not preclude their exhibiting interference phenomena in combined direct and alternating fields. The periodic response to steady fields, on the other hand, seems inexplicable as a bulk effect, and the evidence suggests that it arises locally, at contacts or other macroscopic obstacles to motion[10].

The motion of an elastically deformable CDW over pinning centres has been examined by Sneddon et al.[11], who show that when $E \gg E_T$ the pinning can be treated as a perturbation, and the conductivity then falls short of its limiting value by an amount proportional to $E^{(d-4)/2}$, where d is the dimensionality of the distortion around a pinning site. No analytic solution is available for $E \approx E_T$, but numerical studies by Matsukawa and Takayama[12] suggest that the current I_c carried by the CDW then rises linearly with $E-E_T$. Although this is less rapid than is predicted by rigid-CDW models (which give $I_c \propto (E-E_T)^{\frac{1}{2}}$, so that $I_c'(E) \to \infty$ as $E \to E_T$), an even slower variation, with $I_c''(E) > 0$ over a substantial range beyond E_T, is observed.

It has been shown by Fisher[13] that, given sufficient pinning, I_c is initially proportional to $(E-E_T)^{3/2}$ if, at each pinning site, the distortion of the CDW relaxes towards its mean value for all such randomly-distributed sites. A model due to Sneddon[14], of a chain having infinite-range interactions and pinned by a sinusoidal potential with which it is incommensurate, behaves similarly. Since the couplings within the CDW apparently need to be of infinite range, it is not obvious these mean-field models represent Frohlich conduction at all closely. An indication of how they might do so is provided, however, by Tua and Zawadowski[15], who arbitrarily divide the moving CDW into rigid 'segments' around each pinning site, and a rigid 'frame' to which they are elastically coupled. Their model gives $I_c''(E) > 0$ for E just above E_T, and resembles Fisher's in that the segments relax towards a mean phase set by the frame. In all these models, E_T and the form of $I_c(E)$ depend on the strength of the pinning relative to an elastic modulus of the CDW.

As the elastic behaviour of CDWs is apparent also in electrical memory effects, their study makes it possible to examine the various models experimentally, and perhaps to reveal details of the pinning. This paper summarises the preliminary findings of some experiments undertaken with that aim. Because of its probable freedom from discommensurations, and continuity over macroscopic distances[16], the CDW forming in $NbSe_3$ at 144K was chosen for study. At present measurements have been completed only in a narrow temperature range, but already suggest that the force exerted by the pinning on the slowly-moving CDW comes less from the impurities over which it passes, than from heavily-pinned regions at whose boundaries phase-slip must occur.

Measurements of continuous conduction

The conduction in $NbSe_3$ has been measured at temperatures T between 144K and 59K, where the second CDW appears, using pulses (unidirectional, length > 1 μs) to reduce heating. Contacts were of indium pressed to the crystal surface. A bridge detected departures from linearity, and a pulse-sampling system allowed the Frohlich component I_c of the applied current I to be recorded as a function of field. As there was no sign of the switching phenomena seen at lower temperatures[17], or of the slow changes in E_T observed in some crystals[18], it is assumed that the measurement was of the response to a steady field.

As already reported[16], the threshold field increases as the length ℓ through which the current I flows is reduced. This effect of the finite force needed to initiate phase-slip at or near the current terminals is apparent in figure 1, where the relation between I_c and E is shown for four different lengths of the same crystal. A 4-terminal arrangement was used, passing I between the inner pair and measuring the voltage V developed between the outer; I_c was taken to be I-V/R, with R = V/I for $E < E_T$. As R was roughly proportional to the distance between the inner edges of the relevant contacts, that was accepted as ℓ.

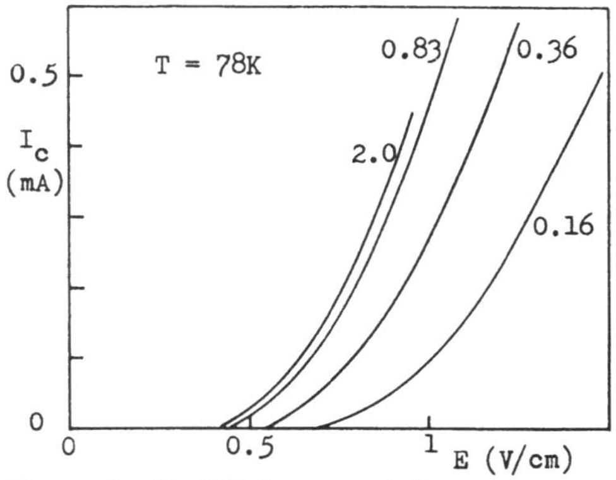

Figure 1. Frohlich current I_c, versus applied field E, in one $NbSe_3$ crystal. Driven lengths ℓ are indicated, in mm.

The threshold field is expressed adequately by

$$E_T(\ell) \quad = \quad E_P \quad + \quad V_s/\ell \tag{1}$$

where E_P comes from pinning distributed over the length ℓ , V_s corresponds to the force required to initiate phase-slip, and the dependence on ℓ^{-1} indicates that the slippage occurs at the current terminals. The slight divergence of the curves in figure 1, as E increases, shows that the force absorbed in maintaining phase-slip increases only slowly with I_c.

The mean-field prediction that I_c varies as $(E-E_T)^{3/2}$ near threshold is followed very closely when T < 100K and ℓ is large enough for V_s to contribute only a small part of E_T. In the example shown in figure 2, I_c is of the expected form until $E \approx 1.5\ E_T$, beyond which it rises more rapidly, evidently because motion then commences in some further part of the crystal. That the variation as $(E-E_T)^{3/2}$ is not a fortuitous result of superposing contributions, perhaps linear in E, from many independent domains has been demonstrated by using the interference phenomenon when direct and alternating fields are combined, to measure the frequency ν at which the CDW advances through wavelengths[20]. Records of dV/dI, obtained by the usual modulation method but with current of the form I + c sin ωt, show anomalies when ν coincides with $\omega/2\pi$ or multiples thereof. The values of ν measured by this means appearing in figure 2 remain proportional to $(E-E_T)^{3/2}$ beyond $E \approx 1.5\ E_T$, though the appearance of anomalies corresponding to unrelated frequencies shows that motion in other domains is then occurring.

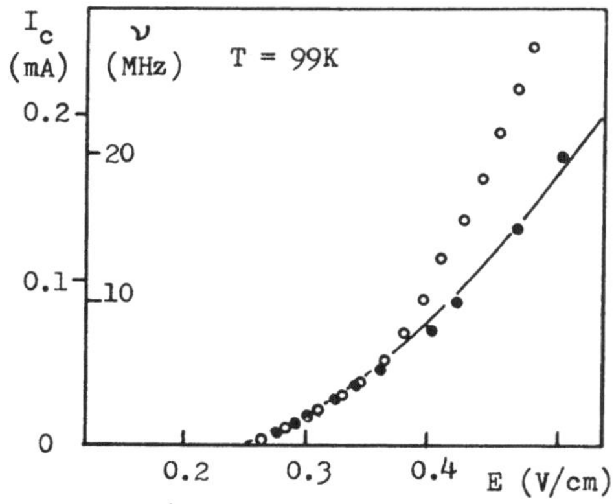

Figure 2. Measured values of I_c(○) and ν(●), compared with a variation as $(E-E_T)^{3/2}$, shown by the continuous line. Crystal as for figure 1: ℓ = 2 mm, cross-sectional area S = 4.2 x 10^{-7} cm^2.

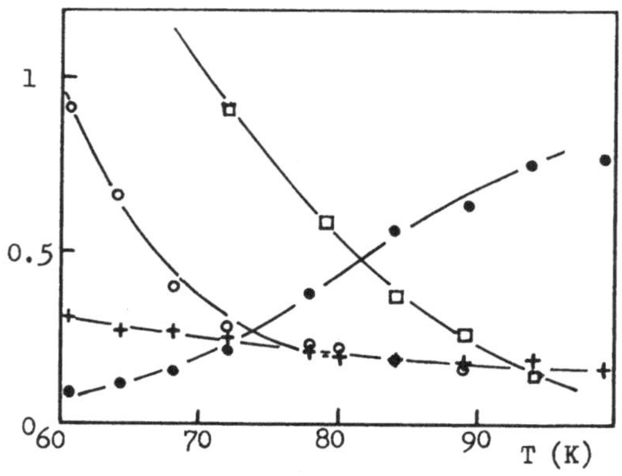

Figure 3. Dependence on temperature T of E_P (+), V_s (□), A (●) and α (○), measured in a crystal having ℓ = 0.39 mm, S = 5.3 x 10^{-7} cm^2. Unit on ordinate axis is for E_P, 1 V cm^{-1}; for V_s, 10 mV; for A, 2000 A $cm^{-2}/(V\ cm^{-1})^{3/2}$; and for α, 2000 J cm^{-3}. Curves for guidance only.

Figure 3 shows the dependence on T of E_P, V_s, and the quantity $A = (I_c/S)(E-E_T)^{-3/2}$ giving the conduction just above threshold. Value of A were derived from data for I_c with E usually less than 1.5 E_T, estimating the cross-sectional area S of the crystal from its ohmic resistance, taking the room-temperature resistivity as 3 x 10^{-4} Ω cm. Measurement of A became unreliable with T > 100K, as the effect of domains was evident with E close to E_T; the determination of A from measurements of ν has not yet been attempted. The quantity α also shown in figure 3 is defined later.

The approach to a limiting conductivity when $E \gg E_T$ has not yet been examined in detail. A few measurements at 78K show, however, that the limit is approached as E^{-n}, with n between 1 and 3/2, rather than 1/2 as would be the case for pinning by point defects (others[21] have found $n \approx 1$). This suggests that the pinning occurs on lines, or more probably on planes, rather than at points, especially as a spread of threshold fields between domains might be expected to reduce the observed value of n.

Memory phenomena and elastic properties

Of the several electrical memory effects known in CDW conductors, that of concern here is the so-called 'overshoot' or 'pulse-sign memory' phenomenon[8] illustrated in figure 4. A temporary increase in Frohlich conduction, evidently arising from the transition of the CDW from one metastable distorted state to another, follows application of a field in the opposite sense to that most recently present. The relevant distortions are not likely to be those which, in mean-field theory, decide the form of $I_c(E)$. The latter will be mainly on the smallest scale capable of giving metastable states,

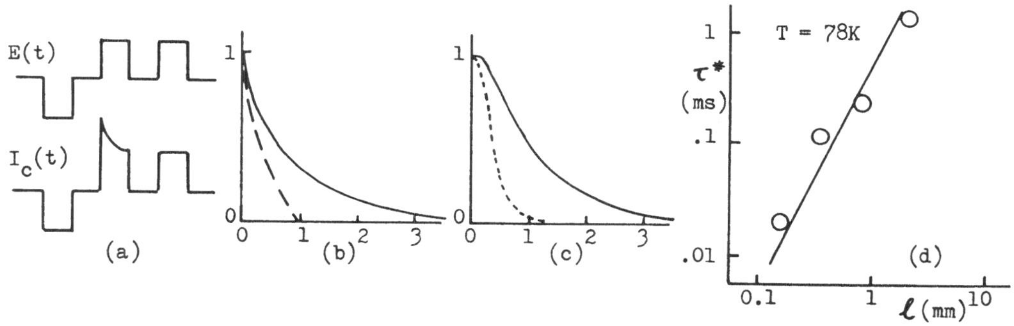

Figure 4. The 'overshoot' phenomenon. The response $I_c(t)$ to a pulsed field E(t) is shown in (a). The behaviour of the voltage transient v(t), according to the model outlined in the text, is shown for voltage terminals in (b), outermost; and in (c), close to the point midway between the current terminals. The curves show v(t)/v(0) as ordinate, versus t/τ as abscissa, for $(I-I_P)/(I_P-I_T) \leq 1$ (———), 1.5 (— — —), and ∞ (-------). The decay times τ^* observed in four lengths ℓ of one crystal are shown in (d); the line corresponds to $\tau^* \propto \ell^2$.

of the order of the characteristic length defined by Lee and Rice[22] and seldom vastly greater than the separation between pinning sites; it is difficult to see how, in the mean-field approximation, they can exhibit directional properties. The overshoot phenomenon, however, is expected to arise predominantly from distortions on the longest scale, namely the distance between places where phase-slip interrupts the continuity of the CDW, because the transfer of charge by the transient conduction corresponds to a mean displacement of the CDW which, for a given magnitude of strain, is proportional to the length over which the distortion occurs. For the CDW now under discussion, that length would appear to be the distance ℓ between current terminals.

That such macroscopic distortion of the CDW is indeed responsible for the transients seen below 120K is confirmed by the experiments next outlined. Their results are accounted for quite well by a linearised model whereby the CDW between the current terminals is assumed to distort elastically, its motion being subject to velocity damping and to a frictional force, corresponding to E_P, which stabilises the distortion after the driving field is removed. Deviations of the wavevector q from its equilibrium value q_0 are assumed to be small (changes ~0.1% are observed), and limited by the commencement of phase-slip. Motion of the CDW in the x-direction, driven by a field $\underline{E}$, is then described by

$$\gamma v = \rho_c(E - E_P) \quad - \quad \alpha\, q'/q_0 \qquad (2)$$

where v is the drift velocity, ρ_c the charge density transported, α the elastic (Young's) modulus, and γ a damping coefficient. The wavevector q satisfies the diffusion equation

$$\dot{q} \quad = \quad (\alpha/\gamma)\, q'' \qquad (3)$$

subject, at the current terminals, to the boundary conditions $v = 0$ unless

$|q-q_0| = \rho_c V_s q_0/2\alpha$, and $|q-q_0| \ngtr \rho_c V_s q_0/2\alpha$. Relation (1) between E_T and ℓ follows at once from these conditions and (2), as also does the prediction of a partial relaxation of the distortion, following removal of $E > E_T$, when $E_T > 2E_P$. An observation of the current due to such relaxation has been reported elsewhere[23].

It is usual to observe the overshoot phenomenon as a small transient v(t) in the voltage V which appears in response to a current I, applied at time t = 0 and steady thereafter. If I flows between terminals at $x = \pm\ell/2$, and V is developed between x_1 and x_2, where $-\ell/2 \leqslant x_1, x_2 \leqslant +\ell/2$, v(t) is proportional to the transient part of $q(x_1) - q(x_2)$. The relevant $q(x,t)$ are available from equation (3) and may be expressed

$$q(x,t) = q_0 + b_0(x/\ell) + \Sigma\, b_n \sin(n\pi x/\ell)\, e^{-n^2 t/\tau} \qquad (4)$$

where the characteristic time τ is equal to $\ell^2(\gamma/\pi^2\alpha)$, and the coefficients b_n (n is an integer $\geqslant 0$) depend on I and the initial conditions, and have to be adjusted when phase-slip commences at $x = \pm\ell/2$.

Figure 4 (b,c) shows the form of v(t) expected when I is applied after a steady state has been established by an equal current in the opposite direction[23]; I_P and I_T are the values corresponding to E_P and E_T. Because the CDW is distorted by the current terminals, v(t) depends on the configuration used to observe it. With voltage terminals outermost (and assuming $\ell >> V_s/E_P$, so that one may neglect motion beyond the current terminals), v(t) decays at first linearly with t, and if $I > I_T$ vanishes when phase-slip commences. In the more usual arrangement with voltage terminals innermost $\dot{v}(t)$ is initially zero, and v(t) always approaches zero asymptotically, though at a rate which is faster when phase-slip releases the CDW at the current terminals. In either case the charge Q transferred by the transient Frohlich current is maximum when $I = I_T$, being then $\rho_c^2 SV_s/6\alpha$ for a crystal with voltage terminals outermost, and cross-sectional area S.

To a substantial extent the predictions of this simple model accord qualitatively with experiment. The features shown in figure 4 are all observed, except that v(t) does not reach zero as sharply as figure 4(b) suggests, presumably because phase-slip does not limit $q(\pm\ell/2)$ as effectively as has been assumed. A tendency of τ to decrease as I rises is attributed to the reduction of γ by the nonlinear dependence of I_c on E. The clearest evidence that the scale of distortion is set by the distance ℓ comes from the duration of the transient. This is shown in figure 4(d); the quantity τ^* is the time taken for v(t), measured with voltage terminals outermost and $I \approx I_T$, to decay to 1/e of its initial value. That $\tau^*(\approx 0.8\tau$ in the model) varies approximately as ℓ^2 is taken to confirm that the distortion of the CDW extends over the entire distance between the current terminals.

With the origin of the transients established, the elastic modulus α becomes accessible to measurement. Its determination from the maximum value of Q, simple in principle, is difficult at low temperatures (when the large E_T leads to heating problems) and perhaps places undue reliance on the linearised model. The alternative adopted was to measure $dQ/dE = \rho_c^2 \ell^2 S/6\alpha$, where Q is transferred on reversal of a field $\underline{E}$ with $E_P < E < E_T$. Some values of α obtained by this means are included in figure 3. The specimen, which also provided the values of the other quantities shown, was short (ℓ= 0.38mm) in order to allow measurement over an adequate range of E; Q was found to vary linearly with $E - E_P$ until E_T was approached. The values of α in figure 3 were based on a charge density $\rho_c \equiv 1.9 \times 10^{21}$ electrons cm^{-3}, and a cross-section S for the whole crystal, with no allowance for domains not contributing to Q.

Discussion

In figure 3, the elastic modulus increases rapidly as the temperature T falls below 90K. While not unexpected, in that a simple theory of the Peierls transition predicts such behaviour, this is difficult to reconcile, in terms of present models of CDW motion, with the accompanying increase in the 'pinning' threshold field E_P. The mean-field models are based on the view of Lee and Rice[22], that E_P corresponds to the force needed to move the CDW over impurities to which, by deforming so as to mimimise total energy, it has become pinned. Unless compensated by increased coupling to the impurities, an increase in elastic modulus then ultimately reduces E_P.

The model proposed by Fisher[13] demonstrates this. At any pinning site j the phase ϕ_j of the CDW, relative to some undistorted stationary state, is assumed to satisfy

$$\Gamma \dot{\phi}_j = \eta(\bar{\phi} - \phi_j) - H \sin(\phi_j - \beta_j) + \rho E \qquad (5)$$

in which Γ specifies motional damping; η the elastic force tending to restore ϕ_j to the mean value $\bar{\phi}$ at all pinning sites; H the strength of the pinning; ρ the coupling to the applied field E; and β a randomly-distributed 'preferred' phase. Denoting H/η by h, and allowing the same values of η and H to apply at all j, the self-consistent solution of (5) gives

$$E_P = (H/\rho)\, f_1(h) \qquad (6)$$

for the threshold field due to pinning, and

$$v = B\,(E - E_P)^{3/2} \,, \qquad (7)$$

where $B = (2\pi/q_0)\,\rho^{3/2}\eta^{-1/2}\Gamma^{-1}.\,f_2(h)$, for the drift velocity just above threshold. The functions $f_1(h)$ and $f_2(h)$ are shown in figure 5; $f_1(h)$, and therefore E_P, vanishes for $h < 1$. The range of E/E_P over which v approximates to (7) becomes smaller as h increases: v is reduced to 90% of the value given above when E/E_P reaches about 1.1 when h = 50, 1.5 when h = 10, and 2 if h = 5.

Except when $E \gg E_p$ (when the model is not expected to apply), the observed form of $I_c(E)$ for $T < 100K$ corresponds to a value of h probably between 2 and 5. As η will be the ratio between some elastic modulus of the CDW and a length dependent on the spatial extent of the distortion around a pinning site, it is not unreasonable to expect it to increase, as does the modulus α, as T falls. To account for the slight rise in E_p a substantial increase in the pinning strength H is then needed. If one assumes η to be proportional to α, increasing by a factor 5.5 between 90K and 60K, then the observed rise in E_p (by a factor 1.7) implies that H increases by a factor 3.0 if h(90K) = 5, or 4.4 if h(90K) = 2. Such a pronounced variation with temperature in a narrow range far below the onset temperature for the CDW would be surprising if H represented coupling to impurities. The quantity A, which should be nearly proportional to B and in figure 3 decreases by a factor ≈ 6 between 90K and 60K, presents a further problem. As f_2 is nearly constant for $2 < h < 5$, $\Gamma\eta^{\frac{1}{2}}$ is required to increase roughly sixfold, of which the observed change in α contributes only 2.3, leaving a further factor 2.6 to be attributed to Γ. Clearly if Γ does behave thus, it cannot refer to the damping which limits the high-field conductivity, for that is almost independent of T below 100K.[21]

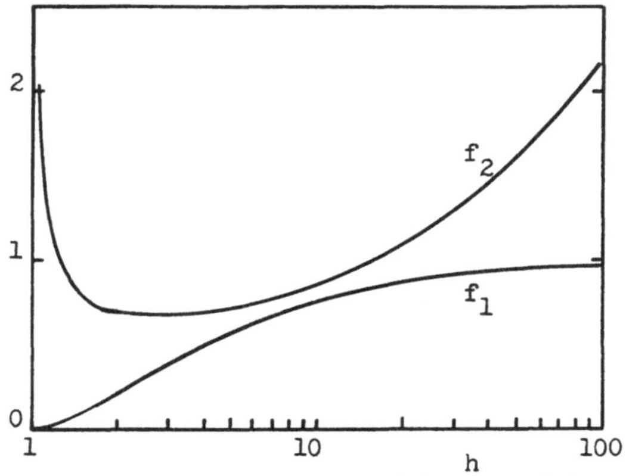

Figure 5. The functions $f_1(h)$ and $f_2(h)$.

A dependence on temperature similar to those ascribed to H and Γ is shown, however, by the quantity V_s associated with phase-slip at the current terminals. It is suggested, therefore, that H and Γ might also refer to phase-slip, in this case at boundaries lying parallel to the current flow or otherwise not obstructing it fully.

The following model is proposed. The CDW is regarded, as Tua and Zawadowski[15] suggest, as composed of a frame, and segments in which the pinning is concentrated. Here, however, the segments are assumed to be so heavily pinned that they do not join in the continuous motion of the frame, which therefore leads to phase-slip at their boundaries. While no sustained motion of the segments occurs, their coupling to the frame results in their elastic deformation, and thus in a limited displacement relative to the pinning. The force F_j exerted by the frame on the segment j is assumed to contain a term periodic in their relative displacement, and also a viscous term proportional to their relative velocity. Although in exerting F_j the frame also distorts, the segment is assumed small enough to relax, when F_j changes, before any extensive rearrangement of the frame can take place. This allows one to treat the frame as though it were rigid, and makes a mean-field approach possible.

Ignoring the direct effect of the applied field on the segments, and neglecting all damping other than that arising from the phase-slip, the equations of motion for segments and frame may then be written

$$\eta V_j \psi_j \quad = \quad F_j \quad = \quad HV_j \sin(\phi_F - \psi_j - \beta_j) + \Gamma V_j(\dot{\phi}_F - \dot{\psi}_j) \qquad (8)$$

and

$$\rho_c V_F E \quad = \quad \Sigma_j F_j \quad = \quad \eta V_s \overline{\psi_j} \qquad (9)$$

where ψ_j expresses (in terms of phase) the displacement of segment j from equilibrium, ϕ_F similarly expresses the position of the frame, V_j and V_F are the respective volumes, and $V_s = \Sigma_j V_j$ is the total volume occupied by segments. The parameters η, H and Γ specify respectively the elastic restoring force, the strength of the coupling between frame and segments, and the viscous damping, all per unit volume of segment. The preferred phase β is distributed randomly, and ρ_c is the density of charge moving with the CDW. With ϕ_j defined as $\phi_F - \psi_j$, and ρ taken to be $\rho_c V_F/V_s$, these equations reduce at once to the form (5) whose solution has already been outlined.

The success of the mean-field approach in accounting for the form of $I_c(E)$ near threshold, together with the observed dependence on temperature below 90K, thus seem explicable in terms of phase-slip at the boundaries of heavily-pinned regions. The microscopic processes which determine η, H and Γ remain to be investigated, as also does the nature of the heavily-pinned segments themselves. One may speculate, in view of the behaviour when $E \gg E_T$, that the pinning is associated with surfaces, possibly dividing the CDW into separate domains, and perhaps including the internal planar features seen in some electron micrographs, as well as the external surface of the crystal. It is interesting that for the CDW discussed here, the threshold field tends to be greater in thin specimens, and has been observed to increase following thermal shock[24].

If one supposes that surface pinning prevents layers of thickness L_s, on either side of a frame of half-thickness $L_F \gg L_s$, from joining in the continuous motion, and allows the segments to be regions of these layers between centres of pinning, then η will be of the order of α/q_0L^2, where L is the segment length. An alternative, perhaps more attractive and giving the model a closer resemblance to that proposed by Sneddon[14], would be for the layers to be pinned over the whole surface, but slightly incommensurate with the CDW of the frame. Equations (8) and (9), with β now expressing the incommensurability, still apply, and η becomes of the order of $\alpha_s/q_0L_s^2$, where α_s is the shear modulus of the layer. In neither case are the measured values of α in obvious conflict with the estimates of η available from $E_P = \eta h f_1(h) L_s/\rho_c L_F$. For example, with $L_F = 3\ \mu m$ (roughly the half-thickness of the present crystals), and letting L_s be 0.2 μm, consistency is achieved if either $L \approx 2\ \mu m$ or, as the relative values of V_s and $L_F\ E_P$ suggest, α_s is about a hundred times smaller than α. Further experiments may perhaps decide which, if either, of these possibilities approximates to the truth.

Finally, mention is made of the possibility that the total force, $\Sigma_j F_j$, exerted by the frame may become sufficient to dislodge the surface layers from their pinning, so that the whole CDW moves together. This will happen if, at low temperatures, E_P rises above the threshold for complete depinning, which presumably is then almost independent of T. No phase-slip at the boundaries of the layers then occurs. It is also possible, because $\Sigma_j F_j$ increases with E (at least until the pinning and damping forces omitted from equation (9) become important), for phase-slip to begin when $E = E_P$, and give way at some higher threshold field to complete depinning. Certain switching phenomena [17] seen in $NbSe_3$ below 50K may perhaps arise in this way.

References

1. H. Frohlich, Proc. Roy. Soc. A 223, 296 (1954).
2. see, e.g., the reviews by G. Gruner, Comments on Solid State Physics 10, 183 (1983), and N. P. Ong, Can. J. Phys. 60, 757 (1982).
3. J. Bardeen, Phys. Rev. Lett. 42, 1498 (1979), and 45, 1978 (1980).
4. S. G. Chung, Phys. Rev. B 29, 6977 (1984).
5. G. Grüner, A. Zawadowski and P. M. Chaikin, Phys. Rev. Lett. 46, 511 (1981).
6. J. Richard, P. Monceau and M. Renard, Phys. Rev. B 25, 948 (1982).
7. J. C. Gill, Solid State Commun. 39, 1203 (1981).
8. K. K. Fung and J. W. Steeds, Phys. Rev. Lett. 45, 1696 (1980).
9. T. Tamegai, K. Tsutsumi, S. Kagoshima, Y. Kanai, M. Tani, H. Tomozawa, M. Sato, K. Tsuji, J. Harada, M. Sakata and T. Nakajima, Solid State Commun. (in press).
10. N. P. Ong, G. Verma and K. Maki, Phys. Rev. Lett. 52, 663 (1984); also J. C. Gill and A. W. Higgs, Solid State Commun. 48, 709 (1983).
11. L. Sneddon, M. C. Cross and D. S. Fisher, Phys. Rev. Lett. 49, 292 (1982).
12. H. Matsukawa and H. Takayama, Solid State Commun. (in press).
13. D. S. Fisher, Phys. Rev. Lett. 50, 1486 (1983).
14. L. Sneddon (preprint).
15. P. F. Tua and A. Zawadowski, Solid State Commun. 49, 19 (1984).
16. J. C. Gill, Solid State Commun. 44, 1041 (1982).
17. A. Zettl and G. Grüner, Phys. Rev. B 26, 2298 (1982); also R. P. Hall and A. Zettl, preprint.
18. J. C. Gill, Mol. Cryst. Liq. Cryst. 81, 791 (1982).
19. I. Batistić, A. Bjeliš and L. P. Gorkov (preprint) consider the case of current terminals covering the ends of the crystal, and predict that phase-slip will occur some distance inside it, giving a contribution to E_T varying approximately as $\ell^{-1.23}$. Whether the dependence in the present experiments, where contact was to one side of the crystal, is significantly different from this is uncertain, as there are indications that V_s is somewhat contact-dependent.
20. P. Monceau, M. Renard, J. Richard, M. C. Saint Lager, H. Salva and Z. Z. Wang, Phys. Rev. B 28, 1646 (1983), conclude that the frequency observed corresponds to passage through half-wavelengths. The values of I_c and ν in figure 2 for E less than 1.5 E_T give $I_c/S\nu \approx 21$ A $cm^{-2}MHz^{-1}$, which appears to agree with that conclusion. However, it is evident that current then flows only through part of the cross-sectional area S, and that the true ratio of current density to ν is greater, possibly by a factor 2.
21. M. Oda and M. Ido, Solid State Commun. 44, 1535 (1982).
22. P. A. Lee and T. M. Rice, Phys. Rev. B 19, 3970 (1979).
23. J. C. Gill, Proceedings of the International Symposium on Nonlinear Transport and Related Phenomena in Inorganic Quasi One-Dimensional Conductors, Hokkaido University, 1983: p. 139. The concentration of distortion near the current terminals noted in this paper resulted from the use of pulses shorter than the characteristic time τ, the large value of which in long specimens was not then appreciated.
24. J. C. Gill, J. Phys. F 10, L81 (1980).

BISTABLE CONFIGURATIONS OF THE PINNED CHARGE DENSITY WAVE: RANDOM-FIELD-MODEL DYNAMICS OBSERVED IN REARRANGEMENT PRIOR TO DEPINNING

N.P. Ong, D.D. Duggan, C.B. Kalem and T.W. Jing
Department of Physics, University of Southern California, Los Angeles, Ca 90089-0484

P.A. Lee
Department of Physics, Massachusetts Institute of Technology, Cambridge, Ma 02139

The pinned charge density wave (CDW) has two stable states A and B. Conversion between them occurs when the applied field changes sign. During conversion the resistance changes logarithmically with time. The total time for conversion varies with temperature (T) and field (E) as exp (const./ET), in striking agreement with theories worked out for Random-Field-Ising-Models. This enormous time variation (10^6 to 10^{-6} s) dominates all transient and ac responses of the pinned CDW. Experimental results supporting this picture are derived from measurements of the dc resistance, pulsed transience experiments and rf ac impedance measurements.

I. INTRODUCTION

The random field due to impurities is known to play a crucial role in the ground state configuration of the pinned charge density wave (CDW)[1]. The problem of explaining why weak impurities can pin the CDW condensate was solved by extending an original argument due to Imry and Ma[2]. This domain approach which predicted a c^2 (c is the impurity concentration) dependence for E_T was quickly confirmed by experiment[3]. The situation with regard to the dynamics of the pinned CDW is less clear. In this paper we examine processes which precede depinning of the CDW and show that these phenomena are related to the conversion of the CDW between two bistable states.

There have been two trends in the study of the general random field problem. In the first trend many workers have shown that the CDW has a rich and puzzling list of dynamic properties in the pinned state[4-8]. Most of these properties may be described roughly as "relaxation". The second trend is in the study of Random Field Ising Models (RFIM)[9]. Recently, theoretical progress[10,11] has been made in understanding the neutron scattering results and critical behaviour of $Rb_2Mn_{0.5}Ni_{0.5}F_4$ and $Fe_cZn_{1-c}F_2$ which correspond to two (2D) and three-dimensional (3D) RFIM's respectively. The theoretical consensus is that domain walls in these systems move logarithmically in time in both 2D and 3D so that equilibration times may approach 10^3 years. We argue that the concept of logarithmic equilibration times may be seen to explain many (we shall argue all) of the puzzling features observed in CDWs prior to depinning.

The picture of the pinned CDW that has emerged[12] from our recent experiments on $NbSe_3$, TaS_3 and $K_{0.3}MoO_3$ using dc, pulse and high frequency techniques is as follows. At a finite temperature T the pinned CDW has two truly stable configurations which we call "pristine" (defect-free) states A and B. The only way to attain the state A (B) is by reducing E from above to below E_T with the CDW initially moving in the right (left) direction. These states cannot be approached by changing T and waiting for the system to equilibrate (unless one is prepared to wait for years). All other configurations are mixtures of these two. Depending on the direction of E conversion to a pristine state occurs logarithmically in time t. We will call such configurations logarithmically unstable. (This immediately implies that ordinary measurements of the Ohmic resistance vs. T sample a succession of logarithmically unstable states[6].) Depinning occurs only when conversion is complete (i.e. the sample is in a pristine state). This appeares to be a necessary though insufficient condition for depinning. Calling t_0 the time needed for a particular fraction of the sample to convert for fixed E we have established[12] that

$$t_0 \ (E,T) \sim \exp(A/ET), \qquad (1)$$

where A is some constant. This result is in striking agreement with some arguments recently presented by Grinstein and Fernandez[10] (GF) for the RFIM. From eq. 1 the value of t_0 can be very long indeed, particularly at low T and E. Thus hidden in the CDW problem is a single process (conversion) with time scales spanning 10^6 s to 10^{-6} s. Various experimental techniques will be sensitive to particular windows of this enormous range of time scales. For e.g. ac lock-in amplifiers will see the window 10^2 to 1 s; pulse transient techniques see the window 10^{-3} to 10^{-6} s; and rf ac impedance techniques are sensitive to 10^{-2} to 10^{-8} s. We believe that all three are glimpses of the same phenomenon.

II. ELECTRICAL HYSTERESIS

Hysteretic behaviour in the dc I-V characteristic of the CDW was first reported in TaS_3 by Gill and coworkers[13]. It has also been seen[5] in $NbSe_3$ and $K_{0.3}MoO_3$. In our experiments[12] high precision ac lock-in measurements of the differential resistance R vs. the dc voltage V are used to study the electrical hysteresis (R-V curves) in high purity $NbSe_3$. Figure 1 shows a typical hysteretic curve with two well-defined resistance states

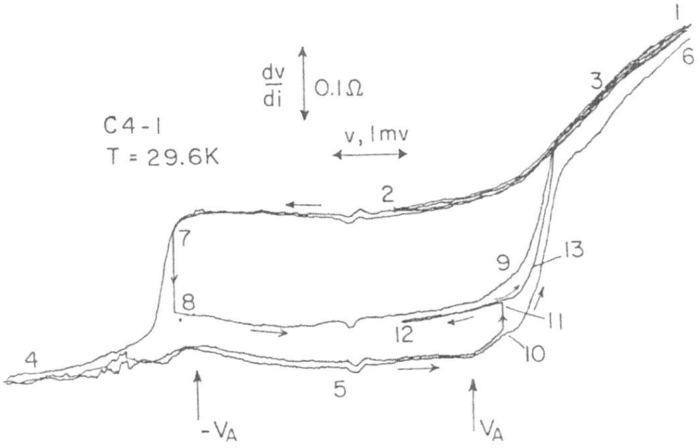

Fig. 1

Electrical hysteresis in $NbSe_3$ shown as differential resistance R vs. voltage V. The numbers indicate the sequence of pen positions. Conversion between pristine states becomes visible when V exceeds V_A. The vertical lines 7-8 and 11-12 represent conversion with V held constant.

R_A and R_B corresponding to the pristine states A and B. (The numbers indicate the time sequence of the recorder pen position.) The vertical lines 7-8, 10-11 represent evolution of R vs. t when V is held constant. If the voltage is held at 7 R changes until it attains a value R_s between R_A and R_B determined by V (and T). (If V is reduced before R attains R_s ("field quenched") R remains at this value until V is restored to its previous value or is reversed in sign. The same is true if V is changed with R at R_s, 10-11.) Figure 2 shows several such relaxations of R vs. log t for different values of V. (All curves are normalized so that the full excursion of R is 1.) For small t R varies as $(\log t)^2$ while for large t R goes as log t until it is <u>interrupted</u> at the value $R(t_0)$ at

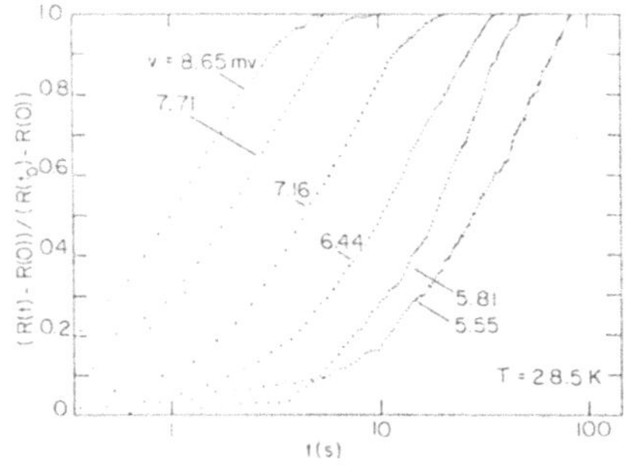

Fig. 2

Variation of differential resistance R vs. log t for various fixed voltages V in $NbSe_3$. For small t R varies as $(\log t)^2$; for large t R goes as log t. R is normalized to the total change in resistance.

time $t=t_0$ (Fig. 3). Clearly t_0 is a function of V. In fact a plot log t_0 vs. 1/V is linear, in agreement with Eq. 1 (Fig. 3, inset). If the ex-

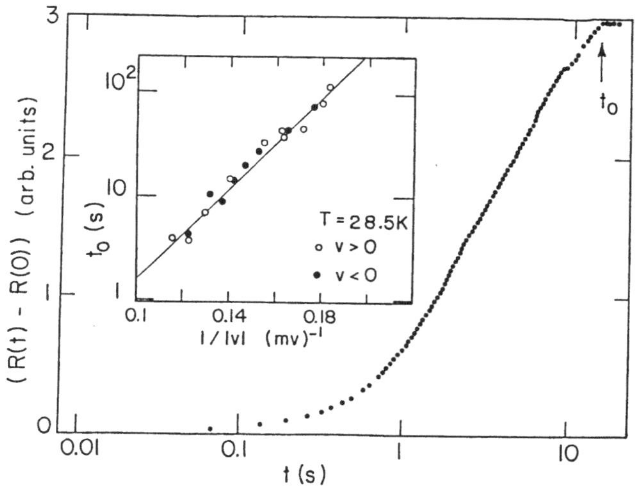

Fig. 3

(Main panel) Variation of R vs. log t for a fixed V in $NbSe_3$ corresponding to the vertical line 7-8 Fig. 1. After time t_0 (which depends on V and T) the logarithmic decay is interrupted. (Inset) Plot of t_θ vs. the voltage V showing that t_0 varies as $\exp(V_0/V)$. The open (solid) circles are for positive (negative) values of V.

periment is carried out at a higher T all time scales are reduced exponentially as anticipated by Eq. 1.

How do we interpret these hysteretic curves? First, the two values R_A and R_B are very well-defined and it is natural to associate them with the true (bi)stable configurations. The intrinsic nature of R_A and R_B is revealed by cycling V so that it exceeds V_T (depinning threshold) in both directions. Thus these are the only two stable values that the system assumes when V = 0; the trajectory passing through R_A and R_B may be called the limit cycle in the R-V space. If the system is perturbed in any way such as changing T to a new value it ends up in a mixed state with R quite different from either R_A and R_B. Logarithmic decay (helped along by applying a finite E) will bring R towards one of these value.

When V is reduced from positive values exceeding V_T in Fig. 1 the system is in state A. As long as V does not change sign pinning and depinning can be reversibly accomplished without changing R (curve 1-2-3). (We ignore the quadratic background caused by ordinary Joule heating of the sample.) As V is made negative conversion from A to B proceeds but does not become visible in this technique until it exceeds $-V_A$ (annealing field). Then conversion proceeds on an ever faster time scale until it is complete at 4 (just prior to depinning). (A more detailed discussion of the paths 8-9, 11-12 is in Ref. 12.) As we emphasized above conversion proceeds whenever V changes sign but - depending on the technique - it is visible only over a narrow window of V (recall Eq. 1). Thus the abrupt increase in R when V just exceeds V_A simply reflects the rapid traversing of this time window. For this reason the definition of V_A as an annealing field is somewhat artificial, though convenient for some discussions.

In our picture conversion represents the growth of the favoured domains at the expense of the unfavoured ones (as determined by field direction). If the favoured domain grows by ΔL the field energy gain is $E\Delta L$ while the random field energy cost is $u_i\sqrt{(L)}$. (u_i is average impurity random field energy.) Treating this as a particle moving in a one-dimensional path with random barriers $u_i\sqrt{(L)}$ much larger than the driving force E we find that the time to cross barriers in a length L is inversely proportional to the thermally activated attempt frequency[10]. Thus we obtain

$$t(L,E,T) = \omega_0^{-1}\exp(A'Lu_i^2/ET) \tag{2}$$

where A' is a constant and ω_0 is some characteristic frequency.

Experimentally we find that the maximum distance the domain wall diffuses L_s is a function of E. When L reaches L_s further conversion halts. Furthermore, the conversion is <u>irreversible</u>, in the sense that if E is reduced in magnitude no change in L can be detected on the same time scale. Equation 2 implies that while the system is converting (for fixed V and T) L which is proportional to $R-R_B$ (or $R-R_A$) changes as log t in agreement with Fig. 3. In Grinstein and Fernandez's model L grows as $(\log t)^2$ when the favoured domains are very small so that the driving term E dominates the impurity barriers $u_i\sqrt{(L)}$. This may describe the quadratic variation in our data for very early times (Figs. 2 and 3). The interruption process at t_0 is not anticipated by GF.

III. PULSED TRANSIENCE

Gill[4] first pointed out that the CDW has a T dependent transient behaviour when the applied E or current I is pulsed. Furthermore, the system responds to a train of unipolar pulses differently than to bibolar pulses.(This has been called the "pulse memory effect). Fleming and Schneemayer[5] (FS) has studied this problem in $K_{0.3}MoO_3$ and shown that the decay time for the transience can exceed 1 ms. FS[5] and MM[6] point out the close resemblance of their results to spin-glass behaviour where logarithmic decay is widely encountered. Mihály, Hutiray and Mihály[7] (MHM) have also pointed out the need for "conditioning pulses" to get consistent conductivity data. They have invoked strains and distortion of the pinned CDW to explain phenomena observed in pulse transience experiments, as has Gill[13]. FS's interpretation[5] is closer in spirit to the model described here.

We will show that these pulsed transience phenomena are identical to the results for $NbSe_3$ discussed in Sec. II, except that the observation time scales are much faster. In our picture the response of the CDW to unipolar voltage pulses is always fast because depinning from a pristine state does not involve conversion (see 1-2-3 in Fig.1). This is true whether V_T is exceed or not. However, when bipolar pulses are applied the system which is in, say, state B after the last pulse, is forced to convert to state A by the new pulse. The capacitive lag is a manifestation of the logarithmic times needed for conversion. Thus the pulse memory effect simply reflects the true stability of the states A and B. When left in one pristine state the system stays in that state until an opposite pulse is applied. The conversion process manifests itself as a logarithmic transient decay. The metastable field of MHM is similar to our annealing field and, like it, is not an intrinsic quantity. The conditioning pulse of MHM[7] simply ensures that measurements are always performed on one of the pristine states and not a mixed configuration after changing T.

To tighten up the argument we have carried out three measurements. First we have digitized the response to bipolar pulses in TaS_3 and shown that it also displays an <u>interrupted</u> logarithmic time dependence as in Figs. 2 and 3. (Also see FS's results[5] on $K_{0.3}MoO_3$.) Next holding T fixed at 140 K we measured the value of t_0 (the time when the log evalution is interrupted) as a function of the voltage across the sample (the applied pulse height). In agreement with Fig. 3 log t_0 is proportional to 1/V. Lastly, we measured t_0 while varying both V and T. In Fig. 4 we show the verification of Eq. 1 over the temperature range 110 K to 140 K. The data in Fig. 4 suggest that one expression for t_0 (Eq. 1) universally applies to all T and V ($<V_T$). These three procedures demonstrate that a single process with time scales given by Eq.1 <u>dominates</u> the transient response of the pinned CDW to changes in the applied field.

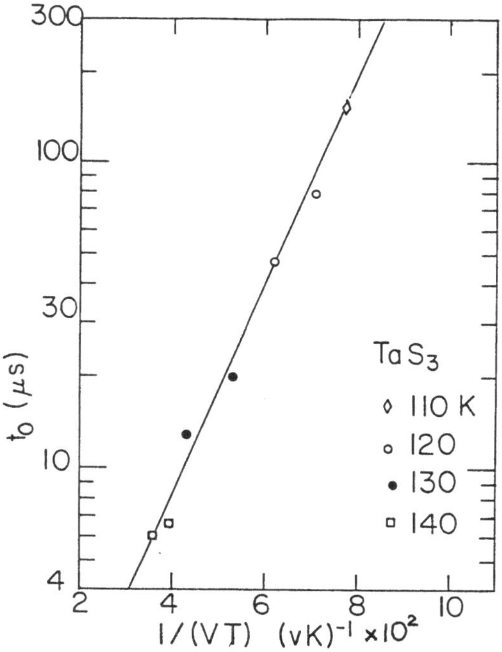

Fig. 4

Variation of log t_0 vs. the reciprocal of the product of voltage and temperature $(VT)^{-1}$ in TaS_3. The time t_0 is measured from the voltage pulse ceasing edge to the point when the logarithmic transience is interrupted in the "pulse memory" experiment. At each T two values of V are used to obtain two values of t_0.

IV. RADIO-FREQUENCY IMPEDANCE MEASUREMENTS

Cava et al.[15] have performed extensive low frequency (12 Hz to 13 MHz) measurements on $K_{0.3}MoO_3$ and found good agreement with the Cole-Cole, Cole-Davidson[16] dielectric expression (ω is angular frequency)

$$\varepsilon(\omega) = 1/(1 + (i\omega\tau_0)^{(1-\alpha)})^{\beta} \tag{3}$$

Equation 3 is the relaxation response of oscillators with a certain distribution of damping constants. The ac conductivity of TaS_3 has also been studied by the Wu et al.[17]. At 120 K they obtain the result

$$\sigma(\omega) = (i\omega)^{-\alpha'} \tag{4}$$

The divergent ε_1 (as ω goes to 0) implied by Eq. 4 would appear to be in conflict with Eq. 3 which has ε_1 going to a finite value as ω goes to 0.

Using a novel techniqe applicable to samples with impedance up to 100 kΩ we have performed dielectric measurements on three samples of orthorhombic TaS_3 at frequencies 100 kHz to 500 MHz and temperature 60 K to 260 K. Some results are shown in Fig. 5. In brief we see behaviour consistent with both the Bell labs.[15] and UCLA[17] experiments. In Fig. 5 the frequency dependence of ε_1 is depicted at various T. At the lowest T (123 K)

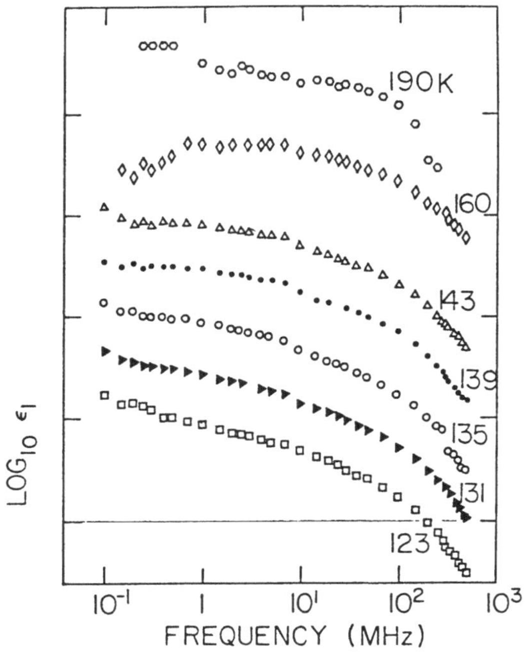

Fig. 5

The frequency (f) dependence of the real part of the dielectric constant (ε_1) in TaS_3 for various temperatures. The curves have been displaced by roughly 1/2 decade each for clarity. Note that at 123 K ε_1 diverges at low f as $f^{-\alpha'}$ (α = 0.28). At temperatures above 139 K this divergence is suppressed and ε_1 is frequency independent below 10 MHz.

behaviour identical to Wu et al's data is obtained. As ω decreases ε_1 gently diverges as $\omega^{-\alpha'}$ where α' appears to be T independent and equals 0.26 and 0.28 for two samples. However, as T is increased there is a noticeable flattening of the dispersion until at high T it becomes insensitive to frequencies below 10 MHz. This behaviour is consistent with the Cole-Davidson formula, Eq. 3. The conclusion we draw from this study is that Eq. 3 very likely describes the low frequency dielectric dispersion of the CDW system (i.e. there is no zero frequency divergence in ε_1). (As noted by Wu et al. it is inadequate in the high frequency range exceeding 100 MHz where ε_1 shows a break in slope indicating a new process.)

How should we interpret the apparent divergence of the UCLA data[17] which extends to 12 Hz at 120 K? An attractive hypothesis is that the time scale τ_0 in Eq. 3 is precisely the same as t_0 in Eq. 1. In fact a numerical simulation using Eq. 3 with τ_0 equal to 0.1 s (and α = 0.1, β = 0.9) mimics Eq. 4 (and Wu et al's data) exactly (expect above 100 MHz). The value for t_0 of 0.1s is quite consistent with the 120 K data in Fig. 4 if we recall that the ac experiment uses much smaller ac fields to probe the sample. The implication is that the divergence in Wu et al's data is cut-off at a frequency much lower than 12 Hz, just as the divergence in our data in Fig. 5 is cut-off at 1 MHz at 139 K.

It should be noted that Cava et al.[15] find the average relaxation time τ_0 in Eq. 3 to be thermally activated in $K_{0.3}MoO_3$ in agreement with Eq. 1. Furthermore, if indeed $(\omega\tau_0) >> 1$ for frequencies above 12 Hz in TaS_3

at 120 K then a log-log plot will show a T-independent slope, as found by Wu et al. and confirmed by our measurements.

Our conclusion is that the anomalous low frequency behaviour is yet another manifestation of the logarithmic conversion time scale summarized in Eq. 1.

Acknowledgement This research is supported by the U.S. National Science Foundation through Grant DMR 81-09971. We thank P. Horn and R.J. Birgeneau for helpful discussions on the RFM and J.C. Eckert for growing the samples. One of us (N.P.O.) wishes to thank the Alfred P. Sloan Foundation for support.

REFERENCES

1. P.A. Lee and T.M. Rice, Phys. Rev. B 19, 3970 (1979)
2. Yoseph Imry and Shang-keng Ma, Phys. Rev. Lett. 35, 1399 (1975)
3. J.W. Brill, N.P. Ong, J.C. Eckert, J.W. Savage, S.K. Khanna and R.B. Somoana, Phys. Rev. B 23, 1517 (1981)
4. J.C. Gill, Solid State Commun. 39, 1203 (1981); J.C. Gill, Solid State Commun. 44, 1041 (1982); J.C. Gill, Mol. Cryst. Liq. Cryst. 81, 791 (1982)
5. R.M. Fleming and L.F. Schneemeyer, Phys. Rev. B 28, 6996 (1983)
6. G. Mihály and L. Mihály, Phys. Rev. Lett. 52, 149 (1984)
7. G. Mihály, Gy. Hutiray and L. Mihály, Solid State Commun. 48, 203 (1983); Gy. Hutiray, G. Mihály and L. Mihály, Solid State Commun. 47, 121 (1983)
8. G. Mihály, G. Kriza and A. Jánossy, Phys. Rev. B 30, (1984) in press.
9. R.J. Birgeneau, H. Yoshizawa, R.A. Cowley, G. Shirane and H. Ikeda, Phys. Rev. B 28, 1438 (1983)
10. G. Grinstein and J.F. Fernandez, preprint.
11. R. Bruinsma and G. Aeppli, Phys. Rev. Lett. 52, 1547 (1984); J. Villain, prepring.
12. D. Duggan, N.P. Ong and P.A. Lee, unpublished.
13. J.C. Gill, Proceedings of the International Symposium on Nonlinear transport phenomena and related phenomena in inorganic quasi-one-dimensional conductors. Hokkaido University, Sapporo, 1983.
14. G. Grüner, L.C. Tippie, J. Sanny, W.G. Clark and N.P. Ong, Phys. Rev. Lett. 45, 935 (1980)
15. R.J. Cava, R.M. Fleming, P.B. Littlewood, E.A. Reitman and L.F. Schneemeyer, Phys. Rev. B, to appear.
16. K.S. Cole and R.H. Cole, J. Chem. Phys. 9, 341 (1941); D.W. Davidson and R.H. Cole, J. Chem. Phys. 19, 1484 (1951)
17. Wu, Wei-yu, L. Mihály, George Mozurkewich and G. Grüner, Phys. Rev. Lett. 52, 2382 (1984)
18. C.B. Kalem and N.P. Ong, unpublished.

ELECTRIC FIELD INDUCED RELAXATION OF METASTABLE STATES IN TaS_3

G. Mihály, A. Jánossy and G. Kriza
Central Research Institute for Physics
H-1525 Budapest, P.O.B. 49, Hungary

The spontaneous decay of thermal quench induced metastable states gives rise to a time dependent ohmic conductivity: $\Delta\sigma/\sigma_o=-\alpha \ln t/\tau$. Detailed experimental results are presented on relaxation of metastable conductivity measured on pure and irradiated orthorhombic TaS_3 in presence of an electric field. The electric field accelerates the relaxation, the time dependence of the conductivity, however, still follows a logarithmic low with an increased value of α. The relaxation is accelerated well below the threshold field indicating a field induced CDW rearrangement in the ohmic conducting range.

In several charge density wave (CDW) systems the low field ohmic conductivity depends on the thermal history of the sample[1-4]. The deviation $\Delta\sigma=\sigma-\sigma_o$ from the stable conductivity σ_o is generally attributed to the presence of metastable CDW configurations frozen by impurity pinning[1,5]. The metastable state can be rearranged by depinning the CDW condensate for a short time, i.e. by application of high field pulses $E>4E_T$; E_T is the threshold field for nonlinear conduction. After this so called "conditioning" the ohmic conductivity is brought to a well defined, stable value, σ_o[2].

Investigations on orthorhombic TaS_3 revealed that the metastable conductivity in absence of an electric field is weakly time dependent and the relaxation towards the stable value follows a logarithmic law: $\Delta\sigma/\sigma_o=-\alpha \ln t/\tau$, with a temperature independent slope, $\alpha=3\times10^{-3}$[6].

In this paper new experimental results are presented on the relaxation of metastable conductivity in presence of an electric field. Measurements on pure and electron irradiated o-TaS_3 demonstrate that i., compared to the spontaneous decay of metastable states electric fields induce a considerably faster relaxation; ii., the acceleration of the relaxation appears for fields well below the threshold field for nonlinear conduction, and iii., comparing samples of different purities and relaxation rates at different temperatures we find the enhancement is determined by E/E_T and not by the normal current.

The threshold field of high purity TaS_3 crystals investigated was E_T=0.5 V/cm (110 K). Crystals of the same bach were irradiated by 2.5 MeV electrons with a dose of 0.1 C/cm^2 creating $3x10^{-4}$ defects per Ta[7]. The threshold field of irradiated crystals is larger by a factor of ten; E_T=5 V/cm (110 K), while the normal conductivity for such a low defect concentration is unchanged in the Peierls distorted phase[8]. Conductivity measurements were performed in a four probe configuration with simoultaneous measurement of voltage V between the inner contacts of distance ℓ and current. In the measurement of the relaxation under field the voltage was kept constant on the outer contacts and the current changed as the resistivity varied. Most measurements were done at 110 K, the long time temperature stability was better than 0.05 K.

Pulse sequences used in the experiments are shown on Fig. 1. The first pulse of $U_h \sim 10\ U_T$ was used to heat and suddenly cool the sample. As the pulse ceases the temperature of the crystal weighting only a few μg is quenched to the temperature of the sample holder. Oscilloscope traces of the sample resistance showed that the sample is heated up by $\Delta T \sim 50$ K and the time constant of the initial exponential cooling is about 2-5 ms (depending on the sample geometry), in agreement with previous studies[6]. This heat treatment results in a well reproducible metastable state; for a given time t following the quench $\Delta\sigma(t)/\sigma_o$ becomes independent of ΔT if $\Delta T > 20$ K[2]. On the other hand fast cooling enables measurements shortly after the heating pulse.

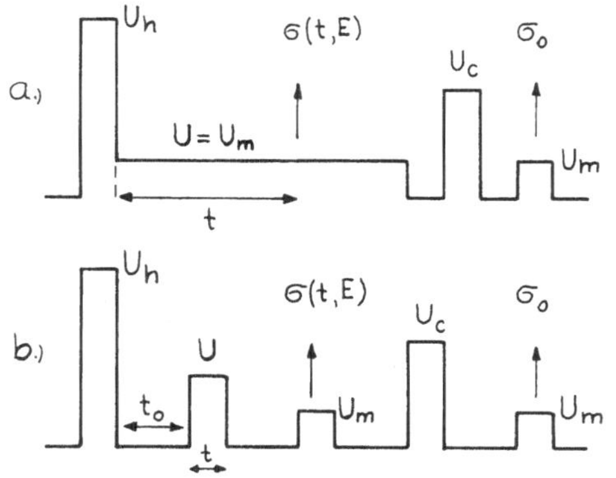

Fig. 1. Pulse sequences applied in relaxation measurements. The pulse U_h heats up the sample by about 50 K. Measurement starts after thermalization time t_o=100 ms. The relaxation of the metastable states created by the thermal quench occurs under the field E produced by pulse U. The conductivity $\sigma(t,E)$ is measured by pulse U_m and after conditioning (U_c) is compared to σ_o. a/ "continuous" measurement used in time interval $10^{-1}<t<10^4$ seconds, b/ "pulsed" measurement for interval $10^{-3}<t<10^0$ s.

The heating pulse U_h is followed by a constant voltage U under which the relaxation is measured (Fig. 1a). The time dependence of the conductivity $\sigma(t,E)$ was determined at various fields E=V/ℓ and the times t were measured from the switch off edge of the heating pulse. Data were registered from times larger than 100 ms when the sample become already thermalized. This method was applied to measure $\sigma(t,E)$ in the range of 10^{-1}-10^{4} seconds. Following the measurement of $\sigma(t,E)$ a conditioning pulse $U_c=4U_T$ was applied and the conditioned value of the conductivity, σ_o, was determined. Values of $\Delta\sigma(t,E)/\sigma_o$ vs time measured by this method are indicated by crosses on Fig. 2.

For fields below 0.4 E_T the relaxation is unchanged. Above $E\sim 0.5\ E_T$ the relaxation accelerates so much that a considerable part of the conductivity decrease occurs during the first hundred milliseconds. This required an extension of the measurement to times shorter than a 100 ms. The pulse sequence used for this purpose is shown on Fig. 1b.

After the heating pulse no field was applied for $t_o\sim 100$ ms and the sample is thermalized to the stable temperature of the holder. We assume that during the first 100 ms the spontaneous relaxation in absence of a field is negligible compared to the conductivity change induced by the following relatively large voltage pulse U ($V/\ell\geq 0.5\ E_T$), even if U is as short as t=1 ms. The pulse U gives rise to the conductivity change of interest. The conductivity $\sigma(t,E)$ was measured by a third pulse U_m of length ~100 ms and amplitude less than $U_T/10$ which again gives negligible relaxation compared to pulse U. Thus when this method is applied we take the time t in $\sigma(t,E)$ as the length of the pulse U since only during this time interval is the relaxation important. Finally a conditioning (U_c) pulse was applied and $\sigma(t,E)$ was compared to the stable value σ_o. The metastable conductivity $\Delta\sigma(t,E)=\sigma(t,E)/\sigma_o$ measured by this pulse sequence characterizes the relaxation under the second pulse U of length t as confirmed by the observation that it is independent of thermalization time t_o (in the range of 50 ms-1 sec) if $E>0.5\ E_T$. The $\Delta\sigma(t,E)/\sigma_o$ vs time curves for fixed field values were taken by repeating the pulse sequence of Fig. 1b with increasing width t of pulse U. This method was used to cover the time range of 10^{-3}-10^{0} seconds. On Fig. 2 circles indicate the values measured in this way. The overlapping range of relaxation curves measured by the two different methods also supports the validity of the method used for short time relaxation experiments. All curves were taken by a computer controlled measuring system enabling us to follow small changes in conductivity over seven orders of magnitude in time.

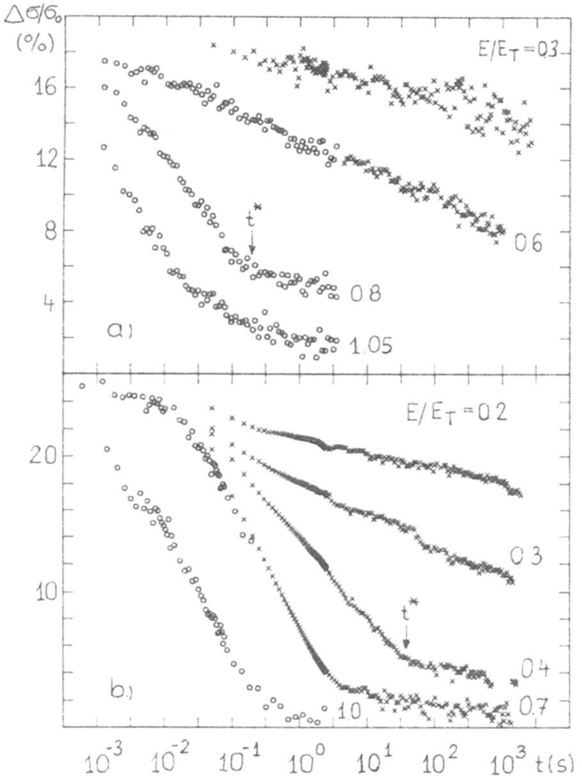

Fig. 2. Time dependence of metastable conductivity under different electric fields at T=110 K. a/ pure crystal, b/ irradiated sample ($3x10^{-4}$ defects/Ta atom).

Fig. 2 summarizes the main effects of an electric field on the relaxation by showing typical curves at various fields measured on both pure and irradiated TaS_3 crystals. At low fields the relaxation is logarithmic; $\Delta\sigma(t,E)/\sigma_o=-\sigma(E)\ln t/\tau$ with $\alpha(E)=\alpha_s\sim 4x10^{-3}$, in good agreement with a previous study of spontaneous relaxation[6]. Above $E\sim 0.4\ E_T$ the relaxation is faster as reflected by an increase of the slope $\alpha(E)$. At even higher fields the relaxation is initially fast until a sharp change in the slope of the $\Delta\sigma/\sigma_o$ vs lgt plot is observed at $t=t^*$. For $t<t^*$ the slope $\alpha(E)$ is field dependent in contrast to times $t\geqq t^*$ when it corresponds to the slope of spontaneous relaxation, α_s, measured in absence of a field. With increasing field t^*, the time from when the relaxation strongly slows down, decreases.

In the irradiated sample the field dependence of the relaxation is similar. Here, however, for fields close to but below threshold for an initial short time interval the relaxation is appearently inhibited (Fig. 2b). This rather special behaviour is discussed later.

Fig. 3 shows the parameters characterizing the field induced relaxation; the field dependence of the initial slope $\alpha(E)$ and the variation of time t^* when relaxation changes from a fast to a slow rate as a

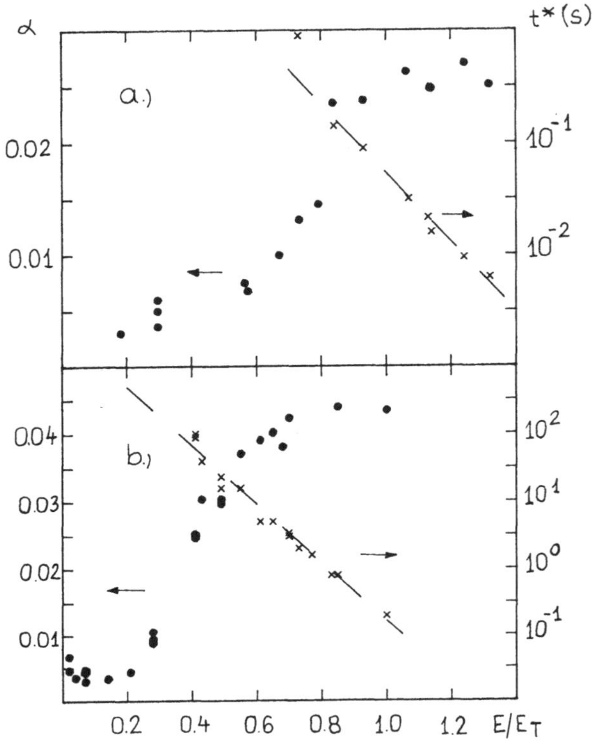

Fig. 3. Electric field dependence of the initial slope of the logarithmic relaxation $\Delta\sigma(t,E)/\sigma_o = -\alpha(E)\ln t/\tau$, and of t*. a/ pure sample, b/ irradiated sample. T=110 K.

function of E. In spite of the order of magnitude difference between the threshold fields of the pure and irradiated samples these parameters show a rather similar behaviour when plotted againts E/E_T. This indicates that the field relative to the threshold is the main factor determining the relaxation enhancement. It should be emphasized that the factor of 6 difference between $\alpha(E_T)$ and α_s (in case of the pure crystal) reflects an enormous difference in the relaxation rates; at $E \sim E_T$ the stable value is reached within ~100 ms, while at low fields, if the logarithmic variation with slope α_s can be extrapolated, several thausand years are needed to complete the relaxation. We also call attention to the linear variation of lnt* with applied field, which is valid over several orders of magnitude in time t*.

As described before for $E<0.5\ E_T$ the experiments were performed by the pulse sequence of Fig. 1a. Since in this case the measuring field and the field inducing the relaxation are the same, $U_m=U$, lowering the field increases the scatter in data. At low measuring levels the determination of $\alpha(E)$ is somewhat ambigous (especially for the pure sample where E_T is small). To demonstrate the existence of a field independent relaxation range at low fields and to determine the field level above which the relaxation is enhanced we performed a more accurate measurement on a pure sample. The pulse sequence of Fig. 1b was applied but the voltage pulse U giving rise to field induced relaxation was increased in small steps and had always the same width of 1 second. This experiment gives a vertical cross section at a fixed time t=1 sec of relaxation curves shown on Fig. 2a.

Results are shown in Fig. 4 for two temperatures.
$\beta(E)=|\sigma(E=0,\ t=1s)-\sigma(E,t=1s)|/\sigma_o$ is zero in the field independent regime and deviations indicate a field induced enhancement of relaxation. We found that below 0.4 E_T application of field does not affect the relaxation - there is a spontaneous relaxation range. More surprisingly even such low fields as 0.4-0.5 E_T - far in the ohmic range - give rise to a change in the metastable conductivity, and thus lead to CDW rearrangement.

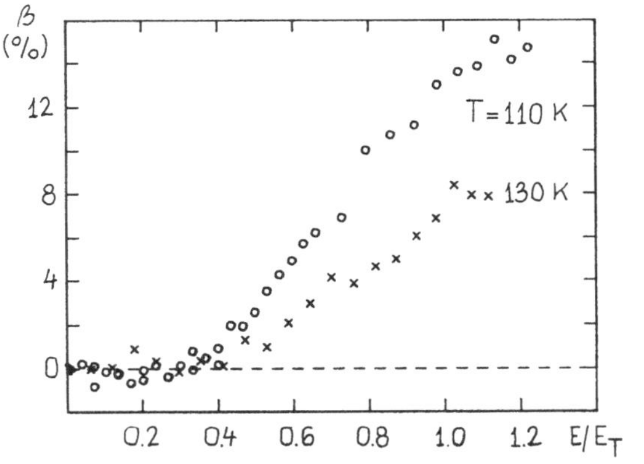

Fig. 4. Enhancement of the relaxation of metastable conductivity by an electric field in pure TaS_3 at two temperatures. β is the difference between conductivities 1 sec after the quench with and without electric field.

At constant fields the normal current in o-TaS_3 increases by a factor of 3 when the temperature is elevated from 110 K to 130 K. In spite of this difference in current the relaxation enhancement appears above the same field level at both temperatures, demonstrating that the normal current intensity does not play any role in the relaxation enhancement.

To describe the logarithmic time dependence of the spontaneous decay of metastable states a model based on an analogy with spin glasses has been proposed[6]. In that model a wide distribution of pinning energies was assumed which leads to the deceleration of CDW rearrangement for longer times when the number of small barriers still to be overcome decreases. With reasonable assumptions a logarithmic time dependence was obtained. This model could be extended to account for the relaxation under electric field (e.g. by assuming a modified barrier distribution in presence of field) and a formal description of the present results might be obtained.

However, we believe, such an extension would prove to be insufficient to describe the rather complex phenomenon of relaxation under field. In the case of fields comparable or larger than threshold the polarized nature of the steady state must be taken into account. Fields larger than threshold give rise to a polarized state in which physical properties like the conductivity vary from one end to the other of the sample[9]. We believe that fields below threshold also polarize the material but the inhomogeneity in this case appears on a smaller scale. We present an experiment which may be interpreted by this assumption.

We show that the field induced relaxation at a field near threshold is inhibited if the field instead of being applied continuously for a relatively long time is applied for the same total time but in short pulses. The experiment was performed on the irradiated TaS_3 crystal. The field applied was $E=0.7\ E_T$, in one case it was applied continuously for 1000 ms while in the other case in 1000 pulses each of 1 ms duration and separated by 100 ms. The continuous field wipes out all the thermal induced deformations, after 1000 ms the conductivity reached already its steady state value (Fig. 2b). On the other hand short pulses have no influence on the relaxation rate: the conductivity decrease after 1000 1 ms long pulses is only ~2 %, the same amount as found for spontaneous relaxation with no field present.

We tentatively interpret this finding in the following way. As seen on Fig. 2b for $E=0.7\ E_T$ in the first 10 ms there is apparently no relaxation. We suggest that during this time the sample becomes polarized in small domains determined by defects with strong pinning potentials. Within each domain the CDW deformation in fact increases during the first 1 ms, a time less than needed to polarize the domain to its largest possible value. The resistivity becomes inhomogeneous but the total resistance is unchanged. If the field is switched off after 1 ms the inhomogeneous state returns to a homogeneous one and by the time the field is switched on again the sample is almost in the same state as at the beginning of the first pulse. If the field is kept continuously on an irreversible fast relaxation takes place at the domain boundaries once the polarization is sifficiently high.

In conclusion we have shown that the relaxation of metastable conductivity is enhanced by electric fields. The field induced enhancement of the rearrangement of metastable states occurs already for fields where the conductivity is ohmic. It depends on E/E_T, the field relative to threshold and not on the ohmic current.

Under electric field the relaxation is logarithmic with a field dependent slope $\alpha(E)$ for $t<t^*$. For $t>t^*$ the relaxation has the same slope α_s as in absence of a field. A common description of spontaneous and field induced relaxation of metastable states has to take into account the existance of an initial reversible process which is, in our belief, related to the polarization of the CDW within domains.

References

1. A.W. Higgs and J.C. Gill, Solid State Commun., 47, 737, (1983)
2. Gy. Hutiray, G. Mihály and L. Mihály, Solid State Commun. 47, 121, (1983)
3. L. Mihály and G. Grüner, present volume
4. J.C. Gill, Mol. Cryst. Liq. Cryst., 81, 73, (1982)
5. G. Mihály, Gy. Hutiray and L. Mihály, Solid State Commun. 48, 203, (1983)
6. G. Mihály and L. Mihály, Phys. Rev. Lett. 52, 149, (1984)
7. H. Mutka, S. Bouffard, G. Mihály and L. Mihály, J. Physique Lett. 45, L-113 (1984)
8. G. Mihály, L. Mihály and H. Mutka, Solid State Commun. 49, 1009, (1984)
9. L. Mihály, G. Mihály and A. Jánossy, present volume

REMANENT DEFORMATION OF CDWS

L. Mihály,[†] G. Mihály and A. Janossy
Central Research Institute for Physics,
H-1525 Budapest, P.O. Box 49, Hungary

We review experimental evidences demonstrating that CDW materials show metastable states in the electronic properties. The latest observations indicate that application of currents above the threshold of nonlinear conduction brings the CDW's into an asymmetric state. The results are interpreted in terms of a phenomenological model where the deformation of CDW wave number accounts for the metastability.

I. Introduction

There is a considerable amount of experimental data indicating that charge density wave materials exhibit metastability in their electrical properties. After the first reports by J. C. Gill on thermal memory[1] and pulse sign memory[2] effects in $NbSe_3$, the Budapest group discovered that the thermally induced metastable conducitivity of TaS_3 can be changed by application of currents above the threshold of nonlinear conduction[3] and, if left unperturbed, it relaxes towards equilibrium following a logarithmic time dependence.[4] The potassium blue bronze $K_{0.3}MoO_3$ also shows temperature hysteresis and long time relaxation processes.[5] Most recently Dumas and Schlenker reported similar phenomena in $Rb_{0.3}MoO_3$.[6] It was demonstrated that mechanical stress leads to metastable states in TaS_3 and $NbSe_3$.[7] Fleming and co-authors related the time dependent strandlike structures, observed in electron microscope studies to the long time relaxation induced by sudden changes in the sample current.[8]

In addition to the long time processes mentioned above there is a wide range of experiments related to the different switching and hysteresis effects in the onset of nonlinear conduction.[9] In this talk we concentrate on the behavior of CDW systems after the CDW current is switched off and we assume that the spontaneous relaxation of the relevant metastable states happens on a time scale much longer than the duration of the experiment.

Several microscopic and phenomenological models can be proposed to explain the different aspects of experimental observations. One class of theories treat the CDWs as a system of coupled domains.[8,10,11] Other attempts consider space charge pile-up due to the change in the amplitude of CDWs.[12]

One of the most fruitful microscopic model of CDWs was proposed by Fukayama and Lee.[12] If treated properly,[13] the model can account for various observations, including metastibility.[14] The experiments discussed here were initiated by a phenomenological approach to the deformable CDW model of Fukayama and Lee. The main points are summarized as follows:

a) The metastable state of the CDWs is associated with the changes in the wave number k. The deformation of the CDW is characterized by the deviation from the equilibrium value k_o. In general $\Delta k = k - k_o$ may depend on the position x. For distances much larger than the spacing between the pinning centers or the size of the CDW domains, $\Delta k(x)$ can be approximated by a smooth function. The relaxation $\Delta k \rightarrow 0$ is hindered by the pinning centers.

b) The deviation Δk has a maximum value; the pinning centers cannot withhold arbitrarily strong deformations.

c) The single particle gap is the function of the distortion of the CDWs. It is a local quantity in the sense that it may vary over distances much larger than the lattice constant. The predominant factor in determining the local gap $\varepsilon_g(x)$ is the deformation of the CDW, $\Delta k(x)$. In a first approximation $\varepsilon_g(\Delta k)$ is presumed to be linear in Δk, $\Delta\varepsilon_g = \frac{d\varepsilon_g}{dk}\,\Delta k$.

d) Application of mechanical stress or temperature change may introduce uniform CDW deformation if the equilibrium wave number k_o depends on these parameters. On the other hand, electric fields close to or above the threshold field of the CDW current induce asymmetric deformations by pushing the CDWs to the end of the sample. The role of the electrical contacts is important in this case and will be discussed later.

The low field resistance of the sample is a sensitive probe to detect CDW deformations. The total resistance of the sample $R = \int dx\,r(x) = \int dx\,r^* e^{-\varepsilon_g(x)/kt}$ where $r(x) = \rho(x)/A$ ($\rho(x)$ is the local resistivity, A is the cross section of the sample) $r^* = \rho_o/A$ and the temperature dependence of the resistivity is approximated by $\rho(T) = \rho_o e^{-\varepsilon_g(x)/kT}$, $\varepsilon_g(x)$ being the position dependent single particle gap. By applying this formula we assume that ρ_o is independent of the deformation and the resistivity change is entirely due to the change in the number of excited electron-hole pairs.

The model presented above has specific predictions on the single particle gap and low field resistance in the metastable state. We will focus on the case when an asymmetric CDW deformation is generated by application of electric field, and we will present experimental evidence demonstrating that when a CDW current is switched off, the sample is left in an asymmetric state.[16]

II. Asymmetry in sample resistance

Conventional silver pain contacts act as extremely strong pinning centers as demonstrated by the narrow band noise spectra of $NbSe_3$ segments[17] and by the broadening of Peierls transition[18] in short segments of TaS_3. The silver paint effectively short circuits the portion of the sample covered by it and significantly decreases the local electric field, so here the CDW remains pinned. The charge carried by a CDW current entering the contact must boil into normal carriers in a region near the contact.[19] This region creates a barrier to CDW conduction and a long range deformation of the CDW system appears with opposite signs at the two contacts. When the current is switched off this deformation, or a part of it, is stabilized by the pinning centers. Corresponding to the model described above the sample is left in an asymmetric state similar to that presented in Fig. 1a,b: one half has an increased resistance while in the other half the resistance is decreased relative to the equilibrium.

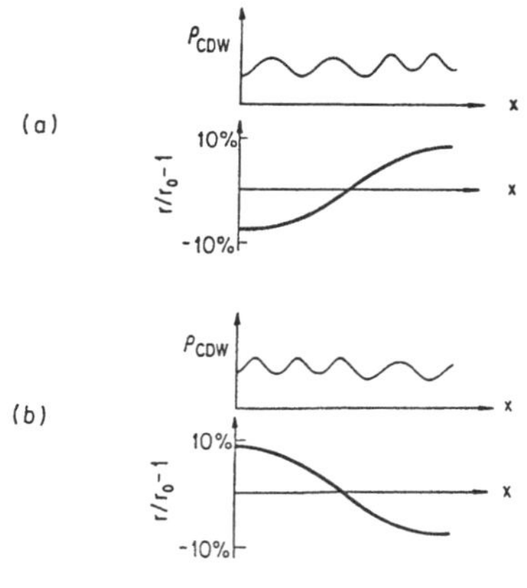

Fig. 1

Deformed charge density wave and inhomogeneous resistivity schematically drawn. a and b differ in the sign of the pulse I_c above threshold preceding measurement. ρ_{CDW} is the oscillating part of the CDW. r is the local resistivity, r_o is the resistivity of the undeformed state.

In order to verify this prediction a five probe TaS_3 sample, shown in Fig. 2 was prepared. Contacts 1,5 are current leads

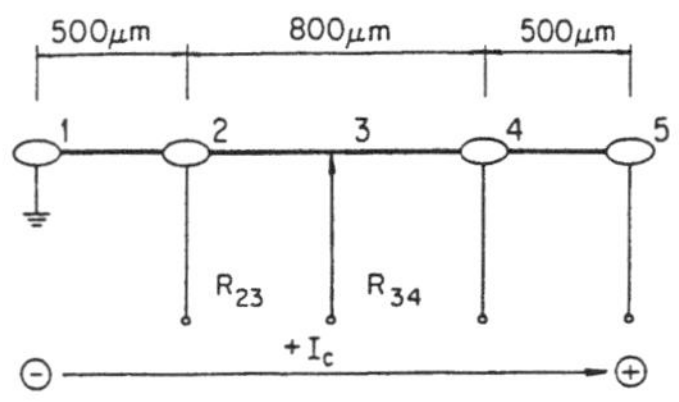

Fig. 2

Five probe sample arrangement to study the current induced asymmetry of the resistivity. Contacts 1, 2, 4 and 5 are silver painted, contact 3 is a touching gold wire. The conditioning current I_c is considered positive if contact 5 is at a higher potential than contact 1.

while contacts 2, 4 serve to measure the potential. Contact 3 is a special, low perturbation probe; it is prepared by touching a ϕ10µm gold wire to the sample. The resulting high ($\sim$ 200 kΩ) contact resistance has a negligible effect on the current flow.

The sample was placed in liquid N_2 in order to maintain a homogeneous stable temperature. Conditioning pulses I_c are applied alternately in both directions between 1 and 5. Following each pulse I_c the low field resistances between segments 2, 3 and 3,4 (and as a check between the silver paint contacts 2, 4) was measured. We denote the measured low field resistances of segments 2,3 by $R_{23}(+I_c)$, $R_{23}(-I_c)$ and similarly for segment 3,4 by understanding that these are measured by currents much smaller than I_c and reflect the metastable resistances <u>after</u> a pulse I_c or $-I_c$ was applied.

Rapid cooling of the sample from room temperature to 77 K gave rise to a non-equilibrium resistance due to a temperature induced deformation.[3] Application of a current pulse $I_c = 4\ I_T$ increased the resistance R_{24} between contacts 2,4 by 7%. After this initial increase no more change was observed in R_{24} with application of further current pulses of any magnitude or sign. Our experiment shows, however, that positive and negative current pulses above threshold do not leave the sample in the same state despite the fact that resistance R_{24} between two strongly perturbing silver paint contacts remains unchanged for this sample. We find that if I_c is larger than threshold and positive (flowing from 1 to 5) then R_{23} is increased and R_{34} decreased compared to the values they have following a negative current pulse $-I_c$. Figure 3 displays the relative resistance differences $\delta_{23} = \Delta R_{23}/R_{23}$ and $\delta_{34} = \Delta R_{34}/R_{34}$ where $\Delta R_{23} = R_{23}(+I_c) - R_{23}(-I_c)$ and $\Delta R_{34} = R_{34}(+I_c) - R_{34}(-I_c)$ respectively. For conditioning currents I_c below threshold δ_{23} and δ_{34} are zero. For I_c above threshold the resistances depend on the preceding current pulse. The relative deviation increases with I_c until about $I_c = 2\ I_T$, above this it becomes constant. This means that once a current pulse $I_c > 2\ I_T$ is applied the resistance change with the polarity of I_c is independent of the magnitude of I_c at least up to 6 I_T. The resistance increase in one half is compensated by

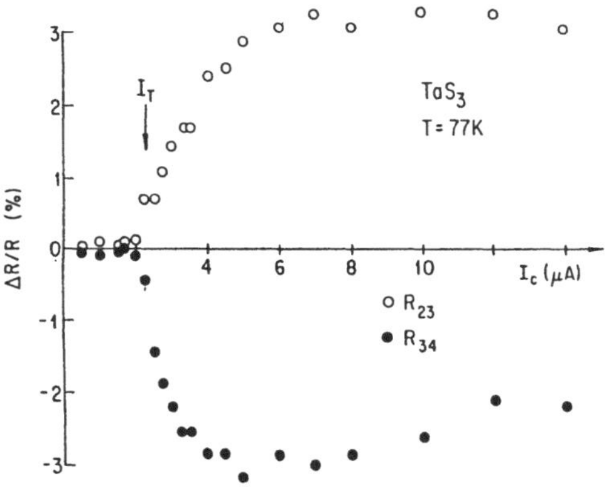

Fig. 3

Resistance changes of segments 2,3 and 3,4 as a function of preceding conditioning currents I_c. Circles correspond to $\Delta R/R = [R_{23}(+I_c)-R_{23}(-I_c)/[R_{23}(+I_c)+R_{23}(-I_c)]$, full dots denote the same quantity for R_{34}.

a resistance decrease of the other half and that is why R_{24} is not changing.

The model described in the introduction predicts a correlation between this effect and the thermal hysteresis of the resistance: the maximum value of the asymmetric deformation in Fig. 1a should be equal to the temperature induced homogeneous deformation. Although we have no direct information on the shape of the deformation function we can make a simple estimate by assuming linear variation. At 77 K the temperature induced deviation in the resistance, corresponding to the homogeneous deformation, is 7%; the resistance variation in the asymmetric case is expected to be 7/3%. This is comparable to the observed relative resistance change $\delta_{23}^{max} = \delta_{34}^{max} = 3\%$.

III. Asymmetry in the single particle gap

To demonstrate that the resistivity changes in the metastable state are really due to changes in the single particle gap, experiments in the presence of a temperature gradient were performed.

The scheme of the experiment is shown on Fig. 4. Silver paint contacts to gold wires were used and the voltage leads were attached to heaters producing the temperature gradient. The low field ohmic resistance of different TaS_3 samples was measured by applying small current pulses I_m in the usual four probe configuration. Each pulse I_m is preceded by a large conditioning current pulse I_c. As demonstrated in Section II the created inhomogeneity, antisymmetric around the center of the sample, does not change the total resistance if the temperature is constant along the sample. On the other hand, if the sample is homogeneous, the application of temperature gradient does not lead to a resistance change either. However, corresponding to our model, if the single particle gap is a monotonic function of position (similar to the r(x) function in Fig. 5) then a resistance change is expected in presence of a temperature gradient:

$$\Delta R/R = \frac{\Delta T}{T} \frac{\Delta \varepsilon_g}{kT} \alpha \left(1 + \frac{\varepsilon_g}{2kT}\right)$$

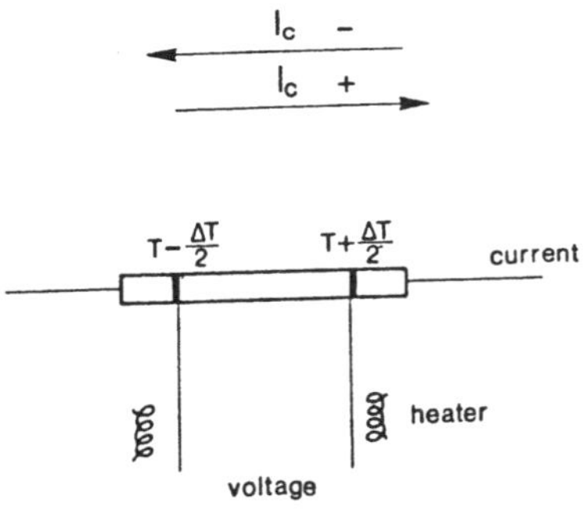

Fig. 4

Experimental setup to demonstrate the effect of conditioning current pulses I_c, $-I_c$. A temperature difference ΔT is set at the potential contacts.

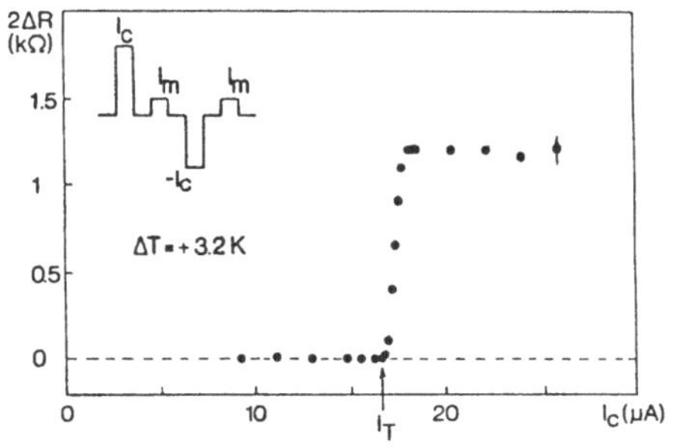

Fig. 5

The resistance difference $2\Delta R$ arising by reversing the direction of the conditioning current I_c as a function of the magnitude of I_c. Sample length ℓ = 820 μm, average temperature T = 90K, total resistance R = 33 kΩ , threshold current I_c = 17 μA, conditioning current I_c = +20 μA (o) and I_c = -20μA, (o), temperature difference ΔT = 3.2K, measuring current I_m = 5 μA. Only a depinned CDW current can change the sign of inhomogeneous deformation.

where ΔT and $\Delta\varepsilon_g$ is the temperature and gap difference between the sample ends and α depends on the actual shape of the $\varepsilon_g(x)$ function. The direction of the last conditioning pulse determines the sign of $\Delta\varepsilon_g$. Therefore the sign of ΔR is expected to depend on the relative signs of ΔT and I_c. As a check that the effect is not due to an asymmetry of the contacts not inherent to the sample, ΔT is varied from positive to negative to interchange the role of the contacts.

A pure and a slightly electron irradiated (defect concentration 3×10^{-4} defects per Ta) orthorhombic TaS_3 single crystal was measured. Fig. 5 shows $\Delta R = R(+I_c)-R(-I_c)$ versus the magnitude of conditioning current I_c for the irradiated sample. The average temperature of the sample was 90K and a ΔT = 3.2K temperature difference was applied at the potential contacts. Resistance change was generated only if I_c is larger than the threshold indicating that depinned CDW currents really change the $\varepsilon_g(x)$ function.

Although this measurement clearly shows that the sample is in a different state after positive or negative CDW currents are switched off, the resulting change in the resistance may be due to the combined effect of the current induced asymmetry and the asymmetry generated by the contact or major lattice imperfections in the sample. Indeed, several authors reported a few percent resistivity change of TaS_3 under similar conditions but without a temperature gradient[20] and this type of behavior is common for blue bronze samples.[21] Therefore it is essential to demonstrate that the resistance is linear in ΔT, corresponding to the prediction of the model based on the variation of the single particle gap. Fig. 6 shows ΔR versus ΔT for positive and negative high current conditioning pulses. The results indicate that ΔR is proportional to ΔT, it reverses sign with reversing ΔT and remains unchanged if both ΔT and I_c are reversed. This is in complete accordance with the model.

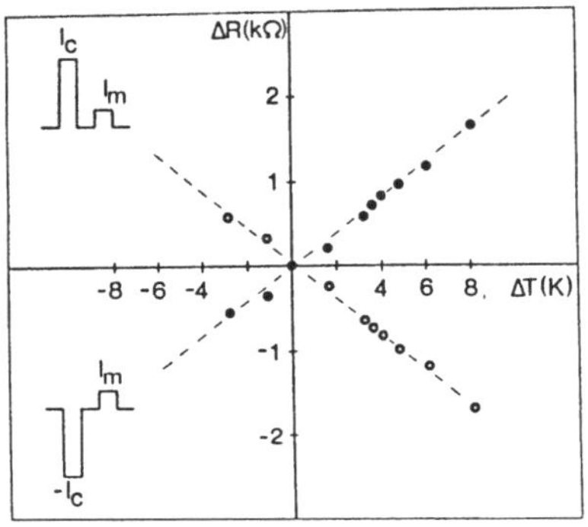

Fig. 6

Resistance difference ΔR induced by conditioning current I_c as a function of the temperature difference ΔT. ΔR is proportional to sign (I_c). ΔT.

We compared the pure and electron irradiated sample at T = 90K and we found $\Delta R/R\Delta T = 6.3 \times 10^{-3} k^{-1}$ for the irradiated TaS_3 sample and $\Delta R/R\Delta T = 1 \times 10^{-3} k^{-1}$ for the pure one. The irradiation increased the threshold field by a factor of 6, thus it seems that the current induced remanent deformation really scales with the pinning strength.

IV. Conclusion

We have shown that a sliding CDW current induces asymmetry in the sample and this asymmetry remains frozen after the current is switched off. The asymmetry is carried by the single particle gap ε_g and for the pure TaS_3 a variation of 0.5÷1% along the sample is estimated at 90K. The position dependent gap leads to position dependent resistivity. A deviation from the equilibrium value which is homogeneous along the sample and bears the same magnitude, may account for the resistivity change observed after temperature cycling.[3,19] We believe that the single particle gap is influenced by the deformed CDW states stabilized by the impurity pinning. The deformation is approximated by a smooth function along the sample although one may not exclude a domain structure with characteristic lengths much smaller than the sample length. We predict that the wave number of CDWs is changing along the sample in the asymmetric state.

The presence of deformed CDW states has important consequencies on the dynamics of the CDW motion. First, it prevents the creation of homogeneous electric field along the sample; below threshold, when the local field is determined by the normal current density and low field conductance, differences as large as 20% may appear. Moreover, the deformed states are metastable and they may lead to complicated time dependent phenomena.[9]

The authors are indebed to Professor G. Gruner, G. Kriza and Gy. Hutiray for valuagle discussions. The work at UCLA was supported by NSF grant DMR84-06896.

References

† Present address: UCLA Department of Physics, Los Angeles, CA 90024

1. J.C. Gill, Mol. Cryst. Liq. Cryst. 81, 73 (1982), see also, J.C. Gill and A.W. Higgs, Solid State Comm. 48, 709 (1983).

2. J.C. Gill, Solid State Commun. 39, 1203 (1981).

3. Gy. Hutiray, G. Mihaly and L. Mihaly, Solid State Commun. 48, 227 (1983).

4. G. Mihaly and L. Mihaly, Phys. Rev. Lett. 52, 149 (1984).

5. L. Mihaly and G. Gruner, abstract submitted to this conference, J. Dumas and C. Schlenker, abstract of invited talk.

6. J. Dumas and C. Schlenker, Phys. Rev. B (1984 Aug.).

7. R.S. Lear, M.J. Skove, E.P. Stillwell and J.W. Brill, Phys. Rev. B 29, 5656 (1984).

8. R.M. Fleming and L.F. Schneemeyer, Phys. Rev. B28, 6996 (1983) and references therein.

9. A. Janossy, G. Mihaly and L. Mihaly, paper presented at this conference and references therein.

10. B. Joos and D. Murray, Phys. Rev. B 29, 547 (1984).
W. Kinzel, Phys. Rev. Lett. 51, 1787 (1983).

11. L. Mihaly and G. Gruner, Solid State Comm. 50, 807 (1984).

12. R. Bruinsma, private communication.
F. Beleznay, private communication.

13. H. Fukayama and P.A. Lee, Phys. Rev. B17, 535 (1977).

14. L. Sneddon, M.C. Cross and D.S. Fisher, Phys. Rev. Lett. 49, 292 (1982).
L. Sneddon, Phys. Rev. B 29, 719 (1984).
D.S. Fisher, Phys. Rev. Lett. 50, 1486 (1983).

15. D.S. Fisher, preprint.
P. Littlewood, abstract of invited talk.
J.R. Schrieffer, private communication.

16. A. Janossy, G. Mihaly and G. Kriza, Solid State Comm. 51, 63 (1984).
L. Mihaly and A. Janossy, Phys. Rev. B, in press.

17. N.P. Ong and G. Verma, Phys. Rev. B27, 4495 (1983).

18. G. Mihaly, Gy. Hutiray and L. Mihaly, Phys. Rev. B28, 4896 (1983).

19. N.P. Ong, G. Verma and K. Maki, Phys. Rev. Lett. 52, 663 (1984).

20. A.W. Higgs and J.C. Gill, Solid State Comm. 47, 737 (1983); N.P. Ong, private communication.

21. H. Mutka, S. Bouffard, J. Dumas and C. Schlenker to be published in Journal de Physique.

RELAXATION OF THE DEFORMED CDW STATE: ELECTRIC AND THERMAL HYSTERESIS

A. Jánossy, G. Mihály and L. Mihály*

Central Research Institute for Physics
H-1525 Budapest, P.O.Box 49, Hungary
*University of California at Los Angeles
Los Angeles CA 90024 USA

An analysis of experimental results on electric and thermal hysteresis of orthorhombic TaS_3 is given in the frame of a deformable charge density wave model. We discuss in detail i/ the relaxation of the deformation created by a sudden temperature change: ii/ the reversal of the polarity of the electric field induced CDW deformation upon reversal of current; iii/ time delayed switching between ohmic and CDW conducting states. These phenomena are shown to be related.

Metastable states

The appearance of metastable states is well documented in a number of charge density wave (CDW) systems showing sliding CDW conductivity[1-8]. Metastability has been revealed mainly by the dependence of electric conductivity on thermal or electric history. Although theoretical studies[9-11] stress the importance of CDW deformations due to impurities a full microscopic understanding of the origin of metastability is still lacking. In this paper we shall assume that metastable states may be characterized by a position dependent out of equilibrium CDW wavevector. In metastable states the charge transfer to the conducting chains is also out of equilibrium, small deformations of the lattice may give rise to a charge transfer from non-metallic components or may represent a redistribution of charge within non-equivalent metallic chains.

We use the term "deformed CDW" in the sense of an out of equilibrium CDW wavevector which is neccesarily accompanied by an out of equilibrium CDW amplitude. The material remains neutral on scales larger than determined by conduction electron screening. Under experimental conditions of interest in this paper this screening length is small compared to sample dimensions.

The assumption[5,6,9] that metastable states differ in the distribution of the CDW wavevector q is a plausible one but has not yet been proven experimentally. A temperature dependent q observed in o-TaS_3[13] and blue bronze[14] may account in this model for the thermal hysteresis. Slight changes in q observed[14] in blue bronze may indicate an electric field induced deformation as discussed below.

In our view the main support for the idea is that the normal resistance of o-TaS_3 may be both larger or smaller than its equilibrium value dependent on thermal history[4]. The resistivity of this material is determined by the Peierls gap, E_g, which for small deviations is assumed to vary linearly with the CDW wavevector q. The non equilibrium resistance has opposite signs depending on whether the deformed state is reached by a sudden increase or decrease of the temperature from an equilibrium resistance state (Fig. 1).

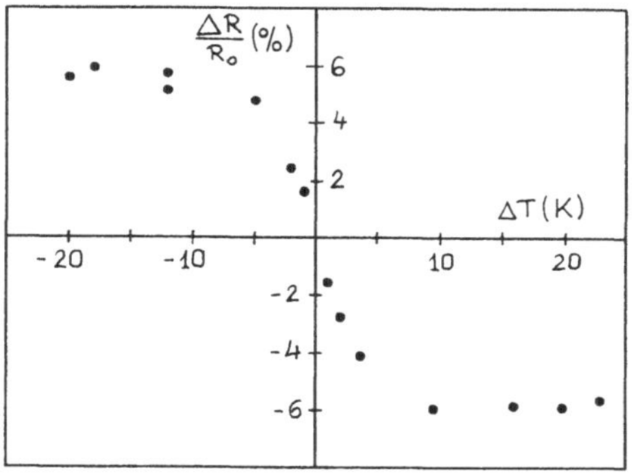

Fig. 1. Non-equilibrium resistance $\Delta R=R-R_o$ following a thermal quench relative to the equilibrium resistance R_o in o-TaS_3 at T=120 K. ΔR is measured a few minutes after the temperature was changed by ΔT from an equilibrium resistance state. Note that ΔR changes sign with changing sign of ΔT and saturates for $|\Delta T|>10$ K.

In this paper the relaxation of metastable states created both by thermal and electric means is discussed. The relaxation of the resistivity following a sudden temperature or electric field change is a consequence of a time dependent CDW wavevector. In absence of an electric field the displacement of the CDW accelerates the relaxation[16]. In all cases when the average q changes a conversion between normal and CDW electrons takes place in terms of a two fluid model. Regions near electrode contacts probably play a special role in CDW conduction[17] and thus in the relaxation of deformations under high fields. It seems that at these regions the CDW to normal electron conversion may be much faster than in the interior.

We first briefly discuss the relaxation of thermally induced states and then the relaxation of deformation induced by an electric field.

The main goal is to show that thermally and electrically induced deformations are similar in nature.

Relaxation of thermal quench induced metastable states

The simplest way to induce metastable states in our model compound, o-TaS_3, is by suddenly decreasing its temperature. In typical experiments[15] - when quenching by $\Delta T \approx 50$ K to temperatures around 100 K - the new thermal equilibrium is reached within a few milliseconds, the resistance, however, is still increasing after several hours. The presence of an electric field - if sufficiently high - increases the relaxation rate[16] (Fig. 2).

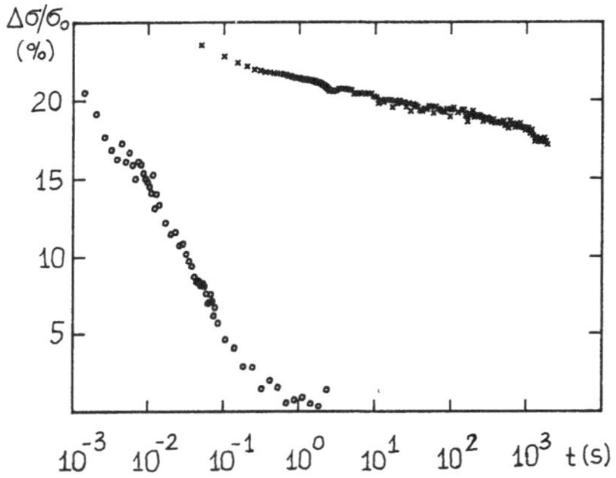

Fig. 2. Relaxation of the conductivity following a thermal quench. X: spontaneous decay in negligible electric field ($E=E_T/5$). O: accelerated relaxation in presence of an electric field ($E=E_T$). $\Delta\sigma/\sigma_0 \approx -\Delta R/R_0$

In absence of an electric field the quench induced deformation is homogeneous on scales larger than the average impurity - impurity distance. The decay of the total resistance, measured by sufficiently small electric fields, measures the time variation of the average CDW wavevector. For not too large deformations the time variation of the resistance R and that of the average wavevector are proportional:

$$\frac{dR}{dt} \sim \frac{dq}{dt} \qquad (1)$$

Figure 2 shows a typical resistance relaxation curve[15] for o-TaS_3. Similar results were found in blue bronze[18]. Within the range of observation (10^{-1} to 10^4 s) the resistance increases with the logarithm of time:

$$R(t)-R_o=\alpha \ln t/\tau \qquad (2)$$

where α and τ are constants. A simple model has been proposed[15] assuming the rearrangement of the CDW to be a thermal process determined by a broad distribution of impurity pinning potentials.

Experimentally relation (2) holds within a large time range. It must break down for very short and very long times. Within reasonable times equilibrium cannot be reached. R_o, the equilibrium value of the resistance, however, may be determined by applying appropriate electric field pulses[4]. The initial resistance characterizing the initial deformation is also difficult to estimate. It seems there is a critical deformation[9] depending on the defect concentration[4,19] above which no deformations can occur. This is indicated on Fig. 1 where the non-equilibrium resistivity $\Delta R=R-R_o$, measured a few minutes after the quench from an equilibrium state is plotted vs. the temperature change, ΔT of the quench. ΔR does not increase without limit with ΔT but saturates above a relatively small value of $\Delta T=10$ K. The simplest explanation is that above some critical value deformations cannot be held by impurities. The remaining initial non-equilibrium resistance depends on temperature[4] and impurity concentration[19]. In high purity o-TaS_3 10^{-1} s after the quench a value of $\Delta R/R_o=0.16$ is found, the non-equilibrium resistance change is far from negligible!

CDW-s deformed by an electric field

Electric fields polarize the CDW system. The polarization due to a rigid collective displacement gives rise to the low field dielectric constant[20] and is well described by the classical model[21]. A much larger polarization may be obtained by the deformation of CDW-s under large fields[22,23]. CDW-s remain pinned at boundaries (near electrode contacts) even in presence of relatively large currents[24] for which the electric field is larger than threshold along the sample. Due to this pinning the electric field compresses the CDW condensate in one direction thus increasing its amplitude and wavevector at one end while decreasing it at the other. Switching off the field the CDW freezes into a polarized state which decays only slowly.

In the polarized state the normal resistivity and the temperature dependence of the resistivity (Peierls gap, E_g) become inhomogeneous as shown by experiments[22,23]. The resistivity compared to the center (x=0) varies with q along the sample:

$$r(x) - r(0) \sim q(x) - q(0) \tag{3}$$

The experiments[22,23] proving the appearance of the polarized state measured the resistivity in the frozen in state after an electric field larger than threshold was switched off. We believe this giant polarization gives rise to important effects with the field switched on also. In our view the anomalous thermoelectric power under large currents found in $NbSe_3$[25] and TaS_3[26] is not related to thermoelectricity but rather is a consequence of the inhomogeneity of the polarized state.

Relaxation of the polarized state

In absence of an electric field a previously induced polarized state relaxes towards a homogeneous steady state. This relaxation is not observed as a time variation of the resistance between boundaries (contacts) as the increase of resistivity with time at one end is compensated by the decrease at the other.

A current pulse much larger than threshold wipes out a thermal quench induced deformation by removing the extra CDW charge. The resulting state is not the equilibrium one - despite the time independent total resistance - after the current is switched off the system remains in a polarized state slowly relaxing towards the homogeneous equilibrium.

Reversal of polarity of field induced CDW deformations

If a field above threshold is reversed a relatively fast transient conductivity increase is observed[1] (Fig. 3 pulse B). This effect - which we call sign memory charging - reflects the displacement current of the deformed CDW system as its polarization is reversed[9,27]. A hint of what is happening during the polarization reversal is given by the time variation of superlattice Bragg reflections of blue bronze following a reversal of a current larger than threshold[14]. The X-ray diffraction of the total sample with current flowing through it was investigated together with the transient conductivity. It was found that during the transient the superlattice reflection narrows substantially in q space. We suggest this to be due to a temporary increase of the homogeneity of q along the sample. In the steady states of opposite polarity q is inhomogeneous, it decreases along the sample giving rise to an increased linewidth. During the reversal the distribution of q(x) narrows and the suparlattice reflections temporarily narrow also. The polarization reversal is observed above a critical current which is, however, somewhat smaller than the critical current for an observable steady state sliding CDW conduction[7]. In our view this is related to the special role of contact regions in sliding CDW conduction. To reverse the polarization a displacement of charge redistributing q(x) without changing the average q is sufficient. In a two fluid model no CDW charge is generated or converted to normal. Thus sign memory charging becomes observable once the electric field is larger than threshold E_T in at least part of the sample. Since CDW to normal conversion is effective only at contacts a field larger than E_T along the whole sample is neccesary for the onset of an observable steady state CDW current.

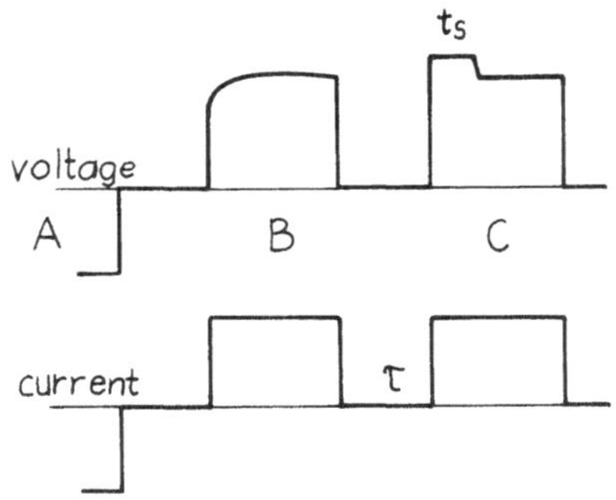

Fig. 3. Pulse sequence to observe transient states. Sign memory charging appears on voltage response B to currents near threshold preceded by pulse A of opposite sign. Delayed switching from normal to CDW conducting states appears on voltage response C preceded by pulse B of the same sign.

There is a simple reason for an inhomogeneous electric field in the polarized state[27]: the current I is constant along the sample and the electric field follows the variation of the resistivity:

$$E(x) = I\ r(x) \tag{4}$$

It seems observable CDW current flows only if $E(x) \geq E_T$ all along the sample.

Delayed switching from a normal to a CDW conducting state

A curious transient is observed[28] in the voltage response to a current pulse if this pulse is preceded by a pulse of the same sign (Fig. 3 pulse C). Experimental details of results on o-TaS_3 are presented in Ref. 29. The transient appears as a voltage corresponding to the ohmic value for some time t_s. At t_s the voltage switches to a smaller value corresponding to a non-ohmic steady state. The delay t_s for switching between the normal and CDW conducting states depends on the current I, it rapidly decreases as I increases above a critical value. Some dependence on the time interval between current pulses, τ, is also found, for short intervals the delay t_s becomes also short. The experimental data discussed below were taken with long enough τ ($\approx$ 1 s) where this dependence is no more observed.

We suggest that delayed switching is due to the relaxation of a transient state in which the polarization of the CDW is larger than the steady state polarization with current I. Our interpretation is different from that of Joos and Murray[30]. We believe the delay t_s is uniquely determined by the current pulses.

We assume that for some - presently not known - reason the polarization, just after the current is switched on, is larger than in steady state. As the relaxation towards steady state begins the resistivity $r(+\ell/2)$ at end $x=+\ell/2$ decreases with time while at the other end $r(-\ell/2)$ increases. By symmetry $r(0)$ remains constant. If the current I is larger than critical, but not too large, the electric field which follows the resistivity at end $x=-\ell/2$ may be initially less than threshold. It reaches threshold after time t_s. During the delay t_s no observable CDW current flows since - as discussed above - it is inhibited if $E<E_T$ in part of the sample. At time t_s the electric field becomes equal threshold at the part at which it is smallest:

$$E(-\ell/2,\, t_s)=r(-\ell/2,\, t_s)I=E_T \qquad (5)$$

and a switching to a CDW current carrying state is observed. By measuring the delay t_s as a function of current I the relaxation of the resistivity $r(-\ell/2,\, t_s)$ at the end blocking the CDW current is measured. During the delay the total resistance is constant, the resistivity increase at $x=-\ell/2$ is balanced by the decrease at $x=+\ell/2$:

$$\frac{dr(-\ell/2)}{dt} = - \frac{dr(+\ell/2)}{dt} \tag{6}$$

An attempt to observe the simultaneous increase and decrease of the resistivities at the ends preceding switching is described elsewhere in this volume[29].

Equivalence of thermal quench and electric field induced deformations

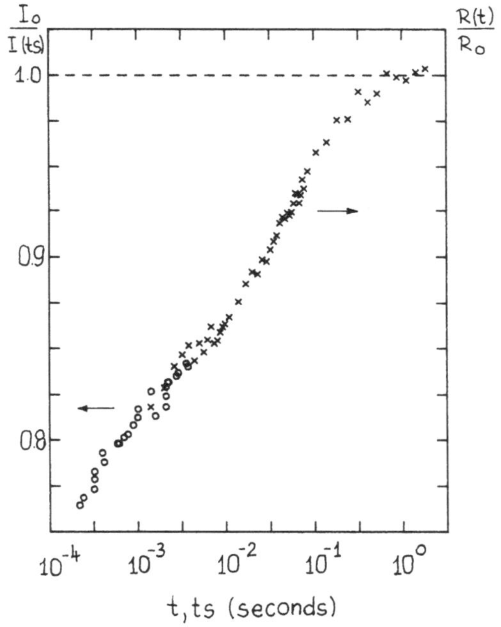

Fig. 4. Equivalence of thermal and electric hysteresis. x: relaxation of resistance R(t) following a thermal quench, 0: dependence of delay for switching, t_s, from normal to CDW conducting states on current I. According to the model I_o/I is proportional to $r(-\ell/2,\ t_s)$ the resistivity at the blocking end.

Our explanation for delayed switching is supported by the observation that the relaxation of the resistance R(t) of a deformed state induced by a thermal quench has the same relaxation rate as the relaxation of the blocking resistance $r(-\ell/2,\ t_s)$ of the polarized state. On Fig. 4 the current ratio I_o/I vs. the delay t_s for switching from a normal to CDW conducting state is shown together with the time dependence of the total resistance following a thermal quench, $R(t)/R_o$. In the two experiments the same o-TaS_3 sample was measured at the same temperature. I_o is the only fitted parameter, it determines the threshold field, $E_T=I_o r(0)$. The relaxation of the total resistance R(t) is measured under a current approximately equal to I_o. Within our model:

$$r(-\ell/2,\ t_s)/r(0)=I_o/I. \tag{7}$$

Figure 4 shows the equivalence of relaxations following a sudden thermal or electric change, we find:

$$r(-\ell/2,\ t_s)/r(0) \approx R(t)/R_o \text{ for } t_s = t \qquad (8)$$

within the range of overlap. It is remarkable that the slopes of the two resistivity time dependence curves are equal indicating that this equivalence holds in a broad time range.

In conclusion we attempted to understand some experimental observations on the thermal and electrical hysteresis of sliding CDW systems by a qualitative model.

We are indebted to Gy. Hutiray, A. Zawadowski, G. Grüner, N.P. Ong and A. Zettl for useful discussions.

References

1. J.C. Gill; Solid State Commun. 39, 1203 (1981)
2. R.M. Fleming; Solid State Commun. 43, 167 (1982)
3. J.C. Gill; Mol. Cryst. Liq. Cryst. 81, 73 (1982)
4. Gy. Hutiray, G. Mihály, L. Mihály; Solid State Commun. 47, 121 (1983)
5. G. Mihály, Gy. Hutiray and L. Mihály; Solid State Commun. 48, 203 (1983)
6. A.W. Higgs and J.C. Gill; Solid State Commun. 47, 737 (1983)
7. G. Mihály and L. Mihály; Solid State Commun. 48, 449 (1983)
8. J. Dumas, C. Schlenker, J. Marcus and R. Buder; Phys. Rev. Lett. 50, 757 (1983)
9. J.C. Gill; Proceedings of the International Symposium on "Non linear transport and related phenomena in inorganic quasi one dimensional conductors" Sapporo (Japan) 1983. p. 139.
10. H. Fukuyawa and P.A. Lee; Phys. Rev. B17, 535 (1977)
11. L. Sneddon, M.C. Cross and D.S. Fisher; Phys. Rev. Lett. 49, 292 (1982)
12. D.S. Fisher; Phys. Rev. Lett. 50, 1486 (1983)
13. Z.Z. Wang, H. Salva, P. Monceau, M. Renard, C. Rouceau, R. Ayroles, F. Levy, L. Guemas and A. Meerschaut; J. Physique 44, L311 (1983)
14. T. Tamegai, K. Tsutsumi, S. Kagoshima, Y. Kanai, M. Tani, H. Tomozawa, M. Sato, K. Tsuji, J. Harada, M. Sakata and T. Nakajima; to be published

15. G. Mihály and L. Mihály; Phys. Rev. Lett. 52, 149 (1984)
16. G. Mihály, A. Jánossy and G. Kriza; in present volume
17. N.P. Ong, G. Verma and K. Maki; Phys. Rev. Lett. 52, 663 (1984)
18. L. Mihály and G. Grüner; in present volume
19. Gy. Hutiray and G. Mihály; in present volume
20. G. Grüner, L.C. Tippie, J. Sanny, W.G. Clark and N.P. Ong; Phys. Rev. Lett. 45, 935 (1980)
21. G. Grüner, A. Zawadowski and P.M. Chaikin; Phys. Rev. Lett. 46, 511 (1981)
22. A. Jánossy, G. Mihály and G. Kriza; Solid State Commun. 51, 63 (1984)
23. L. Mihály and A. Jánossy; Phys. Rev. B in press
24. N.P. Ong and G. Verma; Phys. Rev. B27, 4495 (1983)
25. R.H. Dee, P.M. Chaikin and N.P. Ong; Phys. Rev. Lett. 42, 1234 (1979)
26. J.P. Stokes, A.N. Bloch, A. Jánossy and G. Grüner; Phys. Rev. Lett. 52, 372 (1984)
27. G. Mihály, A. Jánossy and G. Kriza; Phys. Rev. B Sept. 1984.
28. A. Zettl and G. Grüner; Phys. Rev. B26, 2298 (1982)
29. G. Kriza, A. Jánossy and G. Mihály; in present volume
30. B. Joos and D. Murray; Phys. Rev. B29, 1094 (1984)

THERMAL HYSTERESIS IN THE THERMOPOWER OF $o\text{-}TaS_3$

A.W. Higgs*

H.H. Wills Physics Laboratory

Tyndall Avenue, Bristol BS8 1TL

England

A thermopower study has been performed on the charge-density wave (CDW) material orthorhombic tantalum trisulphide ($o\text{-}TaS_3$), for temperatures in the range 300K to 55K. Above ∿63K, the thermopower S is positive and varies in a manner which, although qualitatively similar to that reported by other workers, differs in detail. In particular, for temperatures between T_p∿215K at which coherent CDW formation occurs and ∿55K, S displays a hysteretic temperature dependence. Similar and obviously related behaviour has also been observed in the longitudinal conductivity of $o\text{-}TaS_3$. A discussion of these phenomena is given along with their possible origin.

1. Introduction

The quasi-one dimensional material orthorhombic tantalum trisulphide ($o\text{-}TaS_3$) undergoes a Peierls transition at temperature T_p∿215K, to a charge-density wave (CDW) state[1]. Recent studies of the low-field transport properties of this material [2-4] revealed that the longitudinal conductivity σ, along the c-axis, exhibits thermal hysteresis in the temperature range 55K to 205K. This behaviour was attributed to variability in the CDW wavevector which is incommensurate with the underlying lattice throughout this range[2,3,5]. It was thus the intention to see whether the thermopower S also exhibits hysteretic behaviour. The results in this paper show that this is the case, and they should also resolve the controversy[6-9] over the reported sign of S.

2. Experimental

The thermopower of $o\text{-}TaS_3$ was measured using a standard uni-directional temperature gradient technique. The sample under investigation, typically ∿8mm long and a few μm in cross-section[2,3], was mounted on a sheet of MgO supported between two copper thermal reservoirs. Electrical connections to the sample were made via copper leads attached to its ends using silver paint. The temperature difference across the sample ΔT (typically ∿½K) was provided by a small heater and was determined using a Au(0.03%Fe)-chromel thermocouple.

The thermal voltage produced by the sample $V_s(\Delta T)$ was measured using a passive bridge (Fig.1). $V_s(\Delta T)$ is given by $V(\Delta T)-V(0)$, where V the voltage measured with switch S1 down, is that which is required to balance the bridge with S1 up. The current source supplies the input bias current of the Keithley 177 DVM. With this system S could be measured with a precision of about 1μV/K.

Since it was envisaged that the thermopower of $o\text{-}TaS_3$ may exhibit hysteretic behaviour, all measurements whether on cooling or warming were taken with monotonic

---

* In receipt of S.E.R.C. Fellowship.

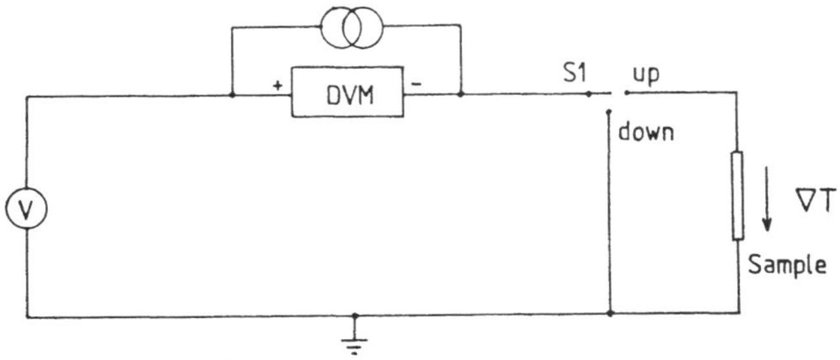

Figure 1. Passive bridge circuit used to measure the thermal voltage produced by the sample.

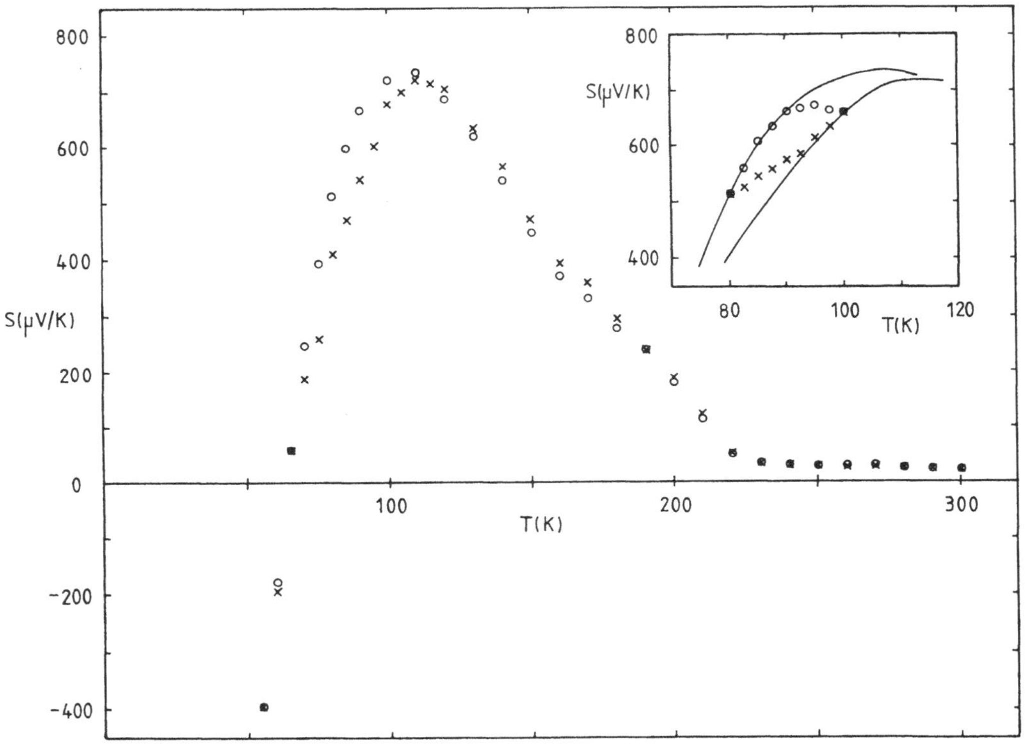

Figure 2. The thermopower of o-TaS_3 S versus temperature T. Inset is a minor hysteresis loop. Results taken on warming are shown (x), and those on cooling (o). The inset extremal curves are shown (—).

changes in temperature. In particular, for measurements on the cooling cycle, ΔT was always established at $\sim$1K above the temperature of interest, before cooling to that temperature.

3. Results

The thermopower S for one sample examined is given in Fig.2, and is similar to that of three other samples studied. From 300K where $S \sim +24\mu V/K$ down to $\sim$63K, S is positive then becomes negative at still lower temperatures. As a check on these signs the thermopower of $NbSe_3$ was determined at room temperature, and was found to be $-11\mu V/K$, in agreement with existing values[10].

As well as the above features, S was found to display a hysteretic temperature dependence from $\sim T_p$ down perhaps to $\sim$55K. On reversal of the direction of temperature change, S moves monotonically from one extremal hysteresis curve to the other, along a path contained within the main loop (see inset in Fig.2). Also, when the system is on one of the extremal curves, application of a current sufficient to cause continuous motion of the CDW, results in the zero-bias thermopower changing to a value within the loop. The exact value of S in this case depends on whether the previously applied current was parallel to or anti-parallel to the temperature gradient.

At temperatures just above 300K measurements indicated that S continues to decrease almost linearly with increasing temperature[7]. Below 55K, the large sample resistance made thermopower measurements extremely difficult and rather unreliable. However, it would appear that S becomes less negative (perhaps even positive) as T is reduced to 30K.

4. Discussion

From the results displayed in Fig.2 it would appear that there are three distinct temperature regions to be considered: (i) 300K to T_p, (ii) T_p to T_m, and (iii) T_m to 55K, where $T_m \sim 105K$ is that for which S is maximum. In region (i), electron-microscope studies have revealed large fluctuation effects which are believed to result from a pseudo-gap in the electronic spectrum at the Fermi surface[11]. The flat variation with temperature of both σ and S, apparently arises from partial localization of the carriers within the pseudo-gap, as suggested by the low carrier mobility $\sim 2cm^2/Vs$ [12].

Below T_p, the Peierls gap opens fully and a three-dimensionally coherent CDW forms. In regions (ii) and (iii) above, the CDW is incommensurate with the underlying lattice, with q_{c^*} the c* component of the CDW wavevector decreasing from $\sim$ 0.255 at T_p to $\frac{1}{4}$ at $\sim$55K[5,2,3]. This suggests that the system contains phase solitons termed discommensurations (DCs), as indicated by optical[13] and various electrical[2,3,14] measurements. The latter, which revealed the hysteretic behaviour of σ attributed to variability in q_{c^*}, led to the proposal of conduction between the localized electronic states associated with the DCs, for its explanation[2,3]. This conduction may be similar to those reported for polyacetylene[15,16]. Clearly, the mechanism would give a low carrier mobility and a rapidly decreasing but hysteretic dependent carrier density, on cooling from T_p to 55K. All these features are consistent with the behaviour of S, at least above T_m.

For temperatures below T_p, two electrons are liberated for every set of four DCs which disappear as q_c^* tends to ¼. These electrons are localized and their excitation only has a noticeable effect on the transport properties below T_m[2,3,17], hence the negative contribution to S. That band conduction breaks down is evident since a crude calculation suggests that the ratio of electron mean free path to lattice parameter is less than unity below T_m. If this is correct, T_m can be thought of as the cross-over point between the high-temperature predominant DC band conduction, and the low-temperature predominant localization conduction. However, in order to verify the above, it will be necessary to model the system more precisely.

5. Conclusions

In conclusion, the thermopower of o-TaS_3 was measured for temperatures in the range 300K to 55K. Above ∿63K, S is positive whilst below that temperature S becomes negative. In addition, between T_p and 55K, where the CDW becomes commensurate, S displays a hysteretic temperature dependence. A model based on DC band conduction which was proposed to explain the behaviour of σ, would appear to give a qualitative description of the variation of S with temperature.

References

1. T. Sambongi, K. Tsutsumi, Y. Shiozaki, M. Yamamoto, K. Yamaya and Y. Abe, Solid State Commun. 22, 729 (1977).
2. A.W. Higgs and J.C. Gill, Solid State Commun. 47, 737 (1983).
3. A.W. Higgs, Ph.D. thesis (University of Bristol 1983) (unpublished).
4. Gy. Hutiray, G. Mihály and L. Mihály, Solid State Commun. 47, 121 (1983).
5. Z.Z. Wang, H. Salva, P. Monceau, M. Renard, C. Roucau, R. Ayroles, F. Levy, L. Guemas and A. Meerschaut, J. de Phys. Lett. 44,L311 (1983).
6. R. Allgeyer, B.H. Suits and F.C. Brown, Solid State Commun. 43, 207 (1982).
7. B. Fisher, Solid State Commun. 46, 227 (1983).
8. D.C. Johnston, J.P. Stokes, Pei-Ling Hsieh and G. Grüner, in Proceedings of the International Conference on Synthetic Low Dimensional Conductors and Superconductors (Les Arcs), J. de Phys. Colloq. 44-C3, 1749 (1983).
9. J.P. Stokes, A.N. Bloch, A. Janossy and G. Grüner, Phys.Rev.Lett. 52, 372 (1984).
10. P.M. Chaikin, W.W. Fuller, R. Lacoe, J.W. Kwak, R.L. Greene, J.C. Eckert and N.P. Ong, Solid State Commun. 39, 553 (1981).
11. C. Roucau, R. Ayroles, P. Monceau, L. Guemas, A. Meerschaut and J. Rouxel, Phys. Stat. Sol. (a) 62, 483 (1980).
12. N.P. Ong, G.X. Tessema, G. Verma, J.C. Eckert, J. Savage and S.K. Khanna, Mol. Cryst. Liq. Cryst. 81, 41 (1982).
13. F. Ya. Nad, in these Proceedings.
14. J.C. Gill and A.W. Higgs, Solid State Commun. 48, 709 (1983).
15. S. Kivelson, Phys. Rev. B25, 3798 (1982).
16. A. Feldblum, R.W.Bigelow, H.W. Gibson, A.J. Epstein and D.B. Tanner, Mol. Cryst. Liq. Cryst. 105, 191 (1984).
17. S.K. Zhilinskii, M.E. Itkis and F. Ya. Nad, Phys.Stat.Sol.(a) 81, 367 (1984).

DELAYED SWITCHING BETWEEN NORMAL AND CDW CONDUCTING STATES IN o-TaS_3

G. Kriza, A. Jánossy and G. Mihály

Central Research Institute for Physics
H-1525 Budapest, P.O.B. 49, Hungary

Delayed switching, a transient in the voltage response to current pulses near threshold for sliding charge density wave conduction is measured in pure and electron irradiated o-TaS_3. The results are in agreement with a model taking into account electric field induced polarization of CDWs. It is experimentally shown that during the transient the resistivity of the two halves of the sample varies in opposite sense.

Among the many unusual phenomena observed near threshold field for non-linear conductivity in CDW systems perhaps the most puzzling is the time delayed switching from normal to sliding CDW state first observed by Zettl and Grüner[1] on $NbSe_3$. If a rectangular current pulse slightly larger than threshold is applied to the sample the voltage response rises initially to a value corresponding to the ohmic conductivity and drops back to the nonlinear value only after a time delay t_s. In this paper we report investigations of this switching phenomenon on pure and electron irradiated o-TaS_3. Measurements of the dependence of the time delay, t_s, on the current pulse amplitude are described first, then an experiment measuring the conductivity during switching in different segments of the sample is presented. Our data support a model based on the deformability of CDWs expounded elsewhere in this volume[2].

Pure samples had a threshold field for non-linear conductivity of E_T=0.5 V/cm at 130 K. In a series of samples defects were created in a well controlled way by high energy electrons[3]. Samples with defect concentration of 10^{-4} defects/Ta atom (threshold field for non-linear conductivity E_T=1.5 V/cm at T=130 K) and $3x10^{-4}$ defects/Ta (E_T=5 V/cm, T=130 K) were investigated. Switching was observable on almost all specimen investigated, although there are substantial differences in the sharpness and magnitude of the voltage step as well as in the temperature range of observability. The phenomenon is more easy to observe in irradiated samples, nevertheless. Switching can only be detected well below the Peierls transition temperature, we have never observed it above 130 K.

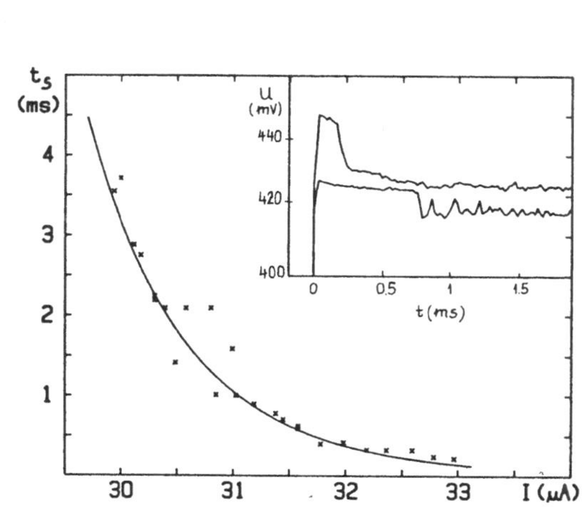

Fig. 1. The dependence of switching delay on current pulse amplitude. Insert: typical voltage responses with switching.

The switching delay measurements were performed in the usual four probe configuration. The voltage response to current pulses of different amplitude were digitalized and stored in computer memory. The dependence of switching delay on current amplitude for a slightly irradiated ($3x10^{-4}$ defects/Ta, E_T=5 V/cm) sample is shown on Fig. 1. Some typical voltage responses are shown in the insert of Fig. 1. Below a threshold the response is ohmic, no transient is observed. Increasing the current pulse amplitude slightly above threshold a switching appears delayed for long time. The longest delay we have observed was about 10 ms. During the delay the voltage is approximately constant, especially for long t_s, and its value is equal to the low field ohmic resistance multiplied by current amplitude. For larger currents the delay t_s decreases rapidly until it becomes unresolved due to the limited resolution of electronic equipment (a few microseconds).

The switching delay also depends on the duration τ of the pause between two pulses: the smaller the pause τ is the shorter the delay will be. At a given pulse separation τ, however, the current pulse amplitude uniquely determines the switching time.

For an irradiated sample at T=110 K we found that increasing the separation between two pulses of the same amplitude the switching delay on the second pulse increases rapidly in the τ=0-~3 ms regime but saturates at a well defined value if τ>10 ms. The time delay date shown in Fig. 1 were taken with a separation τ=200 ms, where the delay is independent of τ.

In some samples the differential resistance calculated from voltage values following switching is negative in a regime above threshold[4] where periodic oscillations with huge amplitude[4] are also present besides the usual narrow band noise.

In an earlier experimental study Zettl and Grüner[1] found on $NbSe_3$ that the switching time at a given current pulse amplitude can be described by a probability distribution. Based on these findings Joos and Murray[5] suggested a theory introducing a probability for the depinning of coupled CDW segments. In o-TaS_3 within our measuring accuracy we attribute all scatter in delay values to instrumental instability.

A model discussed in detail in another paper of this volume[2] is based on the observation[6] that a current higher than threshold induces strongly inhomogeneous deformation of the CDWs. As a result of the deformation the resistivity increases towards one electrode contact and decreases towards the other. Although the average electric field may be higher than threshold it is inhomogeneous along the sample as it follows the variation of resistivitiy. The field at the low-resistivity contact remains under threshold and blocks the CDW current. The inhomogeneous deformation relaxes slowly towards a homogeneous state causing an abrupt switching on of the CDW current when the field at the blocking contact reaches threshold. The full line in Fig. 1 corresponds to the predicted current dependence of the delay time with parameters determined by measuring the relaxation of thermally induced metastable states on the same crystal[2,7].

We stress the importance of the separation τ between pulses. Once the current is switched off, the CDWs relax towards a homogeneous state different from the polarized steady state in the presence of CDW current. The shorter the pause without electric field before the next pulse is the less the system relaxes from the polarized state. Thus, if τ is short the system will be close to the final steady state as the current is switched on and the switching delay is also short.

The finding that the total resistance R_T of the sample remains unchanged in the time interval, t_s, between turning the current on and switching is interpreted in the model by the balance of the resistance decrease at one end of the sample and the increase at the other one during the relaxation of CDW deformation. We performed an experiment by the "touching contact" method[8] to check directly the validity of this hypothesis. The arrangement of electrode contacts is shows in Fig. 2. In addition to the four silver paint contacts a thin gold wire was pressed against the middle of a pure TaS_3 sample. As demonstrated recently[8] such a contact exerts

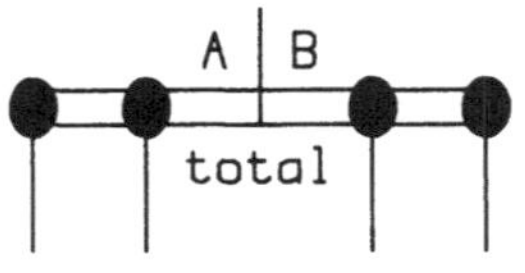

Fig. 2. The arrangement of electrode contacts.

a much smaller perturbation on the CDW system than conventional silver paint contacts encompassing the sample and permits to draw information on the distribution of CDW deformations without significantly changing them.

The voltage responses of A and B sample segments and the total sample to slightly higher than threshold current pulses were recorded and are plotted schematically in Fig. 3. A sign "+" designates voltage responses to electron currents entering the sample at segment A while sign "-" refers to an opposite current direction. Switching with characteristics as described above was observed on the total sample i.e. between the silver paint contacts. A simultaneous rise and fall of the voltages on segments A and B respectively preceding the switching at "+" current direction is in agreement with the model. In this time interval the electric field at the blocking contact of segment A rises to the threshold value as the CDW deformations relax. Simultaneously, the resistivity of segment B is decreasing so that the voltage on the total sample remains constant. Reversing the current direction the role of A and B segments are interchanged. For current direction "-" the voltage decreases on A and increases on B. The sharpness of the switching and other characteristics, however, depend on current direction. This is probably due to a geometrical asymmetry of the sample.

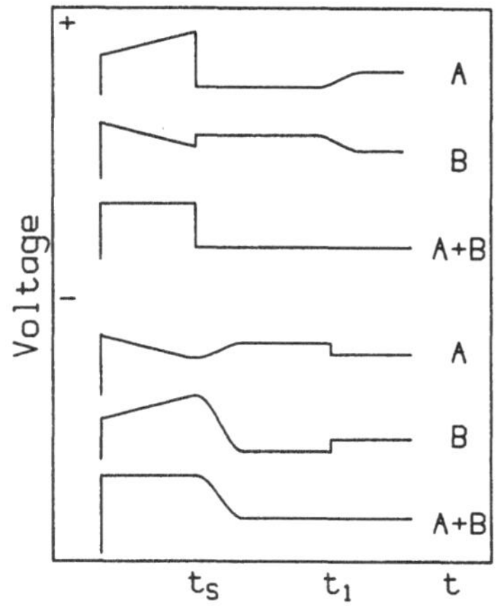

Fig. 3. Schematic plot of the voltage responses of different sample segments.

Another simultaneous switching-like phenomenon occurs at time t_1 (Fig.3). This second switching does not show up in the voltage of the total sample; the upward switching on segment A has the same magnitude as the downward switching on segment B. The origin of these voltage jumps is not known at present.

We presented data on the delayed switching from normal to CDW conducting state in o-TaS_3. A model explaining the current dependence of the switching delay assuming a current induced inhomogeneity of the resistivity is further supported by a direct observation of the resistance of two segments of a sample separated by a weakly perturbing contact.

References

1. A. Zettl and G. Grüner, Phys. Rev. B 27, 4495 /1982)
2. A. Jánossy, G. Mihály and L. Mihály, present volume
3. H. Mutka, S. Bouffard, G. Mihály and L. Mihály, J. Physique Lett. 45, L-113 (1984)
4. L. Mihály and G. Grüner, Solid State Commun., 50, 807 (1984)
5. B. Joos and D. Murray, Phys. Rev. B 29, 1094 (1984)
6. L. Mihály, G. Mihály and A. Jánossy, present volume
7. G. Mihály, G. Kriza and A. Jánossy, Phys. Rev. B, 15. Sept. 1984
8. L. Mihály and A. Jánossy, Phys. Rev. B, 15. Sept. 1984

THE EFFECT OF UNIAXIAL STRAIN ON METASTABLE STATES IN TaS_3

V.B.Preobrazhensky, A.N.Taldenkov
Kurchatov Institute of Atomic Energy, 123182, Moscow, USSR

The uniaxial strain is found to suppress the metastable states in orthorhombic TaS_3 restoring a stable state, in which the small field conductivity doesn't depend on thermal prehistory of the sample.

A distinctly pronounced thermal hysteresis of low field conductivity observed recently by Gy.Hutiray, G.Mihály and L.Mihály [1] and A.Higgs and J.Gill [2] in a Pierls state of TaS_3 proves the existence of metastable CDW states (MS) in 60 to 200 K temperature range. Though a detailed picture of MS is highly unclear up to now there are two hypothesis on how the CDW MS can influence the low field conductivity. Both are based on a recent finding that the period of CDW λ is temperature dependent in the range under discussion, while in MS due to impurity pinning λ can-depending on the thermal prehistory of the sample - be larger or smaller than a stable state value $\lambda_0(T)$. Following the first hypothesis [1] the variations of are accompanied by the changes of the gap; as a result the free carrier concentration is changed. The alternative hypothesis referes the σ changes to the different kink concentration in MS and in a stable state.

A special device, known as a strain transformer [3] was used to study the effect of strain on MS in orthorhombic TaS_3 (Fig. 1). A specimen (typically 3 x 0,01 x 0,003 mm^3) was mounted across the gap between the upper and lower plates of BeCu sample holder in which a connecting ring provides a spring element. The lower plate of the specimen holder is fixed on the cryostat tail. As a force is applied to the upper plate of the holder the specimen is stretched, the strain being dependent on the spring constant k_1 of the ring and the force applied. The upper plate is connected to a spring (k_2) located outside the cryostat and stretched by a servodrive trough a nut an a low-pitch screw. The disolacement of the upper plate (x) is related to that one of the nut (y) by the ratio: $n = y/x = k_1/k_2 \approx 500$. Thus a large displacement of the nut corresponds to a small displacement of the specimen holder upper plate.

Gaidukov has shown [3], that a spring constant of BeCu stretcher is almost temperature unsensitive, allowing a room temperature calibration of such a device to be used throughout the whole temperature range.

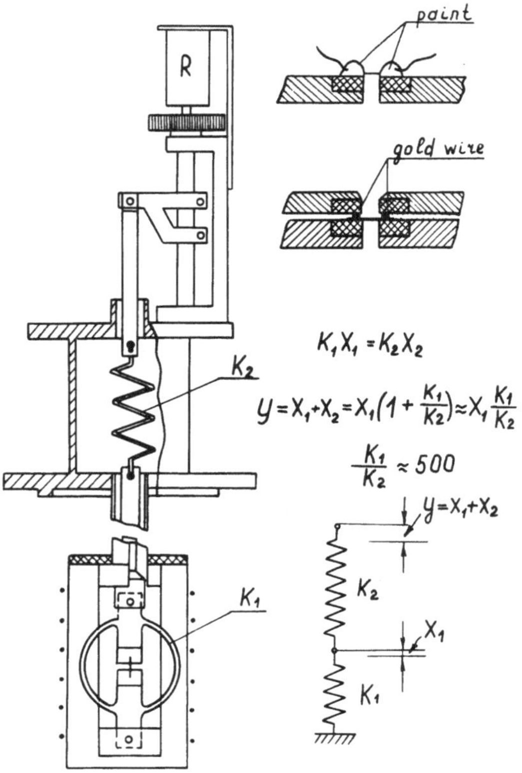

Fig. 1. Schematic of the strain transformer.
R - multiturn potentiometer to provide a signal proportional to the strain.

The sample was clamped between the annealed gold wires served as the electrodes and the sapphire substrates of a stretcher.

Our data on thermal hysteresis of unstretched samples coinside in gross features with that of [1,2].

A uniaxal strain S is found to affect the matastable states in TaS_3 restoring a stable state, where σ_c is intermediate between a cooling and a heating curve. The effect of combined stress and thermal cycling is shown on Fig. 2. Here OC and OH denote overcooled and overheated MS reached at the same temperature T after heating and cooling cycles respectively, ST is a stable one. The a - c b - c curves represent the first run, returning the sample to a stable state ST after heating or cooling cycles. The c - c curve describes a reversible change of the small field conductivity.*

Just the same picture is observed if a sample is subjected to thermal cycling in a stretched state: after the strain is released the sample returns to stable state, in which σ_c doesn't depend on thermal prehistory. This show the MS do exist in a stretched state as well.

A slight discrepancy between a right and a reversal curves is within the limit of our present experimental accuracy.

It is not clear up to now, whether the deformation by itself does produce MS or not. Another question arises: whether a stable state, reached after mechanical cycling is exactly the same, as that

* See a companion paper. This conference.

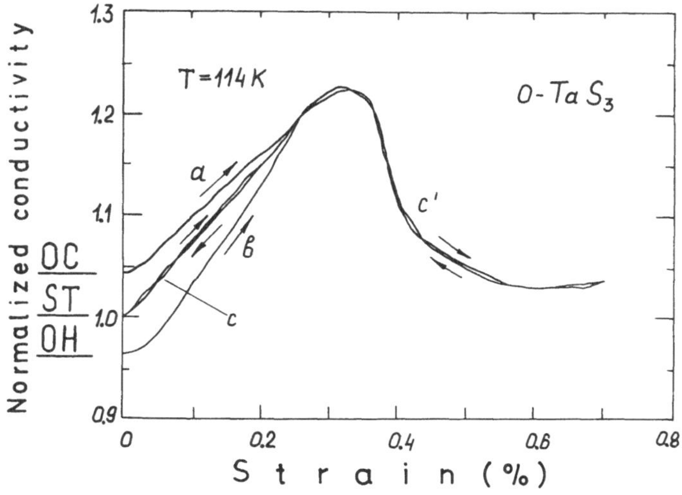

Fig. 2. The combined effect of thermal cycling and strain

OC - overcooled MS state
OH - overheated MS state
ST - stable state

obtained after electric pulse conditioning.

The value of the strain S_m necessary for restoration of a stable state appears to be in agreement with the expected shift of CDW period in MS, provided that a ~20 K overheating or overcooling is reached in MS.

1. Gy.Hutiray, G.Mihály, L.Mihály - Solid State Commun. 47, 121 (1983)
2. A.Higgs, J.Gill - Solid State Commun. 47, 737 (1983)
3. Gaidukov Ju., Danilova N., Tscherbina-Samoilova M. - Pribori i Techn. Eksperim., N 1, 250 (1979)

INFLUENCE OF DEFECTS ON THE METASTABLE STATES OF o-TaS_3

Gy. Hutiray and G. Mihály

Central Research Institute for Physics
H-1525 Budapest, P.O.B. 49, Hungary

We have investigated the influence of defects on metastable states by measuring the hysteresis in the ohmic conductivity as the temperature was cycled. The defect concentration dependence of the width of hysteresis loops taken under the same circumstances is interpreted by an increase of the impurity pinning strength which fixes the otherwise temperature dependent wave number of charge density waves.

The existence of thermally induced metastable states is well known in a number of CDW systems[1-4]. The observation of a temperature dependent wave vector q(T) in o-TaS_3[5] triggered speculations about its possible role in the appearance of metastable states. It has been assumed[3,6] that the freezing of the wave vector q by pinning the phase of CDWs at impurities leads to the out of equilibrium state. We discuss in this framework new experimental results on the conductivity hysteresis of pure and a series of electron irradiated o-TaS_3.

We first discuss results on pure sample. If the temperature is cycled below the Peierls transition a hysteresis of the conductivity σ is observed. Stopping the cycle at any temperature, the relaxation to equilibrium is extremely slow. The steady state conductivity σ_0 can be determined, however, by applying large electric field pulses - called, conditioning pulses - which accelerate the relaxation[2].

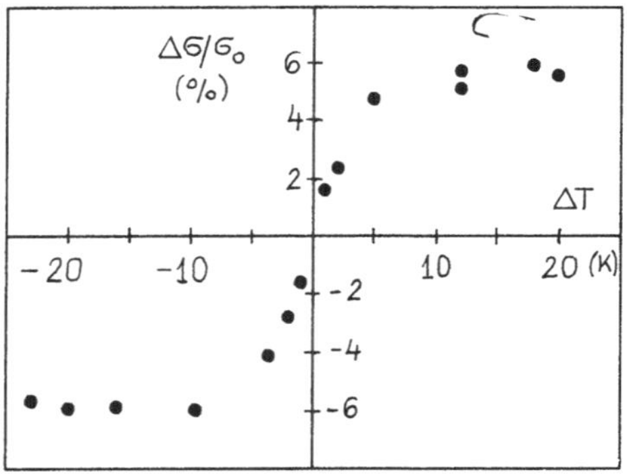

Fig. 1. Deviation of metastable conductivity σ from the stable value σ_0 at T=120 K. $\Delta\sigma/\sigma_0$ characterizes the metastable state reached by cooling/heating from an equilibrium state at temperature T+ΔT.

The metastable conductivity $\Delta\sigma=\sigma-\sigma_0$ measured after the temperature is changed from an equilibrium state at T+ΔT to a temperature T depends on the temperature difference ΔT as shown on Fig. 1. According to the model $\Delta\sigma/\sigma_0$ given in the Figure is measure of $\Delta q/q_0$, the deviation from the equilibrium wave number q_0 at a given temperature T after the sample was cooled/heated from an equilibrium state ($q_o+\Delta q$) at $T+\Delta T$. For small temperature changes the metastable conductivity changes linearly while above $|\Delta T|=10$ K it saturates.

The above experiment is the key in understanding the hysteresis of the conductivity when the temperature is cycled at a constant rate with amplitude ΔT. On Fig. 2 typical hysteresis curves are shown for various cycle amplitudes. For small amplitudes, $\Delta T<10$ K, the hysteresis of the conductivity is negligible small. The material, however, is not adiabatically following its equilibrium state since the slope $\delta\sigma/\delta T$ is much less than observed in usual conductivity measurements without cycling. As the amplitude is increased a hysteresis appears. The width of the hysteresis measured by the maximum conductivity difference between the cooling and heating curves saturates for cycle amplitudes above $|\Delta T|\sim 20$ K.

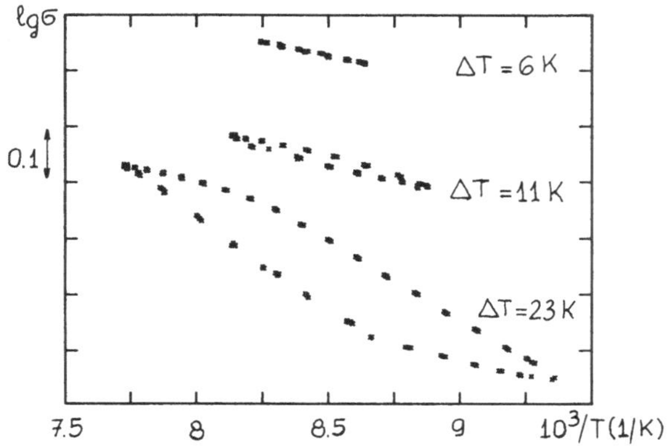

Fig. 2. Typical hysteresis loops of a pure sample for various width of temperature cycling. Curves are shifted by 0.2 for better visiability.

The experiment when the metastable state is reached from an equilibrium value at different temperature (Fig. 1) and the cycle amplitude dependence of the hysteresis loop may be understood if it is assumed - following Gill[7] - that there is a maximum deformation which is still sustained by impurities. We call the corresponding difference from the equilibrium wavevector $\Delta q_{max}=\Gamma$. Thus deformations larger than Γ are rapidly released while relaxation of smaller deformations is slow compared to time interval in experiments above. If the temperature change is small the resulting relatively small deformation ($\Delta q<\Gamma$) is withheld by impurity pinning. The metastable conductivity measured on Fig. 1 is than close to linear with temperature change. Hysteresis cannot be observed if freezing of q is strong enough since q remains all the time during the small cycle constant. The states is, however, out of equilibrium; at the end point of the cycle the wave number differs by $\Delta q=q(T)-q(T+\Delta T)$ from equilibrium. As the cycle is increased the extreme values of Δq increases.

At $|\Delta T|=10$ K the critical deformation ($\Delta q=\Gamma$) is reached at the end points of the cycle. For larger cycle amplitudes the wavevector is no more constant during the cycle, rather it is the difference from equilibrium $\Delta q=q-q_o(T)=\pm\Gamma$ which remains constant in the saturated range and the sign depends on wether the sample is heated or cooled. The hysteresis loop encircles the equilibrium curve measured after conditioning and since relaxation is slow the half width of the hysteresis loop is equal to the saturation value of metastable conductivity given in Fig. 1.

To show the role of pinning in phenomena related to metastable states we varied the strength of pinning by introducing defects through electron irradiation. From the above picture an increase of Γ is expected with increasing defect concentration. The high purity o-TaS_3 was exposed to irradiation by 2.5 MeV electrons which led to the displacement of Ta atoms in the lattice. The irradiation dose - defect concentration relation is known from earlier work of the Fontanay aux Roses group[8,9]. Defect induced in this way act as strong pinning centres as in reflected by the linear increase of threshold for sliding CDW conductivity E_T with concentration c[10].

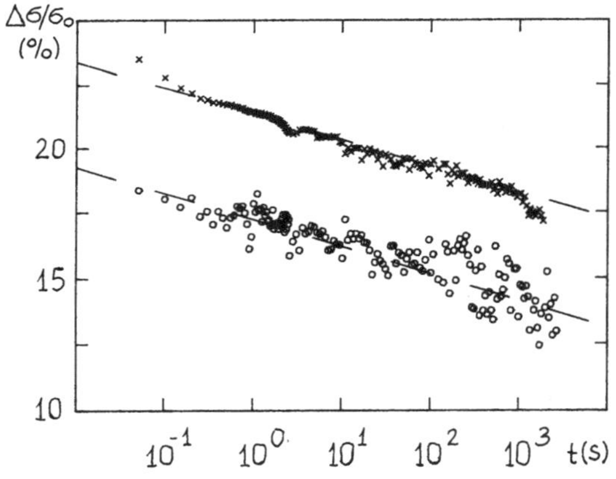

Fig. 3. Spontaneous relaxation of metastable conductivities for pure (0) and irradiated (x) crystals with defect concentration of c=0.03 %.

Fig. 3 shows the metastable conductivity following a sudden change of temperature as a function of time for pure and slightly irradiated (c=0.03%) samples. To suddenly change the temperature a "rapid quench" method described elsewhere was applied[11]. The temperature change of ~50 K applied to the sample is much larger than needed to reach saturation. We find that the relaxation of the pure and irradiated samples are similar, both show a logarithmic time dependence. However, the magnitude of deformation - as measured by the conductivity at a given time after the quench - is much larger for the irradiated than the pure crystal. We deduce from this that the maximum deformation still sustained by impurities, $\Gamma(c)$, is increasing with increasing defect concentration. This is supported by the increase of the width of the conductivity hysteresis loop with defect concentration in the low concentration range, as shown on Fig. 4. In this experiment it was essential that we applied the same temperature cycle (100-150 K) and the same heating/cooling rate (0.1 K/min) for every concentrations. For small concentrations we found that the half width of the hysteresis loop equals to the saturation value of metastable conductivity compared to the conditioned, stable conductivity. Unfortunately for high defect concentrations (c>0.1 %) the equilibrium value of the conductivity could not be determined and we have no direct way to measure the absolute value of the deformation.

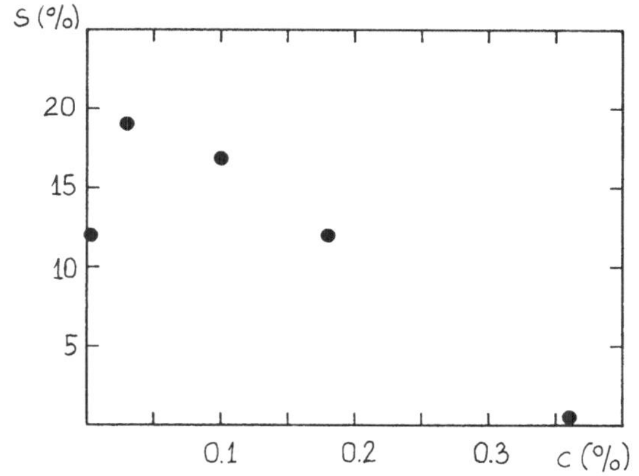

Fig. 4. Half width of hysteresis loops vs. defect concentration. Temperature cycled between 100 and 150 K with the some rate of 0.1 K/min for each concentration.

Fig. 4 shows the half width of the hysteresis loop at T=120 K, versus defect concentration. The initial increase is followed by a gradual diminishing of the hysteresis. For the highest concentrations no hysteresis is found.

It is expected that the maximum deformation which can be withheld by impurities increases with defect concentration, in the same way as the threshold field for nonlinear conduction increases. This behaviour is found at small concentrations where the width of hysteresis loops is proportional to $\Gamma(c)$. On the other hand at high defect concentrations (c>0.3 %) no hysteresis loop is observed. This does not mean - in our view - that the system is not metastable during the cycle. On contrary we believe that for high defect concentrations the maximum deformation Γ possible is large and the wave number of the system remains frozen, close to its value near the Peierls transition temperature, during the whole cycle. For these samples the condition $\Delta q>\Gamma$ is never reached.

In the intermediate concentration range (0.03<c<0.3) the width of the hysteresis loop decreases with defect concentration as espected for the transition from conditions where the change in q reaches the maximum value at relatively small temperature changes to when even the maximum temperature change produces small deformations compared to critical.

In conclusion in the analysis of hysteresis we found a qualitative agreement with a simple model of metastable states. We have shown that metastable conductivity is strongly influenced by impurities even at very low concentration levels. The observed variation of hysteresis loop widths was explained by the simple assumption of a monoton increase of $\Gamma(c)$; the maximum deformation which can be withheld by pinning.

The irradiation of the samples by Drs. S. Bouffard and H. Mutka and discussions with A. Jánossy are thankfully acknowledged.

References

1. J.C. Gill, Solid State Commun. 39, 1203 (1981)
2. Gy. Hutiray, G. Mihály and L. Mihály, Solid State Commun. 47, 121 (1983)
3. A.W. Higgs and J.C. Gill, Solid State Commun. 47, 737 (1983)
4. L. Mihály and G. Grüner, present volume
5. C. Roucau, J. Physique 44, C3-1725 (1983)
6. G. Mihály, Gy. Hutiray and L. Mihály, Solid State Commun. 48, 203 (1983)
7. J.C. Gill, Proceeding of the International Symposium on "Non linear transport and related phenomena in inorganic quasi one dimensional conductors" Sapporo (Japan) 1983. p. 139.
8. G. Mihály, N. Housseau, H. Mutka, L. Zuppiroli, J. Pelissier, P. Gressier, A. Meerschaut and J. Rouxel, J. Physique Lett. 42, L263 (1981)
9. D. Leseur, I. Morillo, H. Mutka, A. Audouard and I.C. Jousset, Rad. Effects 77, 125 (1983)
10. H. Mutka, S. Bouffard, G. Mihály and L. Mihály, J. Physique Lett. 45, L112 (1984)
11. G. Mihály and L. Mihály, Phys. Rev. Lett. 52, 149 (1984)

CHARGE DENSITY WAVE TRANSPORT IN THE BLUE BRONZES $K_{0.30}MoO_3$ and $Rb_{0.30}MoO_3$: METASTABILITY, HYSTERESIS AND MEMORY EFFECTS

J. Dumas and C. Schlenker

Laboratoire d'Etudes des Propriétés Electroniques des Solides, C.N.R.S., B.P. 166, 38042 Grenoble Cedex, France

Non-linear charge density wave transport in the blue bronzes is associated with sharp threshold electric fields and narrow band noise voltage including both high (10-150 KHz) and low (∿ 1 Hz) frequencies as well as metastability, hysteresis and memory effects.

INTRODUCTION

The so-called blue bronzes $K_{0.30}MoO_3$ and $Rb_{0.30}MoO_3$ belong to the class of the ternary transition metal oxides A_xTO_m where A is an alkaline metal and TO_m the highest binary oxide of the transition metal T. $K_{0.30}MoO_3$ and $Rb_{0.30}MoO_3$ are at room temperature metallic conductors. This is due to the charge transfer from the alkaline metal to the conduction band which is built upon a combination of antibonding t_{2g} orbitals from the transition metal and oxygen p_π orbitals.They show a metal-to-semiconductor transition at 180 K[1,3] accompanied with a change of sign of the Hall coefficient[4]. At room temperature, the electrical resistivity[1,2] and the optical reflectivity[5] show anisotropic electronic properties which establish that the blue bronzes are quasi-one dimensional metals. This is well accounted for by the existence in the crystal structure of infinite chains of MoO_6 octahedra extending parallel to the monoclinic (high conductivity) $\vec{b}$ axis and sharing corners only[6]. X-ray refinements studies have shown that most of the 4d electron density is found on the so-called Mo(2) and Mo(3) sites which are involved in the infinite chains[7]. Recently, x-ray diffuse scattering studies have established that the metal-to-semiconductor transition is a Peierls transition towards an incommensurate charge density wave (CDW) state, the component of the wave vector of the distortion along $\vec{b}$ being found to be $q_b = (0.74 \pm 0.01)b^*$ at 110 K[8]. In the semiconducting phase, detailed studies of the dc voltage versus current characteristics have shown that the conductivity is non-linear above a sharp threshold electric field E_t with a switching from the Ohmic regime to the non-Ohmic one depending on the samples and on the temperature range[9]. Other properties typical of CDW transport[10] such as ac 'coherent noise voltage' generated in the non-Ohmic state in the frequency range 10-150 KHz has also been reported[11]. More recently, metastability[12], hysteresis in the low field resistance[13] and memory effects in quenched samples[14] have been found. The conductivity has been found very sensitive to the frequency and the dielectric constant has been well-accounted for by assuming a distribution of relaxation times[15]. Very recently, reflectivity measurements in the far-infrared region have shown a very high reflectivity peak at low energy which has been attributed to the phase mode of the

CDW[16]. Evidence for the amplitude mode has been given by Raman scattering measurements[17].

STRUCTURAL AND PHONONS DATA

Pouget et al[19] had found by x-ray studies precursor effects at room temperature with a q_b value of 0.72 at 300 K. Neutron data by Sato et al.[18] show that q_b increases with decreasing temperature down to ∿ 100 K then keeps a constant value of 0.746 below. Fleming et al.[20] have suggested that a lock-in transition towards q_b = 0.75 could take place at 110 K. This is not inconsistent with recent data by Pouget et al.[19] which obtained a constant value q_b = 0.7495 ± 0.0005 below 100 K. The low temperature value of q_b very likely depends on the stoichiometry and/or the purity of the samples and may be commensurate in the best ones. On the other hand, Tamegai et al.[21] have observed a dependence of the transverse component q_c on the applied electric field above threshold ; this may indicate some distortion of the CDW[21]. Mössbauer spectra of Fe-doped $K_{0.30}MoO_3$ is consistent with a distortion of the MoO_6 octahedra below the Peierls transition[22]. Neutron inelastic scattering studies by Sato et al.[18] and by Pouget et al.[19] have established that a Kohn anomaly takes place at 180 K and corroborate the quasi 1 D nature of the blue bronze.

NON-LINEAR TRANSPORT PROPERTIES

Depending on the samples and, for a given sample, on temperature, three types of behaviour near E_t are found[23] : smooth threshold, switching from the Ohmic to the non-Ohmic regime with precursor voltage pulses, mixed regime with a smooth threshold followed by a switching. Fig. 1a shows switching with precursor pulses, the system hopping between the two regimes. Fig. 1b shows a mixed behavior with two thresholds E_{t1}, E_{t2} together with the onset of the broadband noise at E_{t1} and E_{t2} respectively.

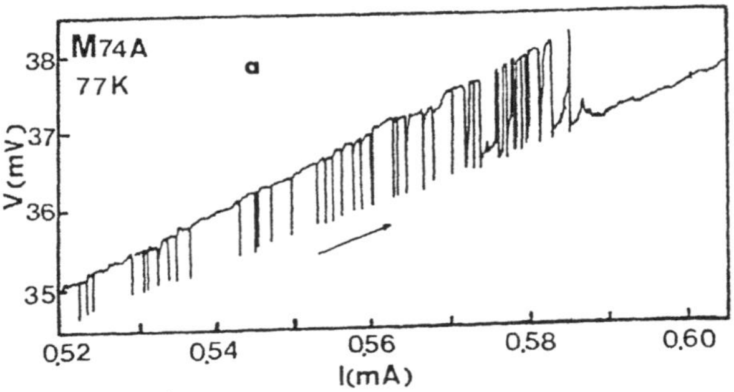

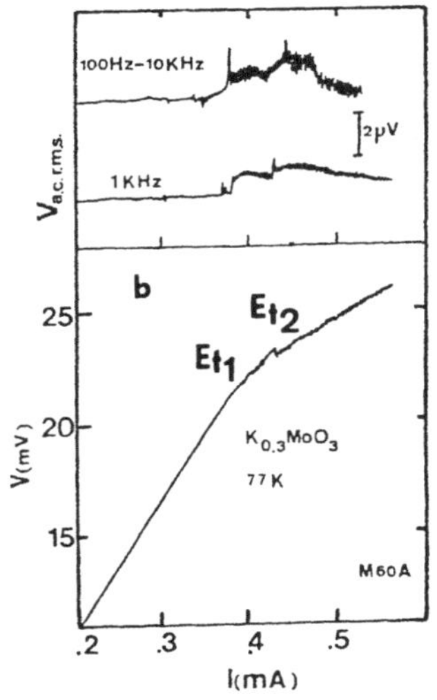

Fig. 1

(a) V-I characteristic of $K_{0.30}MoO_3$ showing precursor pulses ΔV < 0 and switching.

(b) Bottom curve : V-I curve showing two thresholds E_{t1}, E_{t2} ; upper curves : Onset of broad band noise at E_{t1}, then E_{t2}, in the frequency range 100 Hz - 10 KHz and at 1 KHz.

We must also point out that substantial variations in the values of the thresholds fields are found ; E_t lies in the range 50 - 500 mV/cm at 77 K, depending on the samples. No clear correlation between E_t and the resistivity value at 300 or 77 K as well as the resistivity ratio ρ300/ρ77K has been found. These behaviors near E_t suggest that the onset of the non-Ohmic regime is related to two processes : a nucleation of sliding CDW filaments parallel to the high conductivity $\vec{b}$ axis and the transverse propagation of these filaments. The threshold behavior of electron irradied samples may corroborate this picture[24]. The irradiation turns a smooth onset of the non-linearity into a switching process and raises the threshold electric field value. The existence of sliding CDW channels has also been proposed recently by Joos and Murray[25]. In this context, two threshold fields can be defined : a nucleation field E_t^n and a propagation field E_t^p. Depending on the relative magnitude of E_t^n and E_t^p, different behaviors may be expected as it is experimentally found. The data obtained by Hall and Zettl[26] indicate that for $NbSe_3$ both thresholds have different temperature dependences. One should also note that switching with complicated hysteresis has also been observed in $Fe_x NbSe_3$[27]. The temperature dependences of E_t is not fully elucidated. In our samples, E_t reaches a maximum at ∿ 110K while other authors[28] report a continuous rise of E_t. These discrepancies may be due to the fact that the threshold becomes very smooth above ∿ 110 K in our samples and cannot be defined accurately. The abrupt transition at E_t to the non-linear state for those samples exhibiting this behavior is observed only in a limited temperature range ∿ 60 K to ∿ 100 K as shown in Figure 1c.

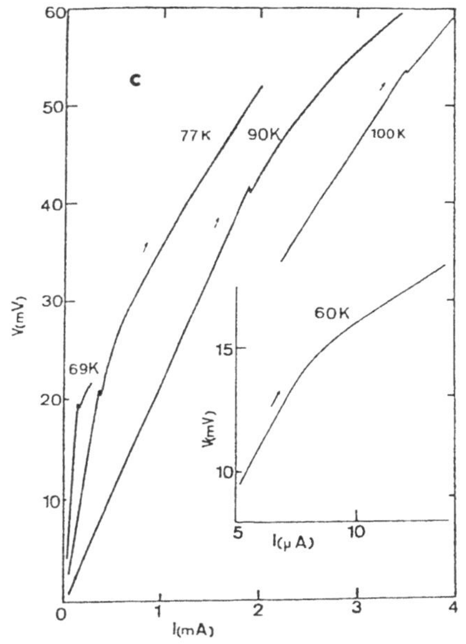

Fig. 1

(c) V-I curves at different temperatures for a $K_{0.30}MoO_3$ crystal exhibiting switching below 100 K.

<u>HIGH FREQUENCY OSCILLATIONS</u>

Fig. 2a shows the Fourier analysis of the noise voltage of $K_{0.30}MoO_3$ for dc currents above E_t. The spectrum consists of one fundamental frequency F_1 defined as the first frequency which appears in the noise spectrum, and its harmonics. As in $NbSe_3$ and TaS_3, the frequency increases linearly with the CDW current[11, 23] with slopes F_1/J_{CDW} = 0.1 MHz/Acm^{-2} for $K_{0.30}MoO_3$ and 1 MHz/A cm^{-2} for $Rb_{0.30}MoO_3$. A.c. conductivity measurements are in agreement with this coherent noise[15]. Monceau et al.[29] have shown that the slope F/J_{CDW} was given by $F/J_{CDW} = 1/ne\lambda$ where n is the concentration of electrons condensed in the CDW and λ the superlattice period. Our results do not seem consistent with this prediction[12].

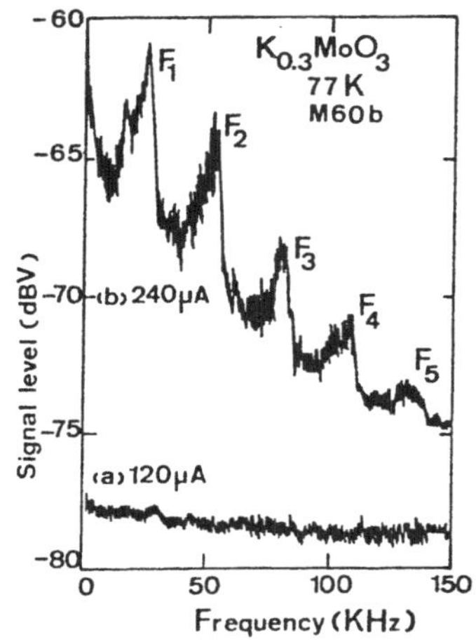

Fig. 2a

Narrow band noise spectrum of $K_{0.30}MoO_3$ at 77K.
Threshold current : I_t = 130 μA.

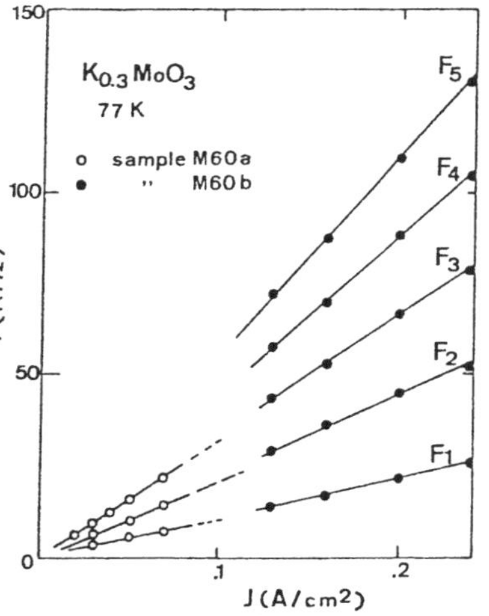

Fig. 2b

Noise frequencies as a function of the excess CDW current at 77K.

LOW FREQUENCY PHENOMENA

At constant current, we have previously reported[12] that pulses of transient voltage corresponding to a sudden decrease of V_{dc} were produced just below and in some samples above E_t. The magnitude of the pulses was ∿ 2 % of V_{dc}. We have attributed these pulses to interactions of CDW domains (boundaries) with defects and indicated that those pulses have some analogies with Barkhausen jumps in ferromagnetic materials. In Figure 3 are shown examples of voltage pulses recorded as a function of time for different values of the current just below the threshold value.

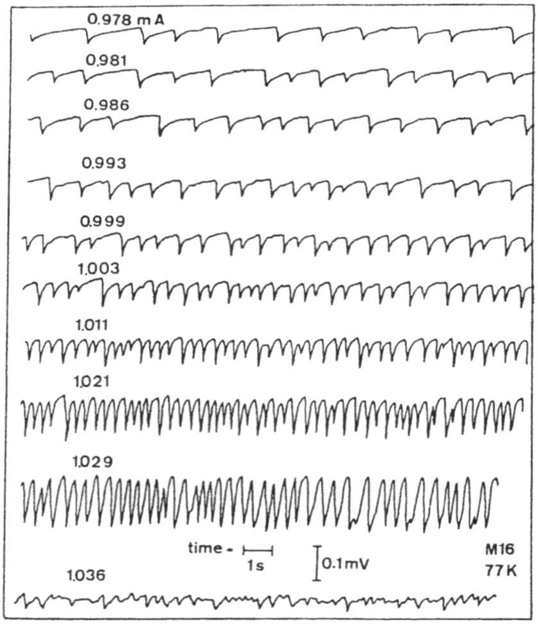

Fig. 3

Voltage pulses as a function of time at indicated currents, for a $K_{0.30}MoO_3$ sample at 77 K. This sample exhibits a switching at I_t = 1.030 mA.

METASTABILITY

In the non-linear state, the V-I curve is not always stable as a function of time[11]. For $K_{0.30}MoO_3$, the excess CDW current may decrease logarithmically with time while for $Rb_{0.30}MoO_3$ and V or Fe doped $K_{0.30}MoO_3$ no significant drift occurs over several hours, as shown in Figure 4.

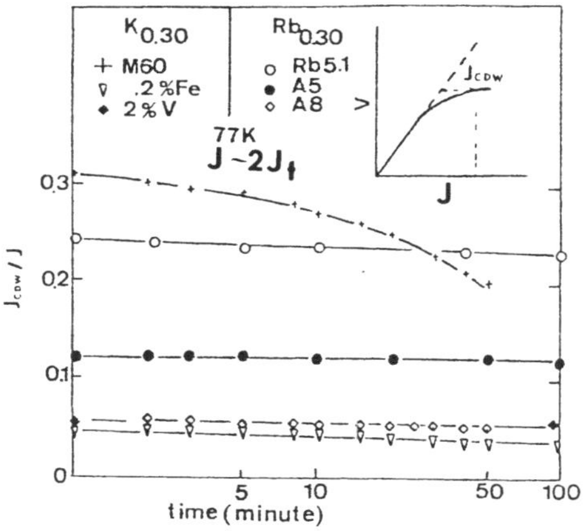

Fig. 4

Ratio of the CDW current density J_{CDW} to the total current density J measured vs time (logarithmic scale). J is kept constant during the experiment ; J_t is the threshold current at 77 K. The upper right inset shows how J_{CDW} is defined.

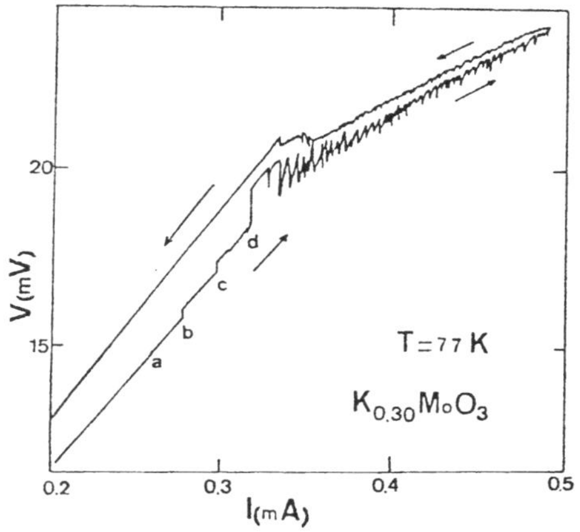

Fig. 5a

dc V-I curve showing steps (a,b,c,d) in the low field resistance.

The drift of the CDW current is not a monotoneous function of time, it exhibits positive and negative steps[11]. These time dependent effects have been attributed to a progressive pinning of CDW domains boundaries as a function of time.
Metastable states have also been found in the low field resistance[13]. Fig. 5a shows that the low field resistance exhibits sudden steps when one sweeps the dc current.

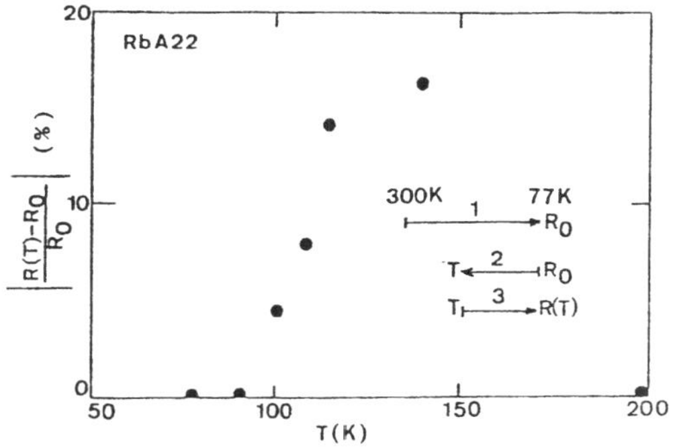

Fig. 5b

Absolute relative change of $\frac{R(T) - Ro}{Ro}$ for successive temperatures cycles for $Rb_{0.30}MoO_3$ (see text). R(T) is measured by ac lock-in technique with currents less than I_t(77K)/100.

Significant changes in the low field resistance values are found when a thermal cycling has been performed. Let Ro be the resistance found after cooling the sample from 300 to 77 K. After heating up to a given temperature T then cooling again to 77 K, a new resistance value R(T) is found at 77 K as shown in Figure 5b. For each point given in this figure, the sample has been heated again up to 300 K before starting a new cycle in order to achieve a virgin state. R(T) is found to increase noticeably when T $\sim$ 100 K ; R(T) is found always smaller than Ro.

For samples showing a switching at E_t, hysteresis is found near the threshold. Figure 6 shows the thresholds $E_{t\uparrow}$ obtained when the current I is swept up to I_{max} and $E_{t\downarrow}$ obtained when I is swept back to zero at the same sweeping rate. If the sweeping rate is much slower when the current is decreased from I_{max}, then $E_{t\downarrow}$ is closer to $E_{t\uparrow}$. On decreasing the current from I_{max}, if one keeps the current constant at a value I_A, the voltage will drift slowly from V_A to $V_{A'}$. Metastability phenomena are more pronounced in electron irradiated samples[24] and W-doped samples[30]. Hysteresis in V-I curves has also been found by Fleming and Schneemeyer[12] in ac experiments.

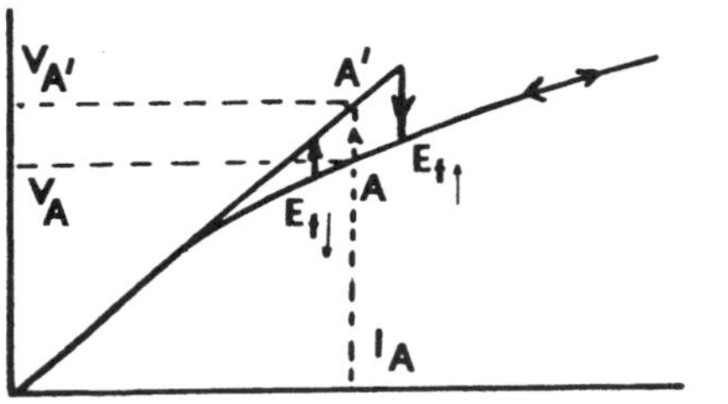

Fig. 6

V-I curve showing hysteresis near E_t. $E_{t\uparrow}$ (resp. $E_{t\downarrow}$) are the threshold values found on increasing(resp. decreasing) the dc current.

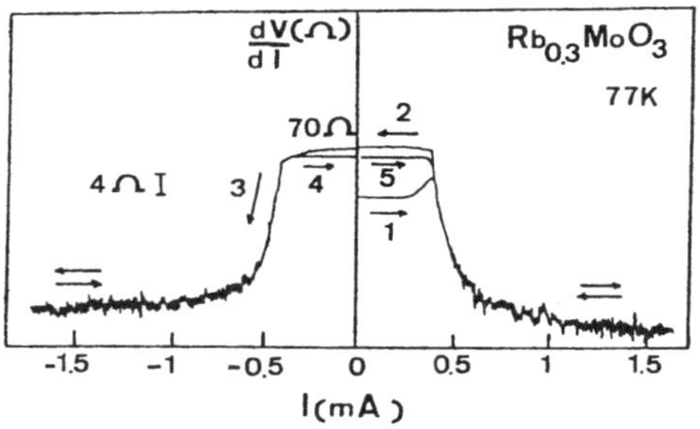

Fig. 7

Differential resistance dV/dI (at 43 Hz) as a function of the dc current for $Rb_{0.30}Mo_3$ at 77 K. 1 refers to the virgin state.

For both $K_{0.30}MoO_3$ and $Rb_{0.30}MoO_3$, the low field resistance depends on the past thermal and electrical history of the sample. Figure 7 shows the differential resistance dV/dI as a function of the current for $Rb_{0.30}MoO_3$. A hysteresis in the low field resistance is found when the current has been swept above the threshold value I_t up to a given value I_{max}. If one labels R_1 the low field resistance in the virgin state and R_2 the resistance after the current has been swept above I_t, one can define an isothermal remanent resistance (IRR) as $\Delta R/R = (R_2 - R_1)/R_1$. After a full cycle (2-5), the following cycles are nearly reproducible if one keeps the same value for I_{max}. R_2 does not seem to drift with time over several hours.

Fig. 8 summarizes the effect of the thermal and electrical history of the sample on the low field resistance. When the sample is cooled from 300 to 77 K with an applied dc current, the Ohmic resistance R_{th} is found larger than the resistance R_1 found with a zero current cooling. One can denote this increase $\Delta R/R = (R_{th} - R_1)/R_1$, the thermoremanent resistance (TRR). The TRR increases noticeably when the current applied during cooling is larger than the threshold current I_t at 77 K. The TRR becomes vanishingly small near 130 K. These results have some similarities with the remanent magnetizations of spin-glasses and also with the results obtained by Tsutsumi et al.[21] on $K_{0.30}MoO_3$ and Hutiray et al. on TaS_3 and by J.C. Gill on $NbSe_3$[31].

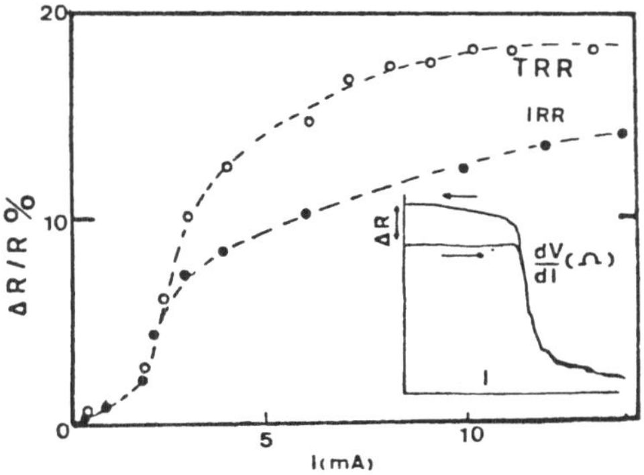

Fig. 8

Thermoremanent (TRR) and isothermal remanent resistance (IRR) of $K_{0.30}MoO_3$ at 77 K. The threshold current I_t at 77 K is 2.2 mA. The inset shows the hysteresis in the differential resistance measurement (1 corresponds to the first sweeping from a virgin state). The horizontal axis corresponds to the current applied during cooling for the TRR and to I_{max} values for the IRR.

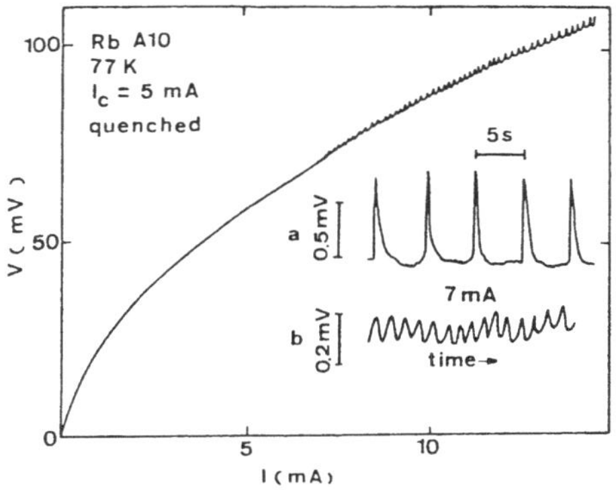

Fig. 9

V-I curve showing voltage fluctuations well above the threshold for a quenched $Rb_{0.30}MoO_3$ sample (I_t = 0.4 mA) ;
inset : (a) voltage pulses for fast cooling ($\Delta V > 0$).
(b) voltage oscillations for slow cooling.
Current applied during cooling I_c = 5 mA.
Measuring current I = 7 mA.

We have found[14] that coherent low frequency ($\sim$ 1 Hz) voltage fluctuations could be generated by quenching the sample with an applied current. Figure 9 shows a V-I curve for $Rb_{0.30}MoO_3$ sample quenched from 300 to 77 K with an applied current. The inset shows quasi-periodic voltage pulses found under fast cooling conditions ($\sim$ 5 s) and voltage oscillations found under slow cooling conditions ($\sim$60 s). For cooling without any current only erratic voltage fluctuations are found. These low frequencies are also found proportional to the excess CDW current. The slope f/J_{CDW} is $\sim$ 0.2 Hz/A cm^{-2} for oscillations in $Rb_{0.30}MoO_3$ and $\sim$ 1 Hz/A cm^{-2} in $K_{0.30}MoO_3$. These results have some similarities with the onset of serrations in stress-strain curves of Al alloys[32].

DISCUSSION

The major role of the metastability in the blue bronzes is now well-established. A possible source of metastability is the existence of crystal defects or impurities. We discuss firstly, in this context, the hysteresis in the low field resistance, then the low frequency voltage oscillations.

The Ohmic resistance is not intrinsic and has to be attributed to non-stoichiometry and/or impurity levels in the Peierls gap. In the 'pure samples', these levels may correspond to localized electrons on Mo^{5+} donor centers. These centers may be located on the Mo sites labeled Mo(2) and Mo(3) in Ref. 6 and have been observed by EPR spectroscopy[33]. After a cooling process, the population of the two corresponding levels would be metastable. If these defects are coupled to the CDW, the motion of the CDW would induce a redistribution of the d electrons population between these two levels. These rearrangements may involve jumps of a 4d electron on neighbouring Mo(2) and Mo(3) sites. Another possibility would be a variable distorsion due to the CDW motion which would then lead to some displacements of the levels in the gap[34]. One also should take into account the effect of the temperature dependence of the superlattice q vector in the metastability phenomena. In this context, different values of the q vector should correspond to different configurations for the CDW domains, possibly for discommensurations. The coupling of mobile defects with discommensurations should then be important. Diffusion of mobile defects in incommensurate structures has already been invoked to account for thermal memory effects[35]. As far as the low frequency voltage fluctuations is concerned, we have proposed[13] that these phenomena could be related to rearrangements of mobile effects, such as Mo^{5+}, under the effect of the applied current during the quenching process : the incoherent voltage pulses would correspond to a random configuration of defects and the coherent ones to a quasi-periodic arrangement. The effect of the quenching would be related to a temperature dependent mobility of these defects, probably involving d electrons jumps between neighbouring sites only.

CONCLUSION

The blue bronzes provide an excellent model compound of the Peierls transition as well as the non-linear transport phenomena attributed to the sliding of CDW. They show most of the properties, including narrow band noise, that had been previously observed on the transition metal trichalcogenides. On top of that, they show both low frequency and metastability phenomena sometimes characteristic of a glass-like behavior. A major interest of these compounds is that the comparatively large size of the crystals allow bulk studies such that optical and inelastic neutron scattering. Direct observation of CDW domains would now be necessary to complete the picture which should emerge in a near future from all the data.

REFERENCES

1. G.H. Bouchard, J. Perlstein, and M.J. Sienko, Inorg. Chem. 6, 1682 (1967) ; W. Fogle and J.H. Perlstein, Phys. Rev. B6, 1402 (1972) ; D.S. Perloff, M. Vlasse, and A. Wold, J. Phys. Chem. Solids 30, 1071 (1969).
2. R.Brusetti, B.K. Chakraverty, J. Devenyi, J. Dumas, J. Marcus, and C. Schlenker, in 'Recent Developments in Condensed Matter Physics', Vol. 2, Ed. J.T. De Vreese, L.F. Leemens, V.E. Van Royen (Plenum 1981) p. 181.
3. P. Strobel and M. Greenblatt, J. Solid State Chem. 36, 331 (1981).
4. E. Bervas, Thesis Docteur-Ingenieur, Université de Grenoble 1984 (unpublished).
5. G. Travaglini, P. Wachter, J. Marcus and C. Schlenker, Solid State Commun. 37, 599 (1981).
6. J. Graham and A.D. Wadsley, Acta Cryst. 20, 93 (1966).

7. M. Ghedira, J. Chenavas, M. Marezio (to be published).

8. J.P. Pouget, S. Kagoshima, C. Schlenker, and J. Marcus, J. Phys. (Paris) Lett. 44, L113 (1983).

9. J. Dumas, C. Schlenker, J. Marcus, and R. Buder, Phys. Rev. Lett. 50, 757(1983).

10. See, for example, P. Monceau, J. Richard and M. Renard, Phys. Rev. B25, 931 (1982) ; R.M. Fleming in 'Physics in One Dimension', Springer Series in Solid State Science 23, Ed. J. Bernasconi and T. Schneider, N.Y. 1981 ; G. Grüner, Physica 8D, 1 (1983); N.P. Ong and G. Verma, 'Proceedings of the International Symposium on Non-Linear Transport and Related Phenomena in Inorganic Quasi-One Dimensional Conductors', Hokkaido Univ, Sapporo (Japan), oct. 1983, p. 115 ; J.C. Gill, ibid. p.139 ; A. Zettl, ibid. p. 41.

11. J. Dumas and C. Schlenker, Solid State Commun. 45, 885 (1983) and in Proc. Sapporo Conf., p. 198.

12. in Ref. 10 and in 'Proc. Int. Conf. on the Physics and Chemistry of Low Dimensional Synthetic Metals' ICSM84, Abano Terme (Italy) (to be published) ; R.M. Fleming and L.F. Schneemeyer, Phys. Rev. B28, 6996 (1983).

13. J. Dumas, A. Arbaoui, J. Marcus, and C. Schlenker, in Proc. ICSM84 ; K. Tsutsumi, T. Tamegai and S. Kagoshima, ibid.

14. J. Dumas, A. Arbaoui, H. Guyot, J. Marcus, and C. Schlenker, Phys. Rev. B30, 2249 (1984).

15. R.J. Cava, R.M. Fleming, P. Littlewood, E.A. Rietman, L.F. Schneemeyer, and R.G. Dunn, Phys. Rev. B, 15 september 1984.

16. G. Travaglini and P. Wachter, Proc. ICSM84, and Phys. Rev. B 30, 1971 (1984).

17. G. Travaglini, I. Mörke, and P. Wachter, Solid State Commun. 45, 289 (1983) ; S.B. Dierker, K.B. Lyons, and L.F. Schneemeyer, Bull. Am. Phys. Soc. 29, 469 (1984).

18. M. Sato, H. Fujishita, and S. Hochino, J. Phys. C, Solid State 16, L877 (1983) and this Conference.

19. J.P. Pouget, C. Escribe-Filippini, B. Hennion, R. Moret, A.H. Moudden, J. Marcus, and C. Schlenker, in Proc. ICSM84 ; C. Escribe-Filippini, J.P. Pouget, R. Currat, B. Hennion, J. Marcus, and C. Schlenker, this Conference.

20. R.M. Fleming and L.F. Schneemeyer, Bull. Am. Phys. Soc. 29, 470 (1984).

21. T. Tamegai, K. Tsutsumi, S. Kagoshima et al., Solid State Commun. 51, 585 (1984). K. Tsutsumi, T. Tamegai, S. Kagoshima in Proc. ICSM84.

22. J.Y. Veuillen, R. Chevalier, D. Salomon, J. Dumas, J. Marcus, and C. Schlenker, this Conference.

23. C. Schlenker, J. Dumas and J.P. Pouget, Proc. ICSM84.

24. H. Mutka, S. Bouffard, J. Dumas, and C. Schlenker, J. Phys. (Paris) Lett. 45, L729 (1984) ; H. Mutka, S. Bouffard, M. Sanquer, J. Dumas and C. Schlenker, Proc. ICSM84. See also : C.H. Chen, L.F. Schneemeyer, and R.M. Fleming, Phys. Rev. B29, 3765 (1984) ; S. Bouffard et al., this Conference.

25. B. Joos and D. Murray, Phys. Rev. B29, 1004 (1984).

26. R.P. Hall and A. Zettl, Solid State Commun. 50, 813 (1984).

27. M.P. Everson and R.V. Coleman, Phys. Rev. B28, 6659 (1983).

28. K. Tsutsumi et al., in Ref. 20.

29. P. Monceau, J. Richard, and M. Renard, Phys. Rev. Lett. 45, 43 (1980).

30. L.F. Schneemeyer, R.M. Fleming, and S.E. Spengler, Bull. Am. Phys. Soc. 29, 357 (1984).

31. Gy. Hutiray, G. Mihaly, L. Mihaly, Solid State Commun. 47, 121 (1983) ; J.C. Gill, Molec. Cryst., Liq. Cryst. 81, 73 (1982).

32. P.G. McCormick, Scripta Met. 4, 221 (1970)

33. G. Bang and G.S. Sperlich, Z. Phys. B22, 1 (1975) ; J. Dumas, C. Escribe-Filippini et al. to be published.

34. A. Janossy, G. Mihaly, G. Kriza, Solid State Commun. 51, 63 (1984).

35. P. Lederer, G. Montambaux, J.P. Jamet, M. Chauvin, J. Phys. (Paris) Lett. 45, L627 (1984).

EFFECTS OF IRRADIATION ON THE BLUE BRONZES $K_{0.30}MoO_3$ AND $Rb_{0.30}MoO_3$

S. Bouffard, M. Sanquer and H. Mutka[+]
SESI, CEN-FAR, B.P. 6, 92260 Fontenay-aux-Roses, France
J. Dumas and C. Schlenker
LEPES CNRS, B.P. 166, 38042 Grenoble Cédex, France
[+] Present address : Technical Research Centre of Finland, SF-02150 Espoo, Finland

The irradiation induced defects act as strong pinning centers at very low defect concentration. The threshold field increases linearly with the defect concentration. The metastable properties are governed by the presence of such defects : a smooth threshold of non linearity becomes a switch associated with large hysteresis. The results suggest that deformable CDW are arranged in domains and trapped in metastable states.

The first investigations of the non-linear properties of the 1D inorganic conductors have been limited to the continuous behaviours[1]. And the theoretical models treated the charge density waves (CDW) as classical particules in a periodic potential[2] or as quantum objects which participate to the conduction by macroscopic quantum tunneling[3]. However, early measurements on some CDW conductors $NbSe_3$ showed a switching phenomena associated with the onset of CDW conduction[4]. Since the discovery of a new non-linear compound (the blue bronze)[5], several unusual transport properties of the CDW state have been more extensively described, namely switching, voltage pulses, memory effects, hysteresis and low-frequency oscillations. These metastable properties have been also observed in the linear chain compounds TaS_3 and $NbSe_3$, but they are less pronounced than in the blue bronzes. It is generally accepted that extend or point defects have the most important role in the pinning of CDW : the threshold field for the onset of CDW current strongly depends on the concentration of pinning centers [6,7] and the transition temperature decreases when the defect concentration increases[6,8]. The radiation experiments permit to introduce in a controlled way these pinning centers. In these contribution, we discuss the effect of the irradiation on the non-linear properties of the blue bronzes and more precisely on the metastable properties.

In the blue bronzes ($K_{0.30}MoO_3$ and $Rb_{0.30}MoO_3$), fast electrons ($E > 200$ keV) produce defects by displacing atoms in elastic collisions with nuclei[8]. However, the electronic excitation at lower energies should be enough efficient to produce defects[9]. The defect concentration can not be directly determined from the damage curves, nevertheless the susceptibility of the d-electron paramagnetic resonance line presenting a Curie like behaviour gives us an idea of the concentration of magnetic centers[8] (about $1 mC/cm^2$ of 2.5 electrons produces 10^{-5} atomic fraction of magnetic centers). The production rate of any non-magnetic defects is surely of the same order of magnitude. The effects of these irradiation induced defects on the classical linear and

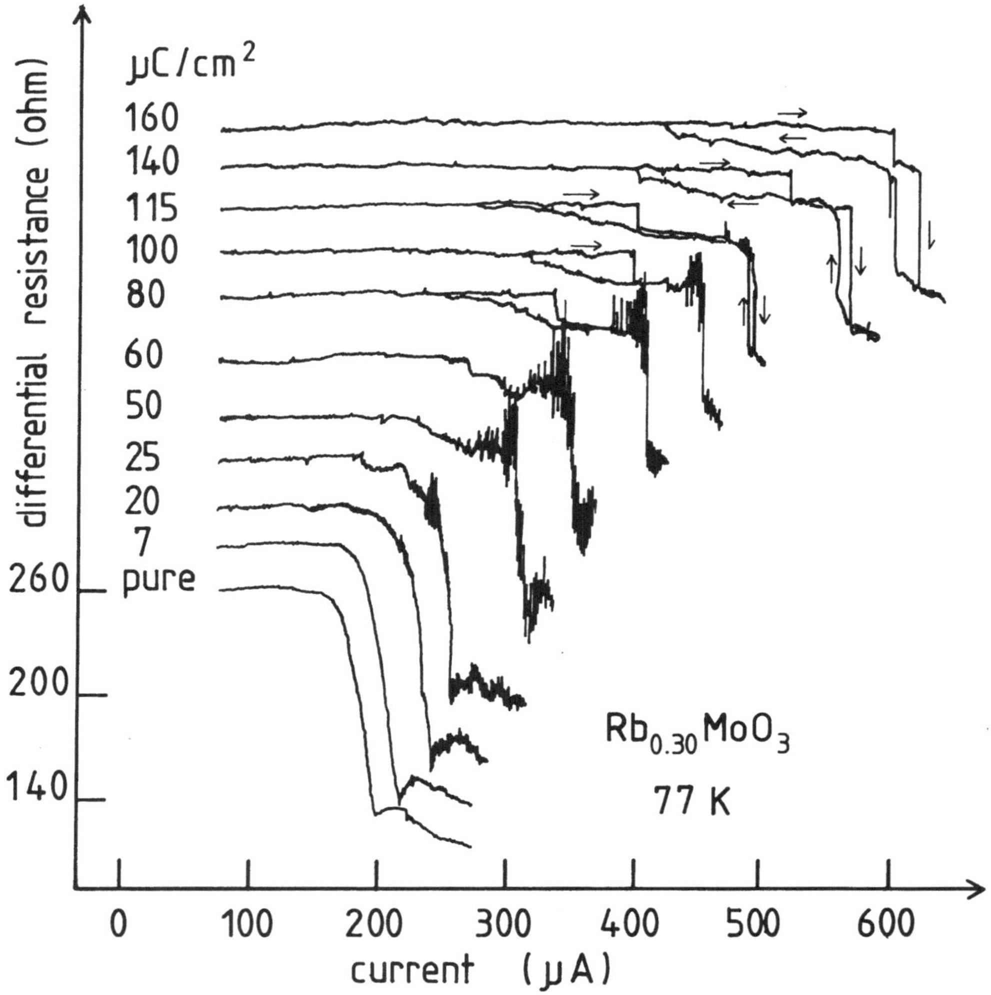

Fig. 1

Differential resistance versus applied current for $Rb_{0.30}MoO_3$. The curves have been shifted to clarify the figure ; the ohmic resistance is not modified by the irradiation. The irradiations (2.5 MeV electrons) and the in-situ measurements have been made in liquid nitrogen.

non-linear properties have been described elsewhere[7,8]. They could be separated in two defect concentration regimes. In the first one when the defect concentration is lower than 10^{-4}, the defects pins the CDW and prevents their gliding : the threshold field increases linearly with the irradiation dose. At higher defect concentration, the average CDW amplitude is affected and, consequently the critical temperature decreases and the ohmic resistance varies.

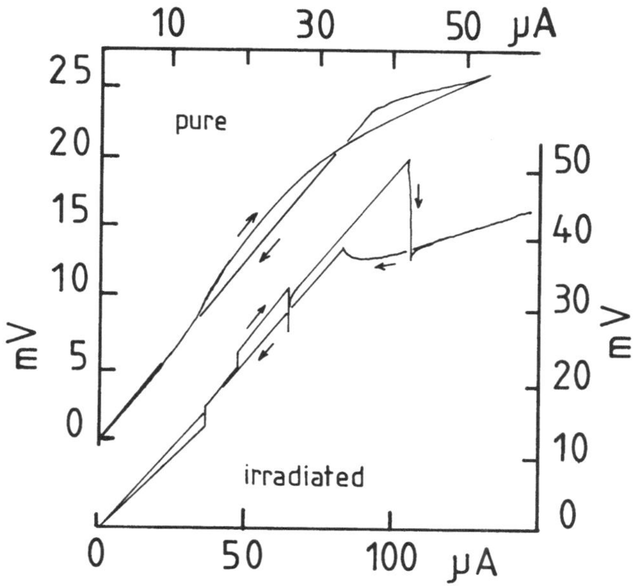

Fig. 2

Voltage versus current for a sample of $K_{0.30}MoO_3$ at 77 K.
The upper curve shows the smooth deviations to the ohmic law which are transformed to switching process by the irradiation (lower curve).

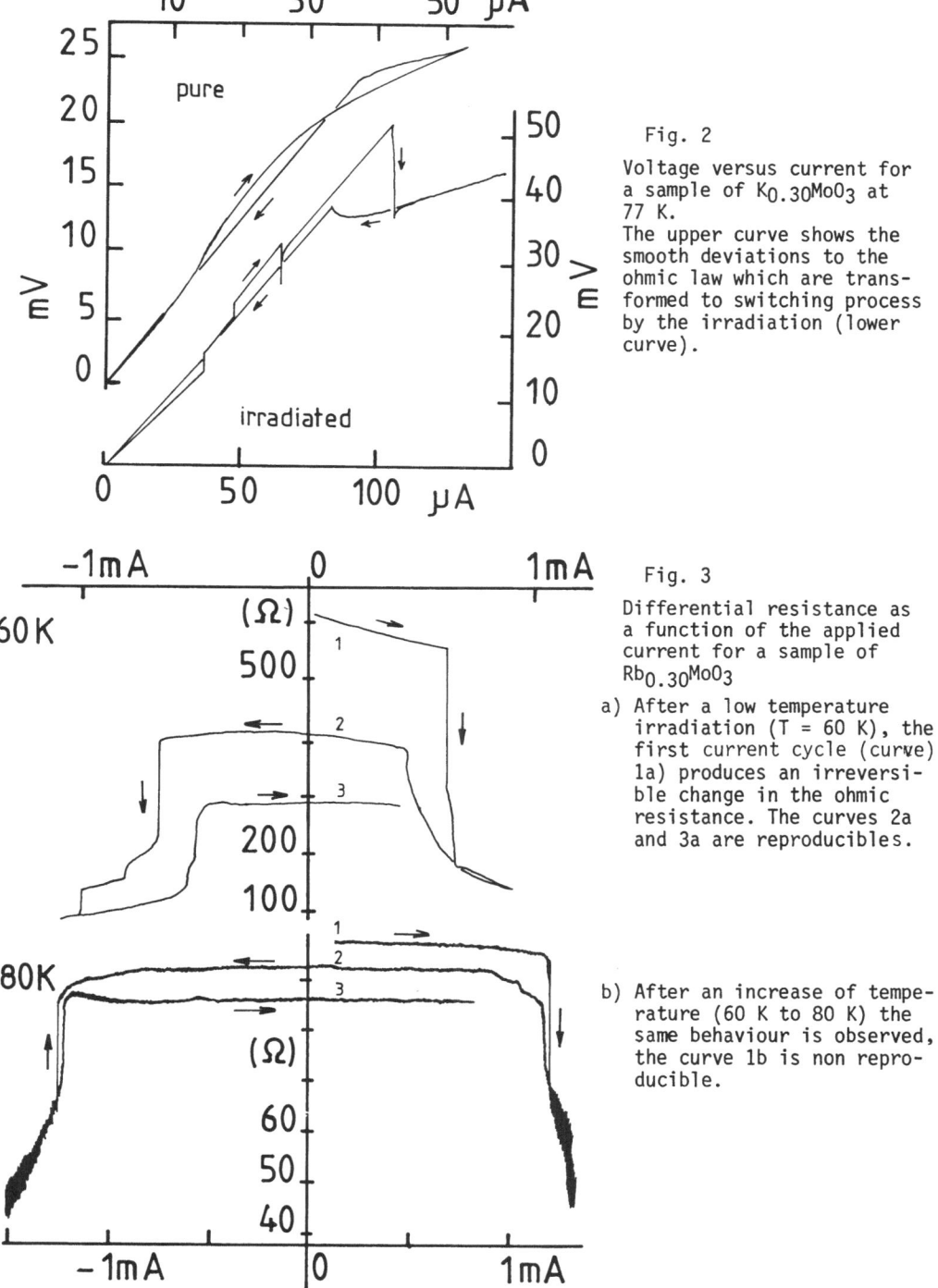

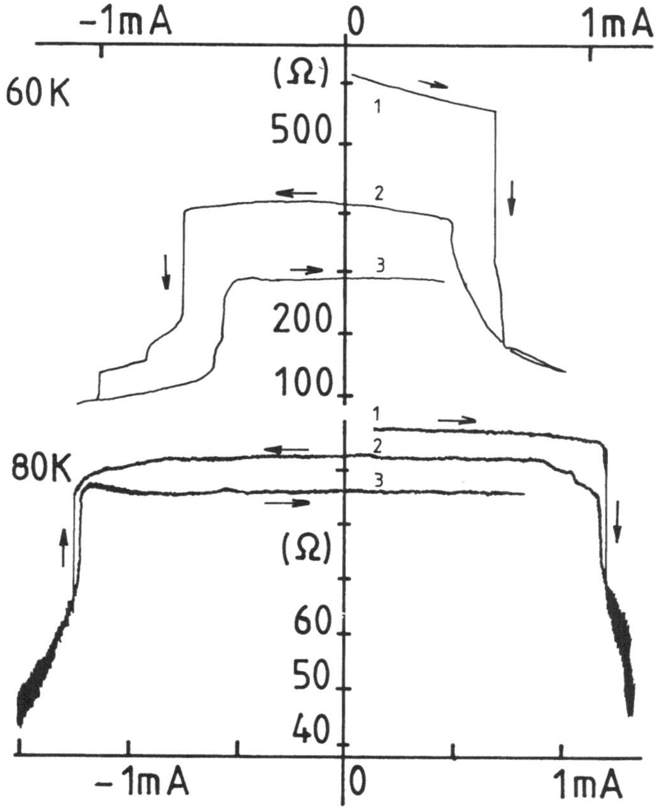

Fig. 3

Differential resistance as a function of the applied current for a sample of $Rb_{0.30}MoO_3$

a) After a low temperature irradiation (T = 60 K), the first current cycle (curve) 1a) produces an irreversible change in the ohmic resistance. The curves 2a and 3a are reproducibles.

b) After an increase of temperature (60 K to 80 K) the same behaviour is observed, the curve 1b is non reproducible.

Some qualitative effects of the irradiation occurs in parallel with these more quantitative results : the shape of the voltage or differential resistance versus current curves is modified by the irradiaton pinning centers. The figure 1 shows the differential resistance as a function of the applied current for a typical sample of $Rb_{0.30}MoO_3$ measured at the liquid nitrogen temperature and for different irradiation doses. In this experiment the current has not been reversed to avoid the hysteresis in the ohmic regime (see below). One should notice the two main effects of irradiation induced defects : the threshold field increases with the dose and the onset of non-linear current is modified. When the pure sample exhibits a smooth threshold field, a very low defect concentration (a few ppm) is sufficient to produce voltage pulses near the threshold field. These voltage pulses grow with the irradiation dose. At higher doses (about 10^{-5}) the pulses change into a switching process between the linear and non-linear state. At the same time an hysteresis appears between increasing and decreasing current curves. These features are sample independent (6 samples measured) and also independent on the alkaline metals (K or Rb). The continuous change of the threshold field shape from a smooth to a switching process through voltage pulses clearly indicates that the pinning centers play the most important role on these phenomenum.

Two other experimental behaviours clearly depend on the presence of irradiation induced defects and are surely associated to the moving and the deformation of CDW. There are the deviation to linearity below the threshold field and the large hysteresis appearing when the current crosses the positive and negative threshold. The figure 2 resumes the former point. In some blue bronzes, a smooth deviation to the ohmic law appears well below the threshold field. This non-linearity is transformed by the irradiation into sudden jumps between two or more ohmic states (at the same time the onset of the CDW current occurs through a switching). The figure 3 shows the second effect. After an irradiation, the ohmic resistance is modified by the first current cycle (curve 1a), while following cycles give reproducible hysteresis and ohmic values. The same behaviours are found after an heating (figure 3b). The first current cycle after an increase of temperature of about 20 K, produces an irreversible change in the ohmic resistance.

These three effects of the irradiation, namely the modification of the shape of threshold field, the deviation to the ohmic law before the gliding of CDW and the irreversible effects (and hysteresis) due to current cycle, have certaintly the same origin : the moving of deformable CDW in presence of domain walls and pinning centers.

Let us introduce a naïve description of these phenomenum. First at all, we have to suppose a certain disorder in the CDW : domain walls and/or discommensuration structures, even if they have not been directly observed. These CDW domains are probably the equivalent of the strandlike domains observed in TaS_3 and $NbSe_3$[10]. The pristine compounds (with a smooth threshold field) contain weak pinning centers :

impurities, dislocations and stoechiometry defects. In this case, the domain structure does not change the depinning process of the CDW (the CDW domains are not decoupled by these defects), however it should induce a certain deformation of the CDW around the pinning centers and consequently a modification of the states in the gap (and of the resistivity). On the other hand, irradiation produces strong pinning centers which are efficient at lower defect concentration[7] (a few ppm). At these defect concentration there is only a few defects in a domain, so the distribution size of the CDW domains implies that the pinning energy depends on the considered domain.
The CDW conduction needing a contineous channel of domains in which the CDW glide, the current carrying by the CDW appears only when all the domains are coupled and depinned[11]. A multiple switching is probably associated to the existence of several conductive channels. On decreasing current, the CDW are trapped in a metastable state which depends on the maximum applied current and on the current direction (i.e. on the deformation of the CDW and/or the domain walls). The different metastable states give rise to different ohmic resistance values (a current lower than the threshold field can induce a jump between two states). In a such picture, the introduction of defects in relaxed CDW domains produces highly metastable configuration which relaxes during the first current cycle (figure 3a). The same behaviour occurs when the new metastable state is created by a temperature run (figure 3b). This latter effect has also been demonstrated in TaS_3[12], however in our case the origin of the thermal effect is not related to a change of the q-vector. However, one cannot exclude that the coupling between CDW and defects induces a redistribution of the localized electrons on Mo^{5+} donor centers between the different Mo sites and consequently a modification of the population on the levels in the gap[13].

This experimental approach demonstrates that the unusual CDW conductivity in the blue bronzes are associated with the presence of strong pinning centers. A switch occurs at the onset of the CDW current when the CDW domains need to be coupled before gliding. At the same time, CDW domains are trapped in metastable states producting different resistance values. A current cycle can induce a jump between two metastable state.

REFERENCES

1 - See for example G. GRUNER in the Proceeding of International Symposium on Non Linear Transport and Related Phenomena in Inorganic Quasi One Dimensional Conductors, Sapporo 20.22 Oct. 1983

2 - L. SNEDDON, M.C. CROSS and D.S. FISHER, Phys. Rev. Lett. 49, 292 (1982)

- G. GRUNER, A. ZAWADOWSKI and P.M. CHAIKIN, Phys. Rev. Lett. 46, 511 (1981)

3 - J. BARDEEN, Phys. Rev. Lett. 45, 1998 (1980)

4 - A. ZETTL and G. GRUNER, Phys. Rev. B26, 2298 (1982)

5 - J. DUMAS and C. SCHLENKER in the Proceeding of International Symposium on Non Linear Transport and Related Phenomena in Inorganic Quasi One Dimensional Conductors, Sapporo 20.22 Oct. 1983

- J. DUMAS and C. SCHLENKER This Conference

6 - H. MUTKA, S. BOUFFARD, G. MIHALY and L. MIHALY, J. Physique Lett. 45, L113 (1984)

7 - H. MUTKA, S. BOUFFARD, J. DUMAS and C. SCHLENKER, J. Physique Lett. 45, L729 (1984)

8 - H. MUTKA, S. BOUFFARD, M. SANQUER, J. DUMAS and C. SCHLENKER, Proceeding of the International Conference on the Physics and Chemistry of Low-Dimensional Synthetic Metals, ABANO TERME, 17-22 June 1984

9 - C.H. CHEN, L.F. SCHNEEMEYER and R.M. FLEMING, Phys. Rev. B29, 3765 (1984)

10- C.H. CHEN and R.M. FLEMING, Phys. Rev. B29, 4811 (1984)

11- C.H. CHEN and R.M. FLEMING, Solid State Comm. 48, 779 (1983)

- B. JOOS and D. MURRAY, Phys. Rev. B29, 1094 (1984)

12- G. MIHALY, G. KRIZA and A. JANOSSY, Phys. Rev. B.(to be published).

13- H. DUMAS, A. ARBAOUI, J. MARCUS and C. SCHLENKER, Proceeding of the International Conference on the Physics and Chemistry of Low-Dimensional Synthetic Metals, ABANO TERME, 17-22 June 1984.

RELAXATION OF METASTABLE STATES IN BLUE BRONZE $K_{0.3}MoO_3$

L. Mihaly[+], Ting Chen, B. Alavi and G. Grüner
UCLA, Department of Physics
Los Angeles, CA 90024 USA

Metastable states, induced by thermal cycling and electric fields were investigated in blue bronze $K_{0.3}MoO_3$. We found a hysteresis in the temperature dependence of the low field resistance. The thermally induced metastable resistivity relaxes obeying a power low time dependence.

The potassium blue bronze $K_{0.3}MoO_3$ exhibits unusual electric properties, most probably related to metastable states in the charge density wave system. The hysteresis in the low field conductivity,[1] similar to that observed in TaS_3[2] and Fe doped $NbSe_3$,[3] demonstrates that positive or negative current above the threshold of the CDW conduction drives the system into different states. The onset of CDW current is accompanied by low frequency oscillation.[4] Long time relaxation processes were observed in the response to sudden current changes.[5]

The study reported here was initiated by the observation of other types of metastable states in the related compound orthorhombic TaS_3. In that material the temperature dependence of the low field conductivity exhibits hysteresis.[6,7] The metastable low field conductivity shows logarithmic or weak power low (exponent close to zero) time relaxation[8] and this relaxation is sensitive to the external field.[9] We observed similar phenomena in $K_{0.3}MoO_3$.

The blue bronze single crystals were grown at UCLA by the well known electrochemical method. Four probe samples were prepared by silver painting gold wires to the previously etched crystals.[10] Typical contact resistances of several ohms were obtained by this method. The threshold field, measured on several samples at 77K varied between 100-200 mV/cm. These values are comparable to the threshold field reported by Cava and coworkers.[11]

The results of the measurement on the temperature dependent low field resistance are summarized in Fig. 1. The resistivity shows considerable hysteresis below the phase transition.[12] We emphasize that this measurement detects thermally induced metastable states only if the spontaneous relaxation of the conductivity is slower than the sweep rate of the measurement.

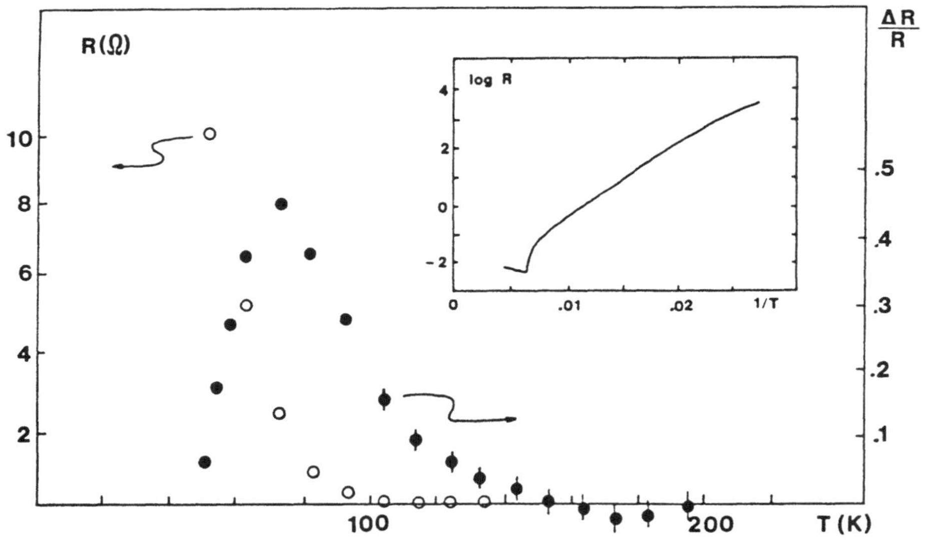

Fig. 1.
Magnitude of the hysteresis in the low field resistance ΔR vs. temperature T. The cooling and the heating rate was 0.4K/min, the resistance variation was calculated by ΔR = R(T) heating - R(T) cooling. The insert shows the temperature dependence of the average resistance R = [R(T) cooling + R(T) heating]/2.

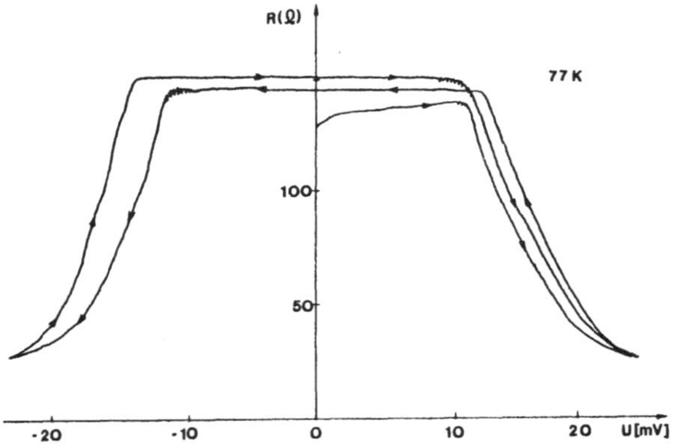

Fig. 2.
Differential resistance vs. voltage drop on the potential contacts at 77K. The current in the sample was swept with a rate of 6 μA/sec. Sweeping to high fields changes the metastable low field resistance induced by the sudden cooling from room temperature to 77K.

Fig. 2. shows the voltage dependence of the differential resistance taken in an LN_2 bath. The sample was quenched in zero electric field from room temperature to 77K. At first the resistance has a lower value. After the first sweep to high fields the curves are reproducible and the hysteresis in the low field resistance, reported earlier by other authors,[1,5] is apparent.

The time dependence of the differential resistance after quenching from room temperature to 77K was investigated on several samples. Fig. 3. demonstrates that in the limited time domain of the experiment the relaxation of the resistance follows a power low time dependence

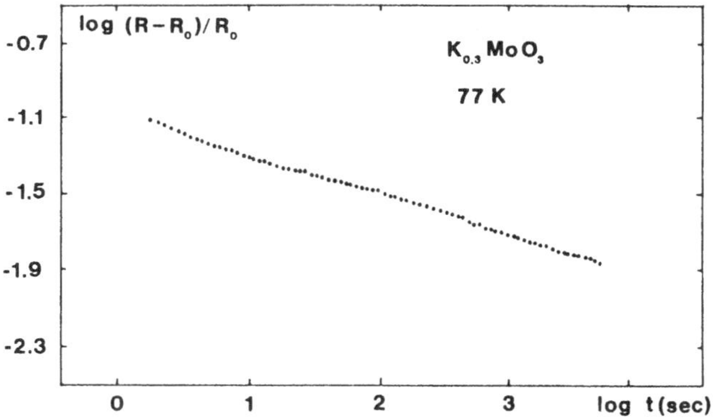

Fig. 3.
Relaxation of low field resistance after quenching to liquid N_2. The time dependence obeys power low with exponent $\alpha=0.2 \pm 0.02$.

$$R(t) = R_o - R(t) = R_o \ (1-At^{-\alpha})$$

where R_o, A and α are fitting parameters. The zero field relaxation was reproductable from sample to sample. Investigations in the presence of constant external fields speed up the relaxation. The process still obeys power low time dependence, but the parameters are sample dependent.

In conclusion, we demonstrated that thermally induced metastable states, similar to those observed in TaS_3 exist in $K_{0.3}MoO_3$. The relationship of this metastable state to the possible domain structure[13] and/or to the temperature dependent wave number[14] of CDWs needs further investigations.

This work was supported by NSF grant DMR84 - 06896.

References

+ Permanent address: Central Research Institute for Physics
H-1525 Budapest, POB 49. Hungary

1. J. Dumas and C. Schlenker, preprint
2. A.W. Higgs and J.C. Gill, Solid State Comm. 47, 737 (1983)
3. M.P. Everson and R.V. Coleman, Phys. Rev. B28, 6659 (1983)
4. J. Dumas and C. Schlenker, Proceedings of the Int. Symp. on Nonlinear Transport, Sapporo, Japan, 1983
5. R.M. Fleming and L.F. Schneemeyer, Phys. Rev. B28, 6996 (1983)
6. J.C. Gill and A.W. Higgs, Solid State Comm. 48, 709 (1983)
7. Gy. Hutiray, G. Mihaly and L. Mihaly, Solid State Comm. 48, 227 (1983)

8. G. Mihaly and L. Mihaly, Phys. Rev. Lett. 52, 149 (1984)
9. G. Mihaly, G. Kriza and A. Janossy, to be published
10. The contacting method was proposed by Professor Martha Greenblatt
11. R. Cava, et al. to be published in Phys. Rev. B
12. K. Tsutsumi, T. Tamegai, S. Kagoshima and M. Sato, present volume
13. C.H. Chen and R.M. Fleming, Solid State Comm. 48, 777 (1983)
14. R. Fleming, private communication, J. Pouget, et al. preprint

RELATED TOPICS

INCOMMENSURATE FERROELECTRICS

R. Blinc
J. Stefan Institute, E. Kardelj University of Ljubljana,
61111 Ljubljana, P.O.Box 53, Yugoslavia

I. Introduction

Systems where the periodicity of the modulation wave is incommensurable with the periodicity of the basic crystal lattice can be divided into two main classes:

a) Conducting compounds where the lattice modulation is the result of an electronic instability.

b) Insulators such as ferroelectric Rb_2ZnCl_4 and thiourea where the lattice modulation is the result of competing interactions between atoms or molecules.

In one-dimensional conductors such as TTF-TCNQ the wave vector of the periodic lattice distortion and the charge density wave is twice the wave vector of the electronic wave function at the Fermi surface and can only vary when the number of electrons in the conduction band varies. In contrast the wave-vector of the modulation is not fixed in insulators and generally varies with temperature or pressure.

There are two conflicting theoretical descriptions of incommensurate insulators. According to the Landau theory[1,2] which is based on the continuum approximation the incommensurate structure consists of nearly commensurate regions which are separated by a regular array of discommensurations (or phase solitons) where the phase of the modulation wave varies rapidly[3]. The soliton width d_o is not critical but the inter-soliton distance x_o diverges at lower temperatures as the incommensurate-commensurate (I-C) transition T_c is approached and the soliton density

$$n_s = d_o/x_o \longrightarrow 0, \quad T \rightarrow T_c^+ \qquad (1)$$

- which is the order parameter of this transition - vanishes. The Landau theory is valid as long as the soliton width is large as compared to the lattice spacing[4]. It predicts that the soliton density and the modulation wave vector vary continuously with temperature.

According to the devil's staircase model[5,6] - which takes the discretness of the crystal lattice explicitely into account - the phase diagram may consist of an infinity of higher order commensurate phases which may or may not be separated by an infinity of truly incommensurate phases. The first of these two cases where the modulation wave-vector varies in steps and "locks in" at an infinity of commensurate values is known as the incomplete and the second as the complete devil's staircase[6].

The complete devil's staircase is expected to exhibit complete pinning of the modulation wave and global wave vector hysteresys whereas the incomplete devil's staircase is - similarly as the Landau theory- connected with the existence of a gapless phason branch[6].

The experimental evidence on the nature of systems exhibiting several stairs in the devil's staircase is rather scarce and very little is known on the dynamics of the modulation wave in the I and higher order C phases.

The following problems are still open:

a) the nature of the modulation wave,
b) the nature of the I-C transition,
c) the observation of phase modes.

II. Landau Theory for Incommensurate Ferroelectrics

The Landau free energy density can be for I ferroelectrics in an external electric field E expressed[7] in terms of the complex order parameter Q, representing the slow one-dimensional modulation along the spatial coordinate x, and the polarisation P:

$$f(x) = \frac{\alpha}{2}|Q|^2 + \frac{\beta}{4}|Q|^4 - i\delta\left(Q\frac{dQ^*}{dx} - Q^*\frac{dQ}{dx}\right) + \frac{\varkappa}{2}\left|\frac{dQ}{dx}\right|^2 -$$

$$- \frac{\gamma}{2}(Q^n + Q^{*n}) + \xi P(Q^p + Q^{*p}) + \frac{P^2}{2\chi_o} - PE \qquad . \qquad (2)$$

It is assumed that $\beta, \gamma, \varkappa, \chi_o > 0$ whereas $\alpha = \alpha_o(T-T_o)$ and $n = 2p$ is even. It equals 6 in Rb_2ZnCl_4 and 10 in $[N(CH_3)_4]_2ZnCl_4$. The Lifshitz term (δ) drives the transition from the paraelectric to the I phase at T_I, whereas the "UMKLAPP" anisotropy terms, i.e. Q^n and PQ^n, represent in lowest order the discretness of the crystal lattice and are responsible for the I-C transition at T_c.

By a minimisation of the average free energy density $F=(1/L)\int_0^L f(x)dx$ with respect to P we get

$$P(x) = -\xi\chi_o(Q^p + Q^{*p}) + \chi_o E \qquad (3)$$

so that P can be eliminated. Introducing the polar representation $Q = A\exp(i\phi)$ the corresponding Euler-Lagrange equations can be written in the constant amplitude approximation $A(x) = A_o$ as

$$\varkappa \phi'' = n\bar{\gamma} A_o^{n-2} . \sin(n\phi) - n\xi\chi_o A_o^{\frac{n}{2}-2} . E . \sin\left(\frac{n}{2}\phi\right) \quad , \tag{4}$$

where $\bar{\gamma} = \gamma + \xi^2 \chi_o$. The above non-linear equation for the phase of the modulation wave is known as the double sine-Gordon equation.

The amplitude of the order parameter A_o varies with temperature for $E = 0$ as

$$A_o = \left[(\alpha_o/\beta)(T_I - T)\right]^{1/2} \quad , \tag{5}$$

where

$$T_I = T_o + \delta^2/(\varkappa\alpha_o) \quad . \tag{6}$$

Close to T_I the solutions of Eq.(4) are of the plane-wave type, i.e. $\phi = (\delta/\varkappa)x$, whereas ϕ is non-linear in x resulting in a multi-soliton lattice in the low temperature part of the I phase.

The I-C transition takes place for E = 0 when

$$A_o^{\frac{n}{2}-1} = \pi\delta/4\,(\varkappa\bar{\gamma})^{1/2} \quad , \quad T = T_c \tag{7}$$

and the inter-soliton distance varies with temperature as:

$$x_o = \frac{d_o}{\pi}\, \ln\left[\frac{4(T_I - T_c)}{T - T_c}\right] \quad . \tag{8}$$

The soliton width d_o at $T = T_c$ equals:

$$d_o = 4\varkappa/(n\delta) \quad . \tag{9}$$

The dielectric susceptibility is sensitive to the soliton density n_s

$$\chi = \chi_o + \frac{\tilde{C} \cdot n_s}{4\pi(T_I - T_c)} \cdot e^{\pi/n_s} \quad , \qquad T > T_c \tag{10}$$

and exhibits a Curie-Weiss law on approaching T_c:

$$\chi = \chi_o + \frac{\approx\!\!\!C}{T - T_c} \quad , \qquad T > T_c \tag{11}$$

whereas it should abruptly drop to χ_o below T_c in the C phase:

$$\chi = \chi_o, \qquad T < T_c \; . \tag{12}$$

III. The Nature of the Modulation Wave in the I Phase

The basic question whether the periodicity of the modulation wave q_I is in fact incommensurable with the periodicity of the basic lattice q_c

$$q_I/q_c \neq \frac{M}{N} \quad ; \text{ M, N ... integers} \tag{13}$$

or whether we deal with an infinity of higher order commensurate phases which are stable over narrow temperature intervals, is hard to answer experimentally as any irrational number can be approximated to any desired degree of accuracy by rational fractions of large numbers. Since however irrational numbers do exist, there is no a priori reason to believe that true I phases should not exist.

In Rb_2ZnCl_4 - which is one of the best investigated incommensurate ferroelectrics with $T_I = 31^oC$ and $T_c = -81^oC$ - and which is well described[8] by the Landau theory - if the mean field exponents are replaced by critical ones - the situation is as follows:

1. Over most of the I phase the modulation is of the plane wave type: $\phi = q_I x$, $q_I = q_c(1 - \delta)$ as shown by X-ray[9] and NMR data[10].

2. According to X-ray and neutron scattering data the "mismatch" continuously decreases on cooling from T_I and vanishes at the "lock-in" transition T_c.

3. The critical exponents[9] $\gamma = 1.26 \pm 0.04$, $\nu = 0.693 \pm 0.005$ and $\beta = 0.345 \pm 0.005$ are those of the n = 2, d = 3 Heisenberg model as predicted.

4. As shown by NMR[11], the modulation wave is over most of the I phase static and pinned.

5. In the high temperature part of the I phase, thermal unpinning has been observed by NMR[11,12] and NQR techniques. The observed magnetic resonance lineshape could be described by

$$f(\nu) = \frac{\text{const}}{\nu_1^2 e^{-\sigma^2} - (\nu - \nu_o)^2} \tag{14}$$

and the effective splitting between the two singularities

$$\Delta\nu = \nu - \nu_o = \pm \nu_1 \, e^{-\sigma^2/2}, \quad \nu_1 \propto (T_I - T)^\beta \tag{15}$$

is reduced as compared with the static case where the mean square phase fluctuation of the modulation wave $\sigma^2 = \langle \phi^2 \rangle$ is zero. The results show that not too close to T_I σ^2 is inversely proportional to the amplitude of the modulation wave

$$\sigma^2 \propto \frac{1}{A^2} \propto (T_I - T)^{-2\beta} \quad . \tag{16}$$

σ was found to vary between 20° at $T_I - T = 7$ K to 65° at $T_I - T = 0.5$ K. The corresponding translations σ/q_I of the incommensurate modulation wave vary between 1.6 and 5 paraelectric cells. The amplitude of this fluctuations is thus too small to average out the incommensurate broadening as observed in Rb_2ZnBr_4[11] but large enough to produce a partial motional averaging of the splitting $\Delta\nu$.

6. In the low temperature part of the I phase a multi-soliton lattice is formed and the temperature dependence of the soliton density n_s has been determined[13]. There is a large thermal hysteresis in n_s between heating and cooling runs even well above T_c demonstrating soliton pinning. It should be noticed that even close to T_c the system is in the "broad" soliton regime ($n_s \approx 0.5$) so that the inter-soliton spacing is only twice the soliton width.

IV. The Incommensurate-Commensurate Transition

At a "classical" I-C transition as for instance in Rb_2ZnCl_4 the average wave-vector mismatch δ is supposed to vanish as the modulation wave "locks in" to the basic lattice. McMillan has shown[14] that the transition is non-trivial and takes place via a formation of a multi-soliton lattice where nearly commensurate (C) regions are separated by discommensurations forming a multi-soliton lattice. The order parameter of the I-C transition is thus a local one (n_s).

Close to T_c the multi-soliton lattice becomes rather soft in view of the increase in x_o. When the soliton-soliton coupling energy

$$V_{s-s} = C \cdot \exp\left(-x_o \pi / d_o\right) \tag{17}$$

becomes of the order of the soliton-discrete lattice pinning energy

$$V_{pinn,L} = K \cdot \exp\left(-n \pi d_o / a_o\right) \tag{18}$$

where a_o is a lattice constant, the periodicity of the soliton lattice is destroyed[14] and the solitons become randomly pinned. The critical inter-soliton distance is

$$\chi_{o,c} = n\, d_o^2 / a_o \tag{19}$$

and the critical soliton density is

$$n_{s,c} = n\, d_o / a_o \quad . \tag{20}$$

For $\chi_o > \chi_{o,c}$ or $n_s < n_{s,c}$ we thus have a "chaotic" phase[14] with no long range order. In the presence of impurities random soliton pinning takes place whenever the soliton-soliton coupling becomes of the order of the soliton-impurity coupling. The incommensurate phase should be thus separated from the commensurate phase by an intermediate "chaotic" phase.

Such an intermediate "chaotic" phase has been indeed observed in Rb_2ZnCl_4 via:

a) the broadening of the incommensurate X-ray sattelite reflections[15] as $T \rightarrow T_c^+$,

b) NMR and dielectric measurements[13] which show the presence of randomly pinned solitons as metastable entities even in the C phase below T_c.

In Rb_2ZnCl_4 the intermediate "chaotic" states seem to be induced by impurities.

V. The Observation of Phason Modes

Whereas amplitudons[16] have been observed in many I systems the observation of the low lying phason branches still represents an experimental challenge and the available data are rather scarce. Clear evidence for propagating phason modes by neutron scattering techniques has been reported so far only for biphenyl[17], $ThBr_4$[18], and K_2SeO_4[19], where the soft mode is underdamped. The data are however inconclusive as to the existence of a gap Δ in the phason.

The difficulty in observing phasons by scattering techniques lies in the fact that in contrast to acoustic modes the phason damping coefficient Γ remains finite in the long wavelenth limit and is comparable with that of the soft mode at T_c.

Conclusive evidence about phason gaps in $[N(CH_3)_4]_2ZnCl_4$, Rb_2ZnCl_4, and Rb_2ZnBr_4 has been recently obtained[20] by NMR spin-lattice relaxation techniques. The method is based on the fact that the variation of the effective spin-lattice relaxation time T_1 over the incommensurate frequency distribution $f(\nu)$, e.g.

$$\frac{1}{T_1} = X^2\left(\frac{1}{T_1}\right)_A + (1 - X^2)\left(\frac{1}{T_1}\right)_\varphi \quad , \quad X = \frac{\nu - \nu_o}{\nu_1} \qquad (21)$$

allows for a separate determination of the amplitudon (T_{1A}^{-1}) and phason ($T_{1\varphi}^{-1}$) induced spin-lattice relaxation rates. The phason contribution $T_{1\varphi}^{-1}$ is in the absence of a gap Δ Larmor frequency (ω_L) dependent

$$(T_1^{-1})_\varphi = \text{Const.} \sqrt{\Gamma/\omega_L} \quad , \quad \omega_L > \Delta \qquad (22)$$

whereas it is Larmor frequency independent in the presence of a gap Δ:

$$(T_1^{-1})_\varphi = \text{Const.}\ \Gamma/\Delta \quad , \quad \omega_L < \Delta . \qquad (23)$$

The T-dependence of $T_{1\varphi}$ thus directly reflects the T-dependence of Δ. The phason gap is of the order of 10^{11} s^{-1} in the I phase and is defect induced. In $[N(CH_3)_4]_2ZnCl_4$ it increases by a factor of two in the $5c_o$ C_1 phase and again by a factor of four in the $3c_o$ C_2 phase as expected due to commensurability effects.

References

1. See, for instance, P.Bak, Rep.Prog.Phys. 45, 587 (1982).
2. R.A. Cowley, Adv.Phys. 29, 1 (1980).
3. W.L.McMillan, Phys.Rev.B 16, 4655 (1977).
4. D.A.Bruce, J.Phys.C 13, 4615 (1980).
5. S.Aubry, Ferroelectrics 24, 53 (1980).
6. S.Aubry, J.Physique 44, 147 (1983).
7. P.Prelovšek and R.Blinc, J.Phys.C 17, 577 (1984).
8. A.Levstik, P.Prelovšek, C.Filipič, ad B.Žekš, Phys.Rev.B 25, 3416 (1982).
9. S.R.Andrews and H.Mashiyama, J.Phys.C 16, 4985 (1983).
10. R.Blinc, B.Ložar, F.Milia, and R.Kind, J.Phys.C 17, 241 (1984).
11. R.Blinc, D.C.Ailion, P.Prelovšek, V.Rutar, Phys.Rev.Lett. 50, 67 (1983).
12. R.Blinc, F.Milia, B.Topič, and S.Žumer, Phys.Rev.B 29, 4173 (1984).
13. R.Blinc, A.Prelovšek, A.Levstik, and C.Filipič, Phys.Rev.b 29, 1508 (1984).
14. P.Bak and V.L.Pokrovsky, Phys.Rev.Lett. 47, 958 (1981).
15. H.Mashiyama, S.Tanisaki, and K.Hamano, J.Phys.Soc.Jap. 50, 2139 (1981); 51, 2538 (1982).
16. J.Petzelt, Phase Transitions 2, 155 (1981).
17. H.Cailleau, F.Mousa, C.M.E.Zeyen and J.Bouillot, Solid State Commun. 33, 407 (1980).
18. L.Bernard, R.Currat, P.Delamoye, C.M.E.Zeyen, S.Hubert, and R. de Kouchovsky, J.Phys.C 16, 433 (1983).
19. J.D.Axe, M.Iizumi, and G.Shirane, Phys.Rev. B 22, 3408 (1980).
20. R.Blinc, J.Dolinšek, D.C.Ailion, and S.Žumer, to be published.

COMMENSURATE AND INCOMMENSURATE PHASES OF A TWO-DIMENSIONAL LATTICE OF SUPERCONDUCTING VORTICES

P. Martinoli, H. Beck, G.-A. Racine, F. Patthey and Ch. Leemann
Institut de Physique, Université de Neuchâtel,
CH-2000 Neuchâtel, Switzerland

Superconducting films whose thickness is periodically modulated in one direction provide an attractive system to study commensurate and incommensurate phases of the two-dimensional vortex lattice. Critical currents, macroscopic quantum phenomena and the dynamic response of the vortices to a small driving rf-field are shown to be sensitive probes of the commensurate-incommensurate phase transition which is triggered by soliton excitations.

1. Introduction

Modulated structures whose period is incommensurable with that of the underlying lattice have been discovered and studied in a variety of condensed-matter systems [1]. They are usually observed in systems showing two competing periodicities as, for instance, rare-gas monolayers adsorbed at the surface of a solid, crystals with two interpenetrating incommensurate sublattices, metallic conductors undergoing a Peierls transition leading to the formation of charge density waves (CDW) and helical or sinusoidal magnetic structures incommensurable with the crystal lattice in certain rare-earth compounds. The very existence of commensurate (C) and incommensurate (I) phases has also been demonstrated for a two-dimensional (2D) lattice of quantized vortices in superconducting films whose thickness is periodically modulated in one direction [2,3]. In this system the grating-like structure of the superconducting layer creates a 1D periodic pinning potential, of wavelength $\lambda_0 = 2\pi/q_0$, for the 2D triangular vortex lattice, whose areal particle density $n_\square = (2/\sqrt{3})a^{-2}$ (a is the lattice parameter) is governed by the transverse magnetic field B which generates the vortices : $n_\square = B/\phi_0$, where ϕ_0 is the superconducting flux quantum. In this paper we review some of the static and dynamic properties of this particular system which shows interesting analogies with the CDW-structures discussed in this Conference.

2. The phase diagram

The phase diagram of 2D crystals exposed to a 1D periodic force field has been studied by Pokrovsky and Talapov [4] and by Martinoli et al. [5]. At low temperatures, where melting phenomena driven by the unbinding of thermally excited dislocation pairs [6-9] are expected to be irrelevant, it is determined by considering only soliton-like topological excitations which trigger the instability of a C-phase with respect to an I-phase. The CI-phase transition is conveniently described in terms of a "mismatch" parameter δ which measures the degree of mismatch between the 2D vortex lattice and the 1D periodic substrate. δ is defined by $\delta(B) = 1 - [g_{mn}(B)/q_0]$, where $\vec{g}_{mn}(B)$ is a reciprocal vortex lattice vector and $\vec{q}_0$ the wave vector of the thickness modulation. Then, by considering the energy of the system, which is the sum of an elastic contribution due to the pinning-induced lattice distortion and of a potential energy contribution due to the periodic pinning force, it can be shown that if δ is less than the critical value :

$$\delta_c = (2/\pi)(\Delta/\mu)^{1/2} \qquad (1)$$

the 2D vortex crystal is in a C-phase. In Eq.(1) Δ and μ are, respectively, the amplitude of the cosine pinning potential and the shear modulus [10] of the vortex lattice. In the ground state of a C-phase (Fig. 1) the vortices are forced to lie in the valley of the periodic potential, i.e. the "phase" field $\phi_e = W_{ex} + \delta x$, which describes the displacement of the vortices with respect to the bottom of the corresponding wells, vanishes everywhere for $\delta < \delta_c$. For an incompressible crystal, as it is the case for a lattice of superconducting vortices, this requires an area conserving homogeneous deformation $\vec{W}_e$ with components given by $W_{ex} = -\delta x$ and $W_{ey} = \delta y$. For $\delta = 0$, i.e. $B = B_{mn} = (\sqrt{3}/2)(\phi_0/\lambda_0^2)(m^2 + n^2 + mn)^{-1}$, the (undistorted) 2D vortex crystal is in perfect registry with the underlying pe-

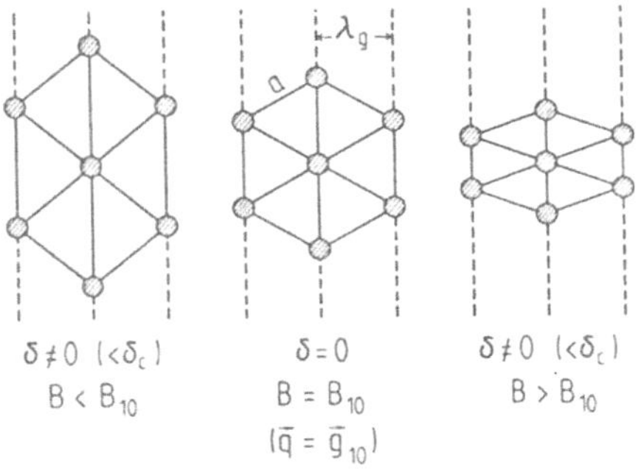

Fig. 1: The fundamental commensurate C_{10}-phase in three different states of deformation.

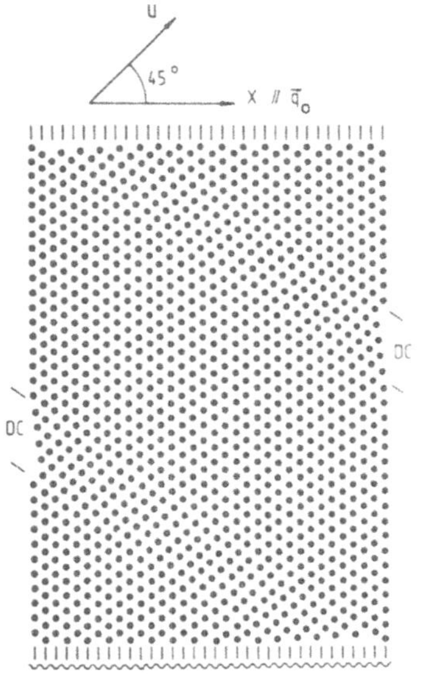

Fig. 2: Incommensurate I-phase for $\delta = 0.13$ ($B/B_{10}=0.76$). Discommensurations (DC) form a 1D soliton superlattice along the u-direction.

riodic pinning structure (matching configuration). At $\delta = \delta_c$ the C-phase becomes unstable against the nucleation of soliton excitations and for $\delta > \delta_c$ the vortex lattice is found in an I-phase (Fig. 2), whose ground state is characterized by a 1D soliton superlattice, of period $P(\delta)$, superposed onto the homogeneous deformation found for the C-phase. For the incompressible vortex lattice in an I-phase $\vec{W}_e$ has therefore components of the form $W_{ex} = -\delta x + \phi_e(u)$ and $W_{ey} = \delta y - \phi_e(u)$, where the stair-shaped phase field $\phi_e(u)$ describes a 1D lattice of solitons varying in a direction u which forms an angle of 45^0 with $\vec{q}_0$. $P(\delta)$ diverges at the CI-phase transition, i.e. at $\delta = \delta_c$.

To determine the complete phase diagram of the system at finite temperatures, several authors [5,9,11-17] have emphasized that one should add to soliton excitations the effect of thermally activated dislocation pairs [6-9] in order to assess the stability of an I-phase against melting into a fluid-like phase. Although important for a deeper understanding of the physics of 2D systems, this aspect of the problem plays only a marginal role in the analysis of the experiments reported later on in this paper. Thus, for simplicity, in determining the shape $\delta_c(T)$ of the CI-phase boundary we rely on an approach based on the so-called Self-Consistent Harmonic Approximation (SCHA) [5] obtaining for $\delta_c(T)$ a result similar to that predicted by renormalization-group theories [4,18]. In the SCHA-scheme the effect of thermal fluctuations enters the problem through a Debye-Waller factor which reduces the amplitude of the periodic pinning potential experienced by the vortices from the "bare" value Δ to the "renormalized" value:

$$\Delta_R = \Delta e^{-(1/2)q_0^2\langle U_{tx}^2\rangle} \quad . \tag{2}$$

The mean square transverse (t) vortex fluctuation $\langle U_{tx}^2\rangle$ in the direction x parallel to $\vec{q}_0$ is easily deduced from the Langevin equation describing the brownian motion of the vortices in the periodic potential. In the limit $\Delta \ll \mu$ one obtains :

$$\langle U_{tx}^2\rangle \approx (k_B T/8\pi\mu)\ln(\Delta_R/\mu) \quad . \tag{3}$$

From Eqs. (2) and (3) it then follows :

$$\Delta_R/\Delta = (\Delta/\mu)^{T/(T_{LU} - T)} \quad , \tag{4}$$

an expression showing that $\Delta_R(T)$ vanishes at a "Locking-Unlocking" temperature T_{LU} implicitly given by $k_B T_{LU} = (4/\pi)\mu(T_{LU})\lambda_0^2$ [4,5]. The phase boundary $\delta_c(T)$ now simply follows by replacing Δ with its renormalized value $\Delta_R(T)$ in Eq. (1). The resulting phase diagram is shown in Fig. 3, where, instead of $\delta_c(T)$,

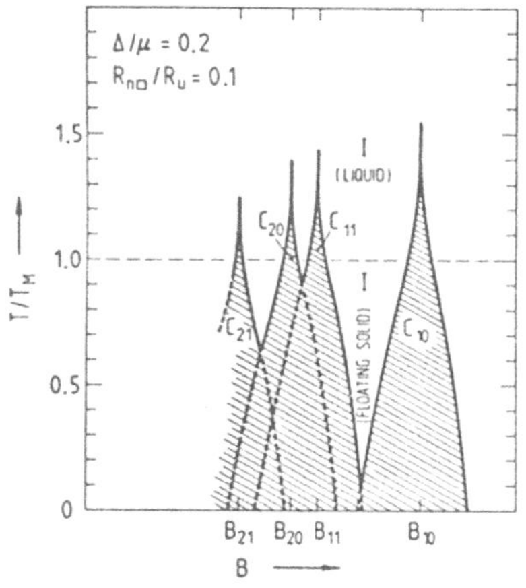

Fig. 3 : Phase diagram in the (B,T)-plane of a 2D vortex lattice in a 1D periodic potential. The ratio T_{LU}/T_M is a function of the normal-state sheet resistance $R_{n\square}$ expressed in units of $R_u = \hbar/e^2$ [5].

we have plotted the related quantity $B_c(T) = B_{mn}\,[1 \pm \delta_c(T)]^2$. Since T is measured in units of T_M, the melting temperature of the 2D vortex lattice [6-9], from Fig. 3 it is seen that, if the CI-transition occurs for $T > T_M$, it is actually a transition from a "locked" C-phase to a fluid-like phase. This is the case for the lower-order C-phases of Fig. 3, where $T_{LU} > T_M$. It can be shown that $T_{LU} = T_M$ for the C_{22}-phase. For C-phases of higher order T_{LU} is always lower than T_M and, consequently, a floating-solid phase separates a C-phase from the liquid phase.

3. Critical Currents

Critical current (I_c) measurements in thickness modulated films [2] prove to be a sensitive probe of the CI-phase transition described in the previous Section. Since a C-phase is "locked" to the periodic substrate, a finite current, flowing parallel to the 1D grooves, is required to depin the vortex lattice and, subsequently, to sustain vortex motion in the dissipative flux-flow régime. An I-phase, on the other hand, is not pinned by the periodic film structure, its energy being independent of the relative position of the solitons with respect to the pinning potential. Thus, as shown by Fig. 4, characteristic peaks reflecting the presence of various C-phases centred at $B = B_{mn}$ show up in the $I_c(B)$-curves. In principle, a detailed comparison of the shape of an I_c-peak with the theoretical model studied by Burkov and Pokrovsky [19] should allow a determination of the CI-phase boundary. In practice, however, pinning effects due to randomly distributed inhomogeneities in real films, which result in a finite contribution to I_c even in an I-phase, and the overlap of adjacent C-phases render the analysis of the peak shape quite difficult. A more accessible experimental quantity to test theoretical predictions is the temperature dependent strength, $I_{CM}(T)$, of a critical current peak. A simple calculation [5] in which the Lorentz force provided by the driving current is balanced against the pinning force experienced by the vortices in a matching configuration ($\delta = 0$) shows that $I_{CM}(T)$ can be written in the form :

$$\frac{I_{CM}(T)}{I_{CM}(0)} = \frac{\Lambda(0)}{\Lambda(T)} \, (\Delta/\mu)^{T/(T_{LU} - T)} , \tag{5}$$

where Λ is the effective penetration depth for 2D superconducting layers [5]. As shown in Fig. 5, good agreement with Eq. (5) is obtained for a reasonable choice of the fitting parameters T_{LU}/T_c (T_c is the BCS-transition temperature) and Δ/μ. For comparison, theoretical curves calculated by neglecting the effect of thermal fluctuations, represented by the last factor in Eq. (4), are also shown.

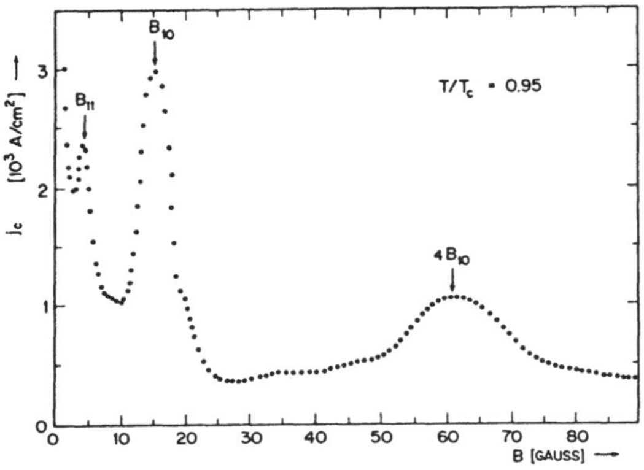

Fig. 4: Critical current density as a function of the transverse magnetic field for a thickness-modulated Al-film (λ_0=1.1μm).

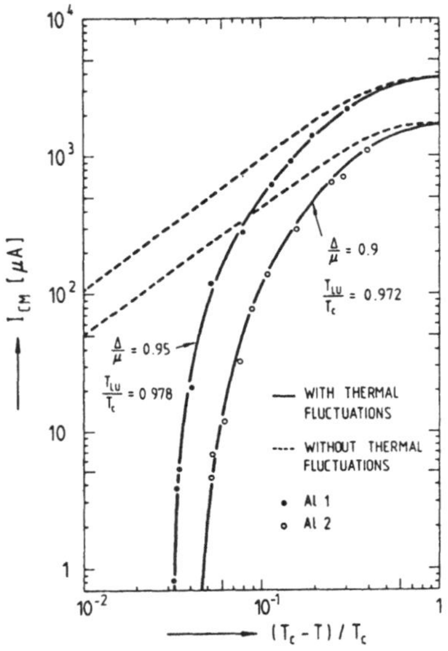

Fig. 5: Temperature dependence of the critical currents of two thickness-molated Al-films in the perfectly registred (δ=0) C_{10}-phase.

4. Macroscopic Quantum Phenomena

The motion of superconducting vortices in a periodic pinning potential shows interestiñg quantum features reminiscent of ac-Josephson phenomena in arrays of superconducting weak links [3,20,21]. When the dc driving current I_{dc} exceeds I_c, a particular flux-flow régime characterizes a C-phase. Dynamic coupling of the 2D vortex lattice with the 1D periodic substrate results, in this case, in a highly coherent velocity oscillation, $\vec{v}(t)$, of the vortices which, in turn, generates a macroscopic oscillating electric field $\vec{E}(t) = -[\vec{v}(t) \times \vec{B}]$, whose presence can be revealed by a sensitive detection system [20]. Since

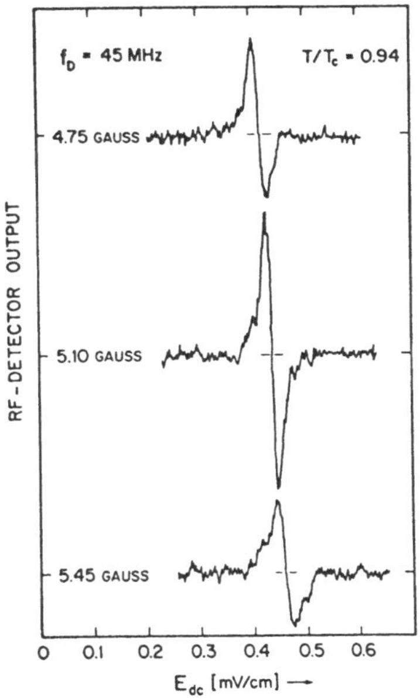

Fig. 6: RF-signals emitted at 45 MHz by a thickness-modulated Al-film (λ_0=1.9 μm) in the C_{10}-phase.

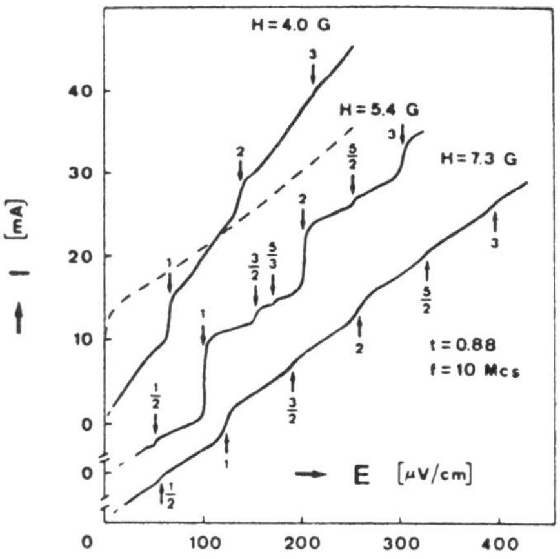

Fig. 7: I-V-curves of a thickness-modulated Al-film (λ_0=1.9 μm) exposed to rf-radiation at 10 MHz in the C_{10}-phase. The dashed line is the unexcited characteristic for B=5.4 G.

$\vec{E}(t)$ has spectral components at frequencies, f_m, given by $f_m = mv_{dc}/\lambda_0 = mE_{dc}/B\lambda_0$, with a detector tuned at f_D signals associated with the quantum oscillation of the vortex lattice are expected whenever $f_m = f_D$, i.e. $E_{dc} = \lambda_0 f_D B/m$. Recorder traces of the detector output are shown in Fig. 6 as a function of E_{dc} for B-values about the matching field B_{10}. It is readily verified that the rf-signals emitted by the modulated film show up at $E_{dc} = \lambda_0 f_D B$, thereby demonstrating that they were generated by the fundamental component of $\vec{E}(t)$.

Indirect evidence for the collective oscillation of the vortices in a C-phase is obtained by exposing thickness-modulated films to rf-radiation. Pinning-induced coupling at rf-frequencies [3,21] between the oscillating motion of the vortex lattice and the applied rf-field gives rise to quantum interference transitions in the current-voltage characteristics (Fig. 7) whenever f_m becomes a multiple of the frequency f of the rf-radiation, i.e. when $E_{dc} = E_{mn} = (n/m)\lambda_0 fB$.

As expected, rf-emission (Fig. 6) and interference effect (Fig. 7) undergoe a rapid degradation as $\delta(B)$ approaches $\delta_c(T)$ and are totally absent when the vortex lattice is in an I-phase [22].

5. Dynamic Vortex Response

Detailed information about the dynamics of the 2D vortex lattice in the various phases it can assume on the 1D periodic substrate can be obtained from a study of the anisotropic response of the vortex medium to a small oscillating driving force of angular frequency ω. To measure the vortex response, we rely on a modified version [23] of the two-coil technique devised by Fiory and Hebard [24]. In the following discussion we shall denote by (α,β) a coil configuration such that the force $\vec{F}_L$ provided by the drive coils acts in the α-direction while the orientation of the receive coil is such that it detects the response arising from the projection of vortex motion along the β-direction.

With the vortex crystal in a C-phase, our experimental method does not excite transverse modes of an infinite lattice. In such a phase coupling to lattice shear modes only arises from the finite size of the sample, from Umklapp (U)-processes and/or from residual random pinning [25]. In an I-phase, on the other hand, the 1D periodic sequence of solitons breaks the translational symmetry of the vortex lattice, thereby allowing intrinsic coupling of the driving field $\vec{F}_L$ to transverse modes of the soliton superlattice. Pronounced structures, reflecting the occurence of the CI-transition at $\delta = \delta_C$, are thus expected in the complex vortex response of modulated layers as B is swept across B_{mn}.

Field derivatives of the vortex response taken at 3 MHz and at different temperatures are shown in Fig. 8 for the (y,y)-coil configuration. Marked structures in both components of $\partial V/\partial B$ emerge from a monotonically varying background around B_{10}. While at low temperatures only minor changes in the shape of the structures

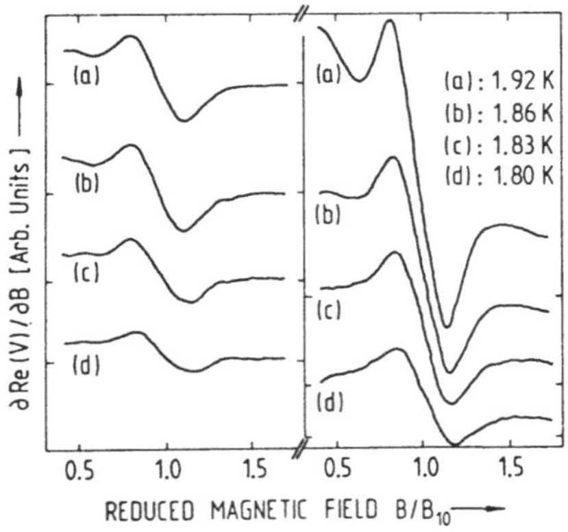

Fig. 8: Field derivatives of the complex vortex response at 3 MHz of a thickness-modulated Al-film (λ_0=0.73μm, T_c=2.05K, $R_{n\square}$=25Ω) measured in the (y,y)-coil configuration.

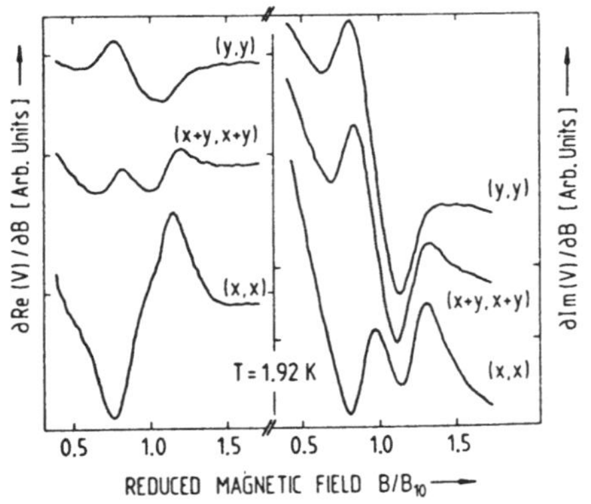

Fig. 9: Field derivatives of the complex vortex response at 3 MHz for three different coil configurations. The film is the same as in Fig. 8.

are observed as the coils are progressively rotated from the (y,y)- towards the (x,x)-configuration, profound modifications in the form of both components of $\partial V/\partial B$, as shown by Fig. 9, occur at higher temperatures. A strongly anisotropic signal, having its maximum strength in the (x,x)-configuration and vanishing in the (y,y)-configuration, emerges as the coils are rotated. The fact that it does not show up at lower temperatures is interpreted as an indication that it cannot be associated with the soliton-triggered low-temperature CI-transition described in Section 2.

To calculate the low-temperature vortex lattice response we rely on a continuum approximation [4,5]. The equation of motion for the deformation field $\vec{W} = \vec{W}_e + \vec{s}$ is linearized in the deviation $\vec{s}$ from the static equilibrium displacement $\vec{W}_e$ (see Section 2). The finite size L of the film is accounted for by allowing for boundary solitons (BS) in a C-phase [19,23]. I-phase and C-phase with BS are matched where $P(\delta) = L$. In practice, the appropriate L may be considerably shorter than the actual sample size, approximately of the order of the mean distance between vortex lattice imperfections which interrupt the long-range coherence of the soliton lattice.

The equation of motion of the dissipative (viscosity η) elastic vortex continuum (Lamé coefficients λ,μ) driven by $\vec{F}_L$ in the periodic potential $U(\phi) = U(W_x + \delta x)$ has the form :

$$\eta \frac{\partial \vec{s}}{\partial t} + \hat{\mathbb{L}}\vec{s} = \vec{F}_L \qquad , \tag{6}$$

where $\mathbb{L}_{\alpha\beta} = -(\lambda + \mu)\nabla_\alpha\nabla_\beta - \mu\nabla^2\delta_{\alpha\beta} + \hat{q}_{0\alpha}U''(\phi_e)\hat{q}_{0\beta}$. Its solution $\vec{s}(\vec{r},t)$ can be expressed in terms of eigenfunctions of the 2x2 matrix operator $\hat{\mathbb{L}}$.

Using London's and Maxwell's equations, the voltage V at the receive coil due exclusively to vortex motion is shown to be given by an integral over Fourier components of the deviation, $\delta\vec{\Phi}$, of the 2D fluxoid field $\vec{\Phi}(\vec{r},t)$ [24,26] from its static value :

$$\delta\underset{\sim}{\vec{\Phi}}(\vec{q},t) = \phi_0(q \times z)[q \cdot \underset{\sim}{\vec{s}}(\vec{q},t)] \qquad . \tag{7}$$

Introducing the solution of Eq.(6) into Eq.(7), V can be approximately expressed by $V = AZ'I_{rf}$, where A describes geometry of drive and receive coils as well as their orientation with respect to $\vec{q}_0$, I_{rf} is the rf-current in the drive coils and the "impedance" $Z'(B,\omega,T)$ contains the vortex dynamics.

For practical calculations we have chosen a piece-wise parabolic potential U. Thus, in the I-phase $U''(\phi_e)$ yields a Kronig-Penney potential with distance $P(\delta)$ between the "spikes". For a C-phase with BS we use periodic boundary conditions, a procedure resulting again in a Kronig-Penney potential with $P(\delta)$ replaced by L. If, as in our experiments, the geometry dependent quantities vary slowly, only small $\vec{q}$ are needed in Eq. (7) and, consequently, $\vec{F}_L$ can be regarded as almost spacially homogeneous. In this case the eigenfunctions of $\hat{\mathbb{L}}$ are Bloch "spinors" $\vec{\Psi}_{\underline{K}}(u)$ varying only along the direction u of the 1D soliton lattice and the relevant Bloch wave vectors $\vec{K}$ are the reciprocal vectors $\vec{G}$ of the soliton lattice. Then, the impedance of the incompressible ($\lambda \to \infty$) vortex lattice can be written in the form :

$$Z'(B,\omega,T) = R_{\square} \sum_{\underline{G}}{}' \langle \Psi_{\underline{G}}(u) \rangle^2 i\omega\tau_{\underline{G}} (1 + i\omega\tau_{\underline{G}})^{-1} \quad , \tag{8}$$

where the sum is over all $\vec{G}$ excluding the $\vec{G}$ = 0-mode and the average of $\Psi_{\underline{G}}(u)$ is over $P(\delta)$ (for the I-phase) or over L (for the C-phase with BS). In Eq. (8) $R_{\square}$ is the flux-flow sheet resistance and the relaxation time $\tau_{\underline{G}}$ is related to the eigenvalues $E_{\underline{G}}$ of $\hat{\mathbb{L}}$ by $\tau_{\underline{G}} = \eta / E_{\underline{G}}$.

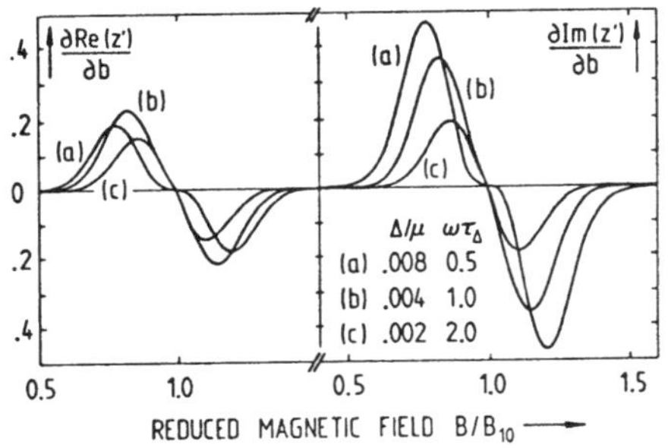

Fig. 10: Theoretical field derivatives of $z' = Z'/R_{\square}$ calculated from Eq. (8) as a function of $b = B/B_{10}$.

Theoretical field derivatives of Z' are shown in Fig. 10 for $L = 30\ \lambda_0$. By replacing Δ with $\Delta_R(T)$ [Eq. (4)], rising temperatures can be simulated by decreasing values of Δ_R/μ and of the relaxation rate $\tau_{\Delta}^{-1} = q_0^2 \Delta_R / \eta$, which, with L, are the parameters of the model. Comparison with the data of Fig. 8 shows good qualitative agreement for the B-dependence as well as for the relative magnitude of the real and imaginary parts of the signal. As observed experimentally, the model correctly predicts a rapid degradation of the structures around B_{mn} as T approaches T_{LU}, the temperature above which the vortex lattice no longer feels the periodic pinning structure. Details of the angular dependence, contained in A, as well as an explanation of the physical origin of the high-temperature anisotropic signal of Fig. 9 need, however, further analysis.

We thank J.R. Clem and V.L. Pokrovsky for stimulating discussions. This work was supported by the Swiss National Science Foundation.

Références

[1] P. Bak, Rep. Prog. Phys. 45, 587 (1982)

[2] O. Daldini, P. Martinoli, J.L. Olsen and G. Berner, Phys. Rev. Lett. 32, 218 (1974)

[3] P. Martinoli, Phys. Rev. B 17, 1175 (1978)

[4] V.L. Pokrovsky and A.L. Talapov, Phys. Rev. Lett. 42, 65 (1979); Sov. Phys. JETP 51, 134 (1980)

[5] P. Martinoli, M. Nsabimana, G.-A. Racine, H. Beck and J.R. Clem Helv. Phys. Acta 55, 655 (1982)

[6] J.M. Kosterlitz and D.J. Thouless, J. Phys. C 6, 1181 (1973)

[7] D.R. Nelson and B.I. Halperin, Phys. Rev. B 19, 2457 (1979)

[8] B.A. Huberman and S. Doniach, Phys. Rev. Lett. 43, 950 (1979)

[9] D.S. Fisher, Phys. Rev. B 22, 1190 (1980)

[10] A.T. Fiory, Phys. Rev. B 8, 5039 (1973)

[11] J.V. José, L.P. Kadanoff, S. Kirkpatrick and D.R. Nelson, Phys. Rev. B 16, 1217 (1977) [Erratum, Phys. Rev. B 17, 1477 (1978)]

[12] S. Ostlund, Phys. Rev. B 23, 2235 (1981)

[13] S. Ostlund, Phys. Rev. B 24, 398 (1981)

[14] J. Villain and P. Bak, J. Physique 42, 657 (1981)

[15] S.N. Coppersmith, D.S. Fisher, B.I. Halperin, P.A. Lee and W.F. Brinkmann, Phys. Rev. Lett. 46, 549 (1981)

[16] T. Bohr, V.L. Pokrovsky and A.L. Talapov, Sov. Phys. JEPT Lett. 35, 203 (1982)

[17] T. Bohr, Phys. Rev. B 25, 6981 (1982)

[18] M.W. Puga, E. Simanek and H. Beck, Phys. Rev. B 26, 2673 (1982)

[19] S.E. Burkov and V.L. Pokrovsky, J. Low Temp. Phys. 44, 423 (1981)

[20] P. Martinoli, O. Daldini, Ch. Leemann and B. Van den Brandt, Phys. Rev. Lett. 36, 382 (1976)

[21] P. Martinoli, O. Daldini, Ch. Leemann and E. Stocker, Solid State Comm. 17, 205 (1975)

[22] P. Martinoli, H. Beck, M. Nsabimana and G.-A. Racine,Physica 107B, 455 (1981)

[23] F. Patthey, G.-A. Racine, Ch. Leemann, H. Beck and P. Martinoli, in Proc. of the 17th Int. Conf. on Low Temp. Phys., Karlsruhe, 1984, Eds. U. Eckern, A. Schmid, W. Weber and H. Wühl (North-Holland, Amsterdam, 1984) p. 573

[24] A.T. Fiory and A.F. Hebard, in Inhomogeneous Superconductors-1979, Eds. D.V. Gubser, T.L. Francavilla, S.A. Wolf and J.R. Leibowitz, AIP 58, 293 (1980)

[25] A.T. Fiory and A.F. Hebard, Phys. Rev. B 25, 2073 (1982)

[26] J. Pearl, Thesis, Polytechnic Institute of Brooklyn, New York (1965), unpublished.

$(TMTSF)_2X$ COMPOUNDS: SUPERCONDUCTIVITY, SPIN-DENSITY WAVES AND ANION ORDERING

H.J. SCHULZ
Laboratoire de Physique des Solides
Université Paris-Sud, 91405 Orsay, France

Abstract The different types of experimentally observed phase transitions in $(TMTSF)_2X$ and $(TMTTF)_2X$ salts (superconductivity, antiferromagnetism, spin-Peierls and anion ordering) are reviewed. Current theoretical models of these transitions and the related low-temperature properties are discussed. Both theoretical arguments and experimental results, mainly from transport and magnetic resonance measurements, suggest important effects of low dimensionality in these compounds.

INTRODUCTION

Most quasi-one-dimensional conductors, whether inorganic (KCP, $NbSe_3$,...) or organic (TTF-TCNQ,...),exhibit Peierls type transitions[1], leading to a low-temperature charge-density-wave (CDW) state. On the contrary, the $(TMTSF)_2X$ compounds (TMTSF = tetramethyltetraselenafulvalene, X an inorganic anion: PF_6, AsF_6, ClO_4, ReO_4,...) and their sulfur analogues $(TMTTF)_2X$ never show Peierls transitions, but rather a variety of other instabilities: most prominently the first example of supraconductivity in an organic substance[2], but one also finds spin-density waves (SDW, = antiferromagnetism) of the Slater-des Cloizeaux-Overhauser type[3], spin-Peierls transitions[4], and order-disorder transitions of the anions. In the present paper I discuss the typical phase diagram of these compounds and its theoretical interpretation, with special emphasis on the electronic phase transitions and associated low-temperature properties. More detailed information can be found in recent conference proceedings[5,6] and review articles[7,8]. For completeness I mention here that supraconductivity has recently also been observed in sulfur-based compounds of the type $(BEDT\text{-}TTF)_2X$, first under pressure[9], and subsequently at ambient pressure[10], with T_c up to 2.5K.

The triclinic crystal structure of the $(TMTSF)_2X$ compounds consists of parallel stacks formed by the planar TMTSF molecules, with the direction of highest conductivity along the stack (a-) direction. The inorganic anions are completely ionized, i.e. there is one hole for two TMTSF molecules, leading to a 3/4-filled band. The bandstructure can reasonably be described by a tight-binding model, with an overall bandwidth $4t_a \approx 1eV$ (t_a is the transfer integral along the chains). From optical and conductivity anisotropies[11] and from calculated values[12] one finds $t_a/t_b \approx 10...20$, $t_b/t_c \approx 20...30$. This leads to an open Fermi surface consisting of two parallel, corrugated sheets. Due to a slight dimerization of the organic stack, there is a gap at the center of the conduction band (approximately t_a below the Fermi level). Thus, strictly speaking, one has a half-filled band, which is important for some aspects of the phase diagram because of the possibility of electron-electron umklapp scattering.

THE PHASE DIAGRAM OF $(TMTSF)_2X$ COMPOUNDS

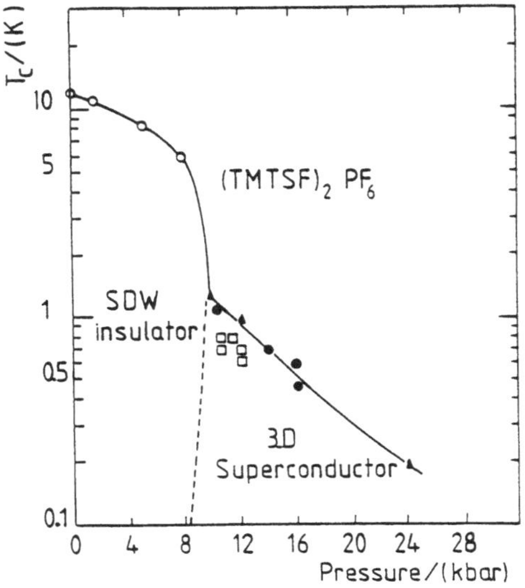

Fig.1: The phase diagram of $(TMTSF)_2PF_6$ (ref. 8). Compounds with other anions have similar phase diagrams (see text).

The generic phase diagram of the $(TMTSF)_2X$ compounds is represented by fig.1. Compounds with different anions have different temperature and pressure scales, the most interesting case being $(TMTSF)_2ClO_4$, where superconductivity exists already at ambient pressure[13]. The existence of superconductivity and its bulk nature are unambiguously shown by resistivity[2], Meissner effect[14], and specific heat[15] measurements. On the other hand, by analogy with TTF-TCNQ-like compounds, the insulating low-pressure state had originally been supposed to be of the Peierls type. However, the absence of any sign of a lattice modulation, and, more directly, the magnetic anomalies at the metal-insulator transition[16] and the observation of antiferromagnetic resonance[17] are evidence for the spin-density wave nature of the low-pressure state. The insulating character of this phase implies a modulation wavenumber of $2k_F$ in the chain direction. However,a detailed determination of the magnetic structure by neutron scattering is lacking so far.

Applying a perpendicular magnetic field to $(TMTSF)_2ClO_4$ or $(TMTSF)_2PF_6$, under conditions with a superconducting ground state at zero field, superconductivity disappears at a critical field $H_{c2}\approx 1$kOe. At considerably higher fields (≈ 50kOe) one finds a transition from the metallic to a semimetallic SDW state[18,19], and for $(TMTSF)_2ClO_4$ with further increasing field a cascade of further transitions is observed[20]. A theoretical explanation has been given by Gorkov and Lebed[21], and has been further worked out by Heritier et al.[22]: due to the finite transverse bandwidth, an SDW transition may leave some electron or hole pockets, which will be narrow tubes in the present case. In a magnetic field it is then energetically favourable to adjust the cross section of the tubes (i.e. the precise value of the ordering wavevector) in such a way that there is always an integral number of occupied Landau levels. This leads to the observed cascade of transitions with increasing field, with a discontiuous change of wavevector at each transition.

In addition to these electronically driven phase transitions, compounds with noncentrosymmetric anions ($X=ReO_4, ClO_4, NO_3, \ldots$) exhibit order-disorder transitions of the anions[23]. However, as manifested by the existence of transitions both with longitudinal wavevector $q_a=2k_F$ and $q_a=0$, and by the absence of any one-dimensional precursor scattering, these transitions are not intrinsic instabilities of the electronic system, but are rather due to a direct anion-anion interaction (of so far unknown type). However, effects of the anion ordering on the electronic structure exist. E.g. , in $(TMTSF)_2ClO_4$ the anion ordering can be suppressed by rapid quenching, and then the superconducting ground state is replaced by an SDW[24]. The anion ordering of the ReO_4 compound (T_c=183K!) disappears under a pressure of about 10kbar, and instead one find superconductivity[25]. A full theoretical understanding of these effects is lacking so far, however some attempts have been made[26].

THEORETICAL INTERPRETATION OF PHASE DIAGRAMS

A basis for the theoretical discussion of the phase diagrams of $(TMTSF)_2X$ compounds is provided by the one-dimensional interacting electron gas ("g-ology") model[27,28], where electron-electron backward and forward scattering are parametrized by the constants g_1 and g_2, respectively. Due to the one-dimensionality, two particularities appear: (i) There are instabilities both of the density wave (CDW/SDW) and superconducting (singlet (SS) and triplet (TS)) type, as manifested by logarithmic divergences ($\ln(E_F/T)$) of density wave (Peierls) and superconducting (Cooper pair) susceptibilities of the noninteracting fermion system. A consistent treatment of interaction effects has to include the coupling between the Peierls and Cooper divergences (in a three-dimensional metal only the Cooper pair susceptibility diverges). (ii) Thermal and quantum fluctuations prevent the occurence of phase transitions or long-range order, however, at low temperatures there can be strong (divergent for $T\to0$) fluctuations of CDW, SDW, SS, or TS type. The divergent fluctuations are described by the correlation function (susceptibilities) R_i representing the different types of order (i=CDW,SDW,SS,TS). Their low temperature behaviour is characterized by power laws:

$$R_i(T) \propto \frac{1}{\pi v_F}\,(T/E_F)^{-\alpha_i} \qquad , \qquad (1)$$

where v_F is the Fermi velocity. Abbreviating $\gamma=(g_2-g_1/2)/\pi v_F$, one finds for $g_1>0$

$$\alpha_{CDW} = \alpha_{SDW} = 1 - \left(\frac{1-\gamma}{1+\gamma}\right)^{1/2} \simeq \gamma \quad ,$$

$$\alpha_{SS} = \alpha_{TS} = 1 - \left(\frac{1+\gamma}{1-\gamma}\right)^{1/2} \simeq -\gamma \quad , \qquad (2a)$$

whereas for $g_1<0$ there are no SDW or TS divergences, and one has

$$\alpha_{CDW} = 2 - \left(\frac{1-\gamma}{1+\gamma}\right)^{1/2} \simeq 1+\gamma \; , \quad \alpha_{SS} = 2 - \left(\frac{1+\gamma}{1-\gamma}\right)^{1/2} \simeq 1-\gamma \; . \qquad (2b)$$

The approximate equalities in (2a) and (2b) are valid for small g_1 and g_2. The regions in the g_1-g_2 plane where there are divergent fluctuations of the different types are shown in fig.2. Another important point is the existence of a gap in the spectrum of spin excitations for $g_1<0$, leading to a thermally activated spin susceptibility, whereas for $g_1>0$ there is no gap, and consequently a Pauli-like (possibly interaction enhanced) susceptibility.

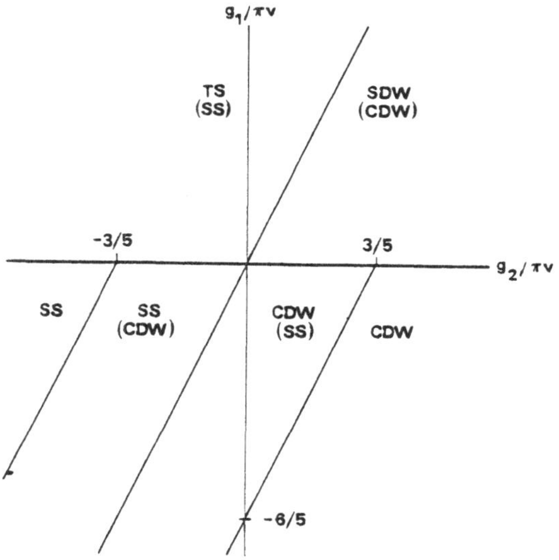

Fig.2: The phase diagram of the interacting one-dimensional electron gas in the absence of umklapp scattering (g_3=0). Parentheses indicate divergences weaker than the leading one. For g_1>0 the amplitudes of these subdominant divergences are quite small (ref.33).

Generally, in a quasi-one-dimensional conductor, with a weak coupling between the chains, a three-dimensionally ordered phase of type i appears at a critical temperature T_c^i given by $\lambda_i R_i(T_c^i) \simeq 1$, where λ_i is the appropriate interchain coupling. From eq.(1) one finds

$$T_c^i \propto E_F(\lambda_i/E_F)^{1/\alpha_i} \quad . \tag{3}$$

Equation (3) can be obtained by a mean-field treatment of the interchain interaction if this interaction couples order parameters on adjacent chains directly[29]. Then λ_{CDW} is mainly due to direct interchein Coulomb interaction, whereas for SDW, SS, or TS interchain exchange interactions contribute. On the other hand, if the coupling is due to single-particle interchain motion, parametrized by an overlap integral $t_\perp$, the situation is less clear. However, a recent calculation[30] shows that even in this case the coupling is due to pairs tunneling at nearby sites, leading again to eq.(3) with $\lambda_i \simeq t_\perp^2/E_F$. For $\alpha_i \lesssim 1$ the thermodynamic relations at T_c^i between the normal state specific heat, the specific heat jump, and the condensation energy are quite close to those of standard BCS theory[31].

An interpretation of $(TMTSF)_2X$ phase diagrams using the g-ology model has been worked out by Barisic, Brazovskii, Emery, and Bruinsma, emphazising umklapp effects[32]. Both the existence of an SDW phase and the Pauli-like susceptibility place these substances in the region g_1>0. Further, due to the weak dimerization of the TMTSF stacks, electron-electron umklapp scattering, parametrized by a constant g_3, is possible. For $g_1<2g_2+|g_3|$ a nonzero g_3 leads to the appearance of a gap Δ_ρ in the charge density excitation spectrum:

$$\Delta_\rho \propto E_F(|g_3|/(2g_2-g_1))^{1/2\gamma} \quad \text{for} \quad |g_3| \ll 2g_2-g_1 \quad ,$$
$$\Delta_\rho \propto E_F \exp(-\pi^2 v_F((2g_2-g_1)^2-g_3^2)^{-1/2}) \quad \text{for} \quad |g_3| \lesssim 2g_2-g_1 \quad , \tag{4}$$

and one has R_{CDW}, $R_{SDW} \propto 1/(\pi v_F)(\Delta_\rho/T)$, i.e. $\alpha_{SDW}=\alpha_{CDW}=1$. For $g_1>2g_2+|g_3|$ there is no gap, and the exponents given in eq.(2) are unchanged. The phase diagram of the $(TMTSF)_2X$ compounds can now be explained assuming that under increasing pressure $g_1-2g_2-|g_3|$ changes from negative to positive. Then, at low pressure $\Delta_\rho \neq 0$ and $\alpha_{SDW}=1$, leading to a relatively high SDW transition temperature, $T_c^{SCW} \simeq \Delta_\rho \lambda_{SDW}$ (one also has $\alpha_{CDW}=1$, but the amplitude of the CDW divergence is usually much weaker than the SDW one[33], and no CDW is expected). With increasing pressure Δ_ρ and therefore T_c^{SDW} decrease, until, beyond some critical pressure $g_1>2g_2+|g_3|$, so that $T_c^{SDW}=0$. Instead superconductivity appears at $T_c^{SC}\propto E_F(\lambda_{SC}/E_F)^{1/\alpha_{TS}}$, however, because $\alpha_{TS}=\alpha_{SS}$ is only slightly larger than zero, one has $T_c^{SC}<<T_c^{SDW}$, as experimentally observed.

The above picture explains qualitatively the phase diagram (fig.1), but some problems and questions remain: (i) The existence of a gap Δ_ρ would imply a low, thermally activated electrical conductivity. Experimentally, the conductivity is very high and metallic and increases with decreasing temperature down to T_c^{SDW}.
(ii) According to fig.2 the superconducting phase should be of the triplet rather than of the usual singlet type. The high sensitivity of the superconductivity to nonmagnetic defects may be an indication of triplet superconductivity[8,24,34], however there is so far no definite experimental evidence on the nature of the superconducting state. (iii) If the long-range (1/r) character of the Coulomb interactions is included in the model, there appears a metallic region in the upper half of fig.2, in between the TS(SS) and SDW(CDW) regions[35]. In a quasi-onedimensional system this implies a metallic ground state, which is not found experimentally.
(iv) In the standard g-ology model there are only instantaneous, Coulomb-type interactions. It is not yet clear how the picture is modified if the retarded electron-phonon interaction is included, though some progress has been made recently[36,37].

Alternative approaches to the phase diagram have been based on mean-field theory[38-40], assuming three-dimensional couplings to be sufficiently strong so that low-dimensional effects can be neglected. These theories make some specific predictions, e.g. a direct transition from a SDW to a <u>singlet</u> superconducting state[38], and its first order character[40]. However, the fundamental question is: Are the effects of the mixing of the density-wave and superconducting istabilities important or not? In the presence of transverse electron motion, the Peierls-type divergence saturates at $T\simeq t_\perp/\pi$. Consequently, one has to ask whether the terms $(g_i/\pi v_F)\ln(t_\perp/E_F)$ are large or small. The effect of Coulomb interactions can be parametrized by a constant C:

$$C = \frac{1}{\varepsilon_\perp}\left(\frac{\omega_{pl}a}{\pi v_F}\right)^2 \quad , \tag{5}$$

where a is the interchain distance, ω_{pl} the plasma frequency, and $\varepsilon_\perp$ the background dielectric constant perpendicular to the chains. For the Coulomb contribution to the g_i's one finds[35] $g_1/\pi v_F\simeq(\pi/8)C$, $g_2/\pi v_F\simeq(\pi/8)\ln(4/C)$. For $(TMTSF)_2X$ one has $C\simeq1$. Even though this estimate is certainly not quantitatively correct, due to the neglect of electron-phonon contributions and the detailed structure of the electronic wave-functions (assumed to be plane waves), it appears that the one-dimensional effects are quite important. A complementary estimate of the importance of one-dimensionality can be obtained from the fluctuation corrections to the mean-field transition temperature, calculated from a Ginzburg-Landau approach. This accounts for thermal fluctuation effects, but the density-wave/superconductivity coupling is completely neglected, and therefore the role of one-dimensionality is certainly underestimated. Nevertheless, using the known bandstructure anisotropies of TMTSF compounds, it appears that an important role of one-dimensionality can by no means be excluded[41]. Evidence more definite than these crude theoretical estimates can be obtained from some experimental results, to be discussed in the following chapter.

THE LOW-TEMPERATURE METALLIC REGIME

In the presence of a finite single-particle interchain tunneling amplitude $t_\perp$ the $2k_F$-density-wave susceptibility of a noninteracting fermion gas ceases to diverge at a crossover temperature $T^x \simeq t_\perp/\pi$, and below this temperature the system appears as three-dimensional (albeit possibly quite anisotropic). However, if the logarithmic interaction terms are important $((g_i/\pi v_F)\ln(E_F/t_\perp)>1)$, self-energy corrections may depress the crossover temperature considerably. A calculation similar to that giving eq.(3) leads to[31,42]

$$T^x \propto \frac{1}{\alpha_i+\lambda_i/E_F} T_c^i \quad . \tag{6}$$

Above T^x, the system has one-dimensional fluctuation properties (however, other properties, e.g. single particle transport, may well be more three-dimensional), whereas below T^x three-dimensional correlations build up, to give finally rise to the three-dimensional phase transition at T_c^i. T^x may well be considerably larger than T_c^i, however the (Ginzburg) temperature region, where critical fluctuations become important near T_c^i, remains quite narrow[31].

Experimentally, evidence for the above discussed behaviour comes from transport and magnetic resonance experiments. Recent NMR measurements[42] of the ^{77}Se spin-lattice relaxation rate $1/T_1$ of $(TMTSF)_2ClO_4$ show, under conditions with a superconducting ground state, a large enhancement over the standard free-electron Korringa relation ($1/T_1T$=const.) below 30K. $1/T_1$ is a direct measure of spin fluctuations. In the one-dimensional regime, a divergent SDW correlation function leads to an enhanced spin-lattice relaxation rate[42]:

$$(T_1T)^{-1} \propto \mathrm{Im}(R_{SDW}(T)) \propto \frac{1}{\pi v_F} (T/E_F)^{-\alpha_{SDW}} \quad . \tag{7}$$

Such a law explains the experiments quite well down to 8K, where the enhancement saturates. This leads to an estimate of $T^x \simeq 8$K.

The dc conductivity of $(TMTSF)_2ClO_4$ is very large at low temperatures[8] ($\sigma(2K) \simeq 10^6 (\Omega cm)^{-1}$), and shows an anomalously large sensitivity to a transverse magnetic field[8,43]. Far infrared studies[44] show that the high conductivity is limited to a narrow (in frequency) mode of width $\simeq 0.5cm^{-1}$. Using the known electron density and mass, a Drude interpretation of this width leads to a conductivity more than 100 times higher than the observed value[44]. The experimental value of the conductivity can be recovered assuming a conducting collective mode with only a fraction of the carriers participating, as is typical for superconducting fluctuation conductivity. Superconducting fluctuations would also explain the absence of saturation of the conductivity of $(TMTSF)_2X$ salts above the superconducting transition and the anomalously high magnetoresistance[45,46]. This picture suggests, at least for $(TMTSF)_2ClO_4$, coexistence between SDW and superconducting fluctuations. This possibility exists in the g-ology picture if spin-dependent interactions are included[28].

The results discussed here, as well as some others[47], clearly demonstrate the importance of collective effects in the $(TMTSF)_2X$ compounds up to temperatures of the order of 30K. The BCS-like specific heat results[15] might be rather taken as evidence for a mean-field-like transition. However, as mentioned in the previous chapter, BCS-like thermodynamics is also expected in weakly coupled chains with strong effects of low dimensionality[31].

$(TMTTF)_2X$ COMPOUNDS

Replacing selenium by sulfur the series of $(TMTTF)_2X$ compounds, isostructural to $(TMTSF)_2X$, is obtained.At ambient pressure these compounds generally show a broad maximum of conductivity around 150...200K, with a subsequent decrease with decreasing temperature[48]. At considerably lower temperaturez (10...20K) one finds phase transitions into magnetic[49] (X=Br,SCN) or nonmagnetic states[48,50] ($X=PF_6$). One also observes anion ordering in compounds with noncentrosymmetric anions ($X=ClO_4$,SCN,NO_3,...), analoguous to those of the TMTSF salts[23,49].

Theoretically, the temperature dependence of the conductivity can be understood by a correlation gap[32,48], induced by a finite g_3 (i.e. $\Delta_\rho \neq 0$). Due both to the larger dimerization and the smaller sulfur atoms (and consequently enhanced Coulomb repulsion) this effect is indeed expected to be stronger than in TMTSF compounds. Below the temperature of the conductivity maximum the electrons become progressively localized (a progressive Mott-Hubbard localization), and the low-temperature phase transitions involve essentially localized spins. The magnetic transitions occur into an antiferromagnetic state (similar to the TMTSF salts, but with stronger Coulomb correlations), whereas the nonmagnetic transitions are thought to be of the spin-Peierls[4,37] type, involving an additional dimerization of the organic chain and the simultaneous formation of singlet pairs by neighboring electrons. Quite different interchain interactions act to stabilize the antiferromagnetic and spin-Peierls states: antiferromagnetism is stabilized by exchange interactions, whereas the spin-Peierls state is stabilized by interchain Coulomb and electron-phonon interactions.

A particularly interesting case is $(TMTSF)_2PF_6$: under pressure of 13kbar the spin-Peierls state is replaced by antiferromagnetism[50], implying a symmetry changing phase transition at some intermediate pressure. This suggests that the $(TMTTF)_2X$ series is the continuation of the phase diagram of fig.1 to the left, with the sequence spin-Peierls→SDW (antiferromagnetism)→superconductivity with increasing pressure. This picture is also supported by the appearance of a metallic low-temperature state[51] in $(TMTTF)_2Br$ under 25kbar. However, up to now superconductivity has not been found in $(TMTTF)_2X$ compounds.

CONCLUSION

In the present article I have discussed the phase diagram and low-temperature properties of the $(TMTSF)_2X$ and $(TMTTF)_2X$ series of compounds. The typical low-T ordered phases of these series are spin-Peierls, SDW, and superconductivity, in order of increasing pressure (or decreasing Coulomb correlation effects). Both theoretical arguments and experimental results, especially from the low-temperature metallic region above the ordered phases, suggest important effects of one-dimensionality. However, a detailed understanding of the relation between simple g-ology model parameters and microscopic interactions (Coulomb versus electron-phonon) is still lacking. Substances with stronger interchain coupling, i.e. less one-dimensionality, may allow higher superconducting transition temperatures.

ACKNOWLEDGEMENTS

I thank my colleagues in Orsay, especially D. Jérome, J. Friedel, and C. Bourbonnais, for many stimulating discussions.

REFERENCES

1. R.E. Peierls, Ann. Phys. (Leipzig) 4, 121 (1930), "Quantum Theory of Solids" (Oxford University Press, London, 1955), p.108.
2. D. Jérome et al., J. Phys. (Paris) Lett. 41, L95 (1980).
3. J.C. Slater, Phys. Rev. 82, 538 (1951); J. des Cloiseaux, J. Phys. Radium (Paris) 20, 607 (1959); A.W. Overhauser, Phys. Rev. Lett. 4, 462 (1960).
4. E. Pytte, Phys. Rev. B 10, 4637 (1974); M.C. Cross and D.S. Fisher, Phys. Rev. B 19, 402 (1979).
5. "Conducteurs et Supraconducteurs Synthétiques à basse Dimension", Les Arcs, France, 1982, J. Phys. (Paris) Colloque 44 (1983).
6. Proceedings of the "International Conference on Low-Dimensional Synthetic Metals", Abano Terma, Italy, 1984, Mol. Cryst. Liq. Cryst., to appear.
7. J. Friedel and D. Jérome, Contemp. Phys. 23, 583 (1982).
8. D. Jérome and H.J. Schulz, Adv. Phys. 32, 299 (1982).
9. S.S.P. Parkin et al., Phys. Rev. Lett. 50, 270 (1883).
10. E.B. Yagubskii et al., Pisma JETP 39, 275 (1984) (JETP Lett. 39, to appear).
11. C.S. Jacobsen et al., Solid State Commun. 38, 423 (1981), Phys. Rev. Lett. 46, 1142 (1981), and in ref.5; J.F. Kwak, Phys. Rev. B 26, 4789 (1982).
12. P.M. Grant, Phys. Rev. B 26, 6888 (1982), and in ref. 5.
13. K. Bechgaard et al., Phys. Rev. Lett. 46, 852 (1981).
14. K. Andres et al., Phys. Rev. Lett. 45, 1449 (1980).
15. P. Garoche et al., J. Phys. (Paris) Lett. 43, L147 (1982).
16. J.C. Scott et al., Phys. Rev. Lett. 45, 2125 (1980); A. Andrieux et al., J. Phys. (Paris) Lett. 42, L87 (1981); K. Mortensen et al., Phys. Rev. Lett. 46, 1234 (1981), Phys. Rev. B 25, 3319 (1982).
17. J.B. Torrance et al., Phys. Rev. Lett. 49, 882 (1982); W.M. Walsh et al., Phys. Rev. Lett. 49, 885 (1982).
18. L.J. Azevedo et al., Physica B 108, 1183 (1981); J.F. Kwak et al., Phys. Rev. Lett. 46, 1296 (1981).
19. T. Takahashi et al., J. Phys. (Paris) Lett. 43, L565 (1982), and in ref. 5.
20. M. Ribault et al., J. Phys. (Paris) Lett. 44, L953 (1983), ibid. 45, to appear, and in ref. 6.
21. L.P. Gorkov and A.G. Lebed, J. Phys. (Paris) Lett. 45, L433 (1984).
22. M. Heritier et al., J. Phys. (Paris) Lett. 45, to appear. See also: J. Friedel, to be published; K. Yamaji, in ref. 6.
23. J.P. Pouget et al., J. Phys..(Paris) Lett. 42, L543 (1982), and in ref. 5.
24. S. Tomic et al., in ref. 5; T. Takahashi et al., ref. 19.
25. S.S.P. parkin et al., Mol. Cryst. Liq. Cryst. 79, 213 (1982).
26. R. Bruinsma and V.J. Emery, in ref. 5; P.M. Grant, Phys. Rev. Lett., (1983).

27. Yu. A. Bychkov et al., Sov. Phys. JETP 23, 489 (1966).
28. J. Solyom, Adv. Phys. 28, 201 (1979); V.J. Emery, in "Highly Conducting One-Dimensional Solids", ed. by J.T. Devreese et al. (Plenum, New York, 1979),p. 247.
29. D.J. Scalapino et al., Phys. Rev. B 11, 2042 (1975); R.A. Klemm and H. Gutfreund, Phys. Rev. B 14, 1086 (1976).
30. H.J. Schulz, to be published.
31. H.J. Schulz and C. Bourbonnais, Phys. Rev. B 27, 5856 (1983); H.J. Schulz, in ref.5.
32. S. Barisic and S. Brazovskii, in "Recent Developments in Condensed Matter Physics", vol. 1, ed. by J.T. Devreese (Plenum, New York, 1981), p. 327; V.J. Emery et al., Phys. Rev. Lett. 48, 1039 (1982).
33. J.F. Hirsch and D.J. Scalapino, Phys. Rev. Lett. 50, 1168 (1983).
34. C. Coulon et al., J. Phys. (Paris) 43, 1721 (1982); S. Tomic et al., in ref. 5.
35. H.J. Schulz, J. Phys. C 16, 6769 (1983).
36. J. Voit and H.J. Schulz, in ref. 6.
37. L.G. Caron and C. Bourbonnais, Phys. Rev. B 29, 4230 (1984).
38. B. Horovitz et al., Solid State Commun. 39, 541 (1981).
39. K. Machida, J. Phys. Soc. Jpn. 50, 2195 (1981); K. Machida and T. Matsubara, ibid. 50, 3231 (1981).
40. K. Yamaji, J. Phys. Soc. Jpn. 52, 1361 (1983), and in ref. 6.
41. H.J. Schulz et al., Phys. Rev. B 28, 6560 (1983).
42. C. Bourbonnais et al., J. Phys. (Paris) Lett. 45, L755 (1984). The crossover exponent determined in this paper differs from eq.(6), probably because different physical quantities are considered.
43. P.M. Chaikin et al., in ref. 5.
44. H.K. Ng et al., J. Phys. (Paris) Lett. 43, L513 (1982), and in ref. 5; T. Timusk and H.K. Ng, in ref. 6.
45. H.J. Schulz et al., J. Phys. (Paris) 42, 991 (1981). The Ginzburg-Landau transport theory used in this paper is certainly at best of some qualitative validity, see also ref. 41.
46. K.B. Efetov, J. Phys. (Paris) Lett. 44, L369 (1983).
47. D. Jérome, in ref. 6.
48. C. Coulon et al., J. Phys. (Paris) 43, 1059 (1982).
49. S.S.P. Parkin et al., Phys. Rev. B 26, 6319 (1982); C. Coulon et al., ibid. 26, 6322 (1982).
50. F. Creuzet et al., in ref. 6.
51. S.S.P. Parkin et al., Mol. Cryst. Liq. Cryst. 79, 605 (1982), and footnote in ref. 5, p. C3-791.

IMPURITY PINNING IN QUASI-1D SUPERCONDUCTIVITY

Hidetoshi Fukuyama
Institute for Solid State Physics, Tokyo University,
Roppongi, Minato-ku, Tokyo 106, Japan

Possible impurity pinning has been investigated in quasi-one-dimensional superconductors, and it is found that a 2π-soliton of Josephson phase can spontaneously be created around an impurity as far as the amplitude of the order parameter of the superconductivity is small enough. The result implies the existence of a novel resistive state just below the critical temperature, which is expected to be highly non-Ohmic.

1. Introduction

Impurity pinning in charge density wave (CDW) has been a subject of extensive investigation. Since the effective mass of CDW is large[1] due to the combined motion of electrons and lattice distortions treated adiabatically, the pinning process is usually examined classically[2]. As the non-adiabaticity of the lattice motion becomes important, the quantum fluctuations have to be treated properly in the pinning process. In the limit of the non-adiabaticity the electron-phonon couplig can be represented as the effective attractive interaction between electrons as in BCS theory for superconductors and then the problem of impurity pinning is reduced to the Anderson localization in interacting electrons. The possibility of interaction induced localization-delocalization transition has been noted by various investigators[3-6].

In this paper we present the coherent description of these problems of the impurity pinning based on the phase Hamiltonian derived by the bosonization of the fermion. Especially we will point out the subtle feature of the role of impurity potential in superconducting state.

We take unit of $\hbar=k_B=1$.

2. Model

In CDW, the classical energy to describe the impurity pinning is given by[7]

$$E_{CDW} = \int dx\ [A(\nabla\theta_s)^2 + V_0 \sum_i \delta(x-x_i) \cos(2k_F x+\theta_s)] \quad , \qquad (1)$$

where $\nabla \equiv d/dx$, $\theta_s(x)$ is the phase variable and is related to spatially slowly varying part of local electronic density, $\rho(x)$, as $\rho(x)=\nabla\theta_s/\pi$. The second term on r.h.s. of eq.(1) represents the potential due to impurities distributed over x_i and k_F is the Fermi momentum.

On the other hand the bosonization[8-12] of the weakly interacting one-dimensional electrons results in the phase Hamiltonian[13-16], by which the charge degree of freedom in the presence of impurity potential[17] is given as follows. (The spin degree of freedom is ignored here for simplicity.)

$$H_1 = \int dx \, [A_+(\nabla\theta_+)^2 + A_-(\nabla\theta_-)^2 + V_0 \sum_i \delta(x-x_i)\cos(2k_F x+\theta_+)] \; , \quad (2)$$

where $\theta_\pm=\theta_\pm(x)$ is the quantum mechanical phase variable and $A_\pm$ is a constant dependent on interaction. The last term on r.h.s. of eq.(2) represents the impurity potential. The two phase variables θ_+ and θ_- do not commute but $[\nabla\theta_+(x),\theta_-(x')]=2\pi i\delta(x-x')$. One of the phase variables, θ_+, is closely related to θ_s in eq.(1) as is seen from the form of the impurity potential in eq.(2). Actually θ_s is the slowly varying classical part of θ_+ and then θ_+ is always written as $\theta_+=\theta_s+\hat{\theta}$, where $\hat{\theta}$ is the quantum fluctuations around θ_s. Hence the classical impurity pinning corresponds to the case of $A_-=0$, while $A_->A_+$ for electrons with attractive interactions. An important fact as seen from eq.(2) is that the impurity pinning of θ_+ in the present phase Hamiltonian is equivalent to the Anderson localization[17]. (It is also shown[18] that the localization problem is equivalent to the classical two-dimensional XY spins in a particular type of random field.) It is reasonable to expect that the pinning force due to impurity potential is reduced as the quantum fluctuations become strong and then the depinning transition will take place eventually at some critical value of A_-/A_+. Such depinning transition, which is examined in detail in the following section, is the localization-delocalization transition in one-dimension caused by mutual interaction. Especially in the present case of attractive interaction it is the transition between the Anderson localized state and the superconducting state.

In reality the inter-chain coupling (i.e. three-dimensionality) of one kind or another ignored in eq.(2) results in various kinds of phase transition. If the system becomes superconductive, such inter-chain interaction will be of Josephson coupling type; i.e.

$$H_3 = -\frac{1}{2}\sum_{i\neq j} J_{ij} \int dx \, \cos(\theta_-(i,x) - \theta_-(j,x)) \quad , \quad (3)$$

where i and j are chain indices. The phase transition and electronic properties of the ordered state will properly be described by treating H_3, eq.(3), in the mean field approximation[19]

$$H_3 \rightarrow - \sum_i \int dx \ \Delta \cos \theta_- (i,x) \ , \qquad \Delta = ZJ < \cos \theta_-(i,x) > \ . \tag{4}$$

Here J_{ij} is assumed only for the nearest neighbour chains whose number is Z.

Consequently the Hamiltonian describing the impurity pinning will be given by

$$H = \int dx \ [A_+(\nabla\theta_+)^2 + A_-(\nabla\theta_-)^2 - \Delta\cos\theta_- + V_0 \sum_i \delta(x-r_i)\cos(2k_F x+\theta_+)] . \tag{5}$$

The new and interesting feature in this H is the coexistence of two different kinds of energy, which favor the fixed values of the phases mutually canonical conjugate.

The conductivity in the Ohmic region is given as follows in terms of θ_+[7,16]

$$\sigma(\omega) = -\omega(\frac{e}{\pi})^2 \int_0^\infty dt \ e^{-i\omega t} \int dxdx' <[\theta_+(x,t), \theta_+(x',0)]> \ . \tag{6}$$

This equation indicates that $\sigma(\omega)\rightarrow 0$ as $\omega\rightarrow 0$ where there exist finite restoring forces for the oscillations of θ_+, i.e. if θ_s is pinned.

3. Transition between Localized and Superconducting States in One-dimension

We treat H_1 given by eq.(2) and ask when the characteristic length, L_0, of the impurity pinning diverges resulting in the depinning. Assuming that $2k_F x_i$ is completely random and noting that we are interested in the region of sufficiently long L_0 near the transition, we can rewrite the last term on r.h.s. of eq.(2) as[20,21]

$$V_0 \sum_i \cos(2k_F x_i + \theta_+(x_i)) \rightarrow \int dx \ V_0 \ \sqrt{n/L_0} \ \cos(\zeta(x) + \theta_+(x)) \ , \tag{7}$$

where n is the average density of impurities and $\zeta(x)$ is the slowly varying random phase resulting from the coarse graining over the distance, L_0, and then $|\nabla\zeta| \sim O(L_0^{-1})$. By eq.(7) we expect that the classical average of $\theta_+(x)$, θ_s, will follow the random phase $\zeta(x)$ to gain energy from impurity potential. But there exist quantum fluctuations, $\hat{\theta}$, around this θ_s as schematically shown in Fig.1. These quantum fluctuations will conveniently be treated by the self-consistent harmonic approximation (SCHA)[15,22,23], i.e.

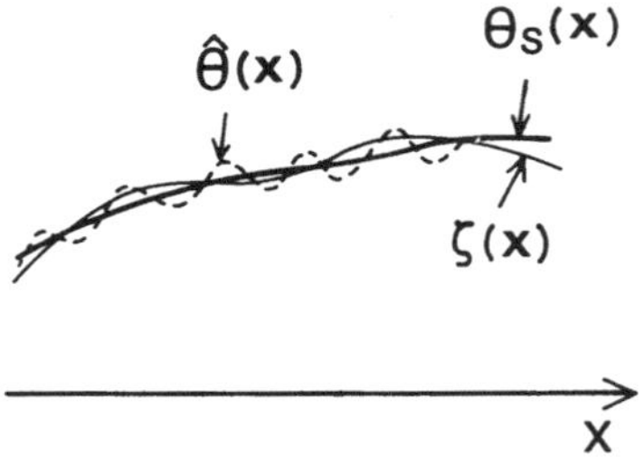

Fig. 1

$$\cos(\zeta(x)+\theta_+(x)) \rightarrow e^{-<\hat{\theta}(x)^2>/2} \cos(\zeta(x)+\theta_s(x))[1-\frac{\hat{\theta}(x)^2-<\hat{\theta}(x)^2>}{2}] \quad , \tag{8}$$

where $<\hat{\theta}(x)^2>$ is the average to be determined self-consistently. Equation (8) implies that the energy gain from impurity potential is of the order of

$$- |V_0| \sqrt{n/L_0}\, e^{-<\hat{\theta}(x)^2>/2} \equiv - |\tilde{V}_0| \sqrt{n/L_0} \quad , \tag{9}$$

per unit length. Here the effective impurity potential, $\tilde{V}_0$, which is the averaged over the quantum fluctuations, is defined by $\tilde{V}_0=V_0\exp[-<\hat{\theta}(x)^2>/2]\equiv V_0\gamma$. The quantity, γ, measures the degree of quantum fluctuations, which are dependent on V_0 as well as $A_\pm$ and will be evaluated explicitly in the following. For this we determine the excitation spectrum, $\omega(q)$, by use of eqs.(7) and (8) and by noting $V_0 \cos(\zeta(x)+\theta_s(x))\simeq-|V_0|$ as

$$\omega(q) = 4\pi\sqrt{A_+A_-}\,(q^2+q_0{}^2)^{1/2} \quad , \quad q_0{}^2 = (|\tilde{V}_0|/2A_+)\sqrt{n/L_0} \quad . \tag{10}$$

In terms of eq.(10), $<\hat{\theta}(x)^2>$ is evaluated as $<\hat{\theta}(x)^2>=\sqrt{A_-/A_+}\,\ln(2\pi/\alpha q_0)$, and then $\gamma=(\alpha q_0/2\pi)^{\kappa/2}$. Here $\kappa=\sqrt{A_-/A_+}$ (> 1) and α is the cut-off parameter of the order of the lattice spacing and $\alpha q_0 \ll 1$ is to be noted. By these equation the self-consistent solution of γ for given L_0 scales with L_0 as $\gamma\propto L_0{}^{-(\kappa/2)/(4-\kappa)}$.

Consequently the energy gain due to impurity potential, eq.(9), is proportional to $-\gamma L_0{}^{-1/2}\propto -L_0{}^{-2/(4-\kappa)}$. On the other hand the cost of elastic energy per unit length is of the order of $A_+(\nabla\theta_s)^2\sim A_+(\nabla\zeta)^2\propto L_0{}^{-2}$. Since L_0 is to be determined variationally by minimizing the total energy, i.e. sum of these two energy, the resultant L_0 is seen to be finite as far as $2/(4-\kappa)<2$, and $L_0\rightarrow\infty$ as $\kappa\rightarrow 3-0$.

Hence the transition between localized and superconducting state occurs at $A_-/A_+=9$.

This result for the critical value of $\kappa_c=3$ is consistent with those by Chui and Bray[4] and Apel and Rice[5,6] for the Tomonaga model[8], in which case the impurity potential couples with θ_+ as $\cos(2k_F x+\sqrt{2}\theta_+)$ instead of $\cos(2k_F x+\theta_+)$ in eq.(1). In the former type of coupling, we have $V=V_0\gamma^2$ and then $\gamma\propto q_0^{\kappa}$ resulting in $\kappa_c=3/2$.

4. Impurity Pinning in Superconducting State

In the preceding section we examined the competion between localized and superconducting ground states in strictly one-dimensional systems. As the three-dimensinality, i.e. interchain electron hopping in this case, is turned on, the actual superconducting transition could happen at finite temperature, T_c, resulting in the long range order with finite order parameter, Δ. We will examine in this section effects of impurity potential on the ground state of such quasi-one-dimesional superconductors[24]. If the interchain coupling is treated within the mean field approximation, our model is given by the Hamiltonian H_3, eq.(5). As in Sec.3 we write θ_- as the sum of slowly varying classical Josephson phase θ_J and the quantum fluctuations $\hat{\phi}$, i.e. $\theta_-=\theta_J+\hat{\phi}$. The SCHA to the Hamiltonian results in

$$H = \int dx\ [A_-(\nabla\theta_J)^2-\bar{\Delta}\cos\theta_J + A_+(\nabla\theta_+)^2 + A_-(\nabla\hat{\phi})^2 + \frac{\bar{\Delta}}{2}\cos\theta_J(\hat{\phi}^2 - \langle\hat{\phi}\rangle^2)] \quad , \tag{11}$$

where $\bar{\Delta}$ is the renormalized order parameter, $\bar{\Delta}=\Delta\exp[-\langle\hat{\phi}(x)^2\rangle/2]=\Delta(\Delta\alpha^2/8A_-)^{1/(4\kappa-1)}$. With finite $\bar{\Delta}$ the classical phase, θ_J, is determined as a solution of the variational equation

$$2A_-\nabla^2\theta_J - \bar{\Delta}\sin\theta_J = 0 \quad . \tag{12}$$

In the ground state θ_J is spatially uniform and there exists the discrete degeneracy, $\theta_J=2n\pi$, n being the integer. This discrete degeneracy is accompanied by the existence of a gap in the excitation of $\hat{\phi}$. As is easily seen this gap in turn leads to the divergent fluctuations of the conjugate variable θ_+, i.e. for any x, $\langle\theta_+(x)^2\rangle=\langle\hat{\theta}(x)^2\rangle=\infty$. This forbids the non-uniform spatial variation of θ_s and hence that of average electron density. This is the characteristic feature of the homogenous superconductivity, where θ_J is uniform in space.

Next we consider an impurity located, say, at the origin. We assume that this impurity potential is weak enough and then it is sufficient to consider the expectation value, E_{imp},

$$E_{imp} = \int dx \quad \delta(x)\ \bar{V}(x)\ \cos\theta_s \quad , \tag{13}$$

where

$$\bar{V}(x) = V_0\ e^{-\langle\hat{\theta}(x)^2\rangle/2} \quad . \tag{14}$$

As is noted $\bar{V}(x)=0$ if the superconducting state is homogenous.

Equation (12) also leads to a 2π-soliton solution where the spatial variation of θ_J is given by

$$\cos\theta_J = 1 - \frac{2}{(\mathrm{ch}\ x/\xi)^2} \quad , \tag{15}$$

where $\xi=(2A\_/\bar{\Delta})^{1/2}$. For such spatially varying θ_J, θ_s is still uniform to make $V_0\ \cos\theta_s=-|V_0|$ but the quantum fluctuation of $\hat{\theta}$ is no longer uniform in space. This spatial dependence is determined in ref.[24] by decomposing $\hat{\phi}$ into the normal modes and the result is

$$\langle\hat{\theta}(x)^2\rangle = \kappa\ [\ell n\ \frac{0.52\xi}{\alpha} + \pi\beta(x)] \quad , \tag{16}$$

where the spatial dependence of $\beta(x)$ is shown in Fig.2. Accordingly $\gamma\equiv\exp[-\langle\hat{\theta}(x)^2\rangle/2]=\gamma(x)$, is now a function of x and is given by

$$\gamma(x) = (\ \alpha\ /\ 0.52\xi)^{\frac{\kappa}{2}}\ e^{-\frac{\pi\kappa}{2}\beta(x)} \quad , \tag{17}$$

or $\gamma(x)/\gamma(0)=\exp[-\frac{\pi\kappa}{2}\beta(x)]\equiv f(x)$. In Fig.3(b) spatial variations of θ_J given by eq.(15), and f(x) are schematically shown, whereas those of homogenous θ_J are shown in Fig.3(a).

Equation (17) implies the non-vanishing value of $\bar{V}(x)$, eq.(14). Especially at the impurity site we obtain

$$\bar{V}(0) = V_0\ (\alpha\ /\ 0.52\xi)^{\kappa/2} \propto (\bar{\Delta})^{\kappa/4} \quad . \tag{18}$$

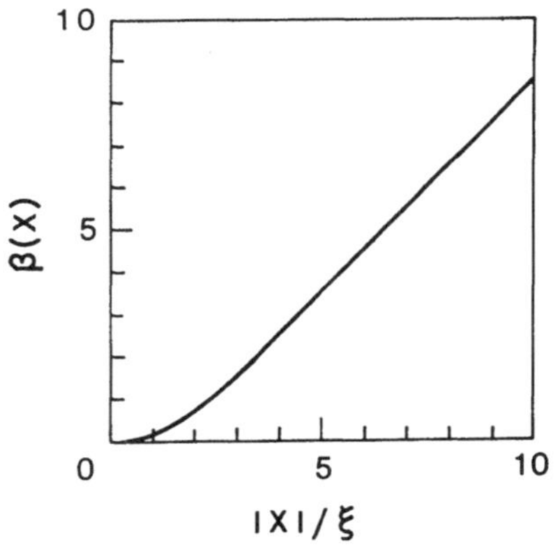

Fig. 2

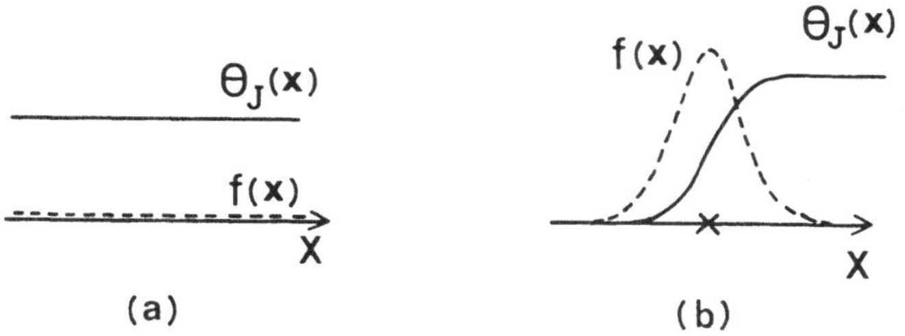

Fig. 3

Hence in this case of the 2 -soliton of the Josephson phase, eq.(15), there exist a finite gain from the impurity potential, which is given by $\bar{V}(0)$, eq.(18). The cost of energy competing with this gain is the formation energy of a soliton. Hence the difference of energy, E_s, of two states shown in Figs.3(a) and (b) is

$$E_s \equiv E\ [\text{Fig.3(b)}] - E\ [\text{Fig.3(a)}]$$
$$= 8\sqrt{2}\ (1 - \frac{1}{4\kappa})\ \sqrt{A_{-}\bar{\Delta}} - \bar{V}(0) \quad . \tag{19}$$

In view of eq.(17) we obtain $E_s<0$ as far as $\kappa<2$ and Δ is small enough, i.e. a 2π-soliton of θ_J is spontaneously created around an impurity.

It is to be noted that there is no current flow, $\langle j(x)\rangle$, in the presence of a soliton, even if θ_J is spatially varying. This is due to $j(x)=(e/\pi)\dot{\theta}_+$ and then $\langle j(x)\rangle=0$ by eq.(12).

In the presence of the finite density of impurities under the similar condition the value of $\theta_s(x)$ will be spatially fixed to gain energy from impurities. This pinning of $\theta_s(x)$ will result in the vanishing Ohmic conductivity given by eq.(6) as in impurity pinning of CDW. However once the order parameter, Δ, gets large enough so that $E_s>0$, the charge fluctuations are no longer pinned and then $\langle\theta_+(x)^2\rangle=\infty$, resulting in infinite conductivity at $\omega=0$, i.e. superconductivity. The strong dependence on the current flow will exist due to the depinning process of the charge density fluctuation by electric field. Although such depinning will share some common features, it can be of quite different nature from CDW.

The discussions so far are confined within independent chains whose interchain interaction has been treated as a mean field. To be more realistic the interchain correlation of θ_J has to be considered and then the soliton should be treated as extended in space, but the qualitative features will remain to be the same. In this context it is to be noted that $\xi \to \infty$ as $\Delta \to 0$, or $T \to T_c - 0$.

5. Summary

In this paper we have investigated various aspects of impurity pinning. Especially we have demonstrated that a 2π-soliton of Josephson phase can spontaneously be created around an impurity and argued about possible impurity pinning process in a macroscopic system. This possibility will be of interest in view of the recent remarkable experiment by Tajima and Yamaya[25], which will be discussed in this conference.

References

1. P.A. Lee, T.M. Rice and P.W. Anderson, Solid State Commun. **14** (1974) 703.
2. For review, e.g. N.P. Ong, Can. J. Phys. **60** (1982) 59; G. Grüner, Comments on Solid State Physics **10** (1983) 183.
3. A. Luther and I. Peschel, Phys. Rev. Lett. **32** (1974) 992.
4. S.T. Chui and J.W. Bray, Phys. Rev. **B16** (1977) 1329, ibid **B19** (1979) 4020.
5. W. Apel, J. Phys. C. **15** (1982) 1973.
6. W. Apel and T.M. Rice, Phys. Rev. **B26** (1982) 7063, J. Phys. C. **16** (1983) L271; ibid L1151.
7. H. Fukuyama, J. Phys. Soc. Jpn. **41** (1976) 513.
8. S. Tomonaga, Prog. Theor. Phys. **5** (1950) 349.
9. A. Luther and I. Peschel, Phys. Rev. **B9** (1974) 2911.
10. A. Luther and V.J. Emery, Phys. Rev. Lett. **33** (1974) 589.
11. D.C. Mattis, J. Math, Phys. **15** (1974) 609.
12. For review, J. Solyom, Adv. in Phys. **28** (1979) 201.
13. K.B. Efetov and A.I. Larkin, Soviet Phys.-JETP **42** (1976) 390.
14. A. Luther, Phys. Rev. **B15** (1977) 403.
15. Y. Suzumura, Prog. Theor. Phys. **61** (1979) 1.
16. For review on the phase Hamiltonian, H. Fukuyama and H. Takayama, Dynamical Properties of Quasi-One-Dimensional Conductors - Phase Hamiltonian Approach in Electronic Properties of Inorganic Quasi-One-Dimensional Compounds, ed. by P. Monceau (D. Reidel Pub. Company).
17. Y. Suzumura and H. Fukuyama, J. Phys. Soc. Jpn. **52** (1983) 2870.
18. T. Saso, Y. Suzumura and H. Fukuyama, LT17 (Karlsruhe, 1984); Y. Suzumura, T. Saso, H. Fukuyama and J. Cardy, ibid.
19. D.J. Scalapino, Y. Imry and P. Pincus, Phys. Rev. **B11** (1975) 2042.
20. H. Fukuyama and P.A. Lee, Phys. Rev. **B17** (1978) 535.
21. H. Fukuyama, J. Phys. Soc. Jpn. **45** (1978) 1266.
22. S. Coleman, Phys. Rev. **D11** (1975) 2088.
23. R. Dashen, B. Hasslacher and A. Neveu, Phys. Rev. **D11** (1975) 3424.
24. H. Fukuyama, Y. Suzumura and T. Saso, J. Phys. Soc. Jpn. **53** (1984) 1206.
25. Y. Tajima and k. Yamaya, J. Phys. Soc. Jpn. **53** (1984) 495.

NUMERICAL STUDIES OF THE EFFECT OF A WALL ON SDW IN A JELLIUM

A. Tagliacozzo
Istituto di Fisica Teorica, Università di Napoli, Napoli, Italy
Gruppo Nazionale di Struttura della Materia, CNR, Italy

INTRODUCTION

We have considered the modifications that occur in the spin density wave (SDW) amplitude of a jellium, near a jellium-vacuum interface. The problem is handled not self-consistently within Hartree Fock (H-F), assuming the one-electron functions to be the ones of the infinite barrier one-electron model on a completely deformable neutralizing background.

Long ago Overhauser[1] showed that the electron gas is unstable in H-F at any density, versus SDW/CDW formation, due to the effect of the exchange interaction. We shall assume result, without questioning how general it is[2,3], as the starting point of our calculation and compute the C/SDW gap parameter of the one-electron spectrum approximately, for a range of electron densities.

Self-consistent calculations of the semi-infinite jellium model have been mostly performed by use of density functional theory and local density approximation applied[4]. Ma and Sahni[5] use physically meaningful density profiles, generated by the linear potential model, to determine the non local surface exchange energy.

Our far from realistic discussion rules out the drastic changes in electronic band structure, that especially occur at transition metal surfaces, which may enhance the local electronic response to modulating potentials[6,7]. Greater importance of the exchange and correlation near the surface should also be taken into account.

Even in our crude approximations, we find that the full $\vec{k}$ dependence of the gap parameter is important in determining the C/S modulation near the surface. Our calculation shows that the excess of local density of states at the Fermi level in vicinity of the surface enhances the SDW amplitude there. This confirms previous results of spin susceptibility in semi-infinite jellium by Perdew[8]. Anisotropic behaviour is found with the direction of the wave vector $\vec{Q}$ of the C/S modulation with respect to the surface plane. If $\vec{Q}$ is perpendicular to the surface, a more pronounced peak appears close to it. The distance inside the sample at which the surface effect is lost and the bulk amplitude recovered, is found to be smaller than the usually defined coherence length.

2. THE GAP EQUATION AND C/S DENSITY

We shall not recall here the standard theory which leads to the H-F self-consistent SDW/CDW gap equation. For the bulk case, a non trivial solution of the H-F equations exists, by pairing electron-hole plane waves, whose wave vectors K differ by $\vec{Q} \simeq 2\vec{k}_f$ (k_f is the Fermi vector), if the self-consistency non linear integral equation is satisfied:

$$\Delta_{\vec{k}} = \sum_{\vec{k}',occ} V^{exch}_{\vec{k}\vec{k}'} \frac{\Delta_{\vec{k}'}}{2\sqrt{\omega^2_{\vec{k}'} + |\Delta_{\vec{k}'}|^2}} \tag{1}$$

Here $\vec{k}$ is the label of the new H-F states[9] and the sum is extended only to occupied states; $\Delta_{\vec{k}}$ is the gap in the one-electron spectrum which opens at $\vec{k} = 0$, that is at k-vectors $\vec{k} = \vec{Q}/2$ of the unperturbed energy parabola, and $\omega_{\vec{k}} = (\varepsilon^-_{\vec{k}} - \varepsilon^+_{\vec{k}})/2 \simeq h^2\vec{k}\cdot\vec{Q}/2m$ is half the difference between the two H-F energies which occur at each $\vec{k}$ value. In the bulk case $V_{\vec{k},\vec{k}'}$ is the usual exchange Coulomb potential $4\pi\ e^2/|\vec{k} - \vec{k}'|L^3$ (L is the length of the sample). If only the singular terms for $\vec{k} \simeq \vec{k}'$ are retained in the spherical harmonics expansion of the kernel of Eq.1, exactly soluble model is obtained[10]. In the extreme case, if Δ is taken independent on $\vec{k}$, the usual BCS form for the gap is recovered: $\simeq \exp(-1/N(0)V)$, with an interaction strength $V = 4\pi\ e^2/L^3k_f^2$, and the free electron density of states at the Fermi energy N(0). Otherwise, the form suggested for the k-dependence by this result is $\approx 1 - (\vec{Q}\cdot\vec{k})^2/k_f^4$.

Our numerical solution of the problem rests on two major approximations:

a) a parameter $\Delta = (\hbar^2 k_f^2/2m)D$, to be determined selfconsistently, is put in the denominator of the r.h.s. of Eq.1, in place of $\Delta_{k'}$, thus reducing Eq.1 to a linear problem to be solved in search of an eigenvalue which should not depend on k but depends on D.
b) the kernel, approximated in this way, and the solution itself, are expanded in spherical harmonics. If we impose the solution to be even with respect to change of sign of $\vec{k}$, only even angular momenta are allowed.

Expressed in terms of the dimensionless parameter D, the integral, equation becomes an universal function of D and the eigenvalue must be equal to $k_f a_B \pi$ (a_B is the Bohr radius). Fig. 1 shows the relation found between k_f and D, if a simplified hydrogen atom like form is assumed for the k

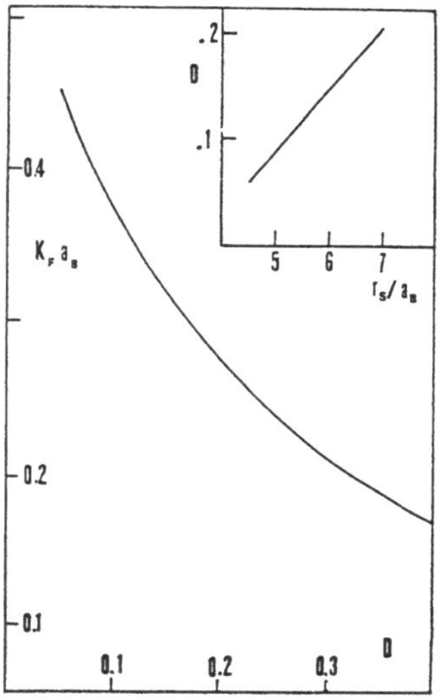

Fig. 1.

Dependence on k_f (on r_s/a_B in the inset) of the magnitude of the gap D.

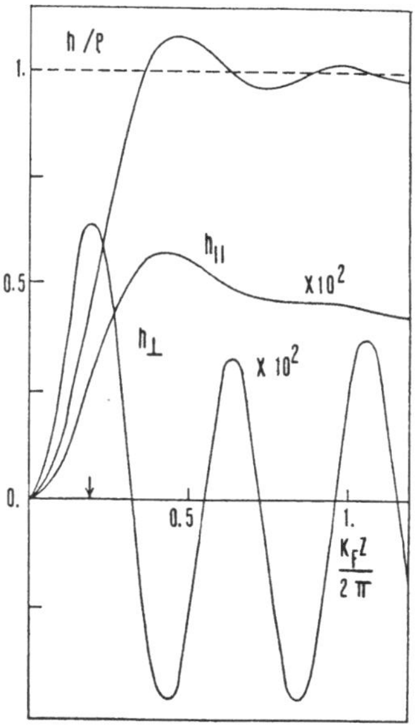

Fig. 2.

Density modulation for $\vec{Q}$ || ,⊥ to surface, versus distance from it. They are scaled by 10^2 with respect to density without SDW, shown for comparison. The arrow indicates the edge of the rigid background.

dependence. Limiting form of Eq.1 for $\vec{k} \to 0$, selects the suitable dependence of Q on k_f, giving a Q vector in the range 2.2 ÷ 2.4 k_f. The resulting dependence of D on r_s $(=(3/4\pi\ \rho)^{1/3}$, ρ is the electron density) is shown in the inset.

To discuss the surface case, let $\underline{k}$ lay on the surface plane, and χ be the component of $\vec{k}$ normal to it ($\chi=\sqrt{(\varepsilon_{\vec{k}} - h^2\underline{k}^2/2m)}>0$). An equation analogous to Eq.1 is obtained, where $V_{k,k'}$ is made of matrix elements of the 2-dimensional Fourier transform of the Coulomb potential integrated over four one-electron functions $f_\chi(z)$ (z normal to the surface). In the semi-infinite jellium, main difference to order 1/L, are interface terms which make the kernel vanish at $\chi = 0$ (or $\chi' = 0$, or $\chi = \chi' = 0$). Thus the boundary condition must be fulfilled, that is $\Delta_{\vec{k}} = 0$ for $\chi = 0$. The parity condition and symmetry arguments show that a solution (if there is any) will be of the form (chosen to be real) $\approx \chi^2$.

The density modulations $n_{\perp,\parallel}(z)$ [11] are plotted in Fig. 2. The occurrence of the maximum close to the surface is certainly favoured by the piling up of charge due to the unphysical confinement of the electrons. Nevertheless, the correct boundary condition for $\Delta_{\vec{k}}$ have striking effect as can be seen from Fig. 3. In Fig. 3, the result of Fig. 2 for Q || to the

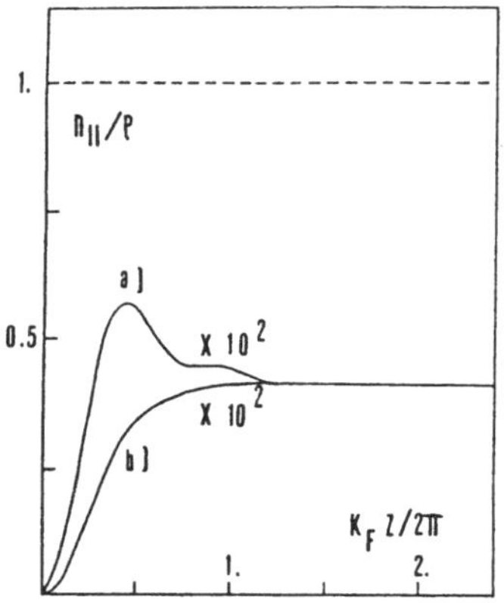

Fig. 3.
The case $\vec{Q}$ || to surface (a) is compared with the corresponding result using the bulk expression for $\Delta_{\vec{k}}$ (b).

surface is compared with the density modulation one would obtain by using the infinite jellium expression for $\Delta_{\vec{k}}$. This is much more like a fall off to be expected in a phenomenological theory.

The peak near the surface is even larger for Q ⊥ to it. This may be because the cutting of the sample doesn't alter the Bragg plane which is this time parallel to the surface. A comparison analogous to the one of Fig. 3 would give for this case a reduction of the first peak of about 1/3.

Surface Friedel oscillations are present, which damp into the bulk like $\sin 2k_f z/2k_f z$, as can be checked analytically.

REFERENCES

1. A.W. Overhauser, Phys. Rev. 128, 1437 (1962)
2. J.P. Perdew and T. Datta, Phys. Stat. Sol. (b), 102, 283 (1980)
3. L.M. Sanders, J.H. Rose and H.B. Shore, Phys. Rev. B21, 2739 (1980)
4. N.D. Lang, in Solid State Physics vol.28, Eds. F. Seitz, D. Turnbull and H. Ehrenreich (Academic Press, New York) (1973)
5. C.Q. Ma and V. Sahni, Phys. Rev. B20, 2291 (1979); V. Sahni and C.Q. Ma, Phys. Rev. B22, 5987 (1980)
6. G. Allan, Phys. Rev. B19, 4774 (1979)
7. D.R. Grempel, Phys. Rev. B24, 3928 (1981)
8. J.P. Perdew, Phys. Rev. B16, 1525 (1977)
9. C. Herring, in Magnetism vol. IV. Eds. G. Rado and H. Suhl (Academic Press, New York) (1966)
10. A. Tagliacozzo, to be published.
11. S.K. Chan and V. Heine, J. Phys. F.: Metal Phys. 3, 795 (1973)

PINNING OF AMPLITUDE SOLITONS IN PEIERLS SYSTEMS WITH IMPURITIES

V.L.Aksenov(a), A.Yu.Didyk(a) and R.Zakula(b)
Joint Institute for Nuclear Research, 101000 Moscow, P.O.Box 79, USSR(a) and "Boris Kidrič" Institute for Nuclear Science, 11001 Belgrade, P.O.Box 522, Yugoslavia(b)

The influence of impurities on properties of amplitude solitons in a one-dimensional model of Peierls systems with nearly half-filled bands is investigated. It is shown that there take place a critical dopant concentration and a depinning temperature at which solitons form an unpinned conducting lattice.

As is well known, in Peierls systems with half-filled bands the ground state is twofold degenerate and in the presence of donor or acceptor impurities the excess electrons or holes initiate the creation of charged solitons. These solitons are amplitude solitons and are described by the φ^4-model. Bak and Pokrovsky[1] have proposed the transition mechanism for the conducting state to occur: the transition takes place when at some critical concentration of excess electrons (solitons), C_{cr}, the soliton-interaction energy becomes equal to the energy of soliton pinning to the lattice. However, a more intensive pinning effect is caused by structural defects among which at least dopant impurities should be taken into account as they lead to the creation of solitons.

In this paper we consider the influence of impurities on amplitude-soliton properties in the one-dimensional φ^4-model of Peierls systems with nearly half-filled bands. Two types of impurities are considered: symmetry-conserving and symmetry-breaking impurities. The modification of the soliton solution in the neighbourhood of the impurity is studied within perturbation theory. The kink-impurity binding energy, a critical concentration at which kinks form an unpinned conducing lattice, and a depinning temperature are calculated.

The equation of motion for displacive fields in φ^4-model with an impurity has in dimensionless variables the form

$$\ddot{\varphi} - \varphi'' - \varphi + \varphi^3 + \gamma\dot{\varphi} = -V(x)\,(\varphi - \varphi_d). \qquad (1)$$

The interaction of kinks with the impurity can be described by the attractive short-range potential: $V(x) = \alpha\,\delta(x - x_d)$, where x_d is the coordinate of the impurity. In the case of conserving-symmetry defect (impurity) its equilibrium position is defined by $\varphi_d(x_d) \equiv \varphi_d = 0$. In the case of breaking-symmetry defects φ_d is different from zero. The damping constant γ describes phenomenologically the stochastic character of kink's motion between their collisions.

We find a solution of Eq. (1) in the form

$$\varphi(x,t) = \varphi_0(x) + u(x,t) \equiv \tanh(x/\sqrt{2}) + u(x,t). \qquad (2)$$

The function $\varphi_0(x)$ is a stationary partial solution of the homogeneous Eq. (2). Fluctuations $u(x,t)$ describe a modification of the soliton solution $\varphi_0(x)$ due to the presence of the impurity and can be represented as

$$u(x,t) = \beta_0(t)\Psi_0(x) + \beta_1(t)\Psi_1(x) + \int_{-\infty}^{\infty} dk\, \beta_k(t)\Psi_k(x), \qquad (3)$$

where $\{\Psi\}$ is the known[2] complete set of eigenfunctions of the self-adjoint linear operator L : $L\Psi \equiv -\Psi'' + [2 - 3\cosh^{-2}(x/\sqrt{2})]\Psi = \omega^2\Psi$.

Substituting representation (2), (3) into Eq. (1) we obtain the system of equations for $\{\beta\}$. In the weak soliton-defect-interaction approximation ($\alpha \ll 1$) we get:

$$\beta_j(t) = \beta_j^{\circ}[1 - e^{-\gamma t/2}(\cos\omega_j t + \gamma \sin\omega_j t/2\omega_j)], \; j = 1, k, \qquad (4)$$

$$\beta_0(t) = \beta_0^{\circ}[t/\gamma - (1 - e^{-\gamma t})/\gamma^2],$$

where

$$\beta_0^{\circ} = \alpha\sqrt{3\sqrt{2}/8}\,[\varphi_d + \tanh((x_0 - x_d)/\sqrt{2})]\cosh^{-2}((x_0 - x_d)/\sqrt{2}),$$

$$\beta_1^{\circ} = -\alpha\sqrt{3\sqrt{2}}\,[\varphi_d + \tanh((x_0 - x_d)/\sqrt{2})]\tanh\frac{x_0 - x_d}{\sqrt{2}}\cosh^{-1}\frac{x_0 - x_d}{\sqrt{2}}\Big/2\omega_1^2,$$

$$\beta_k^{\circ} = \alpha\Psi_k^*(x_d - x_0)[\varphi_d + \tanh((x_0 - x_d)/\sqrt{2})]/\omega_k^2, \quad \omega_1^2 = \frac{3}{2}, \; \omega_k^2 = 2 + \frac{k^2}{2}.$$

Here X_0 is the initial soliton position.

The binding energy of pinned to the impurity kink is given by the energy difference between two configurations[3]:

$$-|E_B| = E(x_0 \to x_d) - E(x_0 \to x_\infty), \qquad (5)$$

where $E \equiv E(x_d) = \frac{1}{T}\int_0^T dt \int_{-\infty}^{x_d} 2H[\varphi(x,t)]dx$ and $H[\varphi(x,t)]$ is the one-particle Hamiltonian the form of which follows from Eq. (1). Using Eqs. (2)-(5) we obtain the expression for pinning energy of kinks to impurities $E_{pin}^{d} = Np|E_B|$:

$$E_{pin}^{d} \approx Np\alpha(\xi_0/2a)[1 + (2 + \sqrt{2}/2)\varphi_d] \quad (\alpha/\gamma \ll 1), \qquad (6)$$

where p is the impurity concentration, ξ_0 the soliton width and a lattice constant.

Besides defects the discreteness of the lattice can hinder to the movement of solitons. According to[1] the pinning energy of kinks to the lattice is

$$E^{l}_{pin} \approx Nc\,\frac{16\pi^4}{3}\left(\frac{\sqrt{2}\,\xi_0}{a}\right)^4 \exp\left(-\frac{\pi^2\sqrt{2}\,\xi_0}{a}\right), \quad \text{for } \xi_0 \gg a. \qquad (7)$$

On the other hand, the interaction between solitons depin them. The soliton interaction energy can be derived by expanding the energy of soliton lattice in a small soliton concentration c (or for $\xi_0 \gg a$) /1,4/:

$$E_{int} \approx Nc\, 4E_0 \exp(-a/c\sqrt{2}\,\xi_0), \qquad (8)$$

where $E_0 = 8\sqrt{2}\,\xi_0/3a$ is the kink energy. A critical soliton concentration c_{cr} is defined by the condition: $E_{int} = E^{l}_{pin} + E^{d}_{pin}$. At $c \leq c_{cr}$ the soliton lattice is pinned while at $c > c_{cr}$ it can be shifted along the chain without any cost of energy.

When only dopant defects are present in the system, the soliton concentration c is equal to the impurity concentration ρ, and by using Eqs.(6)-(8) we obtain:

$$c_{cr} = \frac{a}{\sqrt{2}\,\xi_0} \ln^{-1}\left[\pi^4\sqrt{2}\left(\frac{\xi_0}{a}\right)^3 e^{-\pi^2\sqrt{2}\,\xi_0/a} + \alpha\,\frac{3\sqrt{2}}{64}\,V(\varphi_d)\right], \qquad (9)$$

where $V(\varphi_d) = [1/2 + (1 + \sqrt{2}/4)\varphi_d]$. In the absence of impurities ($\alpha \doteq 0$) and for $\xi_0/a \gg 1$ Eq.(9) goes over to the expression which has been obtained only in the case of lattice pinning/1/: $c_{cr} = a^2/2\pi^2\xi_0^2$. In Fig. 1 the dependence of c_{cr} on the dimensionless kink width $\delta = \xi_0/a$ is shown for conserving - symmetry impurities. For comparison we perform calculations with Eq.(9) for different mechanisms of the pinning: l - only to the lattice ($\alpha = 0$), d - only to impurities ($\alpha = 10^{-5}$), ld - to both the lattice and impurities.

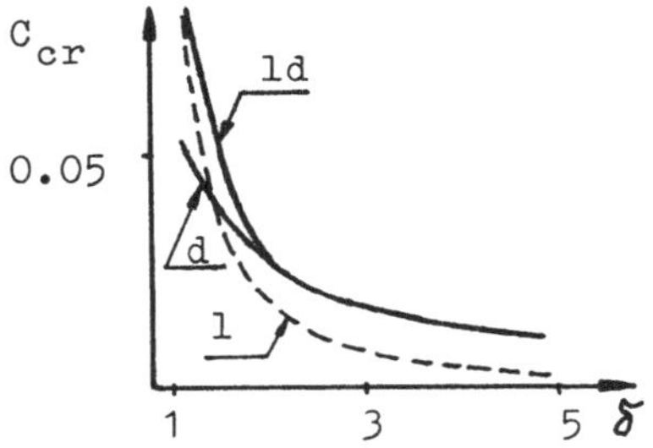

Fig.1. The soliton critical concentration c_{cr} as a function of $\delta = \xi_0/a$

As one can see, the critical concentration and contributions of different mechanisms depend crucially on the kink width. The pinning to the lattice takes place only at small values of δ, while at $\delta \gtrsim 2$ the essential contribution to the pinning comes from impurities. The pinning of kinks to breaking - symmetry impurities is stronger than to conserving-symmetry ones and increases proportionally to φ_d.

So far we were dealing with the system at zero temperature. The enhancement of temperature must lead to the depinning of kinks. Let us estimate a depinning temperature T_d using the relation $\theta = V_0 E_K$, where V_0 is the depth of the one-particle potential and E_K the kink kinetic energy. Thus, $T_d = \theta/V_0$ can be found from the condition :

$T_d = 2(E^l_{pin} + E^d_{pin} - E_{int})/Nc$, if $c < c_{cr}$, and has the form

$$T_d = \frac{32\pi^4}{3}\left(\frac{\sqrt{2}\,\xi_0}{a}\right)^4 e^{-\pi^2\sqrt{2}\,\xi_0/a} + 2\alpha\frac{\xi_0}{a}V(\varphi_d) - \frac{64\sqrt{2}}{3}\frac{\xi_0}{a}e^{-a/c\sqrt{2}\,\xi_0}$$

The δ -dependence of T_d at $c = 10^{-2}$ is shown in Fig. 2 for $\alpha = 10^{-5}$, $\varphi_d = 0$. The dashed part of the curve in Fig.2 corresponds to the previaling contribution of the lattice pinning (note that generally speaking in this region of δ the continuum limit is invalid). $T_d = 0$ means that, for a given δ , $c = 10^{-2}$ is equal to c_{cr} .

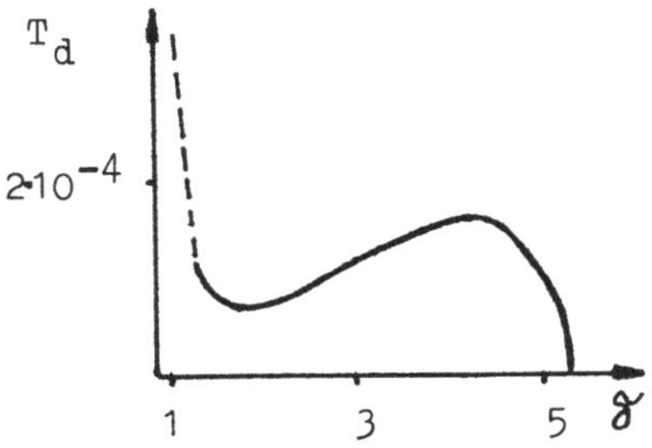

Fig.2. The depinning temperature T_d as a function of $\delta = \xi_0/a$

Our results can be applied for a qualitative description of depinning transitions in CDW-systems and, in particular, the insulator-metal transition in trans--polyacetylene for which, as has been estimated in ref.[1], $2a < \xi_0 < 5a$. Of course, our model of dopant impurities is oversimplified and has to be improved for a detailed comparison with experiment. In a more realistic model it is necessary to include screening effects, electronic correlations, and electron-dopant and interchain interactions.

References

1. P.Bak and V.L.Pokrovsky, Phys.Rev.Lett., 1981, 47, p. 958.
2. W.Hasenfratz and R.Klein, Physics A, 1977, 89, p. 191.
3. T.M.Rice et al., Phys.Rev. B, 1981, 24, p. 2751.
4. B.Horovitz, Phys.Rev.Lett., 1981, 46, p. 742.

NEW RESISTIVE STATE IN LOW DIMENSIONAL SUPERCONDUCTOR $TaSe_3$

K. Yamaya, Y. Tajima and Y. Abe
Department of Nuclear Engineering, Hokkaido University,
Sapporo 060, Japan

From measurements of the effects of the magnetic field and the electric field on the giant resistivity anomaly (GRA) observed within the superconducting (SC) transition region, it is found that the magnetic properties of the GRA are the same as those of the SC phase of $TaSe_3$ and there is a threshold electric field in the nonlinear conductivity in the GRA state. It is concluded that the GRA is one aspect of the SC phenomena and suggested that the origin of the GRA is related with the impurity pinning in the SC state.

It is well known that there are several superconductors such as $TaSe_3$[1], $ZrTe_3$[2] and $Nb_{1-x}Ta_xSe_3$[3], in the transition metal trichalcogenides MX_3. A close relation between charge density waves (CDW) phase and the superconducting (SC) phase generally is taken to be a characteristic feature of the superconductors in MX_3. For example, $NbSe_3$ shows two CDW transitions, but at low temperature this compound becomes superconductive when pressure is applied[4,5] or impurity is doped[6,7]. In $Nb_{1-x}Ta_xSe_3$ Kawabata and Ido proposed that the Ta-concentration dependence of the SC transition temperature T_c is explained in terms of the remaining density of states at Fermi level after two CDW are formed[3]. In $TaSe_3$, however, no anomalies associated with the CDW formation have been observed in the resistivity measurement, and the electron diffraction above 130 K is negative for the CDW formation[8]. Furthermore, the resistive state with the giant resistivity anomaly has been observed within the SC transition region of $TaSe_3$ when the applied current is extremely low[9]. Thus among superconductors in MX_3, $TaSe_3$ is an unique superconductor because no CDW is formed and an anomalous behavior is observed on the SC transition.

In this paper we present the measurements of the effects of current density, magnetic field and electric field on the GRA which was discovered on the SC transition of $TaSe_3$, and discuss an origin of the unusual SC behaviors in comparison with the Josephson-phase soliton model recently proposed by Fukuyama, Suzumura and Saso[10].

$TaSe_3$ was synthesized by the direct reaction of Ta (99.99%) and Se (99.999%). The mixture was sealed in quartz tube in vacuum of $\sim 10^{-5}$ Torr and heated to 650°C for 4$\sim$6 weeks. This heating temperature, which is lower than that reported previously, was set to prevent $TaSe_3$ from decomposing into $TaSe_2$.

Resistance of crystals was measured by the usual four probe dc technique with a current parallel to the b-axis (chain-axis). The electrical contacts were made with silver paint to Te and Au pads evaporated previously on the crystals. Temperature was measured by Ge resistance thermometer. The temperature stability was kept less than 5 mK/min. The electric current was supplied by the constant current sources in the range from 0.1 μA to 1.0 mA with 0.3% accuracy. The sample voltage was measured by a digital voltmeter (YEW model 2501), which has 0.1 μV resolution. Measurements were done by using computer controlled system in which the data were sampled four times and averaged. Thermoelectric voltage across the measurement circuit was cancelled by reversing the current direction.

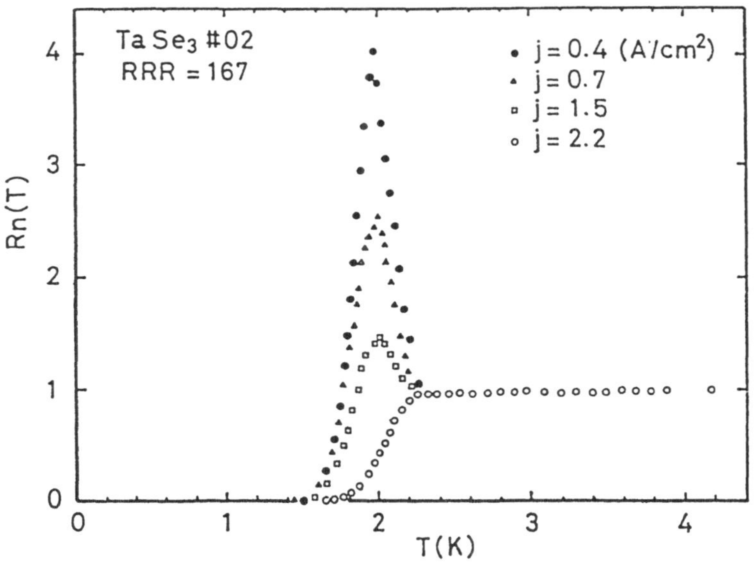

Fig. 1.
Normalized resistance $R_n = R(T)/R(4.2\ K)$ as a function of temperature at several current densities in $TaSe_3$ (ref. 9).

For temperatures above the onset temperature T_c^{onset} of the SC transition, the ohmic resistivity was confirmed in all regimes of the measured current density J. However, a nonohmic resistivity which strongly depends on J was found within the SC transition region. This anomalous behavior was observed in thirteen samples whose residual resistance ratio RRR(=R(300K)/R(4.2K)) were 100∿180. In Fig. 1 is shown a typical temperature dependence of the electrical resistance normalized by the resistance at 4.2K, R_n=R(T)/R(4.2K) for one of samples observed. In relatively high J regime a sharp decrease of R_n to a zero resistance is consistent with the usual SC transition reported previously[1]. While in relatively low J regime (∿ 10^{-1} A/cm^2) R_n increases rapidly below T_c^{onset} and has a peak whose value is almost several times as large as value of R(4.2K), and then decreases rapidly down to zero resistance as same as the usual SC transition. Thus we have found the zero resistance (ZR) transition together with the giant resistivity anomaly (GRA) just within the SC transition region. We call this transition as the ZR_{GRA} transition to distinguish the usual SC transition. No thermal hysteresis is observed in the ZR_{GRA} transition.

From Fig. 1, the effects of J on the ZR_{GRA} transition are summarized as follows. (1) The magnitude of the peak of R_n, R_p is easily reduced by weak J, but the temperature corresponding to R_p, T_p is independent on J. (2) The temperature dependence of R_n in the temperature region from a normal conducting state to the T_p is an activation type. The activation energy is strongly reduced by J. (3) With increasing J, the ZR_{GRA} transition curve coincides with the usual SC transition curve obtained in the high J regime.

The effects of the magnetic field H on the ZR_{GRA} transition have been measured and are compared with those of the usual SC transition. In Fig. 2 is shown a temperature dependence of R_n under H parallel to the b-axis, $H_{/\!/}$ in both regimes of the low and high J. For the high J regime (Fig. 2-b) T_c, which corresponds to R_n = 0.5 in the SC transition curves at various H, shifts to low temperature as H increases. When the SC critical magnetic field H_{c2} is defined as H corresponding to the T_c, the temperature dependence of H_{c2} is consistent with that measured by Yamamoto.[1] It can be seen that the transition width which is defined as the difference of two temperatures, T(0.1) and T(0.9) corresponding to R_n = 0.1 and 0.9 in the SC transition curve, broadens with increasing H. For the low J regime (Fig. 2-a) the ZR_{GRA} transition also shifts to low temperature with the suppression of the peak of R_n as H increases. At

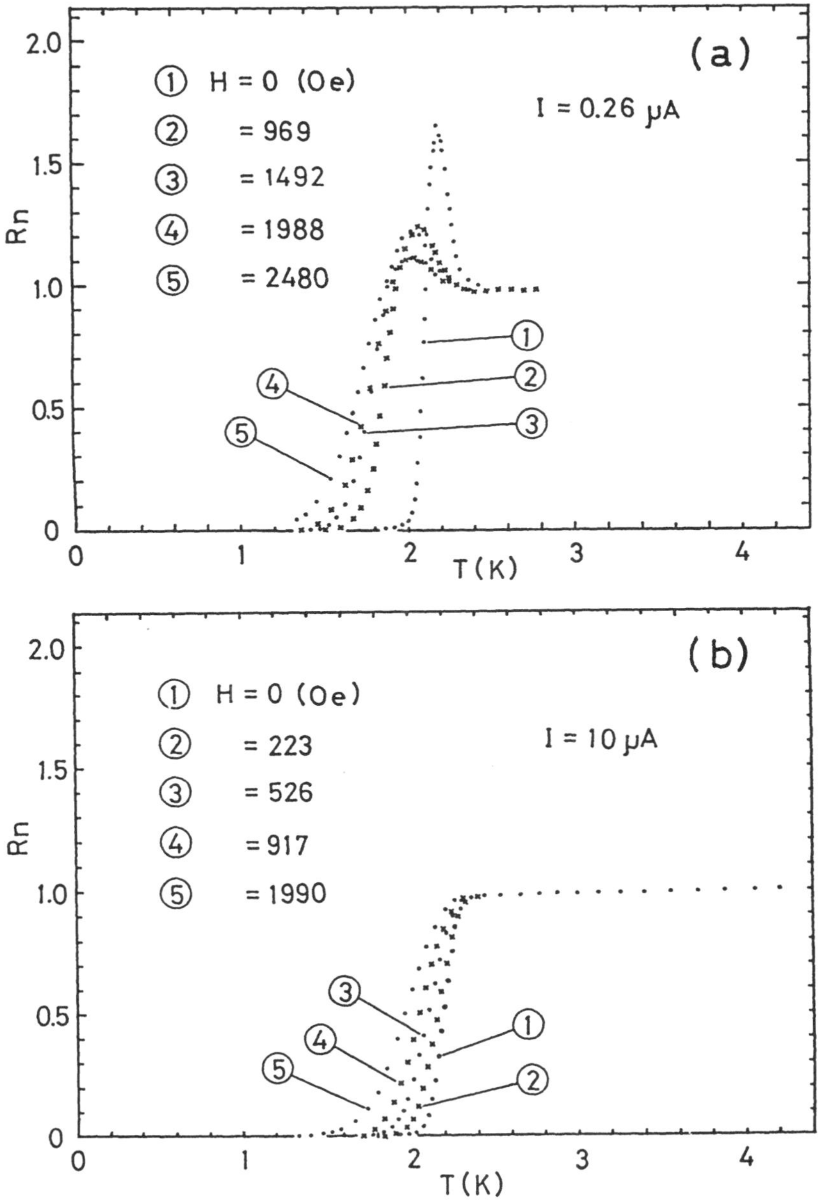

Fig. 2.
Normalized resistance $R_n = R(T)/R(4.2\ K)$ as a function of temperature at several magnetic fields in the low current density regime (a) and the high one (b) in $TaSe_3$.

various H, the T_p nearly equals to the T_c and each temperature of the onset and the end in the ZR_{GRA} transition is almost equal to those in the SC transition, respectively. Therefore, it is found that the magnetic field dependence of the ZR_{GRA} transition follows that of the SC transition.

In order to examine the change of the ZR_{GRA} transition by H, three characteristic magnetic fields, H_{NG}, H_p, and H(low) are defined as the magnetic fields which correspond to three characteristic temperatures in the ZR_{GRA} transition curve. (1) The T_{NG} is defined by the intersection of the linear extrapolation in the transition curve from the normal to the GRA state and the normal conducting value. (2) The T_p is the temperature corresponding to the peak of R_n. (3) T(low) is the temperature corresponding to half value of the resistance in the normal state (R_n = 0.5). In Fig. 3 is shown the temperature dependence of the H_{NG}, H_p, and H(low) under H[11]. At the same time is also shown the temperature dependence of H(0.1), H_{c2}, and H(0.9) which are defined as the magnetic fields corresponding to T(0.1), T_c, and T(0.9) in the SC transition curve, respectively. The temperature is normalized by each characteristic temperature at H = 0 Oe. The temperature dependence of H_{NG}, H_p, and H(low) quite well agree with those of H(0.1), H_{c2}, and H(0.9), respectively. This result shows that the ZR_{GRA} transition is a phenomenon which is closely related with the SC transition although the GRA is contrary to the usual SC phenomena.

It is found from Fig. 3 that the temperature dependence of H_p agrees with that of H_{c2}. The H_{c2} of $TaSe_3$ is known to show a large anisotropy for the directions parallel and perpendicular to the b-axis[1]. In order to obtain the anisotropy of H_p, the temperature dependence has been measured under H perpendicular to the b-axis, $H_\perp$. The result is shown in Fig. 4 with the result obtained for $H_\parallel$ [11]. The temperature is normalized by T_p and T_c at H = 0 Oe. The temperature dependence of the H_p very well agrees with that of the H_{c2} in both directions. This result means that the anisotropy of the H_p is quite the same as that of H_{c2}. The magnitude of the anisotropy of H_p is determined to be 17. This value is nearly equal to that of H_{c2} obtained by Yamamoto, which depends on the samples (10 ∿ 30)[1]. It is found that the magnetic properties of the ZR_{GRA} transition is accompanied with those of the usual SC state of $TaSe_3$ reported previously. Therefore, it is concluded from the effects of the magnetic field on the GRA in $TaSe_3$ that the GRA is a new aspect of the SC phenomena.

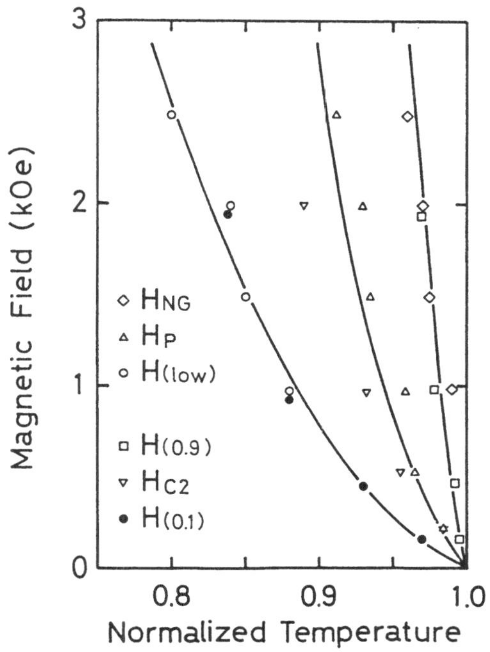

Fig. 3.
The temperature dependence of the H_{NG}, H_p, H(low), H(0.9), H_{c2} and H(0.9) for the direction parallel to the b-axis in $TaSe_3$.

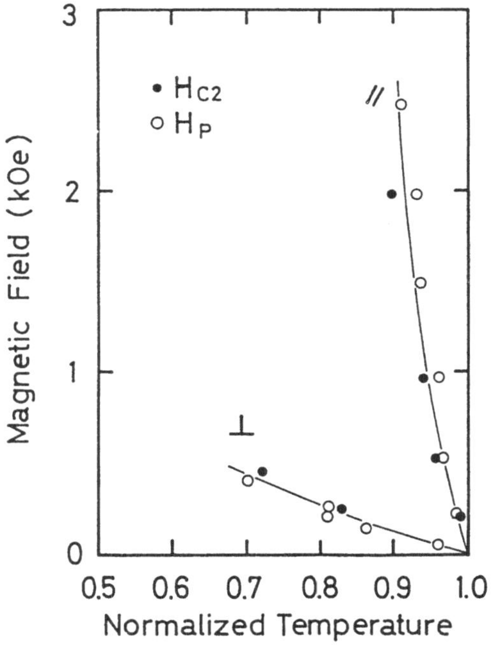

Fig. 4.
The temperature dependence of the H_p and H_{c2} for the direction parallel and perpendicular to the b-axis in $TaSe_3$.

In order to investigate the origin of the GRA, an electric field E dependence of the conductivity σ (= 1/R), has been measured near T_p [11]. The result at 1.95 K is shown in Fig. 5. With increasing E, σ increases rapidly from E $\sim$ 3μV/cm and above E $\sim$ 100 μV/cm saturates to the value of in the SC state ; that is, the nonlinear conductivity is observed. The σ below E $\sim$ 3 μV/cm is uncertain to be constant or not within the experimental accuracy. Therefore it is difficult to determine from the present experimental data (Fig. 5) the threshold electric field E_T, which is the E for the onset of the nonlinear conductivity. However, it is found that the E dependence of σ is well described in a function, $\sigma = \sigma_0 + \sigma_1 \ (1 - E_T/E)$. The second term contributes to the nonlinear conductivity. For the best fit to the experimental data, the value of a parameter, E_T has been obtained to be 3.3 μV/cm. This value agrees well with the value of E ($\sim$3 μV/cm) at which σ increases rapidly as shown in Fig. 5. Thus, it is considered that there is the E_T in the nonlinear conductivity of the GRA state of $TaSe_3$. Since the value of E_T obtained at 1.85 K is $\sim$7 μV/cm, E_T may increase as temperature decreases. On the analogy of the sliding motion of CDW by E above E_T observed in $NbSe_3$ and TaS_3 [12,13], it is considered that the impurity pinning may be responsible. However, the E_T obtained here is more than three orders in magnitude smaller than those obtained in CDW state[12,13].

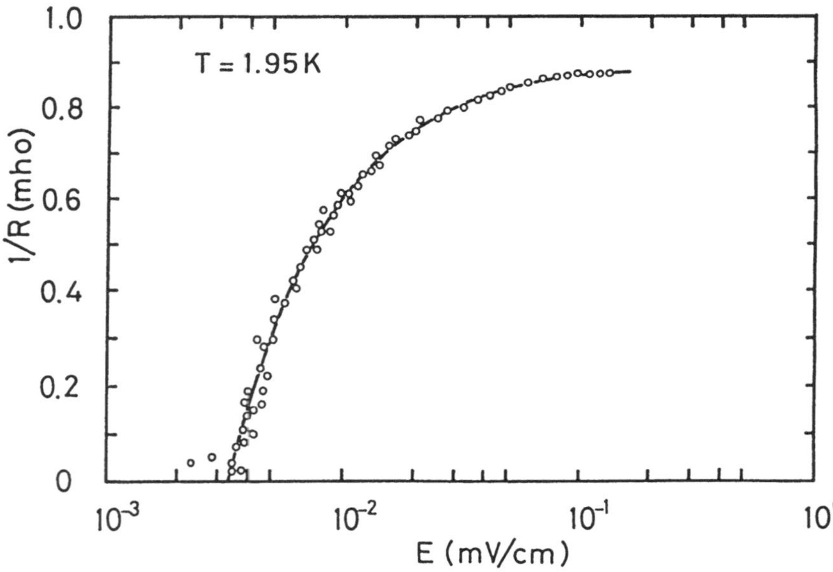

Fig. 5.
The conductivity σ (=1/R) as a function of electric field E at 1.95 K in $TaSe_3$.

The important properties for the GRA in $TaSe_3$ are summarized as follows. (1) The GRA is one aspect of the SC phenomena. (2) There is the threshold electric field in the nonlinear conductivity in the GRA state. (3) The resistivity rise below T_c^{onset} is the activation type. Therefore, the origin of the GRA in $TaSe_3$ is required to explain consistently these properties. One of origins to explain the GRA has been proposed recently by Fukuyama, Suzumura and Saso(FSS)[10]. They have shown that in one-dimensional SC state with an impurity a 2π-soliton of Josephson phase can be formed around an impurity near T_c^{onset}, where the order parameter of the superconductivity begins to grow, and this can lead to a pinning of the charge density fluctuation. They have argued that in $TaSe_3$ the activation type of resistivity rise below T_c^{onset} is due to the impurity pinning and the nonlinear conductivity will be explained by the depinning process of the charge density fluctuation by E. Thus, their theoretical results are qualitatively consistent with our experimental facts. However, the magnitude of E_T and the functional form of the E dependence of σ obtained here have not yet been studied in the FSS model. In particular a detailed investigation of the mechanism of depinning process of the charge density fluctuation will be necessary to compare with our experimental results of the E dependence of σ.

In summary, we have investigated the effects of the the magnetic field and the electric field on the GRA observed within the SC transition region of $TaSe_3$, and found that the magnetic behaviors of the GRA are accompanied with the anisotropy of the H_{c2} of $TaSe_3$, and there is a E_T in the nonlinear conductivity in the GRA state. It is concluded that the GRA is one aspect of the SC phenomena and suggested that the origin of the GRA is related with the impurity pinning in the SC state. Our experimental results may be consistently explained by the Josephson-phase soliton model proposed by FSS. However, much more work, such as detailed investigations of a mechanism of the depinning process in the nonlinear conductivity, a relation between the GRA and the CDW phase in one-dimensional system, and a sample dependence of the GRA will be necessary to make clear the origin of the GRA.

Acknowledgements We would like to thank Prof. T. Sambongi, Prof. M. Ido, Dr. Y. Okamoto, and authors in ref. 10, particularly Professor H. Fukuyama for useful discussions. We also thank Messrs. S. Seto and M. Morita for a technical assistant.

References

1. T. Sambongi, M. Yamamoto, K. Tsutsumi, K.Shiozaki, K. Yamaya, and Y. Abe, J. Phys. Soc. Jpn. 42 (1977) 1421, M.Yamamoto, J. Phys. Soc. Jpn. 43 (1978) 431.
2. S. Takahashi and T. Sambongi, J. Physique 44 supplement C-3 (1983) 1733.
3. K. Kawabata and M. Ido, Proc. Int. Symposium on Nonlinear Transport and Related Phenomena in Inorganic Quasi One Dimensional Conductors, Hokkaido University (1983) 324.
4. P. Monceau, J. Peyrand, J. Richard and P. Molinie, Phys. Rev. Lett. 39 (1977) 161.
5. A. Briggs, P. Moncea, M. Nunez-Reguiero, J. Peyrard, M. Ribout, J. Richard, J. Phys C13 (1980) 2117.
6. W. W. Fuller, P. M. Chaikin and N. P. Ong, Phys. Rev. B24 (1981) 1333.
7. K. Nishida, T. Sambongi and M. Ido, J. Phys. Soc. Jpn. 48 (1980) 331.
8. M. Yamamoto, Ph. D. thesis of Hokkaido University.
9. Y. Tajima, K. Yamaya and Y. Abe, Proc. Int. Symposium on Nonlinear Transport and Related Phenomena in Inorganic Quasi One Dimensional Conductors, Hokkaido University (1983) 292, Y. Tajima and K. Yamaya, J. Phys. Soc. Jpn. 53 (1984) 495.
10. H. Fukuyama, Y. Suzumura and T. Saso, J. Phys. Soc. Jpn. 53 (1984) 1206.
11. Y. Tajima and K. Yamaya, to be published in J. Phys. Soc. Jpn.
12. R. M. Fleming and C. C. Grimes, Phys. Rev. Lett. 42 (1979) 1423, R. M. Fleming, Phys. Rev. B22 (1980) 5606.
13. A. H. Thompson, A. Zettl and G. Gruner, Phys. Rev. Lett. 47 (1981) 64, G. Mihály, Gy. Hutiray and L. Mihály, Solid State Commun. 48 (1983) 203.

SWITCHING IN CDW SYSTEMS AND IN VO_2 - A COMPARATIVE STUDY

B. Fisher

Department of Physics, Technion, Haifa 32000, Israel

Similar phenomena observed in the past in VO_2 and more recently in CDW systems are reviewed. The study and understanding of these phenomena are easier in VO_2 because the electrical instability is accompanied by visible effects observed through an optical microscope. Low frequency oscillations observed in both media are associated in VO_2 with sliding metal-semiconductor domain patterns. Additional effects which bear similarities in the two media are: occurrence of longitudinal domain patterns, mechanical vibrations induced in the samples by passage of current and irreversible changes in the properties of the samples caused by high currents.

INTRODUCTION

VO_2 was extensively studied in the sixties and seventies[1]. It undergoes, upon cooling through T_t = 340 K, a first order phase transition from a tetragonal rutile structure to a monoclinic distorted rutile structure. The vanadium atoms become paired along the rutile axis and the pairs are tilted with respect to that axis. The unit cell is doubled and a gap is opened resulting in a jump in the resistance of more than five orders of magnitude (in pure single crystals). The transition is accompanied by an appreciable change in the optical properties of VO_2 in the visible regime. The multiplicity of ground states of the semiconducting phase with respect to the metallic phase results in a domain structure which depends on the cooling history. The domain pattern is visible under polarized light. In the vicinity of the transition temperature the metallic and semiconducting phases can coexist being separated by sharp boundaries. These domains are easily distinguished under the microscope due to their different colours.

The I-V characteristic of a semiconducting VO_2 needle exhibits a negative resistance (NR) regime and hysteresis loops. This behaviour is driven by the thermal instability[2] of the material. A large variety of metal-semiconductor (MS) domain patterns are observed in the NR regime. Some MS boundaries are static while other slide along the needle in the sense of the electric current (see Fig. 1). It was shown[3] that this drift is a result of Peltier heat - latent heat exchange at the MS boundary. This interpretation is qualitatively supported by the measured value of the drift velocity of the pattern and especially by the sense of its drift which is opposite to the drift of the carriers.

VO_2 and the CDW conductors are very different from the point of view of dimensionality, type of M-S transition and their I-V characteristics in the non-ohmic regime. Therefore it was surprising to find several rather exotic effects that appear in both media.

It may be that some of the similarities are not accidental. These effects are discussed below.

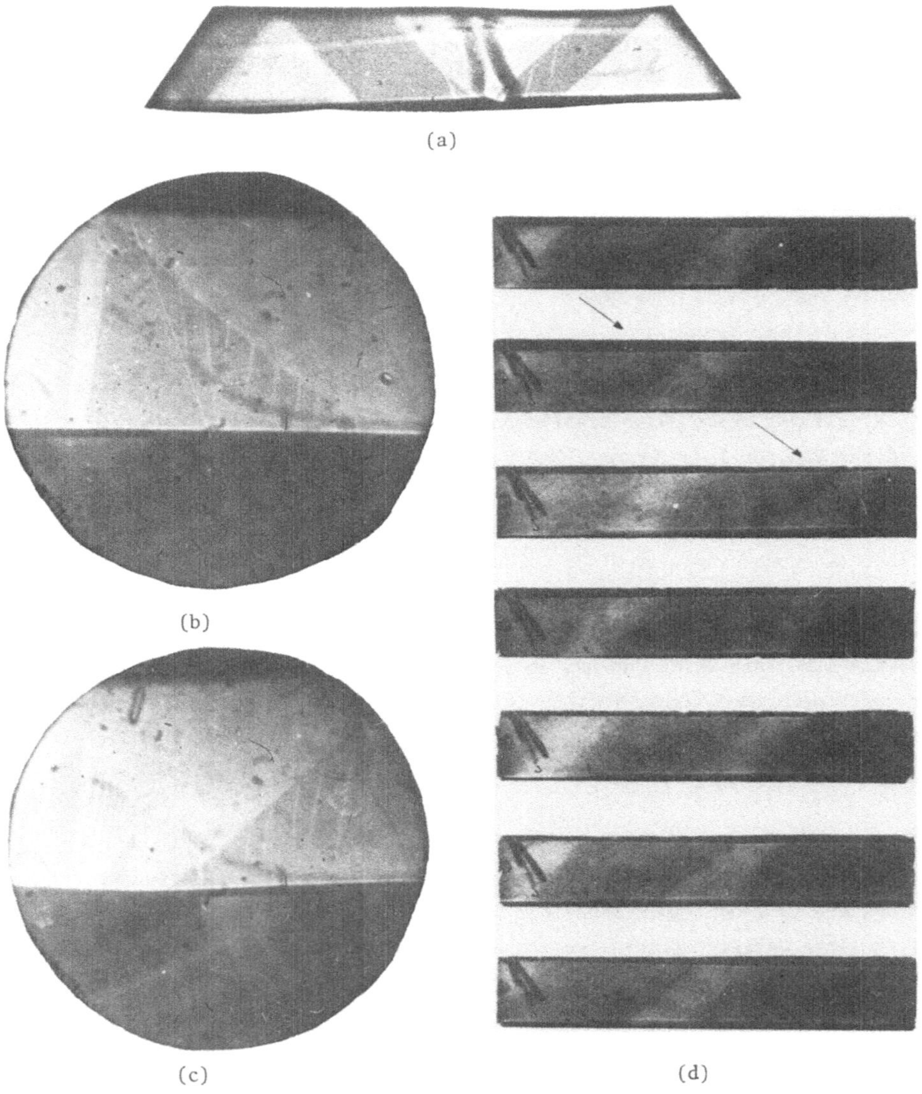

Fig. 1

Typical Metal-Semiconductor domains in switching VO_2 samples. The bright domains are semiconducting, the dark domains are metallic. Widths of samples are: 25 µm (a), 80 µm (b) and (c), 30 µm (d). The arrival of a sliding semiconducting domain to a fixed MS boundary is seen in (b). A sequence of frames from a ciné-film is seen in (d).

LOW FREQUENCY OSCILLATIONS

The sliding periodic pattern of MS domains in a switching VO_2 sample is accompanied by voltage oscillations (see Fig. 2). The frequency of the oscillations is exactly that of the passage of a semiconducting domain (a pair of MS and SM boundaries) through a fixed point in the sample. Though the pattern seen under the microscope can be simple (see Fig. 1 d) the time dependence of ΔV or ΔI is complicated and changes continuously as the applied voltage is varied. The velocity of the domains increases with current. The width of the semiconducting domains decreases with current. No simple

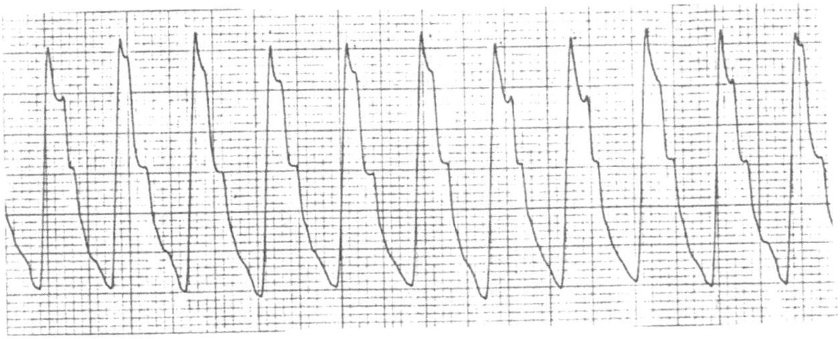

Fig. 2
Low frequency voltage oscillations accompanying sliding metal-semiconductor domain patterns. Horizontal: 1 sec/div.

relation was found between the frequency and the current. The low frequency oscillations (in the Hz regime) were found in the lowest region of the NR part of the IV characteristic. The simultaneous observation of the sliding domains under the microscope and the recording of the voltage oscillations are limited to fairly large M and S domains (tens of microns) and low drift velocities (< mm/sec) that is to low frequencies. Periodic voltage oscillations[4] of higher frequencies appear in small portions of the usually noisy I-V characteristic, at higher currents. A sequence of oscillograms showing a very stable and simple oscillation for a series of increasing currents is shown in Fig. 3. The detailed time dependence of ΔV indicates that the oscillations result from the same mechanism of generation-annihilation of sliding domains. The process is described by the schematical drawing shown in the same figure.

Low frequency quasiperiodic oscillations in the Hz regime have been observed in the blue bronzes of Mo. The interpretation proposed in reference 5 to this slow phenomenon invokes the displacement of some charged boundaries (discommensurations). Low frequency oscillations were observed also in other inorganic CDW conductors.

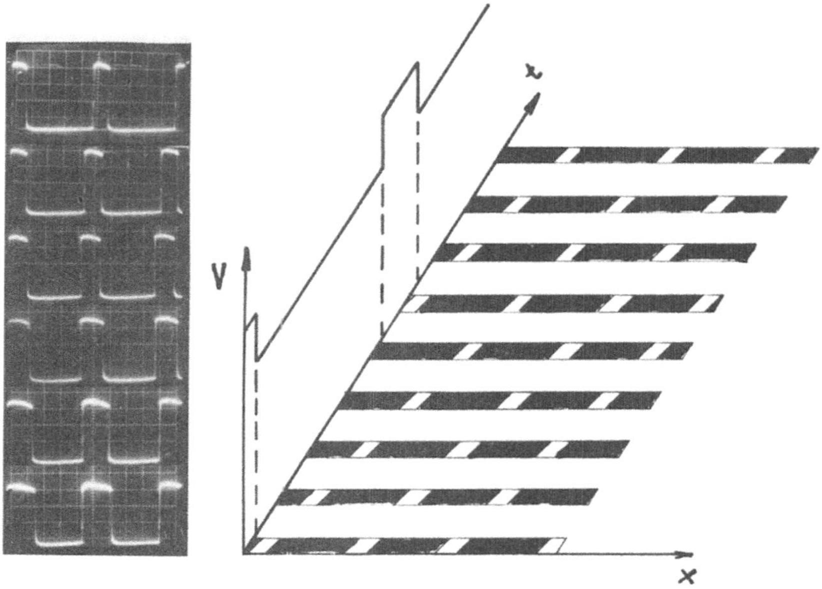

Fig. 3

Stable oscillations in a switching VO_2 sample for various values of increasing current. Horizontal 100µsec/div. The process of generation of such oscillations is described in the adjacent schematic drawing.

STRAND-LIKE DOMAINS

A very uncommon domain configuration[6] observed in a switching VO_2 sample is shown in Fig. 4. Between two tilted MS boundaries the material splits into longitudinal domains of a few µm width. These domains are visible not only due to different reflectivity, but also due to their different heights. The material is strongly wrinkled. A slight change of the electric current causes the boundaries to move, i.e. the wrinkled

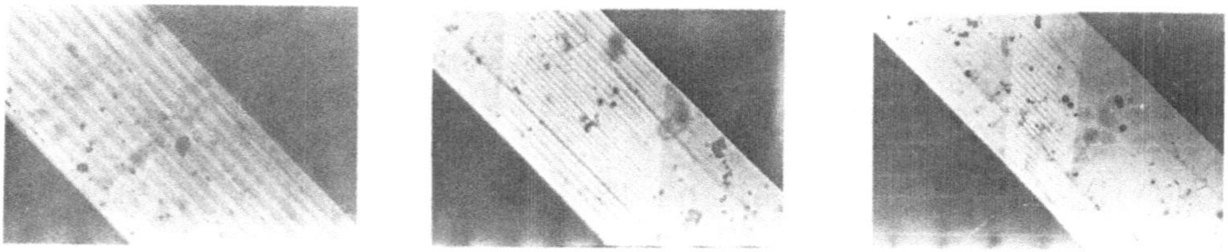

Fig. 4

Strandlike domain in a switching VO_2 sample at different currents. Width of sample is 60 µm.

portion expands or contracts. A faint reminiscence of the wrinkles persists after annealing, indicating the introduction of a permanent damage in the sample. In the presence of such domains the I-V characteristic is very noisy, although the pattern is static.

The pictures in Fig. 4 resemble amazingly that of the strand-like domains[7-9] observed in $NbSe_3$ and in TaS_3. The scale of the pattern found in these materials is smaller than that found in VO_2 by about two orders of magnitude.

MECHANICAL VIBRATIONS

The M-S transition in VO_2 is associated with a large deformation (the net change of the volume of about 0.1% is the result of nearly compensating expansion and contraction of about 1% along transverse axes). The partial transformation of a switching sample is accompanied by large stresses that cause the sample to twist and shake. It is hard sometimes to focus the microscope on such a sample due to its large vibrations. Usually, the vibrations are transient.

It was found[9] that passage of current induces a transient movement in $NbSe_3$. In fact, this effect was used to recognize in a large bundle of filaments exposed to the electro-microscope the fibers which were electrically continuous. The effect was attributed to heating by the current. This result indicates that dilatometric measurements in CDW conductors at various temperatures and electric fields may be of interest.

IRREVERSIBLE CHANGES IN THE SAMPLES

The internal stresses present in switching VO_2 samples often produce cracks, voids and other macroscopic defects. (Therefore, for practical applications thin VO_2 films are preferred rather than single crystals.) Although in most cases destructive, in some cases the switching may improve the properties of the samples. The periodic voltage oscillations described earlier are more easily obtained in samples that have been cycled several times through the NR regime of the I-V characteristic and may last for hours if the external conditions do not change. It is not yet clear if the "improvement" is a bulk or a contact effect.

A "history" dependent improvement of the samples performance was reported for the slow phenomena[5] in the blue bronzes. Recently, stable MHz oscillations[10] were obtained in TaS_3 only after treatment with high current pulses.

FINAL REMARKS

A series of similar phenomena exhibited by two very different media were reviewed. The relatively easy experiments done in VO_2, which leave very little to the imagination, may help in understanding some of the problems encountered in the more complicated CDW systems. Take for example figure 1 b, showing the accommodation of sliding to static MS boundaries. It would suggest a vortex-array model for the accommodation of a sliding CDW phase to a static phase if such a model would not have been invented[11].

A secondary purpose of this talk was to draw attention to the possibility that thermoelectric and thermo-mechanical effects may play an important role in the instabilities observed in the CDW conductors. It should be mentioned that good thermal contact between a sample and a heat bath does not preclude the formation of hot fillaments and hot domains inside the sample while large currents flow through it.

REFERENCES

1. For a review see M. Gupta, A.J. Freeman and D.E. Ellis, Phys. Rev. B 16, 3338 (1977).
2. M. Guntersdorfer, Solid State Electronics 13, 369 (1970).
3. B. Fisher, J. Phys. C 9, 1201 (1976).
4. B. Fisher, J. Appl. Phys. 49, 5339 (1978).
5. Jean Dumas and Clair Schlenker, Proc. Int. Symposium on Nonlinear Transport and Related Phenomena in Inorganic Quasi ID Conductors, Sapporo, Japan 1983, p. 198.
6. B. Fisher, J. Mag. and Mag. Materials 7, 326 (1978).
7. K.K. Fung and J.W. Steeds, Phys. Rev. Letters 45, 1696 (1980).
8. C.H. Chen and R.M. Fleming, Solid State Commun. 48, 777 (1983).
9. R.M. Fleming, C.H. Chen and D.E. Moncton, Journal de Physique, Suppl. 6 44, C3-1651 (1983).
10. B. Fisher - to be published.
11. N.P. Ong, G. Verma, M. Maki, Phys. Rev. Letters, 52, 663 (1984).

THE EFFECT OF VARYING THE BANDFILLING IN A PEIERLS CONDUCTOR

José Carmelo[+] and Kim Carneiro
Physics Laboratory I, University of Copenhagen
Universitetsparken 5, DK-2100 Copenhagen Ø, Denmark

We have considered the effect of changing the electronic bandfilling on the basic properties of the Peierls conductor when it is neither almost empty nor close to half filled. We calculate the zero-temperature gap $|\Delta|_o$ as well as the relation between $|\Delta|_o$ and the transition temperature T_c in the mean field approximation. When necessary we maintain the specific wavevector dependence of the electron-phonon coupling appropriate for the tight binding approximation. Our results are relevant for the interpretation of charge transfer salts with intermediate bandfillings.

Usual investigations of the properties of the Fröhlich Hamiltonian have been restricted to the half-filled band case.[1,2] However, for the quantitative description of one-dimensional conductors with non-integral charge transfers it is important to know the effect of varying the bandfilling.[3] Also the dependence of the electronic wavevector in the electron-phonon coupling is often neglected, but since it changes the electronic energy gain and elastic energy cost of the Peierls instability it may be important to incorporate this effect too. In this paper we address these two features of the Peierls conductor.

We consider the usual Fröhlich Hamiltonian for the electrons on a chain of N ions with lattice spacing a. In the adiabatic approximation where the phonons are treated as classical displacements of wavevector $\pm 2k_F$ with normal co-ordinates $\langle Q_{2k_F}\rangle$ one gets:

$$\begin{aligned} H &= \sum_{s=\pm 1,k} \{\varepsilon_s(k) a^+_{s,k} a_{s,k} + (2M\omega(2k_F)/\hbar N)^{\frac{1}{2}} g_s(k) a^+_{s,k} a_{-s,k} \langle Q_{2sk_F}\rangle\} \\ &= \sum_{s=\pm 1,k} \{\varepsilon_s(k) a^+_{\phi,s,k} a_{\phi,s,k} + |\Delta| f(k) a^+_{\phi,s,k} a_{\phi,-s,k}\} \end{aligned} \tag{1}$$

In (1) $a^+_{\phi,s,k} = \exp(\frac{1}{2} is\phi) a^+_{s,k}$ is the creation operator for an electron of wavevector $k+sk_F$ and phase ϕ, M and $\omega(q)$ are the ionic mass and phonon frequency, and $g_s(k) = isg_o f(k)$ denotes the electron-phonon coupling. $|\Delta|$ is the gap induced in the electronic spectrum at the Fermi level by the phonon displacement $\langle Q_{2k_F}\rangle$. As usual, we treat the electrons in the nearest neighbour approximation by introducing an overlap integral between site j and j+1 of the form:

$$t(j,j+1) = t + t'(u_{j+1} - u_j) \tag{2}$$

where u_j is the displacement of the j'th ion. One then gets:

$$\left.\begin{aligned} \varepsilon_s(k) &= -2t(\cos(k + sk_F)a - \cos(k_F a)) \\ g_o &= 2|t'|(2\hbar/M\omega(2k_F))^{\frac{1}{2}}\sin(k_F a) \\ f(k) &= \cos(ka) \end{aligned}\right| \tag{3}$$

Also $\varepsilon_F = 2t(1-\cos(k_F a))$ and the density of states at the Fermi level $N(0) = 1/2\pi t\sin(k_F a)$. Considering the electron-phonon coupling only within the first Brillouin zone (1) may be diagonalized as follows:

$$\left.\begin{aligned} H &= \sum_{\ell=\pm1,k} \varepsilon_\ell(k) c^+_{\phi,\ell,k} c_{\phi,\ell,k} \\ \text{where} \\ \varepsilon_\ell(k) &= \varepsilon_+(k) + \ell\{\varepsilon_-^2(k) + |\Delta|^2 f(k)^2\}^{\frac{1}{2}} \\ \varepsilon_+(k) &= \tfrac{1}{2}\sum_{s=\pm1} \varepsilon_s(k) = 2t\cos(k_F a)\{1-\cos(ka)\} \\ \varepsilon_-(k) &= \tfrac{1}{2}\sum_{s=\pm1} s\varepsilon_s(k) = 2t\sin(k_F a)\sin(ka) \\ c_{\phi,\ell,k} &= 1/\sqrt{2}\{(1+\ell\alpha(k))^{\frac{1}{2}} a_{\phi,1,k} + \ell(1-\ell\alpha(k))^{\frac{1}{2}} a_{\phi,-1,k}\} \\ \alpha(k) &= \mathrm{sign}(k)\{1 + (|\Delta| f(k)/\varepsilon_-(k))^2\}^{-\frac{1}{2}} \end{aligned}\right| \tag{4}$$

In (4) $\ell = 1$ corresponds to the conduction band and $\ell = -1$ to the valence band separated by the gap $2|\Delta|$. In order to find the temperature dependence of the gap we minimalize the free energy with respect to $|\Delta|$ in the mean field approximation after introducing $k_B T = 1/\beta$ and $\lambda = g_o^2 N(0)/\{\hbar\omega(2k_F)\}$

$$\frac{1}{\lambda} = \int_0^{\tilde{\varepsilon}(k_F)} d\tilde{\varepsilon} \, \frac{f^2(\tilde{\varepsilon})}{\varepsilon_-(\tilde{\varepsilon})} \alpha(\tilde{\varepsilon})\{1 - \mathcal{O}(\tilde{\varepsilon},\beta,\mu)\}\tanh \frac{\beta\varepsilon_-(\tilde{\varepsilon})}{2\alpha(\tilde{\varepsilon})}$$

where

$$\tilde{\varepsilon}(k) = \{2t\sin(k_F a)\}ak \tag{5}$$

and

$$O(\tilde{\varepsilon},\beta,\mu) = \{1 + \cosh^2(\frac{\beta\varepsilon_-(\tilde{\varepsilon})}{2\alpha(\tilde{\varepsilon})})/\sinh^2(\frac{\beta(\varepsilon_+(\tilde{\varepsilon}) - \mu)}{2})\}^{-1}$$

It is easy to check that (5) agrees with earlier results for the half-filled band when inserting $k_F a = \pi/2$.[2] However, apart from the specific dependence on k_F , (5) also includes the temperature dependent chemical potential μ which is identically zero when $k_F a = \pi/2$, but in general is not. We obtained (5) by performing the ℓ summation and keeping the k-integration. In this way we avoid the usual complications related to the introduction of the renormalized energy density of states when ε_ℓ is choosen as an intermediate integration variable. Compared to earlier treatments of arbitrary bandfilling[3] our results are more correct because they take $\mu(T)$, $\varepsilon_\pm(k)$ and $f(k)$ **into** account.

In the weak coupling limit when $\lambda \ll 1$ we derive the following approximate expression for the gap at zero temperature where $\mu_o = 0$:

$$|\Delta|_o = (8t/e)\sin^2(k_F a)\exp\{-1/\lambda\} \tag{6}$$

We may also express (5) both at $T = 0$ and at the mean field critical temperature T_c where $|\Delta| = 0$ and $\mu_c = (k_B T_c \pi)^2 \cos(k_F a)/12t\sin^2(k_F a)$:

$$0 = \int_0^{\tilde{\varepsilon}(k_F)} d\tilde{\varepsilon}\, \frac{f^2(\tilde{\varepsilon})}{\varepsilon_-(\tilde{\varepsilon})}\{(1 - O(\tilde{\varepsilon},\beta_c,\mu_c))\tanh(\tfrac{1}{2}\beta_c\varepsilon_-(\tilde{\varepsilon})) - \alpha(\tilde{\varepsilon})\}. \tag{7}$$

This yields:

$$\ell n \frac{|\Delta|_o}{1.76 k_B T_c} = C\left(\frac{k_B T_c}{2t}\frac{\cot(k_F a)}{\sin(k_F a)}\right)^2 \tag{8}$$

where the constant C may be calculated to be:

$$C = \int_0^\infty dy\, \frac{\sinh(y)}{\cosh^3(y)}\left(y^3 - \frac{\pi^2}{6}y + \frac{\pi^4}{144}\frac{1}{y}\right) = 1.00 \tag{9}$$

Hence the usual BCS relation between $|\Delta|_o$ and T_c is modified according to (8) with $C = 1$.

These results are valid for intermediate bandfillings (for 1/3 and 1/4 we disregard the small commensurability potential considered in reference[4], because we have that $|\Delta|_o/\varepsilon_F \ll 1$). The result (8) shows that the BCS relation between $|\Delta|_o$ and T_c is little modified when $k_F a \neq \pi/2$. Expression (6) tells how $|\Delta|_o$ and thus the elastic energy $N(0)N|\Delta|_o^2/2\lambda$ depend on $k_F a$, taking the bandfilling correctly into account. This is relevant for the understanding of charge transfer salts which often have bandfillings of 1/4 or close to 1/3. The fact that we have derived an analytical solution to the gap equation makes (5) a convenient starting point for taking other effects into account. In particular it would be interesting to consider fluctuations effects as well as to study the limits $k_F a \to \pi/2$ and $k_F a \to 0$, where our results are not directly applicable. Although the half-filled band case where Umklapp processes are allowed is well understood,[5] the way this is approached via the formation of soliton lattices still has some unclear points;[6] and as $k_F a$ approaches zero the weak coupling limit will eventually be violated which may give rise to interesting effects too.[7] Finally it would be worthwhile to examine the free electron band structure where $f(k)$ is as yet unknown.

+INIC-Fellow from Universidade de Évora, Portugal

References

1. G.A. Toombs, Physics Reports C 40, 181 (1978).
2. M.J. Rice, S. Strässler and W.R. Schneider, Lecture Notes in Physics 34, 282 (1975).
3. J.B. Nielsen and K. Carneiro, Solid State Commun. 33, 1097 (1983).
4. P.A. Lee, T.M. Rice and P.W. Anderson, Solid State Commun. 14, 703 (1974).
5. D. Baeriswyl, to be published in "Theoretical Aspects of Band Structures and Electronic Properties of Pseudo-One-Dimensional Solids" (ed. H. Karmimura, D. Reidel Publishing Company).
6. A. Kotani, J. Phys. S. Japan 42, 416 (1977).
7. P.Y. Le Daéron and S. Aubry, J. Phys. C 16, 4827 (1983).

SOLITONS AND POLARONS IN A SPIN DENSITY WAVE CHAIN

B. Pietrass

Zentralinstitut für Festkörperphysik und Werkstofforschung der Akademie der Wissenschaften der DDR

8027 Dresden, Postfach, German Democratic Republic

Model calculations have been done within the unrestricted Hartree-Fock scheme on the structure of intrinsic defects in a finite spin density wave chain with one electron per site, including only on-site interactions. Results are given for the spin and charge distributions of spin density wave defects of the types neutral soliton, charged soliton and polaron.

Introduction

In quasi-one-dimensional charge density wave (CDW) systems solitons and polarons have been extensively studied, especially in connection with the properties of polyacetylene[1,2]. In quasi-one-dimensional spin density wave (SDW) systems solitons should also exist. They have been considered in connection with $(TMTSF)_2X$ compounds[3,4] and with organic polymers[5]. The solitons in SDW systems should obey the same spin-charge relations as those known for solitons in CDW systems[3,4]. Not much is known from the literature, however, about the charge and spin distributions of SDW solitons.

In order to obtain a more detailed picture of solitons and polarons in a one-dimensional SDW system, unrestricted Hartree-Fock calculations have been done for a finite chain, taking into account only on-site interactions. Within the unrestricted Hartree-Fock scheme the ground state of the chain is a dimerized (bond alternating) structure for weak interactions and an antiferromagnetic spin density wave structure for sufficiently strong interactions[6,7].

Method of calculation

In the unrestricted Hartree-Fock scheme the equations for the LCAO coefficients are

$$\sum_{l'} \left(F^{\alpha}_{ll'} - \varepsilon^{\alpha}_{n} \delta_{ll'} \right) c^{\alpha}_{l'n} = 0 \qquad (\alpha = \uparrow, \downarrow) \qquad (1)$$

where l denotes the site and n the energy level. The non-vanishing

Fock matrix elements are

$$F^{\alpha}_{l,l+1} = -t \tag{2}$$

$$F^{\uparrow,\downarrow}_{ll} = U P^{\downarrow,\uparrow}_{ll} \tag{3}$$

where U is the on-site interaction strength of electrons with opposite spin directions. P_{ll} is a diagonal element of the density matrix

$$P^{\alpha}_{ll'} = \sum_{n}^{occ.} C^{\alpha *}_{ln} C^{\alpha}_{l'n} . \tag{4}$$

The equations (1) to (4) have to be solved selfconsistently. Then the net charge and spin at site l

$$Q_l = P^{\uparrow}_{ll} + P^{\downarrow}_{ll} - 1 \tag{5}$$

$$S_l = \frac{1}{2}(P^{\uparrow}_{ll} - P^{\downarrow}_{ll}), \tag{6}$$

the bond order between sites l and l+1

$$B_l = P^{\uparrow}_{l,l+1} + P^{\downarrow}_{l,l+1} \tag{7}$$

and the total energy of the chain

$$E = -t \sum_{l,\alpha} P^{\alpha}_{l,l+1} + U \sum_{l} P^{\uparrow}_{ll} P^{\downarrow}_{ll} \tag{8}$$

are calculated from the density matrix.

Results and discussion

Selfconsistent solutions of the non-linear equations (1) have been obtained by straightforward iteration for the ideal SDW chain as well as for kink and bag like states, starting with suitable initial values. Results are given in Figures 1 to 3. The ideal SDW state is characterized by alternating spin values (Fig. 1a). According to the Figures the uncompensated charges and spins are localized within finite regions where the ideal SDW is strongly disturbed (SDW defects). The width of the SDW defects decreases with increasing values of U/t. The same is true for the extension of chain end effects.
If the chain length is great compaired to the defect width the shape of the SDW defect does not change with varying chain length. Therefore the SDW defects in the Figures 1 to 3 calculated for finite chains are also the intrinsic defects for long SDW chains.

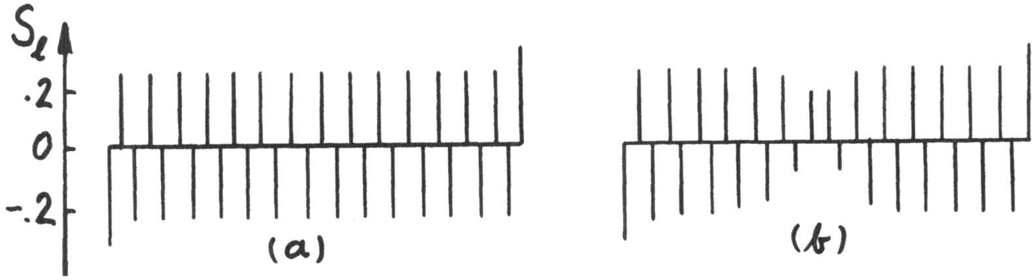

Fig. 1

Spin distribution S_l in a finite spin density wave chain (U/t = 2.5). The charge Q_l is zero at each site. (a) Neutral even chain (30 sites). (b) Neutral odd chain (29 sites). The unpaired spin is localized in the SDW defect at the chain center (neutral domain wall or neutral soliton). $\Sigma_l S_l = 1/2$.

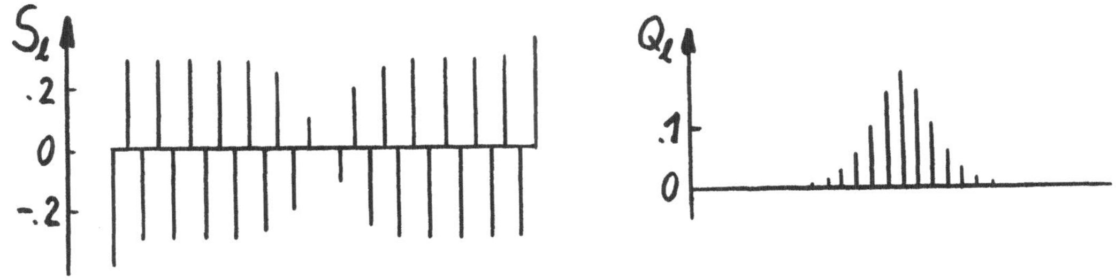

Fig. 2

Spin distribution S_l and charge distribution Q_l in an odd SDW chain (29 sites) with one additional electron. The uncompensated charge is localized in the SDW defect at the chain center (charged domain wall or charged soliton). $\Sigma_l S_l = 0$, $\Sigma_l Q_l = 1$.

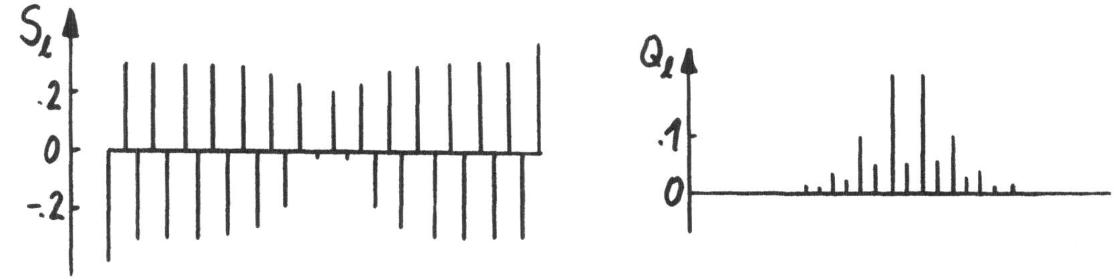

Fig. 3

Spin distribution S_l and charge distribution Q_l in an even SDW chain (30 sites) with one additional electron. The uncompensated spin and charge are localized in the chain center (polaron). $\Sigma_l S_l = 1/2$, $\Sigma_l Q_l = 1$.

Table 1
Defect states in a spin density wave chain

SDW defect	Charge	Spin	Defect Formation Energy (U = 3t)
Neutral Soliton	0	1/2	0.40 t
Charged Soliton	1	0	0.71 t
Polaron	1	1/2	0.87 t

The defect formation energies given in Table 1 are the energy differences between a chain containing a defect and a perfect chain of the same length. Since the perfect SDW structure occurs only in even chains (an odd chain has a neutral soliton in its lowest energy state), the defect formation energy for an even chain is $E_d = E(N) - E_{id}(N)$ (N even), whereas for an odd chain $E_d = E(N) - (N/(N+1))\, E_{id}(N+1)$ (N odd). The charged soliton has a higher formation energy than the neutral soliton. The same has been found to be true in the case of bond alternation solitons when on-site interaction effects are included[6,7].

The localized spin density wave defects are connected with a disturbance of the constant bond order of the perfect spin density wave chains. In the region of a neutral soliton the bond order perturbation is oscillating with maximum bond order weakening at the center. In the regions of charged solitons and polarons the bond order is enhanced.

References

1 W. P. Su, J. R. Schrieffer, and A. J. Heeger, Phys. Rev. B22 (1980) 2099

2 S. A. Brazovskii and N. N. Kirova, Pisma ZETF 33 (1981) 6

3 B. Horovitz, Phys. Rev. Lett. 48 (1982) 1416

4 T.-L. Ho, Phys. Rev. Lett. 48 (1982) 946

5 C. Aslangul and D. Saint-James, J. Phys. (Paris) 44 (1983) 953

6 K. R. Subbaswamy and M. Grabowski, Phys. Rev. B24 (1981) 2168

7 S. Kivelson and D. E. Heim, Phys. Rev. B26 (1982) 4278

CHARGE DENSITY WAVES IN SUPERIONIC CONDUCTORS

N. Plakida

Joint Institute for Nuclear Research
101000 Moscow Head Post Office P.O.Box 79, USSR

For the model of superionic conductor the possibility of CDW formation due to the interaction between fast ions and phonons is considered. On the basis of an equilibrium condition for the lattice it is shown that CDW can exist in temperature range $T_1 < T < T_2$ where $T_1 \leqslant D \leqslant T_2$, D is the activation energy for fast ions.

As has been pointed out in recent experiments (see e.g.,/1-3/), static charge density waves (CDW) may exist in superionic conductors (SC) as a result of a nonuniform distribution of moving ions in the crystal. A phenomenological description of CDW in SC was proposed in/4/, and a microscopic model for the development of CDW due to the interaction of moving-ion density fluctuations with optic lattice vibrations was considered in/5/.

In the present paper the stability condition for lattice of SC based on the model/5/ is analysed. It is shown that CDW can develop only due to interaction of moving-ion density fluctuations with acoustic phonons as proposed in/4/ and CDW can exist for a sufficiently strong coupling in the temperature range $T_1 < T < T_2$ where $T_1 \leqslant D \leqslant T_2$, D being the activation energy for moving ions.

2. Let us consider a model for SC described by the Hamiltonian analogous to/5/:

$$H = \sum_{\bar{K}} E_K a^+_{\bar{K}} a_{\bar{K}} + \frac{1}{2} \sum_{\bar{q} \neq 0} V(q) \rho_{\bar{q}} \rho_{-\bar{q}} + \frac{1}{\sqrt{N}} \sum_{\bar{q} j} g(\bar{q}, j) \rho_{-\bar{q}} Q_{\bar{q} j}$$
$$+ \frac{1}{2} \sum_{\bar{q} j} \{ |P_{\bar{q} j}|^2 + \omega^2_{\bar{q} j} |Q_{\bar{q} j}|^2 \} + \frac{1}{4!} \sum_{q_1 \cdots q_4} V(q_1 q_2 q_3 q_4) Q_1 Q_2 Q_3 Q_4 \qquad (1)$$

where $a^+_{\bar{K}}$, $a_{\bar{K}}$ are Fermi creation and annihilation operators for a moving ion with the momentum $\bar{K}$ and energy $E_k = D + K^2/2M$, where D is the activation energy and M is its effective mass; $\rho_{\bar{q}} = \sum_K a^+_{\bar{K}} a_{\bar{K}+\bar{q}}$. In what follows we consider only the classical limit and the type of statistics (Fermi of Bose) is irrelevant. $V(q) = (1/N)[4\pi(Ze)^2/q^2 \varepsilon v + \beta(q)]$ is the Coulomb and short-range, $\beta(q)$, interaction between moving ions with charge Ze in a crystal with the dielectric constant ε . The number of unit cells in the crystal of volume V is $N = V/v$. Interaction of moving ions with lattice vibrations for branch j is given by the function $g(\bar{q}, j)$ which in the long-wave limit, $q \to 0$, is equal to: $g(\bar{q}0) \approx i\lambda_0/q$ for the optic (0) phonons, $\omega_{q0} \approx \omega_0$, and $g(\bar{q}A) \approx i\lambda_A q$ for the acoustic (A) phonons, $\omega_{qA} \approx cq$. Lattice vibrations are described by the operators $P_{\bar{q}j} = P^+_{-\bar{q}j}$, $Q_{\bar{q}j} = Q^+_{-\bar{q}j}$, $\omega_{\bar{q}j}$ are phonon frequencies and the last term in (1) takes into account the anharmonic interaction of phonons: $1 \equiv q_1 = (\bar{q}_1 j_1)$, etc.

3. Equilibrium conditions for a crystal lattice in the model (1) can be obtained if one equate to zero an average force canonically conjugated to the normal coordinate Q_q:

$$-\langle \frac{d}{dt} P_q(t) \rangle = \frac{i}{\hbar} \langle [P_q, H] \rangle = \omega_q^2 \langle Q_{-q} \rangle + \\ + \frac{1}{\sqrt{N}} g(q) \langle \rho_{-q} \rangle + \frac{1}{8} \sum_{q_1 q_2 q_3} V(q_1 q_2 q_3 q) \langle Q_1 Q_2 Q_3 \rangle = 0 \tag{2}$$

where the statistical average $\langle \dots \rangle$ is taken with the Hamiltonian (1). When CDW exists, $\langle \rho_q \rangle \neq 0$, it produces the corresponding lattice deformation, $\langle Q_q \rangle \neq 0$, for the definite wave vector $\bar{q} = \pm\bar{q}_0 \neq 0$ and branch $j = j_0$: $\pm q_0 = (\pm \bar{q}_0, j_0)$. In order to obtain the necessary condition for existing of a nonzero solution of eq.(2) for the order parameter $y(q_0) = y^*(-q_0) = \langle Q_{q0} \rangle / \sqrt{N}$ one should calculate the density fluctuations $\langle \rho_q \rangle$ in the presence of arbitrary deformation $\langle Q_q \rangle$. To this end we employ the equation-of-motion method for the thermodynamic Green functions. In the lowest order in $g(q)$ we get as in [5] for the nondiagonal Green function

$$\langle\langle a_{\bar{k}+\bar{q}} | a^+_{\bar{k}} \rangle\rangle_\omega = \frac{\lambda(q)}{[\omega - \varepsilon_1(\bar{k})][\omega - \varepsilon_2(\bar{k})]} , \tag{3}$$

with the usual notation for the Fourier-transform of the two-time Green functions. Here we introduced

$$\varepsilon_{1,2}(\bar{k}) \approx \varepsilon_{1,2}(0) = D \pm \varphi(q), \quad \varphi(q) = 2\sqrt{|\lambda(q)|^2}, \\ \lambda(q) = V(q) \langle \rho_q \rangle + g(q) y(q) . \tag{4}$$

In the classical limit of the Boltzmann statistics one obtains from (3):

$$\frac{1}{N}\langle\rho_q\rangle = -\frac{\lambda(q)}{\varphi(q)}\, \mathrm{sh}\frac{\varphi(q)}{T}\, e^{-D/T} \equiv -\lambda(q)\, f(T), \tag{5}$$

where $f(T) \approx (1/T)\exp(-D/T)$ for $\varphi(q) \ll T$.
Now substituting (5) after taking into account definition (4), one gets the equilibrium condition (2) in the form:

$$y(-q)\Big\{\omega_q^2 - \frac{|g(q)|^2 f(T)}{1 + N V(q) f(T)} + \sum_{q'} B(q,q')\langle|Q_{q'}|^2\rangle + B(q)\,|y(q)|^2\Big\} = 0 \tag{6}$$

where we have introduced the anharmonic coupling constant:

$$B(q) = N B(q,q), \quad B(q,q') = 2V(q,-q,q',-q').$$

We point out that this equation has the same form as the equation for the order parameter in the vibronic model of ferroelectric/6/.

4. Nonzero solutions of eq.(6) $y(q) = y(q_0) \neq 0$ can exist only if the screened ion-phonon interaction is sufficiently strong:

$$\tau(q) = \frac{(|g(q)|^2/\omega_q^2)\, f(T)}{1 + N V(q) f(T)} > 1 . \tag{7}$$

The latter condition can be fulfilled in the definite temperature range $T_1 < T < T_2$, where T_1 and T_2 are solutions of the equation

$$\gamma_{ph} - \gamma_c = 1/Df(T) = (T/D)\, e^{D/T} . \tag{8}$$

Here we introduced $\gamma_{ph} = |g(q)|^2/D\omega_q^2$, $\gamma_c = NV(q)/D$.
Eq. (8) has two real solutions $T_1 \lesssim D \lesssim T_2$, when $\gamma_{ph} - \gamma_c > e$. An estimation of the anharmonic term in (6) shows that in the case of a sufficiently small coupling constant $TB(q,q')/\omega_q^2\,\omega_{q'}^2 \lesssim 10^{-2}$, the temperature range where nonzero solutions of eq.(6) exist is given to a high degree of accuracy by the same condition (7) (contrary to the model of vibronic ferroelectric in/6/). Therefore one can estimate the temperature range of the CDW formation as $T_1 < T < T_2$, where T_1 and T_2 are the solutions of eq.(8).

The value of the modulation wave vector $\bar{q}_0$ of CDW depends on the maximum of interaction (7):

$$(d\tau(\bar{q})\ /\ dq)_{q=q_0} = 0 . \tag{9}$$

For optic phonons $|g(\bar{q}0)|^2/\omega^2_{\bar{q}0} \approx \lambda^2_0/\omega^2_0 q^2$ and has its maximum at $q = 0$, therefore CDW of $q \neq 0$ does not exist. For acoustic phonons $|g(\bar{q},A)|^2/\omega^2_{\bar{q}A} \approx \lambda^2_A/c^2$ and the maximum values of $\tau(\bar{q})$ is given by $dV(q)/dq = 0$. Putting $\beta(q) = \beta(0) + Sq^2$ one gets in accordance with /4/: $q_0 = (4\pi\ (Ze)^2/\varepsilon v S)^{1/4}$. Since the dielectric constant ε is SC can be large enough, q_0 may be sufficiently small, and CDW with a macroscopic period can develop.

For strong coupling $\gamma_{ph} - \gamma_c \gg 1$ one finds from eq.(8) $T_1 \ll D$, and CDW can develop in SC just after the phase transition to the superionic state at temperature $T_0 > T_1$. This type of situation is probably observed in Ag_2S /2/ and AgI /3/. In β-$LiAlSiO_4$ /1/ the incommensurate phase with a small q_0 value exist in the temperature range $703 < T < 763$ K that can be obtained as solutions T_1 and T_2 of eq.(8) for the model parameters D = 0.063 eV and $\gamma_{ph} - \gamma_c = 2.720 > e$, that seems to be physically reasonable.

In conclusion we point out that calculation of the phonon Green function $\langle\langle Q_q | Q^+_q \rangle\rangle_\omega$ for the model (1) shows that the frequencies of lattice vibration become unstable in the same temperature range, where eq.(6) has nonzero solutions.

1. W. Press, et al., Phys.Rev., 1980, B21, p. 1250.
2. R.J. Cava, D.B. McWhan, Phys.Rev.Lett. 1980, 45, p. 2046.
3. G. Marriotto, et al., Phys.Rev., 1981, B23, p. 4782.
4. V.N. Bondarev, Fiz.Tverd.Tela (USSR), 1981, 23, p. 2413.
5. A.P. Kovalenko, Fiz.Tverd.Tela (USSR), 1983, 25, p. 1310.
6. N.M. Plakida, G.L. Mailjan, Fiz.Tverd. Tela (USSR), 1977, 19, p. 121.

NUMERICAL STUDY OF IMPURITY PINNING IN ONE-DIMENSIONAL INTERACTING ELECTRON SYSTEMS

T. Saso, Y. Suzumura and H. Fukuyama†
Department of Physics, Tohoku University, Sendai 980, Japan
†Institute for Solid State Physics, Tokyo University, Tokyo 106, Japan

Quantum Monte Carlo calculation of the impurity pinning of charge density waves in one-dimensional interacting electron systems is described. Effects of spin density fluctuation on impurity pinning are also investigated within the classical treatment.

1. Introduction

Growing number of organic and inorganic materials are providing various examples of quasi-one-dimensional interacting electron systems. Much attention has recently been paid theoretically to the role of impurities in such systems[1,2]. Properties of one-dimensional interacting electron systems are most easily investigated by the bosonization method and the use of the phase variables for charge and spin density fluctuation modes.[3] In such an approach, effects of impurities are treated as the pinning of the phase variable for the charge density fluctuation, which may correspond to the Anderson localization of interacting electrons[4,5]. Furthermore, in contrast to the usual impurity pinning problem of CDW, the quantum fluctuation can not be neglected and the spin density fluctuation can play an important role on impurity pinning. The purpose of the present paper is to investigate these problems by applying direct numerical method.

2. Quantum effects on impurity pinning

Impurity pinning of CDW has been investigated by the various methods. Within the classical treatment of CDW, the detailed numerical investigations[6,7] have proved that the theory by Fukuyama and Lee[8] works well even quantitatively. In one-dimensional interacting electron systems, it is expected that the quantum fluctuation of the charge density will weaken the pinning and even lead to the depinning when the magnitude of interactions between electrons is varied[4,5]. We performed a quantum Monte Carlo calculation on the following Hamiltonian on a discrete lattice describing the motion of the phase θ of CDW[7]:

$$H_{1D} = \sum_{i=1}^{N_x} \left[A(\theta_{i+1}-\theta_i)^2 - C \frac{\partial^2}{\partial \theta_i^2} - V \cos(\theta_i-\zeta_i) \right], \qquad (1)$$

where $A = (v_F/4\pi a)(1-g_1/2+g_2)$, $C = (\pi v_F/a)(1+g_1/2-g_2)$, g_1 and g_2 respectively denote the coupling constants for the backward and the forward scattering[3], V denotes the strength of imputiry potential and the random variable $0 < \zeta_i < 2\pi$ simulates the random position of impurities x_i times $-2k_F$. The degree of quantumness is characterized by the parameter $\eta = (1/2\pi)(C/A)^{1/2}$, whereas the parameter $\varepsilon = V/4\pi A$ characterizes the degree of randomness. In this model we disregarded the spin density fluctuations and the Umklapp process. A key to treat eq.(1) is to relate its statistical mechanics with that of the following two-dimensional XY-like classical Hamiltonian with randomness only in x direction and the temperature $T_2 = 4\pi A\eta$:

$$H_{2D} = \sum_{i=1}^{N_x}\sum_{j=1}^{N_y}[\, A(\theta_{i+1,j}-\theta_{ij})^2 + A(\theta_{i,j+1}-\theta_{ij})^2 - V\cos(\theta_{ij}-\zeta_i)]. \quad (2)$$

To see the relationship with H_{1D}, let us calculate the partition function $Z_2(T_2) = \int\delta\theta \exp(-H_{2D}/T_2)$, which by applying transfer integration technique[9,10] leads to $Z_2(T_2) = \mathrm{Tr}\exp(-N_y H_{1D}/4\pi A\eta)$. This is equal to the partition function for H_{1D} with the temperature T_1 replaced by $4\pi A\eta/N_y$. Thus the properties of H_{1D} can be obtained by investigating the classical Hamiltonian H_{2D} by the Monte Carlo method. To know the low temperature properties of H_{1D}, it is necessary to keep the condition[7] $2\pi N_y/N_x \gg 1$. In the following calculation we have chosen $N_x = N_y = N$.

The results are shown in Fig.1. Here, the degree of pinning u is defined as $u = (1/N^2)\Sigma_{ij}\cos(\theta_{ij}-\zeta_i)$ and plotted as a function of η. When the impurities are weak enough, the self-consistent harmonic approximation (SCHA)[3,4] for eq.(1) combined with the Fukuyama-Lee[8] theory results in

$$u = \gamma/\sqrt{N_0}, \quad \gamma = (\varepsilon/2\pi\sqrt{N_0})^{\eta(4-\eta)}, \quad N_0 = 6^{\eta/2(3-\eta)}(\pi/3\varepsilon)^{2/(3-\eta)}, \quad (3)$$

which are also plotted in Fig.1. For the weak pinning case $\varepsilon = 0.1$, the present results show good agreement with SCHA for $\eta < 0.5$. In SCHA, u decreases rapidly for $\eta > 1$ and vanishes at $\eta_c = 3$, showing delocalization transition. The present result decreases slower but do not seem contradicting with $\eta_c = 3$ within the accuracy of the calculation. For the strong pinning case $\varepsilon = 1$, u remains still large at $\eta \sim 3$, indicating that the critical value η_c for delocalization increases as the strength of impurity potential increases.

3. Effects of spin density fluctuation on impurity pinning

In the preceeding section the spin density fluctuations are neglected. However, the impurity potential couples not only with the phase variable of the charge density fluctuation but also with that of the spin density

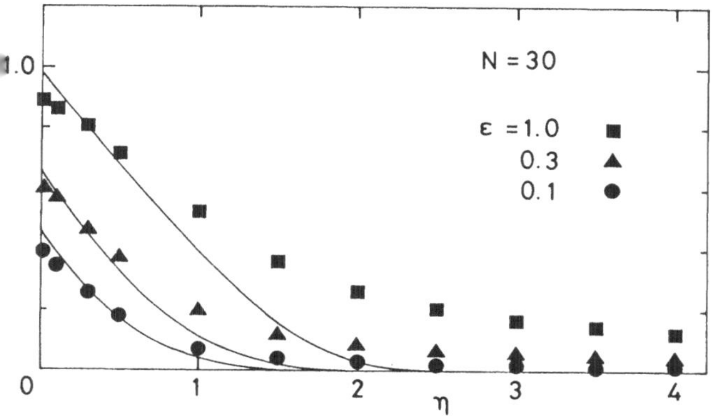

Fig.1 Degree of pinning u as a function of the quantumness parameter η. The solid curves indicate the results of SCHA (eq.(3)).

Fig.2(below) The configuration of the phase variables θ_i (o($\delta=\delta_c-0$) and •($\delta=\delta_c+0$)) and ϕ_i (▲ ($\delta=\delta_c+0$)) are indicated. Also shown by × are the values of θ_i for maximum local impurity gain $\theta_i = -\zeta_i$.

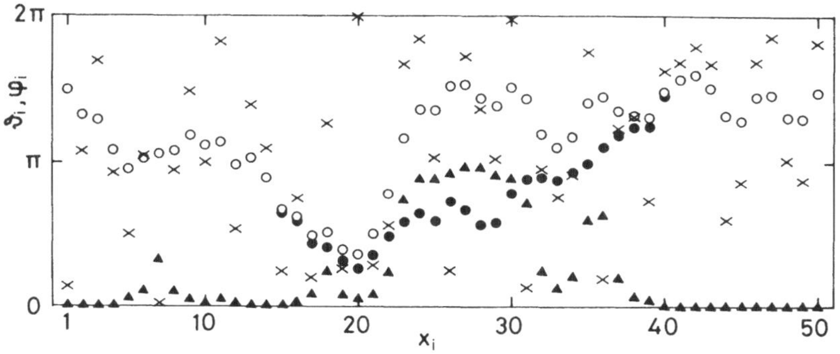

fluctuation[3,4]. To investigate the interplay between these two phase variables, we consider the following model Hamiltonian, ignoring quantum fluctuation for simplicity:

$$H = \sum_{i=1}^{N} [\, A_1(\theta_{i+1}-\theta_i)^2 + A_2(\phi_{i+1}-\phi_i)^2 - V\cos(\theta_i-\zeta_i)\cos\phi_i \,], \qquad (4)$$

where ϕ_i is the phase variable of spin density fluctuation at the site x_i. $A_1 = A$ and A_2 is given as $A_2 = (v_F/4\pi a)(1-g_1/2)$, which is hereafter rewritten as $A_2 = (1-\delta)A_1$ for convenience. We look for a classical solution to minimize eq.(4). For $A_2 > A_1$ ($\delta < 0$), we obtain the solution with ϕ_i = constant = $n\pi$, n = integer, and only the charge density phase variable θ is distorted spatially. In this case, the problem reduces to the usual impurity pinning of CDW. We found for $\delta > 0.6$ a solution with distorted ϕ_i (spin density glass (SDG)!). The transition from constant ϕ_i to SDG seems to occur very sharply and discontinuously for each sample. The critical values δ_c for this transition have a sharp distribution around $\delta_c = 0.6$. For 11 among 20 samples we investigated, δ_c lies between 0.55 and 0.6. Furthermore, the transition occurs by the nucleation of π-domains as shown in Fig.2 at the regions where otherwise the energy gains by impurity pinning of θ were small. After the formation of a domain in ϕ, θ flattens in the same region.

These results can be understood in the following way. First we consider the case $\delta = \delta_c - 0$ and a region of length $N_0 = (\pi/3\varepsilon)^{2/3}$, which corresponds to the characteristic length L_0 in th Fukuyama-Lee theory. The energy per unit length is estimated as $E_0/N_0 \simeq A_1(\pi/N_0)^2 - V/\sqrt{N_0}$. Next, for $\delta = \delta_c + 0$, let us suppose that a π-domain is formed in this region. The energy per unit length may be $E_1/N_0 \simeq A_2(\pi/N_0)2 - V/2\sqrt{N_0}$. The factor 1/2 in the second term appears because $\cos(\theta-\zeta)\cos\phi = (1/2)[\cos(\theta-\zeta+\phi) +\cos(\theta-\zeta-\phi)]$ and only one of the two terms in the square bracket may survive to give a contribution $1/\sqrt{N_0}$. The condition $E_0 = E_1$ gives $\delta_c = 2/3$, which is close to the value obtained above.

The present investigation on the role of spin density fluctuation on impurity pinning of CDW suggests a possible entanglement of electronic and magnetic responses in the quasi-one-dimensional materials. Study on such problems are now in progress.

This work is partially supported by Itoh Science Foundation and the Grant in Aid from the Ministry of Education, Science and Culture.

References

1. S.T. Chui and J.W. Bray, Phys. Rev. B16 (1977) 1329 and B19 (1979) 4020.
2. W. Apel and T.M. Rice, Phys. Rev. B26 (1982) 7063 and J. Phys. C16 (1983) L271.
3. For a review, see H. Fukuyama and H. Takayama, in *Electronic Properties of Inorganic Quasi-One-Dimensional Compounds*, ed. P. Monceau (D. Reidel Pub. Co. 1984).
4. Y. Suzumura and H. Fukuyama, J. Phys. Soc. Japan 52 (1983) 2870.
5. Y. Suzumura and H. Fukuyama, submitted to J. Phys. Soc. Japan.
6. H. Matsukawa and H. Takayama, Solid State Commun. 50 (1984) 283.
7. T. Saso, Y. Suzumura and H. Fukuyama, in *Proceedings of the 17th International Conference on Low Temperature Physics, Karlsruhe, 1984* (North-Holland).
8. H. Fukuyama and P.A. Lee, Phys. Rev. B17 (1978) 535.
9. B. Stoeckly and D.J. Scalapino, Phys. Rev. B11 (1975) 205.
10. Y. Okwamoto, J. Phys. Soc. Japan 49 (1980) 8.

MULTIVALUED CHARGE-DENSITY WAVES

T.V. Lakshmi and K.N. Shrivastava
School of Physics, University of Hyderabad
P.O. Central University, Hyderabad 500 134, India

We propose new type of charge-density waves which occur in a two band model in which one of the bands is a conduction band the other may be a magnetic band. The distortions occur via the interband as well as the intraband electron-phonon interaction so that there are multiple gaps and hence multiple transition temperatures.

In 1954 Peierls[1] invented the lattice distortion in a one-dimensional metal which has only one conduction band. The distortion causes pairing of atoms and hence a redistribution of charges which permits a wave propagation. In a magnetic metal or alloy the electrons in the conduction band as well as those in the valence band interact with the lattice vibrations so that there may be extended distortion of atomic positions due to the charge redistribution in the conduction band to give rise to a charge-density wave as well as local distortions which rearrange the charges in the magnetic valence band. The inter band electron-phonon interaction may transmit the effect of distortions in the conduction band to the valence band and vice versa so that the atoms may get polarized and a third distortion may be encouraged. Corresponding to every one distortion, there is a charge-density wave, so that we find multiple charge-density waves which superimpose to form a pattern quite distinct from that of Peierls.

We write the hamiltonian as

$$\mathcal{H} = \mathcal{H}_o + \mathcal{H}_h + \mathcal{H}_{ep} + \mathcal{H}_s \tag{1}$$

$$\mathcal{H}_o = \sum_{k\sigma} \varepsilon^c_{k\sigma} c^\dagger_{k\sigma} c_{k\sigma} + \sum_{k\sigma} \varepsilon^b_{k\sigma} b^\dagger_{k\sigma} b_{k\sigma} + \sum_q \hbar\omega_q \left(\beta^\dagger_q \beta_q + \beta^\dagger_{-q}\beta_{-q}\right) \tag{2}$$

$$\mathcal{H}_h = \sum_{k\Sigma} V_k \left(c^\dagger_{k\sigma} b_{k\sigma} + b^\dagger_{k\sigma} c_{k\sigma}\right) \tag{3}$$

$$\mathcal{H}_{ep} = \frac{g_1}{\sqrt{N}} \sum_{kq\sigma} \left[c^\dagger_{k+q\sigma} c_{k\sigma} \left(\beta_q + \beta^\dagger_{-q}\right) + c^\dagger_{k\sigma} c_{k+q\sigma} \left(\beta^\dagger_q + \beta_{-q}\right)\right] +$$

$$\frac{g_2}{\sqrt{N}} \sum_{kq\sigma} \left[b^\dagger_{k+q\sigma} b_{k\sigma} \left(\beta_q + \beta^\dagger_{-q}\right) + b^\dagger_{k\sigma} b_{k+q\sigma} \left(\beta^\dagger_q + \beta_{-q}\right)\right] +$$

$$\frac{d}{\sqrt{N}}\sum_{kq\sigma}[c^{\dagger}_{k+q\sigma}b_{k\sigma}(\beta_q+\beta^{\dagger}_{-q}) + c^{\dagger}_{k\sigma}b_{k+q\sigma}(\beta_q+\beta_{-q}) +$$

$$b_{k+q\sigma}c_{k\sigma}(\beta_q+\beta^{\dagger}_{-q}) + b^{\dagger}_{k\sigma}c_{k+q\sigma}(\beta^{\dagger}_q+\beta_{-q})] \qquad (4)$$

$$H_S = U_1\sum_{k\sigma} c^{\dagger}_{k\sigma}c_{k\sigma}\ c^{\dagger}_{k,-\sigma}c_{k,-\sigma} + U_2\sum_{k\sigma} b^{\dagger}_{k\sigma}b_{k\sigma}b^{\dagger}_{k,-\sigma}b_{k,-\sigma} \qquad (5)$$

where $c^{\dagger}_k(c_k)$ are the creation and annihilation operators for the electron in the conduction; $b^{\dagger}_k(b_k)$ describe the electron operators in the valence band and $\beta^{\dagger}_q(\beta_q)$ are the phonon variables.

We calculate the equation of motion for the electron operators as,

$$E << c_{k\sigma}|c^{\dagger}_{p\sigma} >> = <[c_{k\sigma},c^{\dagger}_{p\sigma}]> + << [c_{k\sigma},\mathcal{H}]|c^{\dagger}_{p\sigma} >> \qquad (6)$$

$$E << c_{k\sigma}|c^{\dagger}_{p\sigma} >> = <\delta_{k,p}> + \varepsilon^{c}_{k\sigma} << c_{k\sigma}|c^{\dagger}_{p\sigma} >> + V_{k\sigma} << b_{k\sigma}|\ c^{\dagger}_{p\sigma}>>$$

$$- 2\sum_{k'\sigma'} V_{k'\sigma'} << c^{\dagger}_{k'\sigma'}\ c_{k\sigma}\ b_{k'\sigma'}\ |\ c^{\dagger}_{p\sigma} >> +$$

$$2\sum_{k'\sigma'} V_{k'\sigma'} << b^{\dagger}_{k'\sigma'}\ c_{k\sigma}\ c_{k'\sigma'}\ |\ c^{\dagger}_{p\sigma} >> +$$

$$\frac{g_1}{\sqrt{N}}\sum_{q}[<< c_{k-q,\sigma}(\beta_q+\beta^{\dagger}_{-q})\ |\ c^{\dagger}_{p\sigma} >> + <<c_{k+q,\sigma}(\beta^{\dagger}_q+\beta_{-q})|c^{\dagger}_{p\sigma} >>] +$$

$$\frac{d}{\sqrt{N}}\sum_{q}[<<b_{k-q,\sigma}(\beta_q+\beta^{\dagger}_{-q})|c^{\dagger}_{p\sigma}>> + << b_{k+q,\sigma}(\beta^{\dagger}_q+\beta_{-q})|c^{\dagger}_{p\sigma} >>] -$$

$$\frac{2d}{\sqrt{N}}\sum_{k'q'\sigma'}[<<c^{\dagger}_{k'+q'\sigma'}\ c_{k\sigma}b_{k'\sigma'}\ (\beta_{q'}+\beta^{\dagger}_{-q'})\ |\ c^{\dagger}_{p\sigma} >> -$$

$$<<b^{\dagger}_{k'+q'\sigma'}\ c_{k\sigma}c_{k'\sigma'}\ (\beta_{q'}+\beta^{\dagger}_{-q'})\ |\ c_{p\sigma} >> +$$

$$<<c^{\dagger}_{k'\sigma'}c_{k\sigma}b_{k'+q'\sigma'}(\beta_{-q'}+\beta^{\dagger}_{q'})\ |\ c^{\dagger}_{p\sigma} >> -$$

$$<<b^{\dagger}_{k'\sigma'}c_{k\sigma}c_{k'+q'\sigma'}(\beta_{-q'}+\beta^{\dagger}_{q'})|c^{\dagger}_{p\sigma}>>]+2U_1<<c_{k\sigma}c^{\dagger}_{k-\sigma}c_{k-\sigma}|c^{\dagger}_{p\sigma}>> \qquad (7)$$

and introduce three distortions as,

$$<\beta_q> \;=\; <\beta_q^\dagger> \;=\; \frac{\lambda}{2}\;\delta_{q,Q}\;;$$

$$\Delta_1 \;=\; \lambda g_1/\sqrt{N}\;;$$

$$\Delta_2 \;=\; \lambda g_2/\sqrt{N}\;;$$

$$\Delta \;=\; \lambda d/\sqrt{N}\;;$$

and take the number densities in the two bands as

$$n_{k\sigma} \;=\; <c_{k\sigma}^\dagger\, c_{k\sigma}>\;;$$

$$N_{k\sigma} \;=\; <b_{k\sigma}^\dagger\, b_{k\sigma}>\;;$$

to obtain

$$(E-\varepsilon_{k\sigma}^{c}-2U_1 n_{k,-\sigma}) <<c_{k\sigma}|c_{p\sigma}^\dagger> \;=\; <\delta_{k,}> + V_{k\sigma}(1-2n_{k\sigma})<< b_{k\sigma}|c_{p\sigma}^\dagger>>$$
$$+ \;\Delta_1\;(<<c_{k-Q,\sigma}|c_{p\sigma}^\dagger>>+<<c_{k+Q,\sigma}|c_{p\sigma}^\dagger>>) + \Delta(1-2n_{k\sigma})(<<b_{k-Q,\sigma}|c_{p\sigma}^\dagger>>$$
$$+ <<b_{k+Q,\sigma}|c_{p\sigma}^\dagger>>) \qquad (8)$$

Similarly, we write the equations for $<<b_{k\sigma}|c_{p\sigma}^\dagger>>$, $<<b_{k\sigma}^\dagger|c_{p\sigma}^\dagger>>$, and $<<c_{k\sigma}^\dagger|c_{p\sigma}^\dagger>>$ from which we construct four equations for $<<c_{k\pm Q/2,\sigma}|$, $<<c_{k\pm Q/2,\sigma}^\dagger|$, $<<b_{k\pm Q/2,\sigma}|$, and $<<b_{k\pm Q/2,\sigma}^\dagger|$. The eigen-value equation is then found to be

$$\begin{vmatrix} E_+^c & \Delta_1 & V_{k+Q/2,\sigma}m_+ & m_+\Delta \\ \Delta_1 & E_-^c & m_-\,\Delta & m_-V_{k-Q/2,\sigma} \\ M_+V_{k+Q/2,\sigma} & M_+\,\Delta & E_+^b & \Delta_2 \\ M_-\,\Delta & M_-V_{k-Q/2,\sigma} & \Delta_2 & E_-^b \end{vmatrix} = 0 \qquad (9)$$

where

$$M_\pm \;=\; 1-2\,N_{k\pm Q/2,\sigma}\;;\quad m_\pm \;=\; 1-2\,n_{k\pm Q/2,\sigma}\;;$$

$$E_\pm^c \;=\; \varepsilon_{k\pm Q/2,\sigma}^{c}+2U_1 n_{k\pm Q/2,-\sigma}-E\;;\quad E_\pm^b \;=\; \varepsilon_{k\pm Q/2,-\sigma}^{b}+2U_2\,N_{k\pm Q/2,-\sigma}-E\;; \qquad (10)$$

We take linear Fermi surfaces $E = v_F(|k_F - k|)$, with v_F^b as the Fermi velocity of the electrons in the b-band and v_F^c in the c-band. If we take only the first two elements of the first two rows and set $U_1 = o$, we find that the gap equation occurs in a particularly simple form

$$E = \pm [(v_F k)^2 + \Delta^2]^{\frac{1}{2}}$$

for a one-band ysystem. This is the Peierls case. However, in general there are four solutions of E in (9) so that there occur multiple charge density waves.[2]

References:

(1) R.E. Peierls, Quantum Theory of Solids, p.108, Oxford University Press (1955)

(2) T.V. Lakshmi and K.N. Shrivastava, Solid State Commun. (1984).

Author Index

Lecture Notes in Physics

Vol. 195: Trends and Applications of Pure Mathematics to Mechanics. Proceedings, 1983. Edited by P. G. Ciarlet and M. Roseau. V, 422 pages. 1984.

Vol. 196: WOPPLOT 83. Parallel Processing: Logic, Organization and Technology. Proceedings, 1983. Edited by J. Becker and I. Eisele. V, 189 pages. 1984.

Vol. 197: Quarks and Nuclear Structure. Proceedings, 1983. Edited by K. Bleuler. VIII, 414 pages. 1984.

Vol. 198: Recent Progress in Many-Body Theories. Proceedings, 1983. Edited by H. Kümmel and M. L. Ristig. IX, 422 pages. 1984.

Vol. 199: Recent Developments in Nonequilibrium Thermodynamics. Proceedings, 1983. Edited by J. Casas-Vázquez, D. Jou and G. Lebon. XIII, 485 pages. 1984.

Vol. 200: H.D. Zeh, Die Physik der Zeitrichtung. V, 86 Seiten. 1984.

Vol. 201: Group Theoretical Methods in Physics. Proceedings, 1983. Edited by G. Denardo, G. Ghirardi and T. Weber. XXXVII, 518 pages. 1984.

Vol. 202: Asymptotic Behavior of Mass and Spacetime Geometry. Proceedings, 1983. Edited by F. J. Flaherty. VI, 213 pages. 1984.

Vol. 203: C. Marchioro, M. Pulvirenti, Vortex Methods in Two-Dimensional Fluid Dynamics. III, 137 pages. 1984.

Vol. 204: Y. Waseda, Novel Application of Anomalous (Resonance) X-Ray Scattering for Structural Characterization of Disordered Materials. VI, 183 pages. 1984.

Vol. 205: Solutions of Einstein's Equations: Techniques and Results. Proceedings, 1983. Edited by C. Hoenselaers and W. Dietz. VI, 439 pages. 1984.

Vol. 206: Static Critical Phenomena in Inhomogeneous Systems. Edited by A. Pękalski and J. Sznajd. Proceedings, 1984. VIII, 358 pages. 1984.

Vol. 207: S. W. Koch, Dynamics of First-Order Phase Transitions in Equilibrium and Nonequilibrium Systems. III, 148 pages. 1984.

Vol. 208: Supersymmetry and Supergravity/Nonperturbative QCD. Proceedings, 1984. Edited by P. Roy and V. Singh. V, 389 pages. 1984.

Vol. 209: Mathematical and Computational Methods in Nuclear Physics. Proceedings, 1983. Edited by J. S. Dehesa, J. M. G. Gomez and A. Polls. V, 276 pages. 1984.

Vol. 210: Cellular Structures in Instabilities. Proceedings, 1983. Edited by J. E. Wesfreid and S. Zaleski. VI, 389 pages. 1984.

Vol. 211: Resonances – Models and Phenomena. Proceedings, 1984. Edited by S. Albeverio, L. S. Ferreira and L. Streit. VI, 369 pages. 1984.

Vol. 212: Gravitation, Geometry and Relativistic Physics. Proceedings, 1984. Edited by Laboratoire "Gravitation et Cosmologie Relativistes", Université Pierre et Marie Curie et C.N.R.S., Institut Henri Poincaré, Paris. VI, 336 pages. 1984.

Vol. 213: Forward Electron Ejection in Ion Collisions. Proceedings, 1984. Edited by K. O. Groeneveld, W. Meckbach and I. A. Sellin. VII, 165 pages. 1984.

Vol. 214: H. Moraal, Classical, Discrete Spin Models. VII, 251 pages. 1984.

Vol. 215: Computing in Accelerator Design and Operation. Proceedings, 1983. Edited by W. Busse and R. Zelazny. XII, 574 pages. 1984.

Vol. 216: Applications of Field Theory to Statistical Mechanics. Proceedings, 1984. Edited by L. Garrido. VIII, 352 pages. 1985.

Vol. 217: Charge Density Waves in Solids. Proceedings, 1984. Edited by Gy. Hutiray and J. Sólyom. XIV, 541 pages. 1985.

Selected Issues from

Lecture Notes in Mathematics